LASER TECHNOLOGY

SECOND EDITION

LASER TECHNOLOGY

SECOND EDITION

Lan Xinju *et al.*

Translated by Cao Huamin

CRC Press is an imprint of the
Taylor & Francis Group, an **informa** business

CRC Press
Taylor & Francis Group
6000 Broken Sound Parkway NW, Suite 300
Boca Raton, FL 33487-2742

First issued in paperback 2019

ISBN-13: 978-1-4200-9081-9 (hbk)
ISBN-13: 978-1-138-37276-4 (pbk)

Library of Congress Cataloging-in-Publication Data

Laser technology / editor in chief, Lan Xinju. -- 2nd ed.
p. cm.
"A CRC title."
Includes bibliographical references and index.
ISBN 978-1-4200-9081-9
1. Lasers. I. Xinju, Lan.

TA1675.L37635 2010
621.36'6--dc22 2009031085

Visit the Taylor & Francis Web site at
http://www.taylorandfrancis.com

and the CRC Press Web site at
http://www.crcpress.com

Preface

Since the 1980s, this textbook has been examined and recommended for publication by four sessions of the Educational Steering Committee and was designated by the National Education Committee in 1997 as the National Level Key Textbook of the 9th Five-Year Plan period. Revised on the basis of the first edition, the second edition was designated as the National Level Planned Textbook of the 10th Five-Year Plan period by the Ministry of Education in 2001.

The Editor-in-Chief of this textbook is Professor Lan Xinju with Huazhong University of Science and Technology and the Deputy Editor-in-Chief is Zhu Changhong.

This course is allotted 70 reference class hours, with the whole book divided into 5 parts (8 chapters). Part 1 (Chapter 1) deals with the laser modulation and deflection technology, and is mainly a discussion on the fundamentals and technologies of electro-optics and acousto-optics; Part 2 (Chapters 2–4), Q modulation (Q-switching), deals with ultrashort pulse and amplifying technology, and is mainly a discussion on the basic theories and implementation methods of increasing the power and energy of laser pulse; Part 3 (Chapters 5–6), mode selection technology, deals with frequency stabilizing technology, with emphasis on the physical principles and implementation methods of single mode (transverse, longitudinal) output and stabilizing the oscillating frequency; Part 4 (Chapter 7) deals with nonlinear optical technology by first explaining the physical concept of nonlinear optics, followed by a discussion on the basic principles and methods of implementation of nonlinear optics with the frequency-doubling (second-harmonic-generation) technology as the focus of attention; and Part 5 (Chapter 8) deals with laser transmission technology, and is mainly a discussion on the theory and technology of optical fiber transmission, with a brief introduction to the transmission technology in the atmosphere and underwater.

Compiled in accordance with the syllabus of a specialized basic course in optoelectronic information in the category of electronic information of engineering for institutions of higher education, this textbook is particularly suitable for readers with a basic knowledge about physical optics and the fundamentals of laser. It can also be used as a textbook by undergraduates in optoelectronic technology, optical information technology, technological physics, optoelectronic instrumentation, and applied physics; the principal reference material for graduate students in physical electronics and related subjects; as well as reference material for faculty and students in relevant specialties or engineers and technicians in optoelectronic technology. As the contents of the chapters in this book are basically independent of each other, there is great flexibility in its reading. Hence convenient selection of the parts for use by specific institutions. In addition, despite the fact that stress is placed on the description of basic principles, a definite number of technical methods and examples have been included in certain chapters or sections to facilitate combination of theory with practice since this is a course of rather distinct practicability. Exercises and questions for further consideration are attached to the end of each chapter.

Chapters 1 and 6 are written by Lan Xinju of Huazhong University of Science and Technology, Chapters 2 and 5 by Chen Peifeng of the same university, Chapter 3 by Yao Jianquan and Ning Jiping of Tianjin University, Chapter 4 by Liu Jingsong of Huazhong University of Science and Technology, Chapter 7 by Zhu Changhong of the same university, and Chapter 8 by An Yuying of Xi'an University of Electronic Science and Technology, with Lan Xinju responsible for the whole manuscript. In the course of compilation of this book, not all the works and articles by other authors referred to have been listed in our references one by one. The authors take this opportunity to extend to them our sincere thanks. Owing to our limited scholastic level, criticisms on hard-to-avoid drawbacks and even mistakes are cordially invited.

Editors

Introduction

As one of the major inventions of the 20th century, the laser is playing a unique role in all facets of modern science and technology because of its light emitting mechanism different from that of common light sources and its unusually excellent characteristics, such as very good directionality, high luminance, and good mono-chromaticity and coherence. The impact is keenly felt in industrial, military, telecommunication, medicine, and scientific research. Since the appearance of the first laser in 1960, a diversity of different types of laser, have been invented one after another in the subsequent 40-odd years, mainly including solid lasers such as the Nd:YAG laser, the Nd glass laser, and the ruby laser; gas lasers such as the He-Ne laser, the Ar^+ laser, and the CO_2 laser; semiconductor lasers such as the GaAs laser and the In-GaAsP/InP laser; quasi-molecular lasers such as the XeCl laser and the KrF laser; and free electron lasers. It can be said that each of the great variety of lasers has its own peculiar performance different from others. However, the physical attributes possessed by all lasers are basically fixed and it's impossible for all the above-mentioned characteristics to be ideal. Often, the laser output from an ordinary laser may not necessarily satisfy the requirements of certain applications. In this case, in order to meet the needs of different actual applications, laser techniques intended to improve and enhance the performance of lasers have steadily been investigated and developed with R&D concerning the interaction between laser and material going on at the same time. With these techniques available, laser application has greatly expanded—so much so that a number of brand new physical phenomena have occurred, forming a series of new laser branches and fields of applied technologies, for example, laser physics, nonlinear optics, laser spectroscopy, laser medicine, and information optoelectronic technology.

It is obvious that an ordinary pulse solid laser, with an output optic pulse width of the order of several hundred μs or even ms and a peak power of the level of dozens of kW, is definitely incapable of meeting the requirements of research on precision ranging with laser, laser radar, high-speed photography, high-resolution spectroscopy, to name but a few. It was against such a background that the laser Q modulating technology and mode locking technology were investigated and developed. As far back as 1961, shortly after the laser made its debut in 1960, the concept of Q modulation was put forward, that is, it was assumed that a method could be adopted to compress all the optical radiation into an extremely narrow pulse for emission. In 1962, Hellwarth and Mcclung made the first Q modulation laser with an output peak power of 600 kW and a pulse width of the order of 10^{-7} s. The development in this area was very rapid in the subsequent few years until there appeared a multitude of Q modulating methods, such as electro-optic Q modulation, acousto-optic Q modulation, and saturable absorption Q modulation, with the output power almost abruptly rising. Great progress was also made in pulse width compression. By the 1980s, it was no longer difficult to generate giant pulses with a pulse width of the ns order and a peak power of the GW order using the Q modulating technology. The emergence of the Q modulating technology is an important breakthrough in the history of laser development that has greatly pushed forward the development of the above-mentioned applied technologies. But, constrained by the mechanism of generation, the pulse width could hardly be further narrowed using the Q modulating technology. In 1964, scientists once again proposed and realized a new mechanism of compressing the pulse width and increasing power, known as the mode locking technology. Owing to its capability of shortening the duration of a pulse to the order of picosecond (ps, 1 ps $= 10^{-12}$ s), it is also called ultrashort pulse technology. From the 1960s to the 1970s, the ultrashort pulse technology was rapidly developed. By the beginning of the 1980s, Fork and others had proposed the theory of colliding mode locking. More important, they realized colliding mode locking and obtained a stable 90fs optical pulse sequence. By virtue of its capability of generating ultrashort pulses with a pulse width greater than femtosecond (1 fs$=10^{-15}$ s) and a peak power higher than TW (1 TW $=$ 10^{12} W), the mode locking technology has provided an important means to such disciplines as physics, chemistry, biology, and spectroscopy in learning about the micro-world and ultra-fast process. Owing to the fact that the Q modulating and mode locking technology is capable of making laser radiation highly concentrated in space and time, the monochromatic brightness of laser is enabled to increase by 6 to 9 orders of magnitude over ordinary laser, a new leap in the brightness of the light source. The interaction between lasers of such a high brightness and material has triggered many significant phenomena and novel technologies, making it a powerful instrument in researches in science and technology. If the Q modulating technology is combined with the multilevel amplifying technology, laser of ultra-high power can be generated that can produce extremely high light energy density within an extremely tiny space, thus generating plasmas with a temperature of tens of millions of kilowatt/hours, making it possible to realize the reaction of controllable thermo-nuclear fusion by means of laser ignition.

In addition, for certain laser application fields the laser beam is required to possess very high quality, that is, excellent directionality and monochromaticity. But commonly used lasers often operate in multiple modes (containing the higher-order transverse mode and longitudinal mode), their divergence is rather great, and monochromaticity far from ideal. It's obvious they are incapable of satisfying the requirements of application in precision interferometry, holography, and fine machining. Therefore, in order to improve the beam quality, the mode selecting technology and frequency stabilizing technology have been investigated and developed, the former consisting in selecting the single mode (fundamental transverse mode and single longitudinal mode) from the modes of laser oscillation. Over the years, a multitude of mode selecting methods have been studied and implemented, the selection of the fundamental transverse mode having greatly improved the divergence of the beam. On the other hand, the selection of the single longitudinal mode makes it possible to obtain the single frequency laser output so as to improve the monochromaticity of laser. However, influenced by all kinds of interference from the outside, lasers can only have rather poor frequency stability. That is, the frequency (wavelength) is randomly fluctuating and hence can hardly be applied in precision measurement. Scientists have for years endeavored to study and develop various frequency stabilizing technologies that can enhance the frequency stability of laser and are of practical value. The essence of the frequency stabilizing technology lies in maintaining the stability of the optical path length in the resonator, that is, having the laser oscillation frequency locked at the standard frequency from beginning to end by means of an electronic servo-control system. In the mid-1960s, the center frequency of the atomic spectral line was chosen as the reference standard. For instance, in 1965, the Lamb dip was used as the reference frequency to perform frequency stabilization for a He-Ne laser. The frequency stability obtained was 10^{-9}, but the reproducibility was only 10^{-7}. After 1966, further attempts were made to use the external reference frequency as a standard for frequency stabilization, e.g., the saturated absorption frequency stabilizing method, which consists in using the absorption lines of some molecular gases as the reference frequency. In so doing, the influences of discharge perturbation and pressure broadening could be avoided, thus helping improve the frequency stability. In 1969, making use of CH_4 molecules to perform frequency stabilization for the He-Ne laser's 3.39-μm wavelength, Barger and Hell obtained a frequency stability of 10^{-14} and a reproducibility of 3×10^{-12}. Apart from this, iodine is also a frequently used absorption molecule, e.g., $^{127}I_2$ and $^{129}I_2$. The frequency stabilization of the 633-μm wavelength for the He-Ne laser, too, yielded very high stability and reproducibility rate. During this decade, there emerged the frequency stabilizing technology for other lasers such as Ar^+ and CO_2 lasers. The adoption of the laser mode selecting technology and frequency stabilizing technology has made it possible to obtain high quality beams with excellent frequency stability and extremely small angle of divergence. This will not only satisfy the requirements of such applications as precision measurement and holography, but also, with the appearance of advanced frequency stabilizing technology, the standards for length and time frequency have been unified, that is, in the international measuring standards, the international standard that defines the laser wavelength in "meter" can also be used as a standard for the time frequency "second". Without a doubt, this will exert an extremely far-reaching influence on all fields of physics.

After the laser became commercially available, people immediately began investigating its application in the information technology (the transmission, storage, and processing of information). As laser is a light frequency electromagnetic wave with an extremely high transmission speed and very high frequency and, as a carrier wave, it has a very large content of information capable of providing an excellent information carrier source to applications in such areas as optical communication, optical information processing, etc. Hence the appearance of all kinds of laser modulation technologies one after another. With the unceasing availability of various optical crystal materials, certain physical effects such as the electro-optic, acousto-optic, and magneto-optic effects were successfully utilized to develop a diversity of optical modulation devices and technologies, thereby realizing laser loaded information. In particular, from the end of the 1960s to the beginning of the 1970s, the new conception of the double heterojunction semiconductor laser was put forward by Kressel, Alferov et al., who also succeeded in implementing devices for continuous operation at room temperature. The British scientists of Chinese extraction Gao Kun and Hockham proposed the new concept of light guide fiber based on the principle of total reflection. On the basis of their work, Kapron and others of the American Conning Company successfully developed practical fiber optic products four years later, unraveling the history of vigorous development of the fiber optic communication technology. In addition, in the recent 10-odd years, spatial light modulators have successfully been developed. The very name suggests that they are a kind of device for modulating the distribution of light wave in space. As the devices possess the function of spatially performing real-time modulation of a light beam, they have become the crucial devices in such systems as those of real-time optical information processing, optical computation, optical storage, and optical neural network (ONN), greatly pushing forward the rapid development of the applied technology in those fields.

Prior to the emergence of the laser, the interaction between light and material appeared to be a linear relation while after its appearance, especially following the utilization of the Q modulating and mode locking technology, many highly significant new phenomena and new effects (nonlinear optical effects) ensued, accompanied by the production of a number of nonlinear optical technologies. In 1961, Franken and others observed the second harmonic radiation of ruby laser by focusing the ruby laser beam onto the quartz crystal, which is the phenomenon of frequency doubling. But as the experiment made by Franken and his coworkers was non-phase matched, the conversion efficiency of the second harmonic was very low, about 10^{-8}. By 1962, Kleinman, Giordmaine, and Maker had put forward their phase matching technology, which consists in using the birefringence effect to achieve phase speed matching, thus realizing effective doubling of frequency. In 1965, the theory of nonlinear optics was approaching perfection day by day, with many important nonlinear optical phenomena occurring one after another, for instance, the generation of photomixing (sum frequency, difference frequency), optical parametric amplification and oscillation, multiple photon absorption, self-focusing, and stimulated scattering. With the development of the laser technology and nonlinear optical materials, the above-mentioned nonlinear optical phenomena and effects were extensively applied in expanding the laser band (e.g., the laser frequency converter) and changing or controlling the parameters of lasers (e.g., pulse width, power, frequency, stability). Furthermore, the means of investigating the microscopic properties of material (atoms or molecules) was also provided, thus opening up broad vistas of application for laser. A look ahead makes it clear that the effects of the nonlinear interaction between light and material and researches on its application in various nonlinear optical devices will still be one of the important research directions in the years to come, such as the nonlinear effects of optical fiber in fiber optic communication and the formation and transmission of optical solitons. In a word, the reason lasers can show their magic power in so many fields is because use has been made of the combination and operation of different types of lasers and the relevant laser technologies.

The laser technology involves the theoretical knowledge of multiple disciplines and is itself of diversified kinds and daily updated in terms of development. But as far as its fundamentals are concerned, most are implemented based on the utilization of the physical effects induced by the interaction of light with different kinds of material, mainly a discussion on the electro-optic effect, acousto-optic effect, magneto-optic effect, and nonlinear optical effect as well as the adoption of different forms of use in controlling a certain parameter (energy, power, polarization, mode, line width, and pulse width). Despite the different functions of different laser devices and the steady increase in their assortment, the principle is the same, the basic physical laws are invariable. Therefore, as long as these laws and knowledge are grasped, one will be enabled to learn more from what has been learned by inference and solve numerous technical problems. By the 1980s, the laser technology had already developed to the stage of maturity, and its contents including basic theories and techniques, also, had been so enriched that they had become an important component in the development of the disciplines of optoelectronic technology and optoelectronic information. For this reason, it is the indispensable basic knowledge for people engaged in research on optoeletronic technology and its application in different areas. It is the main purpose of this textbook to give a detailed account of the basic concepts and theories, as well as the principal kinds of laser technology, including the modulation technology, Q modulation technology (Q-switching), ultrashort pulse technology, amplifying technology, mode selection technology, frequency stabilizing technology, nonlinear optical technology, and transmission technology and the fundamentals of the role played by the physical effects in technical devices and implementing methods so that the reader will have a fairly clear and systematic understanding of the physical processes of different laser technologies.

Translated by Cao Huamin

Revised by Lan Xinju

Contents

CHAPTER 1

Laser Modulation and Deflection Technology

1.1 The basic concept of modulation

Laser is a kind of light frequency electromagnetic wave with good coherence. Like the radio wave, it can be used as an information transmitting carrier. As laser possesses very high frequency, as high as $10^{13}\sim10^{15}$ Hz, the frequency band available is very wider and hence a great volume of information is transmitted. Furthermore, owing to the extremely short wavelength and extremely rapid transmission speed possessed by laser, what with the independent transmission characteristics of the optic wave, the 2-dimensional information on a plane can be instantly transmitted onto another with very high resolution, providing conditions for 2-D parallel optic information processing. So, laser is an ideal light source for transmitting information, including speech, language, images, and symbols.

If laser is intended to serve as a carrier of information, it is necessary to solve the problem of loading the information onto laser. For instance, for the laser telephone, it is necessary to load the speech information onto laser to let it "carry" the information through a definite transmission path (atmosphere, optical fiber) to the receiver; then it will be identified by the optical receiver and reduced to the original information so as to achieve the goal of communication by telephone. Such a process of loading information onto laser is called modulation and the device by means of which this process is completed is the modulator, while laser is spoken of as carrier and the low frequency information that plays the role of control is called the modulating signal.

The electric field intensity of the laser optical wave is

$$e_c(t) = A_c\cos(\omega_c t + \varphi_c) \tag{1.1-1}$$

where A_c is amplitude, ω_c angular frequency, and φ_c phase angle. Since laser possesses such parameters as amplitude, frequency, phase, intensity, and polarization, if a certain physical method can be used to change a certain parameter of the optic wave to make it vary in accordance with the law of the modulating signal, then laser is modulated by signal so that the goal of "transporting" information is attained. There are many methods for implementing laser modulation. Depending on the relative relation of the modulator with the laser, they can be divided into external modulation and internal modulation. By internal modulation we mean that the loading of the modulating signal is carried out in the process of oscillation of laser, that is, the modulating signal is used to change the oscillation parameters of the laser so as to change the output characteristics of laser to realize modulation. For instance, the injected semiconductor laser uses the modulation signal to directly change its pump-driven electric current, making the intensity of the laser output modulated (this method is also spoken of as direct modulation). There is another internal modulation method that consists in having the modulation components placed in the laser resonator and using the modulating signal to control the variation of the physical characteristics of the components to change the parameters of the resonator, thereby changing the output characteristics of the laser. The Q modulation technology described in Chapter 2 is in effect a modulation of this type. Internal modulation is currently used mainly in the light source of the injected semiconductor in optical communication. By external modulation we mean placing the modulator on the optical path outside the laser after the formation of laser while changing the physical characteristics of the modulator with the modulation signal. When laser passes through the modulator, a certain parameter of the optic wave will get modulated. External modulation is easily adjusted and will not affect the laser. In addition, the mode of external modulation is not limited by the operation velocity of the semiconductor devices. Hence its modulation velocity is higher than that of internal modulation, about one order of magnitude higher, and the modulation bandwidth is much wide. Therefore, in the application in high speed large volume optical communication and optical information processing in future, it is bound to capture greater attention.

In terms of the properties modulated, laser modulation can be divided into amplitude modulation, frequency modulation, phase modulation, and intensity modulation, the concept of each of which will be briefly discussed below.

1.1.1 Amplitude modulation

Amplitude modulation means the oscillation of the amplitude of carrier as it varies according to the law of the modulating signal. Suppose the electric field intensity of the laser carrier is as shown in Eq. (1.1-1). If the modulation signal is a cosine function of time, then

$$a(t) = A_{\mathrm{m}}\cos\,\omega_{\mathrm{m}}t \tag{1.1-2}$$

where A_{m} is the amplitude of the modulating signal, ω_{m} the angular frequency of the modulation signal. After the laser amplitude is modulated, the laser amplitude A_{c} in Eq. (1.1-1) is no longer a constant, but a function proportional to the modulation signal, the expression of whose amplitude-modulated wave is

$$e(t) = A_{\mathrm{c}}(1 + m_{\mathrm{a}}\cos\omega_{\mathrm{m}}t)\cos(\omega_{\mathrm{c}}t + \varphi_{\mathrm{c}}) \tag{1.1-3}$$

Expanding the above equation with the formula of triangular function, we have the frequency spectrum formula of the AM wave

$$\begin{aligned} e(t) = {} & A_{\mathrm{c}}\cos(\omega_{\mathrm{c}}t + \varphi_{\mathrm{c}}) + \frac{m_{\mathrm{a}}}{2}A_{\mathrm{c}}\cos\{(\omega_{\mathrm{c}} + \omega_{\mathrm{m}})t + \varphi_{\mathrm{c}}\} \\ & + \frac{m_{\mathrm{a}}}{2}A_{\mathrm{c}}\cos\{(\omega_{\mathrm{c}} - \omega_{\mathrm{m}})t + \varphi_{\mathrm{c}}\} \end{aligned} \tag{1.1-4}$$

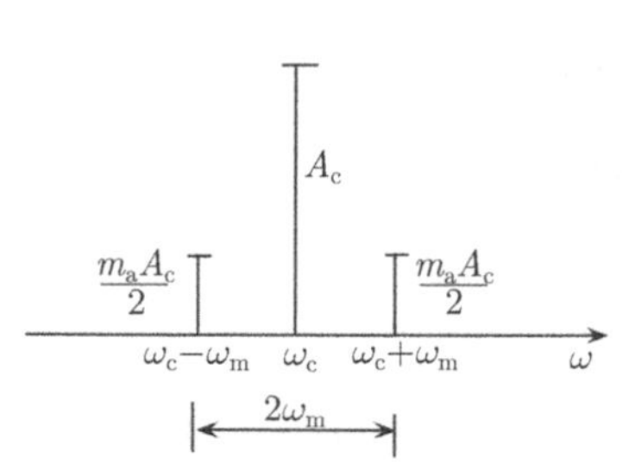

Fig. 1.1-1 The frequency spectrum of the AM wave

where $m_{\mathrm{a}} = A_{\mathrm{m}}/A_{\mathrm{c}}$ is called the AM coefficient. It can be seen from Eq. (1.1-4) that the frequency spectrum of the AM wave is composed of three frequency components, of which the first is the carrier frequency component, while the second and third are generated because of the modulation and are called side frequency components, as shown in Fig. 1.1-1. The above analysis is made on the case of the modulation of the single frequency cosine signal. If the modulating signal is a complicated periodic signal, then the frequency spectrum of the AM wave will be made up of the carrier frequency component and two side frequency bands.

1.1.2 Frequency modulation and phase modulation

Frequency modulation and phase modulation are simply the oscillation of frequency or phase of the optical carrier that varies according to the variation law of the modulating signal. As the two kinds of modulated wave are both manifested as a change in the total phase angle $\varphi(t)$, they are called angular modulation.

For frequency modulation, it should be noted that the angular frequency ω_c in Eq. (1.1-1) is no longer a constant, but varies with the modulating signal, i.e.,

$$\omega(t) = \omega_\mathrm{c} + \Delta\omega(t) = \omega_\mathrm{c} + k_\mathrm{f}a(t) \tag{1.1-5}$$

If the modulating signal is still a cosine function, then the total phase angle of the FM wave is

$$\begin{aligned}\varphi(t) &= \int_0^t \omega(t)\mathrm{d}t + \varphi_\mathrm{c} = \int_0^t [\omega_\mathrm{c} + k_\mathrm{f}a(t)]\mathrm{d}t + \varphi_\mathrm{c} \\ &= \omega_\mathrm{c}t + \int_0^t k_\mathrm{f}a(t)\mathrm{d}t + \varphi_\mathrm{c}\end{aligned} \tag{1.1-6}$$

Substituting Eq. (1.1-6) into Eq. (1.1-1), we have the expression of the frequency modulated wave

$$e(t) = A_\mathrm{c}\cos(\omega_\mathrm{c}t + m_\mathrm{f}\sin\omega_\mathrm{m}t + \varphi_\mathrm{c}) \tag{1.1-7}$$

where k_f is called the proportionality factor and $m_\mathrm{f} = \Delta\omega/\omega_\mathrm{m}$ is called the frequency modulating factor.

Similarly, phase modulation is simply the variation of phase angle φ_c in Eq. (1.1-1) with the variation of the modulating signal; the total phase angle of the phase-modulated wave is

$$\begin{aligned}\varphi(t) &= \omega_\mathrm{c}t + \varphi_\mathrm{c} + k_\varphi a(t) \\ &= \omega_\mathrm{c}t + \varphi_\mathrm{c} + k_\varphi A_\mathrm{m}\cos\omega_\mathrm{m}(t)\end{aligned} \tag{1.1-8}$$

Then the expression of the phase-modulated wave is

$$e(t) = A_\mathrm{c}\cos(\omega_\mathrm{c}t + m_\varphi\cos\omega_\mathrm{m}t + \varphi_\mathrm{c}) \tag{1.1-9}$$

where $m_\varphi = k_\varphi A_\mathrm{m}$ is called the phase modulating factor.

Let's turn to look at the frequency spectrum of the FM and PM wave. Since FM and PM are in the final analysis modulation of the total phase angle, they can be written in a unified form:

$$e(t) = A_\mathrm{c}\cos(\omega_\mathrm{c}t + m\sin\omega_\mathrm{m}t + \varphi_\mathrm{c}) \tag{1.1-10}$$

Expanding Eq. (1.1-10) with the triangular formula, we have

$$\begin{aligned}e(t) = A_\mathrm{c}\{&\cos(\omega_\mathrm{c}t + \varphi_\mathrm{c})\cos(m\sin\omega_\mathrm{m}t) \\ &- \sin(\omega_\mathrm{c}t + \varphi_\mathrm{c})\sin(m\sin\omega_\mathrm{m}t)\}\end{aligned} \tag{1.1-11}$$

Expand $\cos(m \sin \omega_\mathrm{m}t)$ and $\sin(m \sin \omega_\mathrm{m}t)$ in the above equation into

$$\cos(m\sin\omega_\mathrm{m}t) = J_0(m) + 2\sum_{n=1}^{\infty} J_{2n}(m)\cos(2n\omega_\mathrm{m}t)$$

$$\sin(m\sin\omega_\mathrm{m}t) = 2\sum_{n=1}^{\infty} J_{2n-1}(m)\sin[(2n-1)\omega_\mathrm{m}t]$$

When the modulation factor m is known, the values of all orders of the Bessel functions can be found in the table of Bessel functions. Substituting the above two equations into Eq. (1.1-11) and performing expansion, we have

$$\begin{aligned}e(t) &= A_\mathrm{c}\{J_0(m)\cos(\omega_\mathrm{c}t + \varphi_\mathrm{c}) + J_1(m)\cos[(\omega_\mathrm{c} + \omega_\mathrm{m})t + \varphi_\mathrm{c}] \\ &\quad - J_1(m)\cos[(\omega_\mathrm{c} - \omega_\mathrm{m})t + \varphi_\mathrm{c}] + J_2(m)\cos[(\omega_\mathrm{c} + 2\omega_\mathrm{m})t + \varphi_\mathrm{c}] \\ &\quad + J_2(m)\cos[(\omega_\mathrm{c} - 2\omega_\mathrm{m})t + \varphi_\mathrm{c}] + \cdots\} \\ &= A_\mathrm{c}J_0(m)\cos(\omega_\mathrm{c}t + \varphi_\mathrm{c}) + A_\mathrm{c}\sum_{n=1}^{\infty} J_n(m)[(\cos\omega_\mathrm{c} + n\omega_\mathrm{m})t + \varphi_\mathrm{c} \\ &\quad + (-1)^n\cos(\omega_\mathrm{c} - n\omega_\mathrm{m})t + \varphi_\mathrm{c}]\end{aligned} \tag{1.1-12}$$

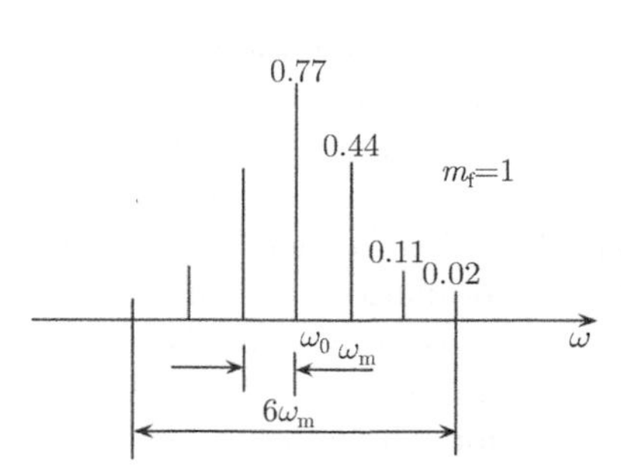

Fig. 1.1-2 The frequency spectrum of the angular modulation wave

It can be seen that during single-frequency sinusoidal modulation, the frequency spectrum of the angular modulation wave is made up of the light carrier frequency and the infinitely many pairs of side frequencies symmetrically distributed on its two sides. The frequency interval between the side frequencies is ω_m and the magnitude $J_n(m)$ of the side frequencies is determined by the Bessel function. If $m = 1$, it will be found from the table of Bessel functions that $J_0(m) = 0.77$, $J_1(m) = 0.44$, $J_2(m) = 0.1$, $J_3(m) = 0.02$, $\cdots$. The frequency distribution is as shown in Fig. 1.1-2. Obviously, if the modulating signal is not a single-frequency sinusoidal wave, its spectrum will be even more complex. In addition, if the angle modulating factor is rather small, i.e., $m \ll 1$, its frequency spectrum has the same form as that of the AM wave.

1.1.3 Intensity modulation

Intensity modulation is the laser oscillation of the optical carrier wave intensity (light intensity) that varies in obedience to the law of the modulating signal, as shown in Fig. 1.1-3. Usually, the form of intensity modulation is adopted for laser modulation since the receiver (detector) in general directly responds to the variation of the light intensity received.

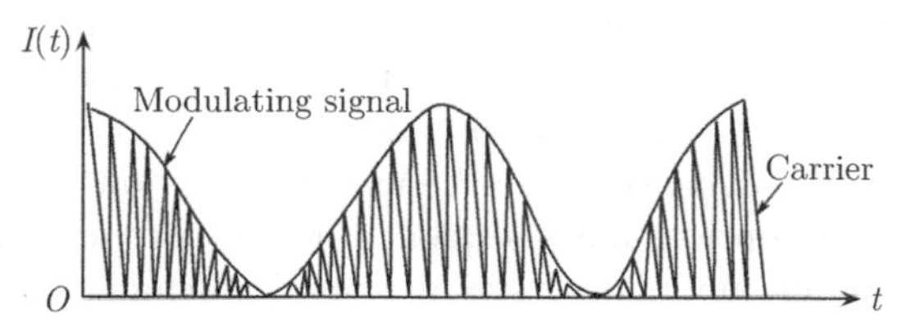

Fig. 1.1-3 Intensity modulation

The light intensity of laser is defined as the square of the electric field of the optical wave, whose expression is

$$I(t) = e^2(t) = A_c^2 \cos^2(\omega_c t + \varphi_c) \tag{1.1-13}$$

Thus, the expression for light intensity of intensity modulation can be written as

$$I(t) = \frac{A_c^2}{2}[1 + k_p a(t)] \cos^2(\omega_c t + \varphi_c) \tag{1.1-14}$$

where k_p is the factor of proportionality. Suppose the modulating signal is a single frequency cosine wave $a(t) = A_m \cos \omega_m t$, which we substitute into Eq. (1.1-14). Letting $k_p A_m = m_p$ (called the intensity modulation factor), we have

$$I(t) = \frac{A_c^2}{2}[1 + m_p \cos \omega_m t] \cos^2(\omega_c t + \varphi_c) \tag{1.1-15}$$

This is a fairly ideal formula for light intensity modulation when the modulation factor $m_p \ll 1$. The frequency spectrum of the light intensity modulating wave can be obtained using a method similar to the one described previously, but the results are slightly different from those of the spectrum for the modulated wave. With respect to the frequency spectrum distribution, apart from the carrier frequency and the symmetrically distributed side frequencies, there is the low frequency ω_{m} as well as the DC component.

In actual application, in order to obtain sufficiently strong anti-interference effects, people often use the secondary modulation mode. That is, first the low frequency signal is used to modulate the frequency of a high frequency subcarrier, then this frequency modulated carrier will be used for intensity modulation, called FM/IM modulation, to cause the intensity of light to vary with the variation of the subcarrier. This is because in the process of transmission, although the atmospheric agitation and other interference wave will be directly superposed onto the optical signal wave, after demodulation, the information will be contained in the frequency modulated subcarrier. So the information will not be interfered with and the original information can be reproduced distortion-free.

1.1.4 Pulse modulation

The modulated waves obtained in the above-mentioned modes of modulation are all continuously oscillating waves, referred to as simulated modulation. In addition, in current optical communication, there are pulse modulation and digital modulation, also spoken of as pulse code modulation, that are extensively adopted for modulation in the discontinuous state. For this kind of modulation, usually electrical modulation (simulated pulse modulation or digital pulse modulation) is performed first, followed by light intensity modulation for the light carrier.

Pulse modulation is a modulating method that is implemented by using an intermittent periodic pulse sequence as the carrier, a certain parameter of which varies in obedience to the law of the modulating signal. That is, first the simulated modulating signal is used to perform electrical modulation for a certain parameter (amplitude, width, frequency, position, etc) of an electric pulse sequence, to make it vary in accordance with the law of the modulating signal, as shown in Fig. 1.1-4, to make it a pulse modulated sequence. Then, this modulated electric pulse sequence should be used to carry out intensity modulation for the light carrier so that the optical pulse sequence of the corresponding variation can be obtained. For instance, if the modulating signal is used to change the time at which each pulse in the electric pulse sequence is generated, then the position of each pulse and that prior to modulation will have a displacement that is proportional to the modulating signal. Such a modulation is called pulse position modulation (PPM), as shown in Fig. 1.1-4(e). Then, by modulating the light carrier wave emitted by the light source, the corresponding optical pulse position modulated wave can be obtained, whose expression is as follows:

$$\begin{aligned} e(t) &= A_{\mathrm{c}}\cos(\omega_{\mathrm{c}}t + \varphi_{\mathrm{c}}) \quad (\text{When } t_{\mathrm{n}} + \tau_{\mathrm{d}} \leqslant t \leqslant t_{\mathrm{n}} + \tau_{\mathrm{d}} + \tau) \\ \tau_{\mathrm{d}} &= \frac{\tau_{\mathrm{p}}}{2}[1 + M(t_{\mathrm{n}})] \end{aligned} \tag{1.1-16}$$

where $M(t_{\mathrm{n}})$ is the amplitude of the modulating signal, and τ_{d} is the retardation of the carrier pulse front edge relative to the sampling time t_{n}. To prevent the pulse from getting overlapped onto the period of the adjacent sample, the maximum retardation of the pulse must be smaller than the period of the sample τ_{p}.

If the modulating signal causes the repeating frequency of the pulse to vary so that the range of frequency shift is proportional to the amplitude of the modulating signal voltage independently of the modulation frequency, then such a modulation is called pulse frequency modulation (PFM). The pulse frequency modulated wave is expressed as

$$e(t) = A_{\mathrm{c}}\cos\left(\omega_{\mathrm{c}}t + \Delta\omega\int M(t_{\mathrm{n}})\mathrm{d}t + \varphi_{\mathrm{c}}\right) \quad (\text{When } t_{\mathrm{n}} \leqslant t \leqslant t_{\mathrm{n}} + \tau) \tag{1.1-17}$$

For both pulse position modulation and pulse frequency modulation, a light pulse of very narrow width can be adopted, with the shape of the pulse unchanged but only the position of the pulse or the repeating frequency varying with the variation of the modulating signal. With a fairly strong anti-interference ability, both modulating methods are widely used in optical communication.

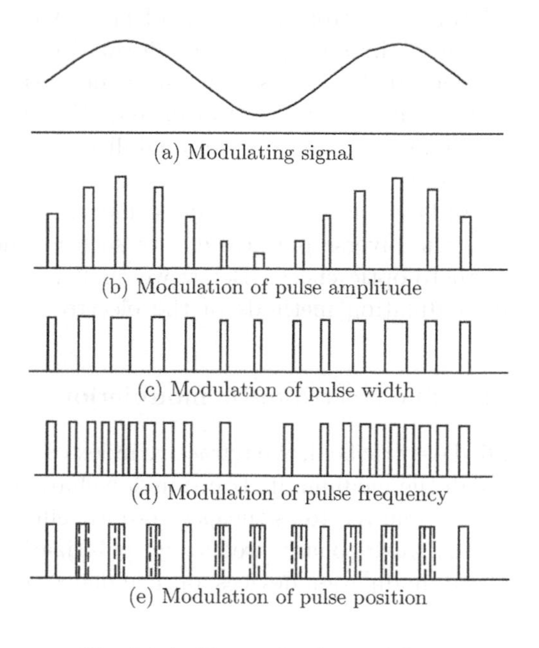

Fig. 1.1-4 Forms of pulse modulation

1.1.5 Pulse coding modulation

This kind of modulation consists in converting the simulating signal into an electric pulse sequence first and then the binary code representing the signal information (PCM digital signal) for intensity modulation of the light carrier prior to performing information transmission.

The implementation of pulse code modulation has to undergo three processes, namely, sampling, quantization and coding.

1. Sampling

By sampling we mean segmenting a continuous signal wave into a discontinuous pulse wave, represented by a pulse sequence of a definite period, and the amplitude of the pulse sequence (called sample value) is in correspondence to the amplitude of the signal wave. That is to say, after sampling, the original simulating signal becomes a pulse amplitude modulated signal. According to the theorem of sampling, if only the sampling frequency is over twofold higher than the highest frequency of the signal transmitted, the waveform of the original signal can be restored.

2. Quantization

By quantization we mean performing level by level round-off treatment of the pulse amplitude modulated wave after sampling by replacing the magnitude of the sampling value with a limited number of representative values. This process is called quantization. So, a sample cannot become a digital signal until it has undergone the process of quantization.

3. Coding

The process of transforming a quantized digital signal into the corresponding binary code is called coding; that is, a group of pulses of equal amplitude and equal width are used as "numerals" and "with" pulse and "without" pulse are used to represent the numerals "1" and "0", respectively. Then the series of electric pulses that reflect the law of the digital signal are added to a modulator to control the output of laser, with the maximum of the laser carrier representing "1" bit of the binary code of information sample amplitude while the zero value of the laser carrier, "0" bit. Thus, different combinations of numerals will be able to represent the information intended to be transmitted. For this form of modulation, both a broader bandwidth and high anti-interference capability are required. So it is widely used in digital fiber communication.

Despite the different kinds of laser modulation, the mechanism of operation of the modulation is invariably based on the diverse physical effects such as the electro-optic effect, acousto-optic effect, and magneto-optic effect. Below we shall have a discussion on each of the basic principles and modulating methods of the electro-optic, acousto-optic, and magneto-optic modulation.

1.2 The electro-optic modulation

Under the action of an applied electric field, the refractive index of certain crystals will vary. When a light wave goes through the medium, its transmission characteristics vary under the influence. Such a phenomenon is referred to as the electro-optic effect, which is widely used to realize control over the light wave (phase, frequency, polarization, and intensity) and developed as a variety of optical modulation devices, optical deflection devices, electro-optic filtering devices, etc.

1.2.1 The physical basis of electro-optic modulation

The law of propagation of light wave in a medium is constrained by the distribution of the medium refractive index which, in turn, is closely related to its medium constant. It has been proved by both theory and experiment that the medium's dielectric constant is related to the distribution of the electrical charges in the crystal. When an electric field is applied on the crystal, a redistribution of the bound electrical charges will ensue and may lead to a slight deformation of the ionic lattice. The result will be a change in the dielectric constant until the change in the refractive index of the crystal. So the refractive index will become a function of the applied electric field and the crystal's refractive index can now be expressed by the power series of the applied electric field E as

$$n = n_0 + \gamma E + hE^2 + \cdots \tag{1.2-1}$$

or

$$\Delta n = n - n_0 = \gamma E + hE^2 + \cdots \tag{1.2-2}$$

where γ and h are constants and n_0 is the refractive index before application of the electric field. In Eq. (1.2-2), γE is the primary term, the variation of the refractive index caused by what is known as the linear electro-optic effect or Pockels effect while the variation of the refractive index induced by the secondary term hE^2 is called the secondary electro-optic effect or Kerr effect. For most electro-optic crystalline materials, the primary effect is more appreciable than the secondary effect and the secondary term can be neglected (only in a centrosymmetric crystal, owing to the absence of the primary electro-optic effect, will the secondary electro-optic effect become rather obvious). So we shall only discuss the linear electro-optic effect.

1. The electrically induced variation of the index of refraction

There are two methods for analyzing and describing the electro-optic effect: the method using the electromagnetic theory and that using geometric figures. The former involves tedious and complicated mathematical derivation while the latter, owing to its use of the refractive index ellipsoid, also called indicatrix, is intuitive and convenient and is therefore preferred in practice.

When no external electric field is applied on the crystal, the refractive index ellipsoid in the coordinate system of the principal axis is depicted by the following equation:

$$\frac{x^2}{n_x^2}+\frac{y^2}{n_y^2}+\frac{z^2}{n_z^2}=1 \tag{1.2-3}$$

where x, y, and z represent the directions of the principal axis of the medium, that is to say, within the crystal the electric displacement D along these directions and electric field intensity are parallel to each other; n_x, n_y, and n_z are the main refractive indices of the refractive index ellipsoid. As this equation can be used to depict the characteristics of the light wave propagation in the crystal, it is possible to infer the influence of the crystal on the law of optical wave propagation after an external electric field is applied on the crystal. An analysis can also be made by the aid of the variation of the factors $\frac{1}{n_x^2}, \frac{1}{n_y^2}, \frac{1}{n_z^2}$ in the equation of the index ellipsoid.

When an electric field is applied on the crystal, "deformation" of the ellipsoid will take place. The corresponding equation of the index ellipsoid will change into the following form:

$$\left(\frac{1}{n^2}\right)_1 x^2+\left(\frac{1}{n^2}\right)_2 y^2+\left(\frac{1}{n^2}\right)_3 z^2+2\left(\frac{1}{n^2}\right)_4 yz+2\left(\frac{1}{n^2}\right)_5 xz+2\left(\frac{1}{n^2}\right)_6 xy=1 \tag{1.2-4}$$

It is known by comparing Eq. (1.2-3) with Eq. (1.2-4) that, owing to the action of the external electric field, the factors $(1/n^2)$ of the index ellipsoid will linearly vary as a result, the quantity of variation being defined as

$$\Delta\left(\frac{1}{n^2}\right)_i=\sum_{j=1}^{3}\gamma_{ij}E_j \tag{1.2-5}$$

where γ_{ij} is called the linear electro-optic factor, i takes values $1\sim 6$, and j takes values $1\sim3$. Equation (1.2-5) can be expressed in the matrix form of a tensor as

$$\begin{bmatrix} \Delta\left(\frac{1}{n^2}\right)_1 \\ \Delta\left(\frac{1}{n^2}\right)_2 \\ \Delta\left(\frac{1}{n^2}\right)_3 \\ \Delta\left(\frac{1}{n^2}\right)_4 \\ \Delta\left(\frac{1}{n^2}\right)_5 \\ \Delta\left(\frac{1}{n^2}\right)_6 \end{bmatrix}=\begin{bmatrix} \gamma_{11} & \gamma_{12} & \gamma_{13} \\ \gamma_{21} & \gamma_{22} & \gamma_{23} \\ \gamma_{31} & \gamma_{32} & \gamma_{33} \\ \gamma_{41} & \gamma_{42} & \gamma_{43} \\ \gamma_{51} & \gamma_{52} & \gamma_{53} \\ \gamma_{61} & \gamma_{62} & \gamma_{63} \end{bmatrix}\begin{bmatrix} E_x \\ E_y \\ E_z \end{bmatrix} \tag{1.2-6}$$

where E_x, E_y, and E_z are components of the electric field along the x, y, and z directions; the 6×3 matrix with elements γ_{ij} is called the electro-optic tensor. The value of each element is determined by the specific crystal and is a quantity that characterizes the strength of inductance polarization. Below we shall make an analysis with the frequently used KDP crystal.

KDP (KH_2PO_4) class crystals belong in the tetragonal system, $\bar{4}$2m point group, and are negative single axis crystals. Hence there is $n_x = n_y = n_{\mathrm{o}}$, $n_z = n_{\mathrm{e}}$, and $n_{\mathrm{o}} > n_{\mathrm{e}}$. The electro-optic tensor of such crystals is

$$[\gamma_{ij}] = \begin{bmatrix} 0 & 0 & 0 \\ 0 & 0 & 0 \\ 0 & 0 & 0 \\ \gamma_{41} & 0 & 0 \\ 0 & \gamma_{52} & 0 \\ 0 & 0 & \gamma_{63} \end{bmatrix} \tag{1.2-7}$$

Furthermore, $\gamma_{41} = \gamma_{52}$. Hence such crystals have only two independent electro-optic factors, i.e., γ_{41} and γ_{63}. Substitution of Eq. (1.2-7) into Eq. (1.2-6) yields

$$\begin{aligned} &\Delta\left(\frac{1}{n^2}\right)_1 = 0, \quad &&\Delta\left(\frac{1}{n^2}\right)_4 = \gamma_{41}E_x \\ &\Delta\left(\frac{1}{n^2}\right)_2 = 0, \quad &&\Delta\left(\frac{1}{n^2}\right)_5 = \gamma_{41}E_y \\ &\Delta\left(\frac{1}{n^2}\right)_3 = 0, \quad &&\Delta\left(\frac{1}{n^2}\right)_6 = \gamma_{63}E_z \end{aligned} \tag{1.2-8}$$

Substituting Eq. (1.2-8) into Eq. (1.2-4), we obtain the equation for the new index ellipsoid:

$$\frac{x^2}{n_{\mathrm{o}}^2} + \frac{y^2}{n_{\mathrm{o}}^2} + \frac{z^2}{n_{\mathrm{e}}^2} + 2\gamma_{41}yzE_x + 2\gamma_{41}xzE_y + 2\gamma_{63}xyE_z = 1 \tag{1.2-9}$$

It can be seen from Eq. (1.2-9) that the applied electric field has led to the appearance of the "crossed" term in the equation for the index ellipsoid, which shows that, upon application of the electric field, the principal axis of the ellipsoid will no longer be parallel to the x^-, y^-, and z-axes. Therefore, we have to find a new coordinate system, so that Eq. (1.2-9) will become the "principal axis" in this coordinate system since only by so doing will it be possible to determine the effect of the electric field on light propagation. For simplicity, make the direction of the applied electric field parallel to the z-axis, that is, $E_z = E$, $E_x = E_y = 0$. Thus, Eq. (1.2-9) becomes

$$\frac{x^2}{n_{\mathrm{o}}^2} + \frac{y^2}{n_{\mathrm{o}}^2} + \frac{z^2}{n_{\mathrm{e}}^2} + 2\gamma_{63}xyE_z = 1 \tag{1.2-10}$$

In order to seek a new coordinate system (x', y', z') so that the equation for the ellipsoid will not contain any crossed term, we have the following form:

$$\frac{x'^2}{n_{x'}^2} + \frac{y'^2}{n_{y'}^2} + \frac{z'^2}{n_{z'}^2} = 1 \tag{1.2-11}$$

where x', y', and z' represent the directions of the ellipsoid's principal axis after application of the electric field. This principal axis is usually called the inductive principal axis; $n_{x'}$,

$n_{y'}$, and $n_{z'}$ are the principal refractive indices in the new coordinate system. As x and y in Eq. (1.2-10) are symmetrical, the x, y coordinates can be rotated around the z-axis an α angle. Thus the transformation relation from the old coordinate system into the new one is

$$\begin{aligned} x &= x' \cos\alpha - y \sin\alpha \\ y &= x' \sin\alpha + y' \cos\alpha \end{aligned} \tag{1.2-12}$$

Substituting Eq. (1.2-12) into Eq. (1.2-10), we have

$$\begin{aligned} &\left[\frac{1}{n_\text{o}^2} + \gamma_{63}E_z \sin 2\alpha\right] x'^2 + \left[\frac{1}{n_\text{o}^2} - \gamma_{63}E_z \sin 2\alpha\right] y'^2 \\ &+ \frac{1}{n_\text{e}^2} z'^2 + 2\gamma_{63}E_z \cos 2\alpha x'y' = 1 \end{aligned} \tag{1.2-13}$$

Letting the crossed term be zero, that is, cos2α = 0, we find $\alpha = 45°$ and the equation becomes

$$\left(\frac{1}{n_\text{o}^2} + \gamma_{63}E_z\right) x'^2 + \left(\frac{1}{n_\text{o}^2} - \gamma_{63}E_z\right) y'^2 + \frac{1}{n_\text{e}^2} z'^2 = 1 \tag{1.2-14}$$

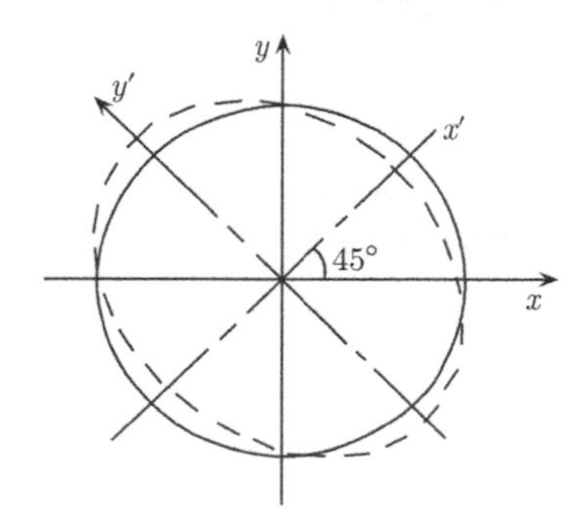

Fig. 1.2-1 Deformation of ellipsoid after application of electric field

This is the new equation of ellipsoid after an electric field is applied on the KDP class crystal along the z-axis, as shown in Fig. 1.2-1. Comparing Eq. (1.2-14) with Eq. (1.2-11), we have

$$\begin{aligned} \frac{1}{n_{x'}^2} &= \frac{1}{n_\text{o}^2} + \gamma_{63}E_z \\ \frac{1}{n_{y'}^2} &= \frac{1}{n_\text{o}^2} - \gamma_{63}E_z \\ \frac{1}{n_{z'}^2} &= \frac{1}{n_\text{e}^2} \end{aligned} \tag{1.2-15}$$

As γ_{63} is very small (about 10^{-10} m/V), in general, $\gamma_{63}E_z \ll \dfrac{1}{n_\text{o}^2}$, using the differential formula $\mathrm{d}\left(\dfrac{1}{n^2}\right) = -\dfrac{2}{n^3}\mathrm{d}n$, that is, $\mathrm{d}n = -\dfrac{n^3}{2}\mathrm{d}\left(\dfrac{1}{n^2}\right)$, we have

$$\left.\begin{aligned} \Delta n_x &= -\frac{1}{2}n_\text{o}^3\gamma_{63}E_z \\ \Delta n_y &= \frac{1}{2}n_\text{o}^3\gamma_{63}E_z \\ \Delta n_z &= 0 \end{aligned}\right\} \tag{1.2-16}$$

So

$$\left.\begin{aligned} n_{x'} &= n_\text{O} - \frac{1}{2}n_\text{o}^3\gamma_{63}E_z \\ n_{y'} &= n_\text{O} + \frac{1}{2}n_\text{o}^3\gamma_{63}E_z \\ n_{z'} &= n_\text{e} \end{aligned}\right\} \tag{1.2-17}$$

It can be seen that, when an electric field is applied on the KDP crystal along the z-axis, it is turned from a single-axis crystal into a bi-axial one, the principal axis of the refractive index ellipsoid having rotated round the z-axis an angle of 45°. This rotational angle has nothing to do with the magnitude of the applied electric field, but the variation of the index of refraction is proportional to the electric field. The Δn value in Eq. (1.2-16) is called the electrically induced refractive index variation. This is the physical basis of implementing such technologies as electro-optic modulation, Q modulation, and mode locking using the electro-optic effect.

2. Electro-optic phase retardation

Below we shall make an analysis of how the electro-optic effect causes phase retardation. In actual application, the electro-optic crystal is always cut up along certain particular directions of the relative optical axis. The electric field is also applied on the crystal along the direction of a certain principal axis. There are two frequently used methods; one consists in having the direction of the electric field coincide with the clear direction, called the longitudinal electro-optic effect, and the other, the transverse electro-optic effect, with the direction of the electric field normal to the clear direction. We shall still use the KDP crystal as an example for our analysis. After an electric field is applied on the crystal along its z-axis, the cross section of its refractive index ellipsoid is as shown in Fig. 1.2-2. If the light wave propagates along the z direction, then its birefringence characteristics depend on the ellipse formed by the intersection of the ellipsoid with the plane normal to the z-axis. In Eq. (1.2-14), letting $z = 0$, we have the equation of the ellipse,

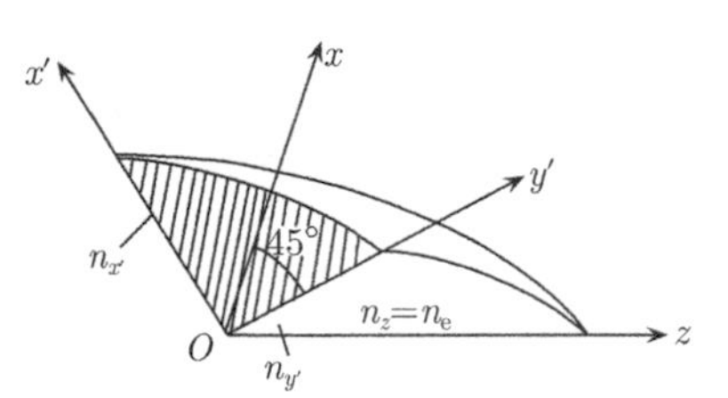

Fig. 1.2-2 The cross section of the refractive index ellipsoid

$$\left(\frac{1}{n_{\text{o}}^2} + \gamma_{63}E_z\right)x'^2 + \left(\frac{1}{n_{\text{o}}^2} - \gamma_{63}E_z\right)y'^2 = 1 \tag{1.2-18}$$

One quadrant of the ellipse is shown as the shaded part in the figure, the long and short half-axes of which overlap x' and y', respectively, x' and y' being the directions of polarization of the two components. The corresponding refraction indices $n_{x'}$ and $n_{y'}$ will be determined by Eq. (1.2-17).

When a beam of linearly polarized light is injected into the crystal along the z-axis direction, it is resolved into two vertical polarized components along the x' and y' directions. Owing to the difference in the refractive index between the two, the propagation speed of the light vibrating along the x' direction is high while that of the light vibrating along the y' direction is low. After a length L, the light paths they have traveled are $n_x L'$ and $n_{y'}L$, respectively. So the phase retardations of the two polarized components are, respectively,

$$\varphi_{n_{x'}} = \frac{2\pi}{\lambda}n_{x'}L = \frac{2\pi L}{\lambda}\left(n_{\text{O}} - \frac{1}{2}n_{\text{o}}^3\lambda_{63}E_z\right)$$

$$\varphi_{n_{y'}} = \frac{2\pi}{\lambda}n_{y'}L = \frac{2\pi L}{\lambda}\left(n_{\text{O}} + \frac{1}{2}n_{\text{o}}^3\lambda_{63}E_z\right)$$

Therefore, when the two beams of polarized light penetrate the crystal, there will be a phase difference

$$\Delta\varphi = \varphi_{n_{y'}} - \varphi_{n_{x'}} = \frac{2\pi}{\lambda}(n_{y'} - n_{x'})L = \frac{2\pi}{\lambda}Ln_{\text{o}}^3\gamma_{63}E_z = \frac{2\pi}{\lambda}n_{\text{o}}^3\gamma_{63}V \tag{1.2-19}$$

It can be seen from the above analysis that this phase retardation is completely induced by the birefringence due to the electro-optic effect and is therefore called the electro-optic phase retardation. $V = E_zL$ in the equation is the voltage applied along the z-axis of the crystal. When the electro-optic crystal and the clear wavelength are determined, the variation of phase difference is only dependent on the applied voltage to vary proportionately.

In Eq. (1.2-19), the voltage applied is called the half-wave voltage when the optical path difference $(n_{x'} - n_{y'})$ between the two vertical components $E_{x'}$ and $E_{y'}$ is half of a wavelength (the corresponding phase difference being π), usually denoted by V_π or $V_{\lambda/2}$. From Eq. (1.2-19), we have

$$V_{\lambda/2} = \frac{\lambda}{2n_o^3\gamma_{63}} = \frac{\pi c_0}{\omega n_o^3\gamma_{63}} \tag{1.2-20}$$

The half-wave voltage is an important parameter for characterizing the quality of the performance of an electro-optic crystal. The lower this voltage is, the better it will be, especially in the broad frequency band and high frequency condition. When the half-wave voltage is low, thc modulating power needed will be low. The half-wave voltage can usually be measured with the static method by applying DC voltage; then the electro-optic factor of the crystal can be found using Eq. (1.2-20). For this reason, accurate measurement of the half-wave voltage is extremely important for research on electro-optic crystalline materials. The half-wave voltage and the γ_{63} values of KDP class crystals found with the static method are listed in Tab. 1.2-1.

Tab. 1.2-1 Half-wave voltage and γ_{63} (for $\lambda = 0.550$ μm) of KDP type ($\bar{4}2$ m crystal class) crystals

Crystal	Formula	n_0	V_π/kV	$\gamma_{63}\times 10^{-10}$/(cm/V)
ADP	$NH_4H_2PO_4$	1.526	9.2	8.4
D-ADP	$NH_4D_2PO_4$	1.521	6.55	11.9
KDP	KH_2PO_4	1.512	7.45	10.6
D-KDP	KD_2PO_4	1.508	3.85	20.8
RbDP	RbH_2PO_4	1.510	5.15	15.5
ADA	$NH_4H_2AsO_4$	1.580	7.20	9.2
KDA	KH_2AsO_4	1.569	6.50	10.9
D-KDA	KD_2AsO_4	1.564	3.95	18.2
RbDA	RbH_2AsO_4	1.562	4.85	14.8
D-RbDA	RbD_2AsO_4	1.557	3.40	21.4
CsDA	CsH_2AsO_4	1.572	3.80	18.6
D-CsDA	CsD_2AsO_4	1.567	1.95	36.6

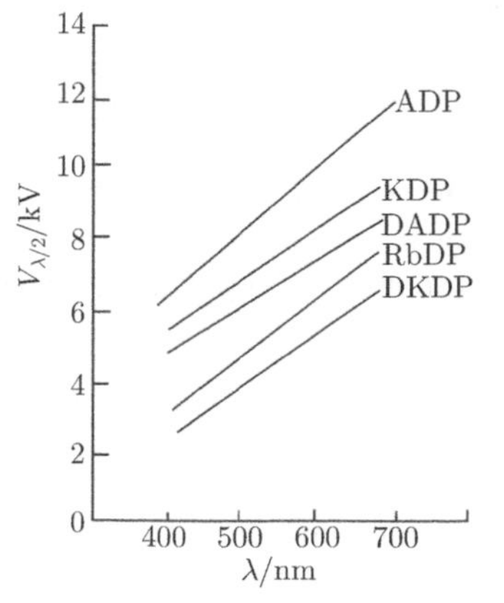

Fig. 1.2-3 The relationship between $V_{\lambda/2}$ and λ of KDP class crystals

The half-wave voltage of a crystal is a function of the wavelength. Figure 1.2-3 shows the relationship between some phosphates in $V_{\lambda/2}$ and wavelength. It can be seen from the figure that within the range measured (400~700 nm), this relationship is linear.

3. Change of the optical polarization state

It is known from the above analysis that the difference between two polarized components in phase speed will cause one component to have a phase difference relative to the other and the action of this phase difference will change the polarization state of the outgoing beam. It is known from "physical optics" that the "wave plate" can be used as a transformer for the light wave polarization state, whose change of the polarization state of the incoming light is determined by the thickness of the wave plate. In general, the outgoing synthetic vibration is an elliptically polarized light, which is expressed mathematically as

$$\frac{E_{x'}^2}{A_1^2}+\frac{E_{y'}^2}{A_2^2}-\frac{2E_{x'}E_{y'}}{A_1A_2}\cos\Delta\varphi=\sin^2\Delta\varphi \tag{1.2-21}$$

Here we adopt a phase retardation crystal that varies proportionately with the applied voltage (equivalent to an adjustable polarization state transformer). Therefore, it is now possible to transform the polarization state of the incoming light wave into the polarization state needed using the electrical method. For illustration, let's first have a look at the change of polarization under specified conditions.

(1) When no electric field is applied on the crystal, $\Delta\varphi = 2n\pi$ $(n = 0, 1, 2, \cdots)$, the above equation is reduced to

$$\left(\frac{E_{x'}}{A_1}-\frac{E_{y'}}{A_2}\right)^2=0$$

or

$$E_{y'}=(A_2/A_1)E_{x'}=E_{x'}\tan\theta \tag{1.2-22}$$

This is a rectilinear equation, showing that the synthetic light that has gone through the crystal is still linearly polarized light and its direction coincides with the polarizing direction of the incoming light. This is equivalent to the action of "a whole wave plate".

(2) When an electric field $(V_{\lambda/4})$ is applied on the crystal so that $\Delta\varphi=\left(n+\frac{1}{2}\right)\pi$, Eq. (1.2-21) can be reduced to

$$\frac{E_{x'}^2}{A_1^2}+\frac{E_{y'}^2}{A_2^2}=1 \tag{1.2-23}$$

This is a positive elliptic equation. When $A_1 = A_2$, its synthetic light will become circularly polarized light. This is equivalent to the action of "a quarter of wave plate".

(3) When the applied electric field $(V_{\lambda/2})$ causes $\Delta\varphi = (2n + 1)\pi$, Eq. (1.2-21) can be reduced to

$$\left(\frac{E_{x'}}{A_1}+\frac{E_{y'}}{A_2}\right)^2=0 \quad \text{or} \quad E_{y'}=-(A_2/A_1)E_{x'}=E_{x'}\tan(-\theta) \tag{1.2-24}$$

The above equation shows that the synthetic light has become linearly polarized light, but the direction of polarization has rotated a 2θ angle relative to the incident light (If $\theta = 45°$, then it has rotated 90°, along the y direction). The crystal has played the role of a half wave plate.

To sum up, suppose a beam of linearly polarized light is injected normal to the x'-y' plane and vibrates along the direction of the x-axis, and the moment it enters the crystal $(z = 0)$ it is resolved into two mutually perpendicular polarized components x' and y'. After traveling a distance L,

$$\text{Component } x' \text{ is} \quad E_{x'}=A\mathrm{e}^{\mathrm{i}[\omega_c t-(\frac{\omega_c}{c})(n_0-\frac{1}{2}n_0^3\gamma_{63}E_z)L]} \tag{1.2-25}$$

$$\text{Component } y' \text{ is} \quad E_{y'}=A\mathrm{e}^{\mathrm{i}[\omega_c t-(\frac{\omega_c}{c})(n_0+\frac{1}{2}n_0^3\gamma_{63}E_z)L]} \tag{1.2-26}$$

The phase difference between the two components at the outgoing surface of the crystal can be obtained from the difference between the indices in the above two equations:

$$\Delta\varphi=\frac{\omega_c n_0^3\gamma_{63}V}{c} \tag{1.2-27}$$

Figure 1.2-4 shows the two components $E_{x'}(z)$ and $E_{y'}(z)$ at a certain instant (to facilitate observation, the two vertical components are plotted separately) as well as the scanned loci of the apices of the light field vectors at different points along the path. At $z = 0$, the phase difference $\Delta\varphi = 0$, the light field vector is linearly polarized light along the x direction; at point e, $\Delta\varphi = \frac{\pi}{2}$, the synthetic light field vector becomes clockwise, rotating circularly polarized light; at point i, $\Delta\varphi = \pi$, the synthetic light field vector becomes linearly polarized light along the y direction, having rotated 90° relative to the incident polarized light. If a polarizer normal to the direction of the incident light polarization is placed at the output end of the crystal, when the voltage applied on the crystal varies in the range $0 \sim V_{\lambda/2}$, the light output from the analyzer is merely a component along the y direction of the elliptically polarized light. Therefore, the variation of the polarization state (polarization modulation) is transformed into one of light intensity (intensity modulation).

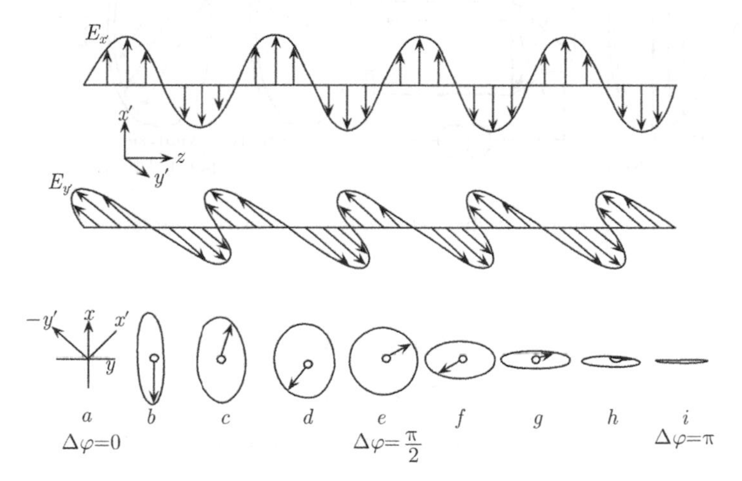

Fig. 1.2-4 Longitudinal application of the variation of the polarization state of light wave in KDP crystals

1.2.2 electro-optic intensity modulation

There are two cases of implementing electro-optic modulation using the Pockels effect. For one, the electric field applied on the crystal is basically uniform spatially but variable temporally. When a beam of light has gone through the crystal, an electrical signal varying with time can be converted into an optical signal, with the information to be transmitted embodied by the intensity or phase variation of the optical wave. This is mainly applied in such fields as optical communication or optical switches. The other case is one in which the electric field applied on the crystal is adequately distributed spatially, with electric field images formed, that is, distributed with the intensity transmissivity or phase of the variation of the x and y coordinates, but invariable temporally or varying slowly, thereby modulating the light wave passing through it. The spatial light modulator to be described later is a case in point, but for this section we shall discuss the former.

1. longitudinal electro-optic modulation

Figure 1.2-5 shows a typical structure of longitudinal electro-optic intensity modulation. The electro-optic crystal (KDP) is placed between two orthogonal polarizer, of which the polarizing initiating P_1's polarizing direction is parallel to the x-axis of the electro-optic crystal while the polarizing direction of the analyzer P_2 is parallel to the y-axis. After an electric field is applied along the z-axis of the crystal, they will rotate 45° to become inductive principal axes x' and y'. Hence the incident light beam along the z-axis becomes linearly polarized light parallel to the x-axis via P_1 and is resolved into two components along the x' and y' directions after entering the crystal ($z = 0$), their amplitude (equal to $\frac{1}{\sqrt{2}}$ that of the incident light) and phase being equal and expressed by

$$E_{x'} = A\cos\omega_c t$$

$$E_{y'} = A\cos\omega_c t$$

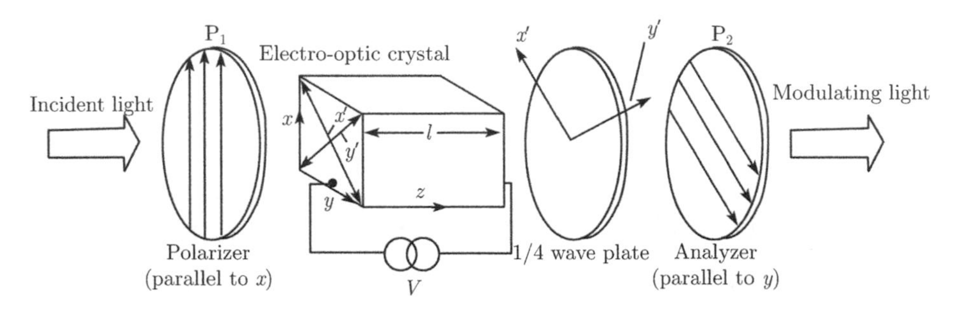

Fig. 1.2-5 Longitudinal electro-optic intensity modulation

or expressed by complex numbers as

$$E_{x'}(0) = A$$

$$E_{y'}(0) = A$$

As light intensity is proportional to the square of the electric field, the intensity of the incident light is

$$I_i \propto E \cdot E^* = |E_{x'}(0)|^2 + |E_{y'}(0)|^2 = 2A^2 \tag{1.2-28}$$

After light goes through length L of the crystal, owing to the electro-optic effect, a phase difference $\Delta\varphi$ is generated between the two components $E_{x'}$ and $E_{y'}$. Thus,

$$E_{x'}(L) = A$$
$$E_{y'}(L) = A\exp(-i\Delta\varphi)$$

So, the total electric field intensity is the sum of the projections of $E_{x'}$ (L) and $E_{y'}(L)$ in the y direction, i.e.,

$$(E_y)_0 = \frac{A}{\sqrt{2}}[\exp(-i\Delta\varphi) - 1]$$

Its corresponding output light intensity is

$$\begin{aligned} I \propto [(E_y)_0 \cdot (E_y^*)_0] &= \frac{A^2}{2}\{[\exp(-i\Delta\varphi) - 1][\exp(i\Delta\varphi) - 1]\} \\ &= 2A^2 \sin^2\left(\frac{\Delta\varphi}{2}\right) \end{aligned} \tag{1.2-29}$$

Comparing the intensity of the outgoing light with that of the incident light and considering the relationship between Eq. (1.2-9) and Eq. (1.2-20), we have

$$T = \frac{I}{I_i} = \sin^2\left(\frac{\Delta\varphi}{2}\right) = \sin^2\left[\frac{\pi}{2}\frac{V}{V_\pi}\right] \tag{1.2-30}$$

where T is called the transmissivity of the modulator. Based on the above-mentioned relationship, the curve of the light intensity modulating characteristics can be plotted as shown in Fig. 1.2-6. It can be seen from the figure that, in general, the relationship between the output characteristics of the modulator and the externally applied voltage is nonlinear. If the modulator is operating in the nonlinear part, the modulating light will get distorted. In order to obtain linear modulation, a fixed phase retardation $\pi/2$ can be introduced to make the voltage of the modulator biased at the operating point with $T = 50\%$. There are two frequently used methods. One consists in applying an additional bias voltage $V_{\lambda/4}$ apart from the signal voltage already applied on the modulating crystal. But this method needs more complicated circuitry and the operating point is not stable. For the other method, a quarter-wave plate should be inserted on the optical path of the modulator, the quick and slow axes forming an angle of 45° with the crystal's principal axis x, causing a fixed phase difference $\pi/2$ to be generated between the two components $E_{x'}$and $E_{y'}$. So the total phase difference in Eq. (1.2-30):

$$\Delta\varphi = \frac{\pi}{2} + \pi\frac{V_{\mathrm{m}}}{V_\pi}\sin\omega_{\mathrm{m}}t = \frac{\pi}{2} + \Delta\varphi_{\mathrm{m}}\sin\omega_{\mathrm{m}}t$$

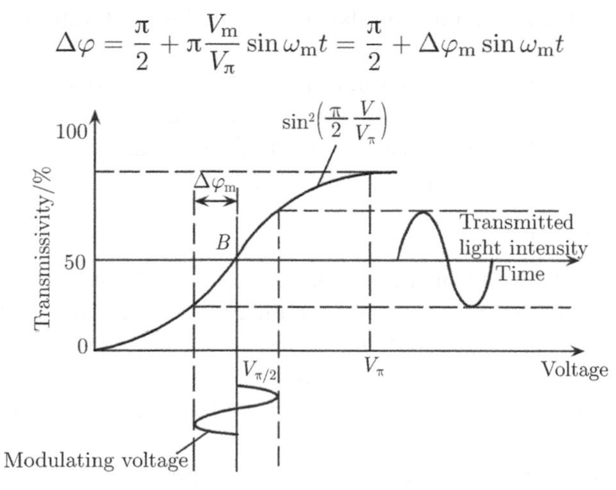

Fig. 1.2-6 Curve of characteristics of electro-optic modulation

where $\Delta\varphi_{\mathrm{m}} = \pi V_{\mathrm{m}}/V_\pi$ is the phase difference corresponding to the applied modulating signal voltage V_{m}. Hence the transmissivity of the modulator can be expressed by

$$\begin{aligned} T = \frac{I}{I_{\mathrm{i}}} &= \sin^2\left(\frac{\pi}{4} + \frac{\Delta\varphi_{\mathrm{m}}}{2}\sin\omega_{\mathrm{c}}t\right) \\ &= \frac{1}{2}[1 + \sin(\Delta\varphi_{\mathrm{m}}\sin\omega_{\mathrm{m}}t)] \end{aligned} \tag{1.2-31}$$

Expanding ($\Delta\varphi_{\mathrm{m}}$ sin $\omega_{\mathrm{m}}t$) in the above equation using the Bessel function identity, we have

$$T = \frac{I}{I_{\mathrm{i}}} = \frac{1}{2} + \sum_{n=0}^{\infty} J_{2n+1}(\Delta\varphi_{\mathrm{m}})\sin[(2n+1)\omega_{\mathrm{m}}t] \tag{1.2-32}$$

It can thus be seen that the output modulating light contains higher order harmonic components that make the modulating light distorted. To obtain linear modulation, it is indispensable to have the higher order harmonic controlled within the allowable range. Suppose the amplitude of the fundamental frequency wave and that of the higher order harmonic are I_1 and I_{2n+1}, respectively; then the ratio of the higher order harmonic to the fundamental frequency wave constituent is

$$\frac{I_{2n+1}}{I} = \frac{J_{2n+1}(\Delta\varphi_{\mathrm{m}})}{J_1(\Delta\varphi_{\mathrm{m}})} \quad (n = 0, 1, 2, \cdots) \tag{1.2-33}$$

If we take $\Delta\varphi_{\mathrm{m}} = 1$ rad, then $J_1(1) = 0.44$, $J_3(1) = 0.02$, $I_3/I_1 = 0.045$; that is, the third-order harmonic is 5% of the fundamental wave. Within this range approximately linear modulation can be obtained, so

$$\Delta\varphi_{\mathrm{m}} = \pi\frac{V_{\mathrm{m}}}{V_\pi} \leqslant 1 \text{ rad} \tag{1.2-34}$$

is taken as the criterion for linear modulation. Now, substitution of $J_1(\Delta\varphi_{\mathrm{m}}) \cong \frac{1}{2}\Delta\varphi_{\mathrm{m}}$ into Eq. (1.2-32) yields

$$T = \frac{I}{I_{\mathrm{i}}} \cong \frac{1}{2}[1 + \Delta\varphi_{\mathrm{m}} \sin\omega_{\mathrm{m}}t] \tag{1.2-35}$$

Hence to obtain linear modulation, the modulating signal is required to be not too big (small signal modulation). Then, the output light intensity modulated wave is just the linear reproduction of the modulating signal $V = V_{\mathrm{m}}\sin\omega_{\mathrm{m}}t$. If the condition of $\Delta\varphi_{\mathrm{m}} \ll 1$ rad cannot be fulfilled (big signal modulation), then the light intensity modulated wave will get distorted.

The above-discussed electro-optic modulator possesses such advantages as simple structure, stable operation, and freedom from the influence of spontaneous birefringence. Its shortcoming is that its half wave voltage is too high, especially when the modulated frequency is rather high, resulting in great power loss.

2. Transverse electro-optic modulation

It has been mentioned that, according to physical optics, there are three different modes of applying the transverse electro-optic effect, namely, ① apply the electric field along the z-axis direction, with the light transmitting direction normal to the z-axis and forming an included angle of 45° with either the x^- or $y-$axis (the crystal being $45° - z$ cut), ② apply the electric field along the x-axis direction (i.e., the direction of the electric field is normal to the optic axis), with the light transmitting direction normal to the x-axis and forming an included angle of 45° with the z-axis (the crystal being $45° - x$ cut), and ③ apply the electric field along the y-axis direction, with the light transmitting direction normal to the y-axis and forming an included angle of 45° with the z-axis (the crystal being $45° - y$ cut). Here the analysis will be made with only the first mode of application of the KDP class crystal as an example.

The transverse electro-optic modulation is as shown in Fig. 1.2-7. As the applied electric field is along the z-axis, similar to the longitudinal application, $E_x = E_y = 0$ and $E_z = E$. The principal axes x, y rotate 45° to x', y', and the three corresponding indices of refraction are as shown in Eq. (1.2-17). But the light transmitting direction is normal to the z-axis and is injected along the y' direction (the polarizing direction of the incident light forms an angle of 45° with the z-axis) and will be resolved into two components vibrating along the x' and z directions after entering the crystal, their refractive indices being $n_{x'}$ and n_z, respectively. If the length of the crystal in the light transmitting direction is L, the thickness (the distance between two electrodes) is d, and the applied voltage $V = E_z d$, then the phase difference between the two outgoing optical waves

$$\Delta\varphi = \frac{2\pi}{\lambda}(n_{x'} - n_z)L = \frac{2\pi}{\lambda}\left[(n_{\mathrm{o}} - n_{\mathrm{e}})L - \frac{1}{2}n_{\mathrm{o}}^3\gamma_{63}\left(\frac{L}{d}\right)V\right] \tag{1.2-36}$$

It can be seen that the γ_{63} transverse electro-optic effect of the KDP crystal causes the phase difference of the light wave after passing the crystal to include two terms. The first term is the phase retardation due to the spontaneous double refraction of the crystal itself unrelated to the applied electric field and has no contribution to the operation of the modulator. Instead, when the temperature of the crystal varies, it will bring about disadvantageous influences. So effort should be made to eliminate (compensate for) it. The second term is the phase retardation generated by the action of the applied electric field, which is related to the applied voltage V and the crystal's dimension (L/d). If the crystal's dimension is appropriately chosen, its half-wave voltage can be reduced.

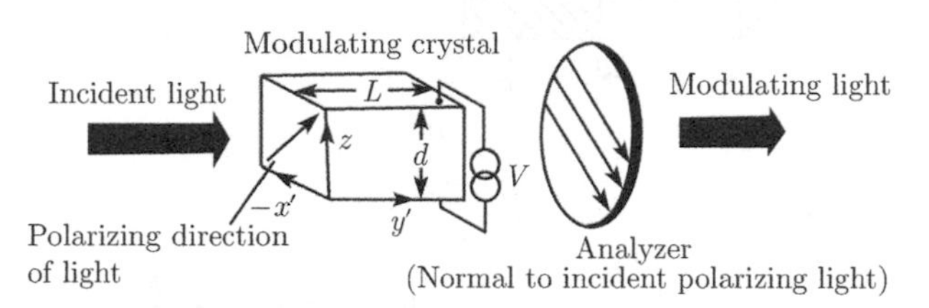

Fig. 1.2-7 Schematic diagram of transverse electro-optic modulation

The main shortcoming of the transverse electro-optic modulation with the KDP crystal is the presence of the phase retardation induced by natural birefringence. This implies that, in the absence of an applied electric field, there exists a phase difference in the two polarized components resolved from the linearly polarized light that enters the crystal. When the temperature of the crystal varies, owing to the difference in the rate of variation with the temperature between the refractive indices n_o and n_e, drift will occur with the phase difference between the two light waves. It has been proved by experiment that the rate of variation with the temperature of the difference between the KDP crystal's two refractive indices is $\Delta(n_o - n_e)/\Delta T \cong 1.1 \times 10^{-5}/°C$. Suppose a KDP crystal of a length $L = 30$ mm is made into a modulator. When it passes through laser of a wavelength of 632.8 nm, the variation of the phase difference induced by the temperature

$$\Delta\varphi = \frac{2\pi}{\lambda}\Delta nL = \frac{2\pi}{0.6328 \times 10^{-6}} \times 1.1 \times 10^{-5} \times 0.03 = 1.1\pi$$

If it is required that the variation of phase not exceed 20 mrad, the accuracy of constant temperature of the crystal must be kept within 0.005°C, which is evidently impossible. Therefore, in the KDP crystal transverse modulator, the influence of the natural birefringence will lead to the modulating light getting distorted, and even to the stop of operation of the modulator. So, in actual application, apart from taking as many measures (e.g., heat dissipation, constant temperature, etc.) as possible to reduce the shift of the crystal's temperature, mainly the structure of a "combined modulator" should be adopted to make the compensation, There are two commonly used methods, one consisting in arranging in series the optical axes of two pieces of crystal of almost completely identical geometric dimensions so that they form an angle of 90° with each other; that is, the y'-axis and z-axis of one piece of crystal are parallel to the z-axis and y'-axis of the other, respectively (Fig. 1.2-8(a)), and the other, having the z-axis and y'-axis of two pieces of crystal arranged parallel to each other in the opposite direction, with a 1/2 wave-plate placed in between (Fig. 1.2-8(b)). The principles of compensation of the two methods are the same. The external electric field is along the direction of the z-axis (optical axis), but in the two pieces of crystal, the electric field is in the opposite direction relative to the optical axis. When the linearly polarized light is injected into the first piece of crystal along the y'-axis, the electrical vector is resolved into light e_1 along the z direction and light o_1 along the x' direction. When they pass through the first piece of crystal, the phase difference between two beams of light is

$$\Delta\varphi_1 = \varphi_{x'} - \varphi_z = \frac{2\pi}{\lambda}\left(n_o - n_e + \frac{1}{2}n_o^3\gamma_{63}E_z\right)L$$

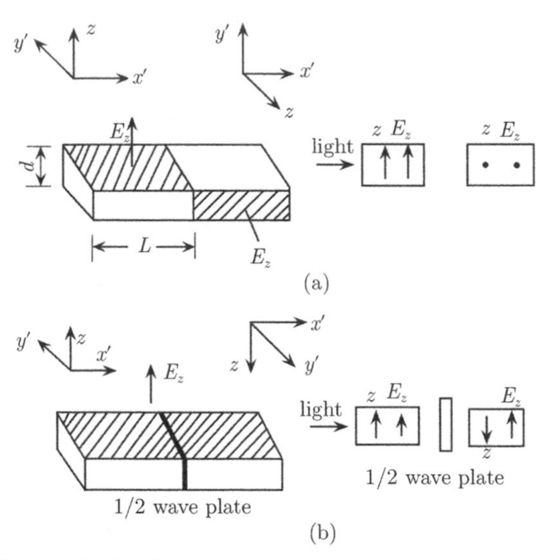

Fig. 1.2-8 Two methods of compensation for γ_{63} transverse electro-optic effect

After the action by the 1/2 wave plate, the directions of polarization of two beams of light each rotate 90° and, after that by the second crystal, the original light e_1 has become light o_2 while light o_1 has become light e_2. So after passing through the second piece of crystal, their phase difference is

$$\Delta\varphi_2 = \varphi_z - \varphi_{x'} = \frac{2\pi}{\lambda}\left(n_e - n_o + \frac{1}{2}n_o^3\gamma_{63}E_z\right)L$$

Thus, the total phase difference after going through two pieces of crystal is

$$\Delta\varphi = \Delta\varphi_1 + \Delta\varphi_2 = \frac{2\pi}{\lambda}n_o^3\gamma_{63}V\left(\frac{L}{d}\right) \tag{1.2-37}$$

Therefore, if the dimensions and performance of and the influences exerted on two pieces of crystal from the outside are completely identical, then the influences of the natural birefringence can be compensated for. According to Eq. (1.2-37), when $\Delta\varphi = \pi$, the half-wave voltage is $V_{\lambda/2} = \left(\frac{\lambda}{2n_o^3\gamma_{63}}\right)\frac{d}{L}$, where what is in the parentheses is the half-wave voltage of the longitudinal electro-optic effect. So

$$(V_{\lambda/2})_t = (V_{\lambda/2})_l\frac{d}{L}$$

It can be seen that the transverse half-wave voltage is (d/L) times the longitudinal half-wave voltage. Reducing d while increasing the length L can reduce the half-wave voltage. But two pieces of crystal have to be used for this method and so the structure will be complicated. Furthermore, the requirements on their dimension processing are extremely high. For KDP crystals, if the difference in length is 0.1 mm, when the temperature undergoes a change of 1°C the variation of the phase difference will be 0.6°C (for a wavelength of 632.8 nm). So, in general, no one will adopt the transverse mode of modulation for the KDP class crystal. In actual application, as both the GaAs crystal ($n_o = n_e$) of the $\bar{4}3$m group and the $LiNbO_3$ crystal (electric field applied along the x-axis, clear along the z direction) of the 3m group are free from the influence of the natural birefringence, the transverse electro-optic modulation is adopted very often.

1.2.3 Electro-optic phase modulation

Shown in Fig. 1.2-9 is a diagram of the principle of electro-optic phase modulation which is composed of an initiating polarizer and a KDP electro-optic crystal. The polarizing axis of the polarizer is parallel to the crystal's inductive main axis x' (or y') and the electric field is applied along the direction of the z-axis. Now the polarized light of the incident crystal will no longer be resolved into two components polarizing along x' and y', but along either the direction of the x'- or y'-axis. So the external electric field does not change the polarization state of the outgoing light, but merely changes its phase, whose variation is

$$\Delta\varphi_{x'} = -\frac{\omega_c}{c}\Delta n_{x'} L \tag{1.2-38}$$

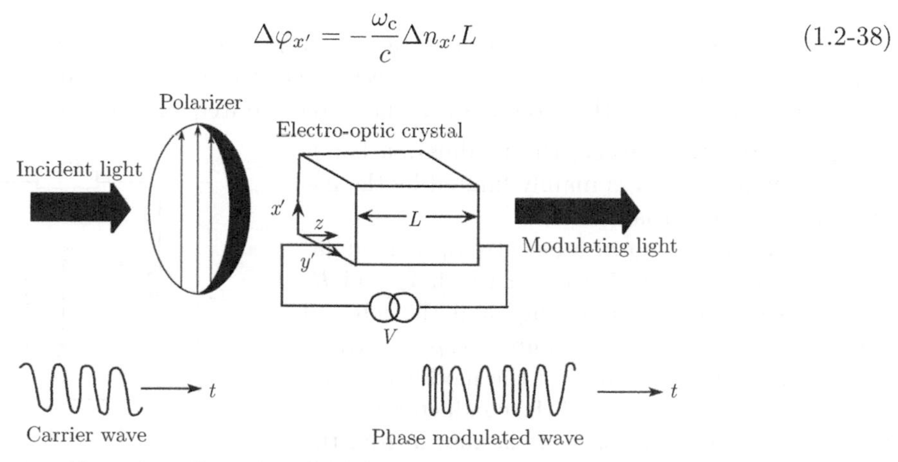

Fig. 1.2-9 The principle of electro-optic phase modulation

As the light wave is polarized only along the x' direction, the corresponding refractive index is $n_{x'} = n_0 - \frac{1}{2}n_0^3\gamma_{63}E_z$. If the applied electric field is $E_{in} = A_c \cos\omega_c t$, then the output light field (at $z = L$) becomes $A_c \cos\left[\omega_c t - \frac{\omega_c}{c}\left(n_0 - \frac{1}{2}n_0^3\gamma_{63}E_m \sin\omega_m t\right)L\right]$. By omitting the constant term of the phase angle, which exerts no influence on the modulating effect, the above equation can be rewritten as

$$E_{out} = A_c \cos(\omega_c t + m_\varphi \sin\omega_m t) \tag{1.2-39}$$

where $m_\varphi = \dfrac{\omega_c n_0^3\gamma_{63}E_m L}{2c} = \dfrac{\pi n_0^3\gamma_{63}E_m L}{\lambda}$ is called the phase modulation coefficient. Expanding the above equation using the Bessel function, we can obtain the form of Eq. (1.1-12).

1.2.4 The electrical performance of the electro-optic modulator

Speaking of the electro-optic modulator, it's only natural to hope for high modulating efficiency and a modulation bandwidth that meets the requirement. Below we shall make an analysis of the operation characteristics of the electro-optic modulator at different modulating frequencies.

In the foregoing analysis of electro-optic modulation, it is always considered that the frequency of the modulating signal is much lower than that of the light wave (i.e., the wavelength of the modulating signal $\lambda_m \gg \lambda$) and λ_m is far greater than the length L of the crystal. Therefore, within the transit time $\left(\tau_d = \dfrac{L}{c/n}\right)$ when the light wave goes through the crystal L, the electric field of the modulating signal is uniformly distributed in all parts of the crystal and the phase retardation obtained by the light wave in all the parts, too, is identical. That is to say, at no moment will the light wave be acted on by a modulating electric field of a different intensity or in the opposite direction. In this case, the modulating crystal with an electrode mounted on it is equivalent to a capacitor; that is, it can be regarded as a lumped element in the circuit, usually spoken of as the lumped parametric modulator, the frequency characteristics of which are mainly influenced by the parameters of the external circuit.

1. Limits to modulated bandwidth by the external circuit

The modulation bandwidth is an important parameter of the light modulator. For the electro-optic modulator, the electro-optic effect itself of the crystal will not limit the frequency characteristics of the modulator as the resonance frequency of the lattice can reach as high as 10^{12} Hz. Hence, the modulation bandwidth of the modulator is mainly limited by the parameters of the external circuit.

The equivalent circuit of an electro-optic modulator is as shown in Fig. 1.2-10, in which V_s and R_s represent the modulating voltage and the internal impedance of the modulating source, respectively, C_e is the equivalent capacitance of the modulator, and R_e and R are the resistance of the conductor and the DC resistance of the crystal. It is known from the figure that the actual voltage acting on the crystal is

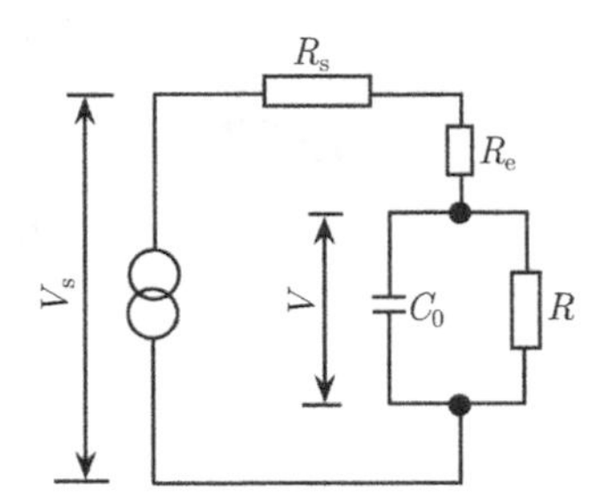

Fig. 1.2-10 The equivalent circuit of an electro-optic modulator

$$V = \frac{V_s\left(\dfrac{1}{(1/R)+\mathrm{i}\omega C_0}\right)}{R_s + R_e + \dfrac{1}{(1/R)+\mathrm{i}\omega C_0}} = \frac{V_s R}{R_s + R_e + R + \mathrm{i}\omega C_0(R_s R + R_e R)}$$

During modulation at a low frequency, there is in general $R \gg R_s + R_e$, and $\mathrm{i}\omega C_0$ is also rather small. Therefore, the signal voltage can be effectively applied on the crystal. But when the modulating frequency is further increased, the impedance of the modulating crystal becomes small. When $R_s > (\omega C_0)$, the greater part of the modulating voltage will drop onto R_s, showing that the impedance is not matched between the modulating source and the crystal loaded circuit. Now the modulating efficiency will greatly decrease and, worse still, operation may stop. To realize impedance matching, an inductor L should be connected in parallel at the two ends of the crystal, whose resonance frequency is $\omega_0^2 = (LC_0)^{-1}$, and a shunt resistor R_L, whose equivalent circuit is as shown in Fig. 1.2-11. When the frequency of the modulating signal $\omega_m = \omega_0$, the impedance of this circuit will be equal to R_L. If we choose $R_L \gg R_s$, then we can make the greater part of the modulating voltage applied onto the crystal. Although this method can increase the efficiency of modulation, the bandwidth of the resonance circuit is limited, whose impedance is rather high only within the range of the frequency internal $\Delta\omega \approx \dfrac{1}{R_L C_0}$. For this reason, to prevent the modulating wave from becoming distorted, its maximum modulation bandwidth (i.e., the frequency bandwidth occupied by the modulating signal) must be smaller than

$$\Delta f_m = \frac{\Delta\omega}{2\pi} \cong \frac{1}{2\pi R_L C_0} \tag{1.2-40}$$

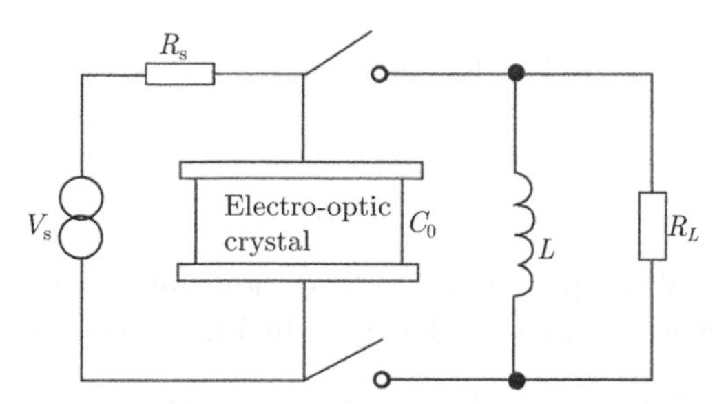

Fig. 1.2-11 The parallel resonance circuit of the modulator

As it is, the modulation bandwidth depends on the specific application requirements. Moreover, definite peak phase retardation $\Delta\varphi_m$ is also required, its corresponding driving peak modulating voltage being

$$V_m = \frac{\lambda}{2\pi n_0^3 \gamma_{63}} \Delta\varphi_m \tag{1.2-41}$$

For the KDP crystal, the driving power needed to obtain the maximum phase retardation

$$P = V_m^2 / 2R_L \tag{1.2-42}$$

By making use of Eqs. (1.2-40) and (1.2-41), the above equation can be further written as

$$P = V_m^2 \pi C_0 \Delta f_m = V_m^2 \pi (\varepsilon A/L) \Delta f_m = \frac{\lambda^2 \varepsilon A \Delta\varphi_m^2}{4\pi L n_0^6 \gamma_{63}^2} \Delta f_m \tag{1.2-43}$$

where L is the length of the crystal, A is the cross sectional area of normal to L, and ε is the dielectric constant. It is known from the above equation that when the crystal's category, dimensions, laser wavelength, and phase retardation required are determined, its modulating power will be proportional to its modulating bandwidth.

2. The influence of transition time during high frequency modulation

When the modulating frequency is extremely high, within the transit time the light wave passes the crystal, the electric field may undergo rather significant changes. That is, the modulating voltage in different parts of the crystal is different, especially when the modulation period ($2\pi/\omega_m$) can be compared to the transition time $\tau_d (= nL/C)$, and the modulating electric field borne by the light wave in various parts of the crystal is different. Hence, the accumulation of the above-mentioned phase retardation will be destroyed. Now the total phase retardation should be obtained from the following integral:

$$\Delta\varphi(L) = \int_0^L aE(t')\mathrm{d}z \tag{1.2-44}$$

where $E(t')$ is the transient electric field, $a = \dfrac{2\pi}{\lambda} n_0^3 \gamma_{63}$. As the time in which the light wave passes through the crystal is τ_d, and $\mathrm{d}z = C\mathrm{d}t/n$, the above equation can be rewritten as

$$\Delta\varphi(t) = \frac{aC}{n} \int_{t-\tau_d}^{t} E(t')\mathrm{d}t' \tag{1.2-45}$$

Suppose the applied electric field is a single frequency sinusoidal signal, i.e., $E(t') = A_0 \exp(\mathrm{i}\omega_m t')$. Thus,

$$\begin{aligned} \Delta\varphi(t) &= \frac{aC}{n} A_0 \int_{t-\tau_d}^{t} \exp(\mathrm{i}\omega_m t')\mathrm{d}t' \\ &= \Delta\varphi_0 \left[\frac{1 - \exp(-\mathrm{i}\omega_m \tau_d)}{\mathrm{i}\omega_m \tau_d} \right] \exp(\mathrm{i}\omega_m t) \end{aligned} \tag{1.2-46}$$

where $\Delta\varphi_0 = \frac{aC}{n} A_0 \tau_d$ is the peak phase retardation at $\omega_m \tau_d \ll 1$; the factor

$$\gamma = \frac{1 - \exp(-i\omega_m \tau_d)}{i\omega_m \tau_d} \tag{1.2-47}$$

characterizes the decrease of the peak phase retardation induced by the transit time and is thus called the high frequency phase retardation reducing factor. The relationship of γ with $\omega_m \tau_d$ is as shown in Fig. 1.2-12. Only when $\omega_m \tau_d \ll 1$, that is, $\tau_d \ll \frac{T_m}{2\pi}$, will $\gamma = 1$, that is, there is no reducing action. This shows that the transit time of the light wave in the crystal has to be much shorter than the period of the modulating signal to be able to prevent the modulating effect from being affected. This implies that, for the electro-optic modulator, there exists a limit to the highest modulating frequency. For instance, if $|\gamma| = 0.9$ is taken as the modulating limit (corresponding to $\omega_m \tau = \frac{\pi}{2}$), then the upper limit to the modulating frequency is

$$f_m = \frac{\omega_m}{2\pi} = \frac{1}{4\tau_d} = \frac{C}{4nL} \tag{1.2-48}$$

For the KDP crystal, if we take $n = 1.5$, length $L = 1$ cm, then $f_m = 5 \times 10^9$ Hz.

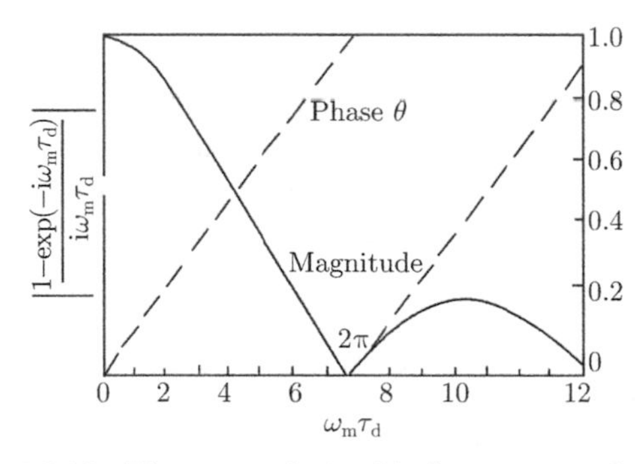

Fig. 1.2-12 The curve relationship between γ and $\omega_m \tau_d$

3. The traveling wave modulator

In order to be able to operate at higher modulating frequencies without being affected by the transition time, a structural form of the so-called traveling wave modulator can be adopted, as shown in Fig. 1.2-13. Its principle consists in having the modulating signal applied on the crystal in the form of a traveling wave to cause the high frequency modulating field to interact with the light wave field in the form of the traveling wave while enabling the light wave and modulating signal to possess identical phase velocity in the crystal from beginning to end. In this way, the modulating voltage borne by the light wavefront in the process of passing through the entire crystal will be identical so that the effects of the transition time can be eliminated. As the high frequency electric field in most transmission lines is transversely distributed, the traveling wave modulator usually adopts transverse modulation.

Below let's see why such a structure can realize high frequency modulation and high modulating efficiency. Suppose the wavefront of the light wave enters the incident face (where $z = 0$) at moment t and is transmitted to where z is at moment t'; then we have $z(t') = \frac{c}{n}(t' - t)$. The phase retardation of the light wave is generated owing to the action of the modulating field

$$\Delta\varphi(t) = a\frac{c}{n}\int_{t}^{t+\tau_d} E[t', z(t')]\mathrm{d}t' \tag{1.2-49}$$

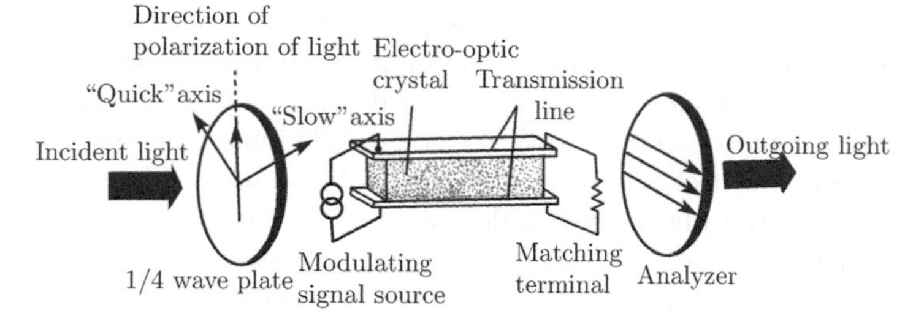

Fig. 1.2-13 The traveling wave electro-optic modulator

where $E\ [t',\ z\ (t')]$ is the transient modulating field. If the traveling wave modulating field is taken as

$$E[t', z(t')] = A_0 \exp[\mathrm{i}(\omega_m t' - k_m z)] = A_0 \exp\{\mathrm{i}[\omega_m t' - k_m (c/n)(t' - t)]\} \tag{1.2-50}$$

where $k_m = \omega_m/c_m$, and c_m is the phase velocity of the modulating field. Substituting Eq. (1.2-50) into Eq. (1.2-49) and after integration, we have

$$\Delta\varphi(t) = \Delta\varphi_0 \exp(\mathrm{i}\omega_m t)\left\{\frac{\exp\{\mathrm{i}\omega_m\tau_d[1 - c/(nc_m)]\} - 1}{\mathrm{i}\omega_m\tau_d[1 - c/(nc_m)]}\right\} \tag{1.2-51}$$

where $\Delta\varphi_0 = aLA_0 = a(c/n)\tau_d A_0$ is the phase retardation generated by the DC electric field equal to A_0. Let

$$\gamma = \frac{\exp\{\mathrm{i}\omega_m\tau_d[1 - c/(nc_m)]\} - 1}{\mathrm{i}\omega_m\tau_d[1 - c/(nc_m)]} \tag{1.2-52}$$

be the reduction factor of phase retardation. This is similar to Eq. (1.2-47) of the aforementioned modulator in form. What is different is simply the replacement of τ_d here with $\tau_d[1 - c/(nc_m)]$. It can be seen that, as long as the phase velocity of the light wave in the crystal is equal to that of the high frequency modulating field, i.e., $c = (nc_m)$, then $\gamma = 1$. Now, whatever the length of the crystal, the maximum phase retardation can be obtained. Similar to the treatment of Eq. (1.2-48), the upper limit of the modulating frequency can be obtained as

$$f_{max} = \frac{c}{4nL[1 - c/(nc_m)]} \tag{1.2-53}$$

A comparison of the above equation with Eq. (1.2-8) shows that the modulation with the traveling wave can raise the upper limit to the modulated frequency $[1 - c/(cn_m)]^{-1}$ times. Currently, the modulation bandwidth of modulators of this type has reached the order of magnitude of dozens of GHz.

1.2.5 The electro-optic waveguide modulator

The various electro-optic modulators discussed in the previous sections are all separate devices of rather large volume and are in general spoken of as "volume modulators". The limitation common to them is the fact that nearly the entire crystal material is acted on by the externally applied electric field. So a powerful electric field has to be applied on the device to change the optical characteristics of the whole crystal in order that the optical wave it passes through will be modulated. With the development of such applied technologies as optical communication, optical information processing, and optical computation, there has emerged a brand new discipline at the same time, viz. integrated optics. The principle of integrated optics lies in having the light wave limited to the characteristic of propagating along a definite direction in the waveguide region of the order of magnitude of μm using optical waveguide to enable optical devices to go plane and optical systems to go integrated. To be specific, active devices such as lasers, modulators, and detectors are all "integrated" on the same substrate and connected by means of passive devices such as waveguides and couplers to constitute an integral mini-optical system. Obviously, the medium optical waveguide is the basic component of the integrated optical technology. The waveguide is mainly divided into two kinds, the plane waveguide and rectangular waveguide, and the former, in turn, is subdivided into plate waveguide and gradually changing refractive index waveguide. The plate waveguide is the most frequently used waveguide of the simplest structure as shown in Fig. 1.2-14. It is composed of a thin film waveguide with a high refractive index inserted between the substrate with a low refractive index and the coverage. n_f, n_s, and n_e are the refractive indices of the waveguide layer, substrate, and coverage, $n_i > n_s \geqslant n_e$. If the coverage is just air, then $n_e \cong 1$. The difference in the refractive index between the thin film waveguide layer and the substrate is in general within the range of $10^{-3}\sim10^{-1}$. The thickness of the waveguide layer is in general of the μm order of magnitude (compared with the length of light wave). Total reflection will occur with the light wave on the upper and lower interfaces of the thin film to make the light wave confined in the film layer propagating as a sawtooth-shaped optical path, that is, to the region whose transverse dimension is only of the magnitude of the light wavelength. If signals are input from the outside to control the light wave propagating in the thin film, optical waveguide devices with different functions will be constructed. Such a device whose guide light is controlled from the outside is called the optical waveguide modulator. As it is mainly only acted on by an external electric field in a very small part of the waveguide region to confine the field to the vicinity of the waveguide thin film, the driving power needed by the optical waveguide modulator is reduced by one to two orders of magnitude lower than the volume modulator. A waveguide modulator of the simplest form needs two kinds of material, at least one of which should meet the modulator's requirements. Furthermore, the two kinds of material have to possess a definite relatively stable refractive index. Not only should a waveguide modulator possess good active properties, but also the material should be capable of combining with another kind of material with a different refractive index to construct a waveguide.

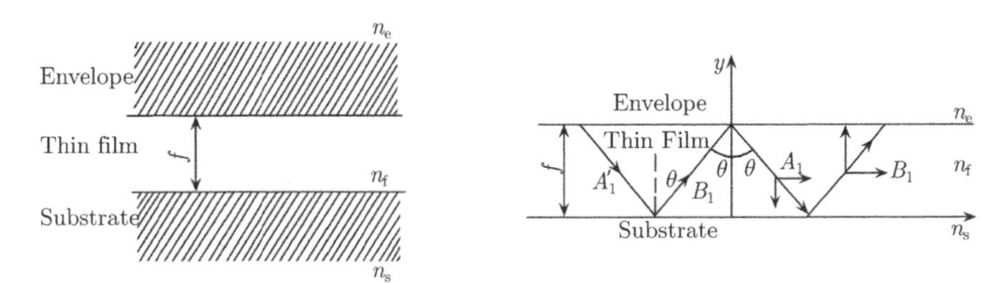

Fig. 1.2-14 The cross section of a plane waveguide

On the one hand, an optical waveguide modulator constructed from a medium is identical to the volume modulator in terms of the process of control over the optical parameters by its physical (electro-optic, acousto-optic, etc.) effects, that is, the ability to cause the dielectric tensor of the medium to undergo slight variation (i.e., variation of the refractive index) resulting in a phase difference between two propagating modes; and, on the other, it is different from the volume modulator in that, as the action of the external field will lead to the variation of the propagating characteristics of the eigenmode (e.g., the TE mode and

TM mode) in the waveguide and the coupling conversion between the two different modes (called the mode coupling modulation), the basic characteristics of the optical waveguide modulator can be depicted using the theory of the medium optical waveguide coupling mode.

1. The principle of modulation of the electro-optic waveguide modulator

The physical basis of realizing modulation by the electro-optic waveguide modulator is the Pockels effect of the crystalline medium. When an electric field is applied on the waveguide, slight variation of the dielectric tensor ε (refractive index) will be generated, which will bring about the variation of the eigenmode propagation in the waveguide or the transformation of power coupling between different modes. In the coordinate system of the waveguide, there is a one-to-one correspondence relationship between the elements $\Delta\varepsilon$ for the electric field to cause the variation of the medium tensor and the coupling between different modes. If what is contained is only the diagonal dielectric tensor element $\Delta\varepsilon_{xx}$ or $\Delta\varepsilon_{yy}$, then self-coupling between the TE modes or TM modes will ensue with only their respective phase varied, thereby generating relative phase retardation, a situation similar to the volume electro-optic phase modulation. However, if in the waveguide coordinate system the variation of the dielectric tensor contains nondiagonal tensor element $\Delta\varepsilon_{xy}$, then mutual coupling between the TE modes and TM modes will result, leading to the transformation of power between modes. That is, the power of an input mode TE (or TM) will be transformed onto the output mode TM (or TE), whose corresponding coupling equation, after proof by inference[1], can be reduced to

$$\frac{\mathrm{d}A_m^{\mathrm{TE}}}{\mathrm{d}z} = -\mathrm{i}\kappa A_l^{\mathrm{TM}} \exp[-\mathrm{i}(\beta_m^{\mathrm{TE}} - \beta_l^{\mathrm{TM}})z] \tag{1.2-54a}$$

$$\frac{\mathrm{d}A_l^{\mathrm{TM}}}{\mathrm{d}z} = -\mathrm{i}\kappa A_m^{\mathrm{TE}} \exp[\mathrm{i}(\beta_m^{\mathrm{TE}} - \beta_l^{\mathrm{TM}})z] \tag{1.2-54b}$$

where A_m^{TE} and A_l^{TM} are the amplitudes of the two modes of the m th and l th order, respectively, β_m^{TE} and β_l^{TM} are the propagation constants of the two modes, respectively, and κ represents the mode coupling coefficient, whose expression is

$$\kappa = \frac{\omega}{4}\int_{-\infty}^{\infty} \Delta\varepsilon_{xy}(x)E_y^{(m)}(x)E_x^{(l)}(x)\mathrm{d}x \tag{1.2-55}$$

Equation (1.2-54) is a description of the coupling between the TE modes and TM modes in the same direction, showing that the variation of amplitude of each mode is a function of the variation of the dielectric tensor (refractive index), the distribution of the mode field, and other mode amplitudes. Suppose the electro-optic material in the waveguide is homogeneous and the distribution of the electric field, too, is homogeneous. When the TE modes and TM modes are completely confined within the waveguide thin film layer and are of the same order ($m = l$), maximum value is taken for the integral of Eq. (1.2-55). Now the field distribution of TE modes and that of TM modes are almost identical, differing only in the direction of the electrical vectors. Moreover, when we have $\beta_m^{\mathrm{TE}} = \beta_l^{\mathrm{TM}} = \beta = k_0 n_{\mathrm{o}}$, then the coupling coefficient κ will approximately be

$$\kappa = -\frac{1}{2}n_0^3 k_0 \gamma_{ij} E \tag{1.2-56}$$

Under phase matched conditions, $\beta_m^{\mathrm{TE}} = \beta_l^{\mathrm{TM}}$, and the light wave is input in the single mode, $A_m = A_0, A_l = 0$. Then the solutions for Eqs. (1.2-54) are

$$A_m^{\mathrm{TE}}(z) = -\mathrm{i}A_0 \sin\kappa z \tag{1.2-57a}$$

[1] See A. Yariv, *Quantum Electronics* (Second Edition). Chap. 19(1975).

$$A_l^{\mathrm{TM}}(z) = A_0 \cos \kappa z \tag{1.2-57b}$$

It can be seen from Eq. (1.2-56) that, if one wants to obtain complete TE → TM power transformation in the waveguide of length L ($Z = L$), it is necessary to satisfy $\kappa L = \frac{\pi}{2}$. Now, the length of the optical waveguide is

$$L = \pi/(2\kappa) \tag{1.2-58}$$

But when the power is transformed into 0, the corresponding waveguide length is

$$L = n\pi/\kappa \quad (n = 0, 1, 2, \cdots) \tag{1.2-59}$$

It can be seen that the conditions for this case are the same as those needed by "on" and "off" of the previously described crystal electro-optic modulator. But, in general, the coupling coefficient κ is smaller than the value given in Eq. (1.2-56), so the EL product needed for achieving complete power transformation should be increased correspondingly.

Example There is a GaAs waveguide modulator that takes $\lambda = 1$ μm, $n_0 = 3.5$, $n_0^3 r = 59 \times 10^{-12}$ m/V, and the externally applied electric field $E = 10^6$ V/m. The coupling coefficient $\kappa = 1.85$ cm^{-1} and the length $L = \frac{\pi}{2\kappa} = 0.85$ cm needed for power transformation are obtained from Eq. (1.2-56).

The ratio of the output light intensity (TM) to the input light intensity (TE) can be obtained from Eq. (1.2-57):

$$I/I_0 = \sin^2 \frac{\Delta\varphi}{2} \tag{1.2-60}$$

while

$$\Delta\varphi = 2\kappa L = \frac{2\pi n_0^3 \gamma}{\lambda}\left(\frac{L}{d}\right)V$$

where d represents the thickness of the waveguide thin film, whose transmissivity curve obtained by experiment is shown in Fig. 1.2-15.

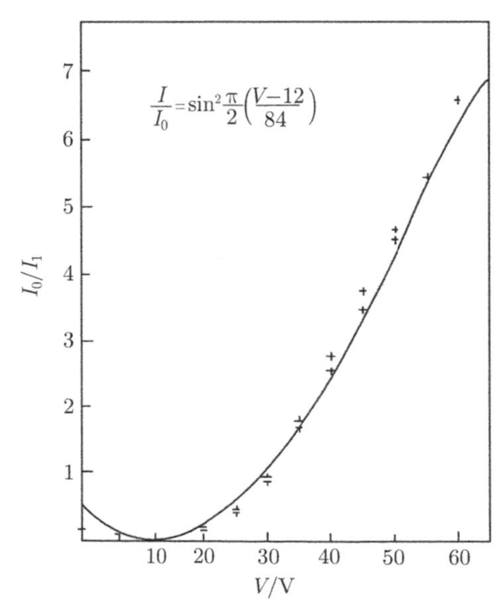

Fig. 1.2-15 The transmissivity of a waveguide placed between intersecting polarizers

2. The electro-optic waveguide phase modulation

Figure 1.2-16 shows a schematic diagram of the structure of an electro-optic waveguide phase modulator. With $LiNbO_3$ as the substrate, Ti diffusion forms the plane waveguide and the sputtering method is used to sediment a pair of thin-film electrodes. In the figure, x, y, z represents the waveguide coordinate system and a, b, c, the orientation of the crystal axes. When a modulating voltage is applied on the electrodes, if what is propagating in the waveguide is the TM mode, the electric field vector will be along the z-axis (corresponding to the c-axis of the crystal), the main electric field component being E_z. Owing to the variation of the waveguide refractive index due to the electro-optic effect, after the guide wave light passes through the electrode region, its phase varies with the variation of the modulating voltage. That is,

$$\Delta\varphi = \pi n_0^3 \gamma_{33} E_z l/\lambda \tag{1.2-61}$$

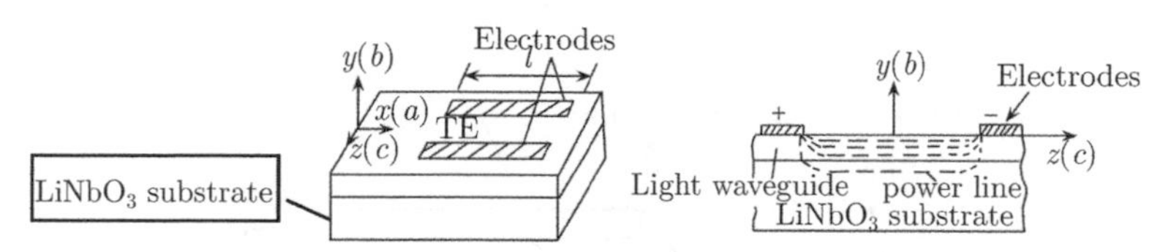

Fig. 1.2-16 Schematic of the structure of an $LiNbO_3$ electro-optic waveguide phase modulator

where E_z is the electric field component along the direction of the c-axis generated by the plane electrode in the crevice, l the length of the electrode, and γ_{33} the electro-optic coefficient.

The electro-optic waveguide phase modulation does not involve the mutual coupling among different modes, whose equation of amplitude is

$$\frac{\mathrm{d}A_\mathrm{m}(x)}{\mathrm{d}x} = -\mathrm{i}\kappa_{\mathrm{min}} A_\mathrm{m}(x) \tag{1.2-62}$$

whose solution is $A_\mathrm{m}(x) = A_\mathrm{m}(0)\exp[-\mathrm{i}\kappa_{\mathrm{min}} x]$. If the E_y incident wave corresponds to the TM mode, its mode field can be expressed as

$$E_y(x, y, z) = A_y(0)E_y(y, z)\exp\{\mathrm{i}[\omega t - (\kappa_{yy} + \beta_y)x]\} \tag{1.2-63}$$

where the self-coupling coefficient is

$$\kappa_{yy} = \frac{\omega}{4}\iint \Delta\varepsilon_{\mathrm{TM-TM}} E_y E_y^* \mathrm{d}y\mathrm{d}z \tag{1.2-64}$$

where $\Delta\varepsilon_{\mathrm{TM-TM}}$ is $\Delta\varepsilon_{22}$. In addition, by introducing the expression of the plane waveguide TM mode power normalization

$$\frac{\varepsilon_0 \omega n^2}{2\beta}\int_{-\infty}^{\infty} E_y E_y^* \mathrm{d}y = 1$$

which is then substituted into Eq. (1.2-64), the self-coupling coefficient κ_{yy} can be determined.

3. The electro-optic waveguide intensity modulation

The device for performing this kind of modulation is similar to the Mach Zehnder (MZ) interferometer. A schematic diagram of the MZ interference type modulator is shown in Fig. 1.2-17. It is composed of the Ti diffusion forking strip-shaped waveguide fabricated by radio frequency sputtering etching on the substrate of the $LiNbO_3$ crystal as the waveguide base-film. Surface electrodes are fabricated in the middle and on the two sides of the strip-shaped waveguide. This device consists in having two branched beams of light get modulated in the region of the electro-optic effect and meet again prior to their coherent synthesis to realize intensity modulation. For instance, stimulate a TE mode at the input end of the waveguide. Then under the action of the externally applied electric field, the guide mode transmitted in the forked waveguide undergoes phase variation of $\Delta\varphi$ and $-\Delta\varphi$, respectively, owing to the action of the electric field E_c of identical dimensions and opposite signs (since the two branched waveguide structures are completely symmetric). Suppose the length of the electrode is l and the distance between two electrodes is d. Then the phase difference between the two guide modes is $2\Delta\varphi = 2\pi n_e^3 \gamma_{33} E_c l/\lambda$. The coherently synthesized light intensity by the two beams of light at the second confluence of forks will be different as the phase difference differs, thereby resulting in intensity modulation.

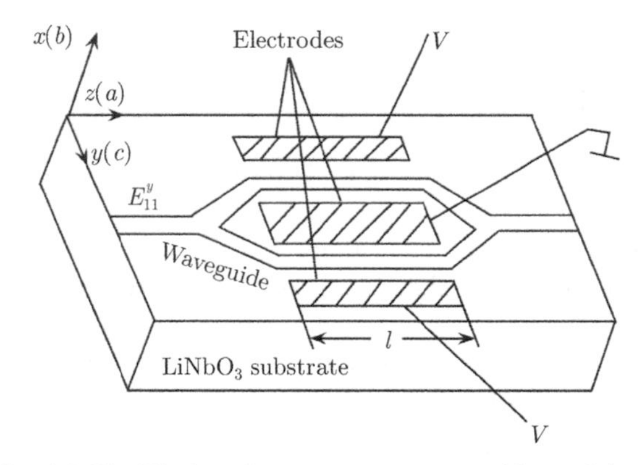

Fig. 1.2-17 The interferometer-type waveguide modulator

In the MZ interferometer-type intensity modulator, in order to increase the modulating depth and reduce insertion loss, it is very important to take the following measures: ① the branch field angle should not be too big (1° or thereabouts in general), as the greater the field angle is, the greater the radiation loss will be; ② the waveguide must be designed as single mode to prevent a higher order mode from getting stimulated; ③ the waveguide and electrodes should be rigorously symmetric structurally so that the fixed phase difference between two phase modulated waves will be equal to zero.

For the MZ interferometer-type modulator made of Ti diffusion $LiNbO_3$ waveguide, its modulating depth can reach as deep as 80%, the half-wave voltage is about 3.6 V, power loss 35 μW/MHz or so, and the modulation bandwidth can reach 17 GHz.

In addition, there are other types of electro-optic waveguide intensity modulators such as the predetermined couple (PC)-type modulator, the electrical absorption (EA)-type modulator, the electro-optic grating modulator, etc. that will not be dealt with here. The reader is referred to books on optical waveguide devices.

1.2.6 Electro-optic deflection

The light beam deflection technology is one of the basic technologies in laser application (such as laser display, facsimile, optical storage, etc.). It can be realized by means of the mechanical deflecting mirror, the electro-optic effect, acousto-optic effect, etc. and is divided into two classes according to the different purposes in its application. One is the analog deflection with the deflection angle of light continuously varying, which can depict the continuous displacement of light, and the other, the discontinuous digit deflection, which consists in making the spatial position of the beam of light "jump vary" in certain specific positions of the chosen space. The former is used in various kinds of display while the latter, mainly in light storage.

1. The principle of electro-optic deflection

The electro-optic deflection is intended to change the direction of propagation of a light beam in space, the principle of which is shown in Fig. 1.2-18. The length of the crystal is L, the thickness is d, and the beam of light is injected into the crystal along the y direction. If the crystal's refractive index is a linear function of the coordinate x, that is,

$$n(x) = n + \frac{\Delta n}{d}x \tag{1.2-65}$$

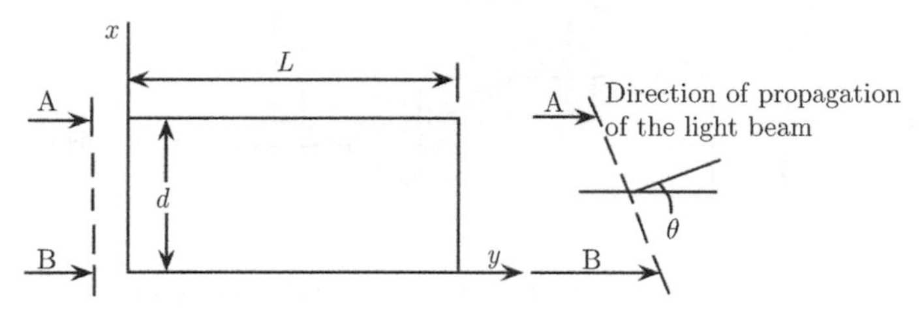

Fig. 1.2-18 The principle of electro-optic deflection

where n represents the index of refraction at $x = 0$ (below the crystal) and Δn the amount of variation of the refractive index on thickness d, then the index of refraction at $x = d$ (on the crystal) is $n + \Delta n$. When a plane wave passes through the crystal, the refractive index "experienced" by the light wave's upper part (ray A) and that by the lower part (ray B) are different, and so are the periods of time needed for passing through the crystal, which are

$$T_{\mathrm{A}} = \frac{L}{c}(n + \Delta n), \quad T_{\mathrm{B}} = \frac{L}{c}(n)$$

respectively. As the time in which one ray passes through the crystal is different from that in which the other passes through, ray A will be a distance behind relative to ray B.

$$\Delta y = \frac{c}{n}(T_{\mathrm{A}} - T_{\mathrm{B}}) = L\frac{\Delta n}{n}$$

This implies that when the light wave reaches the outgoing surface of the crystal, its wave surface has deflected a small angle relative to the axial line of propagation. Its deflection angle (in the output end crystal) is

$$\theta' = -\frac{\Delta y}{d} = -L\frac{\Delta n}{nd} = -\frac{L}{n}\frac{\mathrm{d}n}{\mathrm{d}x}$$

where $\frac{\Delta n}{d}$ is replaced by the refractive index linear variation rate $\frac{\mathrm{d}n}{\mathrm{d}x}$. Then, the deflection angle θ after the beam of light is emitted from the crystal can be found according to the theorem of refraction $\sin\theta/\sin\theta = n$. Suppose $\sin\theta \approx \theta \ll 1$. Then,

$$\theta = \theta' n = -L\frac{\Delta n}{d} = -L\frac{\mathrm{d}n}{\mathrm{d}x} \tag{1.2-66}$$

The minus sign in the equation is introduced by the coordinate system, that is, the turning of angle θ from y to x means negative. It can be seen from the above discussion that if only the refractive index varies along certain directions when the crystal is under the action of the electric field, then the light beam will be made to deflect when it is injected along a specific direction, the size of its deflection angle being proportional to the linear variation rate of the crystal's refractive index.

Shown in Fig.1.2-19 is a dual KDP wedge-shaped prism deflector fabricated on this principle, which is composed of two KDP right angle prisms. The three sides of the prism are along the x'^{-}, y'^{-}, and z-axis directions, respectively, but the z-axes of the two pieces of crystal are parallel to each other in opposite directions while the other two axes have the same orientation.

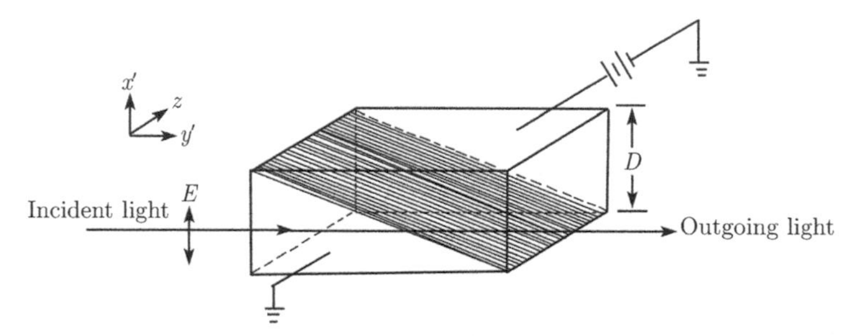

Fig. 1.2-19 A dual KDP wedge-shaped prism deflector

The electric field is along the direction of the z-axis, while the rays propagate along the y' direction and deflect along the x' direction. In this situation, ray A in the upper part propagates completely in the upper prism and the refractive index "experienced" is

$$n_{\mathrm{A}} = n_0 - \frac{1}{2}n_0^3\gamma_{63}E_x$$

whereas in the lower prism, as the electric field is opposite in direction with respect to z, the refractive index "experienced" by ray B is

$$n_{\mathrm{B}} = n_0 + \frac{1}{2}n_0^3\gamma_{63}E_x$$

Thus the difference between the upper and lower refractive indices ($\Delta n = n_{\mathrm{B}} - n_{\mathrm{A}}$) is $n_0^3\gamma_{63}E_z$, which, substituted into Eq. (1.2-66), yields

$$\theta = \frac{L}{d}n_0^3\gamma_{63}E_x \tag{1.2-67}$$

For instance, take $L = d = h = 1$ cm, $\gamma_{63} = 10.5 \times 10^{-12}$ m/V, $n_0 = 1.51$, and $V = 1000$ V, then $\theta = 35 \times 10^{-7}$ rad. It can be seen that the electro-optic deflection angle is too small to meet the requirement for use in practice. To make the angle become bigger while keeping the voltage from becoming too high, it is customary to connect a number of KDP prisms in series to construct a deflector of length mL, width d, and height h, as shown in Fig. 1.2-20. The two pieces at the two ends are isosceles triangular prisms each with an angle of $\beta/2$, and those in the middle with the apex angle of β are isosceles triangular prisms, their z-axes all normal to the face of the plot. The width of a prism is parallel to the z-axis and the z-axes of two prisms adjacent to each other in tandem are in opposite directions. The electric field is along the direction of the z-axis. The directions of the major inductive axes of the prisms are as marked in the figure and the refractive index of the prisms is alternately $(n_0 - \Delta n)$ and $(n_0 + \Delta n)$, where $\Delta n = \frac{1}{2}n_0^3\gamma_{63}E_x$, so after a light beam passes through the deflector, the total deflection angle is m times the deflection angle of every stage (a pair of prisms), that is,

$$\theta_t = m\theta = \frac{mLn_0^3\gamma_{63}V}{dh} \tag{1.2-68}$$

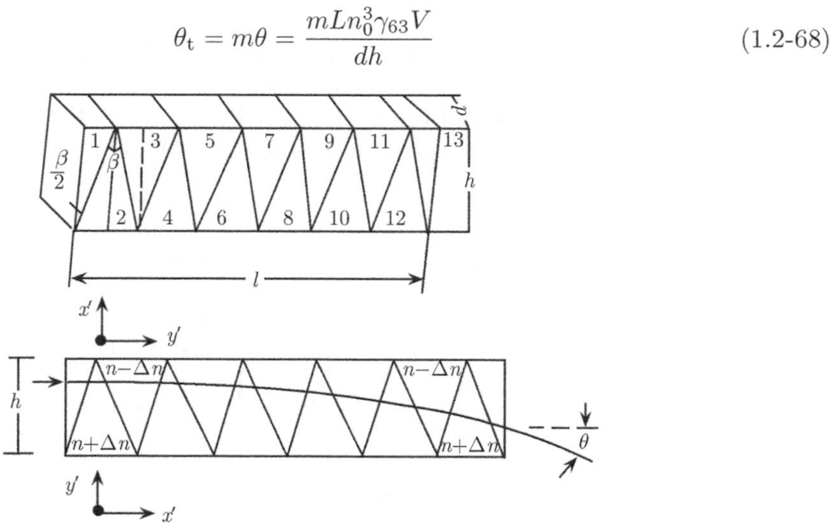

Fig. 1.2-20 A multi-stage prism deflector

In general, m is 4~10, θ_t but a few minutes. The reason why m cannot be infinitely increased is because the laser beam has a definite dimension while the magnitude of h is limited so that it's impossible for the light beam to deflect beyond h.

2. The electro-optic digital deflection

The electro-optic digital deflection is carried out by combining the electro-optic crystal with the birefringence crystal. The principle of its construction is shown in Fig. 1.2-21. In the figure, S is the KDP crystal, and B is the calcite ($CaCO_3$) or nitrate ($NaNO_3$) birefringence crystal (separate prism), the deflection being capable of resolving the linearly polarized light into two beams of light that are parallel to each and whose vibrating directions are normal to each other. The interval b is the degree of splitting, ε is the angle of splitting (also called walk-off angle), and γ is the included angle between the normal direction of the incident light and the optical axis. The x-axis (or y-axis) of the KDP electro-optic crystal S should be parallel to the plane formed by the optical axis of the birefringence crystal B and the crystal face normal. If the polarization direction of a beam of incident light is parallel to the x-axis in S (equivalent to o light* for B), when no voltage is applied on S, the polarization state does not change after the light wave passes through S, then the direction still remains unchanged when it passes through B. When a half-wave voltage is applied on S, then the polarized surface of the incident light will rotate 90° and become e light**. We know that light waves along different polarization directions take different orientations of the optical axis, as do the optical paths of their transmission. So, e light now passing through B has deflected an angle ε relative to the incidence direction and e light emitting from B is b apart from o light. It is already known from physical optics that when n_o and n_e are determined, the corresponding maximum angle of splitting is $\varepsilon_{max} = \text{arctg}\left(\frac{n_e^2 - n_0^2}{2n_e n_0}\right)$. Take calcite as an example. Its $\varepsilon_{max} = 6°$ (in the wave band of visible light and near-infrared light). A stage one digital polarizer is constructed from the above-mentioned electro-optic crystal and birefringence crystal. The incident linearly polarized light occupies one of two "addresses"

* o light (ordinary light).

** e light (extraordinary light).

according to whether a half-wave voltage is or is not applied on the electro-optic crystal, representing state "0" and state "1", respectively. If n such digital deflectors are combined, then n-stage digital deflection can be carried out. Shown in Fig. 1.2-22 is a 3-stage digital deflector as well as how an incident light is resolved into 2^3 deflection points. The short line (|) on the light path represents the polarization surface is parallel to the surface of paper, "·", normal to surface of paper. Of the light emitted the latest, "1" represents that voltage has been applied on a certain electro-optic crystal while "0" represents that no voltage has been applied.

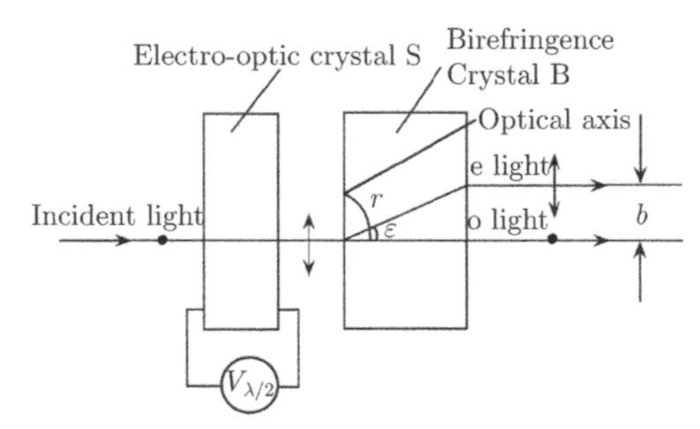

Fig. 1.2-21 The principle of digital deflection

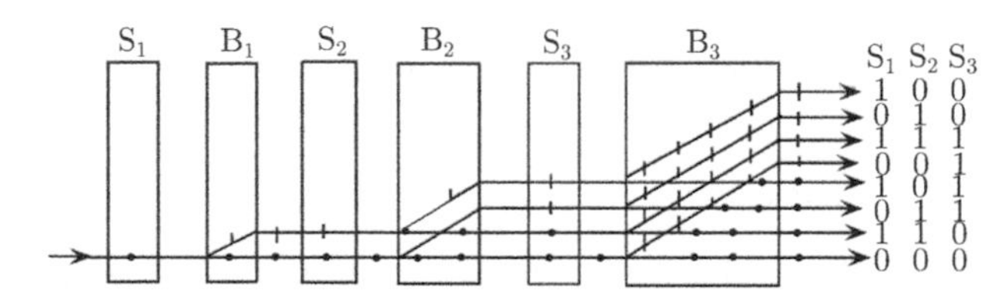

Fig. 1.2-22 A 3-stage digital light deflector

If it is desired that the controllable positions are distributed in the 2-dimensional direction, it is only necessary to combine two n-stage deflectors normal to each other. In so doing, $2^n \times 2^n$ controllable positions can be obtained.

1.2.7 Problems that should be considered in designing the electro-optic modulator

A high quality electro-optic modulator should mainly satisfy the following requirements: ① the modulator should have a sufficiently broad bandwidth to accomplish highly efficient and distortion-free information transmission; ② little electric power is consumed by the modulator; ③ the linear range of the modulation characteristics curve should be wide; and ④ operation is very stable.

1. Selection of the electro-optic crystal material

The material of the modulation crystal plays the crucial role in the effect of modulation. So, a number of factors should be carefully considered when selecting the crystal material. To begin with, the optical performance should be good, the modulated light wave should be highly transparent and the losses in absorption and dissipation should be small, and the refractive index of the crystal should be uniform. The variation of the refractive index should fulfill $\Delta n \leqslant 10^{-4}$/cm, then the electro-optic coefficient should be great as the half-wave voltage of the modulator and the power consumed are each proportional to $1/\gamma_{63}$ and $1/\gamma_{63}^2$. In addition, the modulation crystal should also have fairly good physicochemical performance (mainly the hardness, the threshold of optical destruction, the effects of temperature, air-slake, etc). Table 1.2-2 gives some of the frequently used electro-optic materials.

Tab. 1.2-2 Frequently used electro-optic crystal materials and their physical performance

Name of material	Point group symmetry	Electro – optic coef. $\gamma_{ij}(10^{-12}\text{m/V})$		Refractive indices n_e	n_O	Relative dielectric constants $\varepsilon/\varepsilon_9$
KDP (0.633 μm)	$\bar{4}2$ m	$\gamma_{41} = 8.6$	$\gamma_{63} = 10.6$	1.47	1.51	$\varepsilon//c = 20$ $\varepsilon\perp c = 45$
KD*P (0.633 μm)	$\bar{4}2$ m	$\gamma_{63} = 23.6$		1.47	1.51	$\varepsilon//c \sim 50$ (24°C)
ADP (0.633 μm)	$\bar{4}2$ m	$\gamma_{41} = 28$	$\gamma_{63} = 8.5$	1.48	1.52	$\varepsilon//c = 12$
Quartz (0.633 μm)	32 m	$\gamma_{41} = 0.2$	$\gamma_{63} = 0.47$	1.55	1.54	$\varepsilon//c = 4.3$ $\varepsilon\perp c = 4.3$
CuCl	$\bar{4}3$ m	$\gamma_{41} = 6.1$			$n_O = 1.97$	7.5
ZnS	$\bar{4}3$ m	$\gamma_{41} = 2.0$			$n_O = 2.37$	~10
GaAs (10.6 μm)	$\bar{4}3$ m	$\gamma_{41} = 1.6$			$n_O = 3.34$	11.5
CdTe (10.6 μm)	$\bar{4}3$ m	$\gamma_{41} = 6.8$			$n_O = 2.60$	7.3
$LiNbO_3$(0.633 μm)	3 m	$\gamma_{33} = 30.8$ $\gamma_{13} = 8.6$	$\gamma_{51} = 28$ $\gamma_{22} = 3.4$	2.16	2.26	$\varepsilon//c = 50$
$LiTaO_3$(30°C)	3 m	$\gamma_{33} = 30.3$	$\gamma_{13} = 5.7$	2.18	2.175	$\varepsilon//c = 43$
$BaTiO_3$(30°C)	4 mm	$\gamma_{33} = 23$ $\gamma_{51} = 820$	$\gamma_{13} = 8.0$	2.365	2.437	$\varepsilon\perp c = 4300$ $\varepsilon//c = 106$

2. The method of reducing the power loss of the modulator

The half-wave voltage of the KDP class electro-optic crystal is fairly high. To reduce its power loss, it is advisable to adopt the method of connecting n-stage crystals in series (i.e., series connection on the optical path but parallel connection on the circuit). Figure 1.2-23 shows a longitudinal modulating crystal by connecting four KD*P crystals in series with electrodes of identical polarity linked together. In order to make the four crystals all have the same sign with respect to the phase retardation of the two components of the incident polarized light, the x-axis and y-axis of the crystals are arranged by rotating 90° piece by piece (for example, the x-axis and y-axis of the second crystal are rotated 90° relative to the x-axis and y-axis of the first and the third crystals). The result is the addition of the phase retardation, which is equivalent to a reduction in the half-wave voltage. But it is inadvisable to connect too many crystals in series lest the transmissivity should become too low or the capacitance too high.

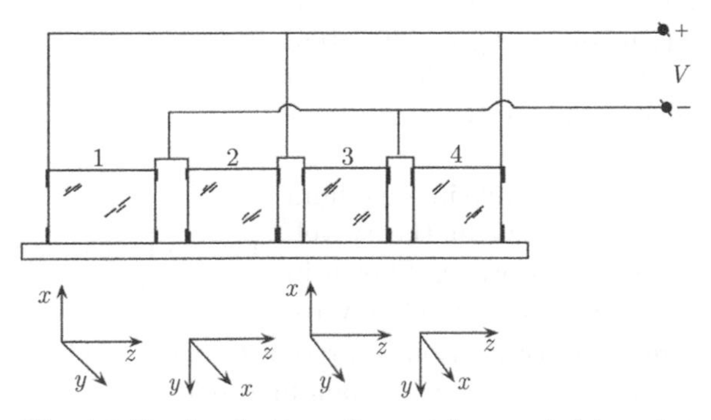

Fig. 1.2-23 An electro-optic crystal connected in series

3. Selection of the modulating voltage

It can be seen from the curve of the modulation characteristics shown in Fig. 1.2-6 that the modulator is already operating at point B. But, if the amplitude of the modulating signal voltage is too high, it will still reach the nonlinear part to cause the distortion of the modulated light. The relationship between the degree of distortion of the modulated light and the amplitude of the modulating voltage has already been analyzed previously. To make the distortion as little as possible, it is necessary to confine the amplitude of higher-order harmonics within the allowable range. When $\eta = 100\%$, $I_{3\omega}/I_{\omega} \approx 0.05$, we find $V \approx 0.383$ $V_{\lambda/2}$, showing that it can be ensured that no distortion will occur.

4. Selection of the dimensions of the electro-optic crystal

By the dimensions of the electro-optic crystal we mean its length and the size of its cross section. In the longitudinal application of the KDP class crystals, although the half-wave voltage is not related to the length of the crystal, an increase in its length will decrease the capacitance of the modulator (because $C_0 = \varepsilon A/L$) to broaden the frequency band. However, the greater the length is, the higher will the requirements be on machining and the accuracy in mounting and modulation. Otherwise, it will be impossible for the optical axis of the crystal to be completely parallel to the direction of light wave propagation. Rather, it will be affected by the spontaneous double refraction of the crystal, thus increasing the instability of phase retardation of the modulator. So L should not be too long. The size of the transverse cross section mainly depends on the requirements of the light transmitting aperture.

1.3 Acousto-optic modulation

1.3.1 The physical basis of acousto-optic modulation

The acoustic wave is a kind of elastic wave (longitudinal stress wave). When propagating in a medium, it causes the medium to generate corresponding elastic deformation, stimulating the particles in the medium to vibrate along the direction of propagation of the acoustic wave and the density of the medium to alternately become thick and thin. Hence the refractive index of the medium, too, will undergo corresponding periodic variation. This part of the action of the ultrasonic field is like an optical "phase grating", whose interval (grating constant) is equal to the acoustic wavelength λ_s. When the light wave passes through this medium, light diffraction will occur, the intensity, frequency, direction, etc. of the diffracted light varying with the variation of the ultrasonic field.

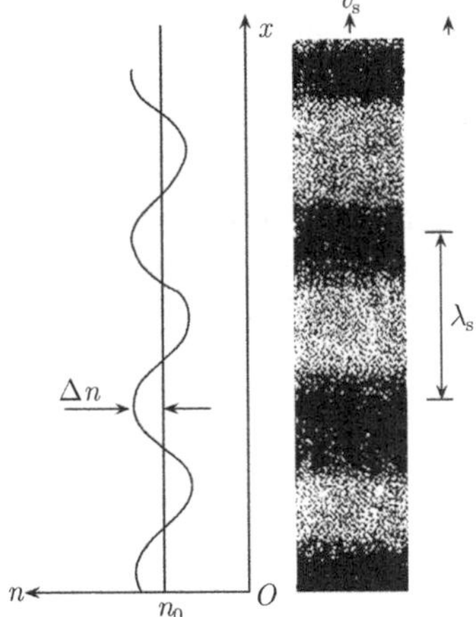

Fig. 1.3-1 The propagation of the ultrasonic traveling wave in the medium

The propagation of the acoustic wave in a medium is divided into two forms, the form of the traveling wave and that of the standing wave. Shown in Fig. 1.3-1 is the case of the ultrasonic traveling wave in a certain instant, where the dark portions represent that the medium is being contracted and the density becomes thick as is the corresponding index of refraction, while the white portions represent that the medium's density has become thin. Under the action of the acoustic field of the traveling wave, the refractive index of the medium alternately varies between increase and decrease and pushes forward with v_s (in general 10^3 m/s). As the acoustic velocity is only one several hundred thousandth of the velocity of light, for the light wave, the moving "acousto-optic grating" can be regarded as stationary. Suppose the angular frequency of the acoustic wave is ω_s and the wave vector is $k_s \left(\dfrac{2\pi}{\lambda_s}\right)$. Then the equation of the acoustic wave is

$$a(x,t) = A\sin(\omega_s t - k_s x) \tag{1.3-1}$$

where a is the transient displacement of the medium particle and A is the amplitude of the displacement of particles. It can be approximately considered that the variation of the medium's refractive index is proportional to the rate of variation of the medium particles displacement along the x direction, i.e.,

$$\Delta n(x,t) \propto \frac{\mathrm{d}a}{\mathrm{d}x} = -k_s A\cos(\omega_s t - k_s x)$$

or

$$\Delta n(x,t) = \Delta n\cos(\omega_s t - k_s x) \tag{1.3-2}$$

Then the refractive index of the traveling wave is

$$\begin{aligned} n(x,t) &= n_0 + \Delta n\cos(\omega_s t - k_s x) \\ &= n_0 - \frac{1}{2}n_0^3 PS[\cos(\omega_s t - k_s x)] \end{aligned} \tag{1.3-3}$$

where S is the strain generated by the medium due to the ultrasonic wave and P is the elasto-optic coefficient of material.

The acoustic standing wave is composed of two beams of acoustic wave of identical wavelength, amplitude, and phase but opposite directions of propagation, as shown in Fig. 1.3-2. Its equation of acoustic standing wave is

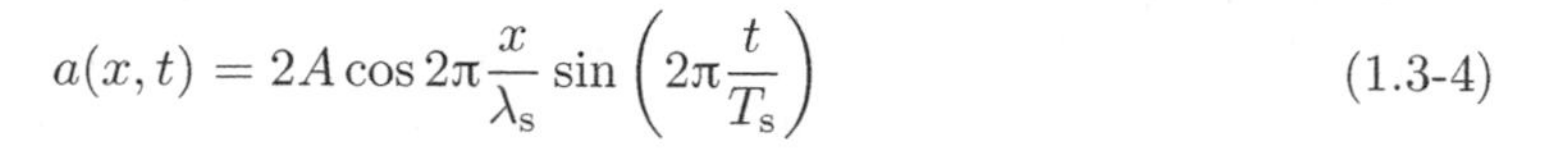

$$a(x,t) = 2A\cos 2\pi\frac{x}{\lambda_s}\sin\left(2\pi\frac{t}{T_s}\right) \tag{1.3-4}$$

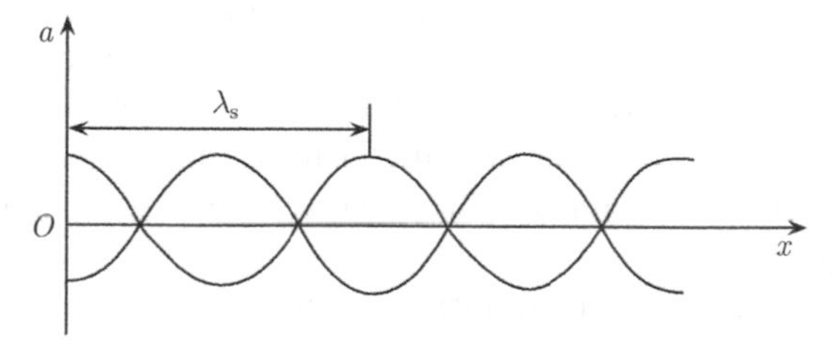

Fig. 1.3-2 The ultrasonic standing wave

The above equation shows that the amplitude of the acoustic standing wave is 2Acos $(2\pi\ x/\lambda_s)$, which is different at different points in the x direction, but the phase $2\pi\ t/T_s$ is the same at all points. Meanwhile, it can also be seen from the above equation that, at the points $x = 2n\lambda_s/4$ $(n = 0, 1, 2, \cdots)$, the amplitude of the standing wave is maximum (equal to $2A$). These points are called the antinodes; the distance between one antinode and another is $\lambda_s/2$. At the points $x = (2n+1)\ \lambda_s\ /4$, the amplitude of the standing wave is zero. These points are called nodes and the distance between two nodes is also $\lambda_s/2$. As the positions of the acoustic standing wave's antinodes and nodes in the medium are fixed, the grating formed by it is also fixed in space. The variation of the refractive index formed by the acoustic standing wave is

$$\Delta n(x,t) = 2\Delta n\sin\omega_s t\sin k_s x \tag{1.3-5}$$

For the acoustic standing wave, the dense and sparse layers appear twice in the medium and the density remains unchanged at the nodes. Therefore, the refractive index varies once every half period ($T_s/2$) at the antinodes, from maximum (or minimum) to minimum (or maximum). In a certain instant during the two times of variation, the indices of refraction in all parts of the medium are identical, being equivalent to a homogeneous medium free from the action of the acoustic field. If the ultrasonic frequency is f_s, then the number of times of appearance and disappearance of grating is $2f_s$. Thus the modulated frequency of the modulating light obtained after the light wave passes the medium will be twofold of the acoustic frequency.

1.3.2 Two types of interaction between sound and light

Depending on the difference in the height of frequency of the sound wave and the length of action between the sound wave and light wave, the interaction between sound and light can be divided into two types, the Raman-Nath diffraction and Bragg diffraction.

1. Raman-Nath diffraction

When the ultrasonic wave frequency is rather low, the light wave is injected parallel to the surface of the sound wave (i.e., normal to the direction of the sound field propagation), and the interaction length L of sound and light is rather short, Raman-Nath diffraction will be generated. As the sound velocity is much smaller than that of the light velocity, the acousto-optic medium can be regarded as a stationary plane phase grating. Furthermore, the length of the sound wave λ_s is much greater than the light wavelength λ. When the light wave passes through the medium parallelly, it does so almost without going through the sound wave surface. So only its phase is modulated, that is, the light wave surface passing through the optically dense (high refractive index) portion will be delayed while its passage through the optically sparse (low refractive index) portion will be advanced. Thus, there is the phenomenon of convexity and concavity in the passage through the plane wave surface of the acoustic-optic medium, resulting in a creased camber, as shown in Fig. 1.3-3. The secondary wave emitted by the subwave sources on the outgoing wave surface will effect the action of coherence to form multistage light of diffraction that is symmetrically distributed with the incident direction, which is the Raman-Nath diffraction.

Below let's make a brief analysis on the direction of light wave diffraction and the distribution of light intensity.

Suppose the sound wave in the acousto-optic medium is a plane longitudinal wave (sound column) of width L and propagating along the x direction; the wavelength is λ_s (angular frequency ω_s), the wave vector k_s pointing to the x-axis, and the incident light wave vector k_i pointing to the direction of the y-axis, as shown in Fig. 1.3-4. The elastic strain field induced by the sound wave in the medium can be expressed as

$$S_1 = S_0 \sin(\omega_s t - k_s x)$$

According to Eq. (1.3-3), there is

$$\Delta\left(\frac{1}{n^2}\right) = PS_0 \sin(\omega_s t - k_s x)$$

or

$$\Delta n = -\frac{1}{2} n^3 P S_0 \sin(\omega_s t - k_s x) \tag{1.3-6}$$

Then

$$n(x,t) = n_0 + \Delta n \sin(\omega_s t - k_s x) \tag{1.3-7}$$

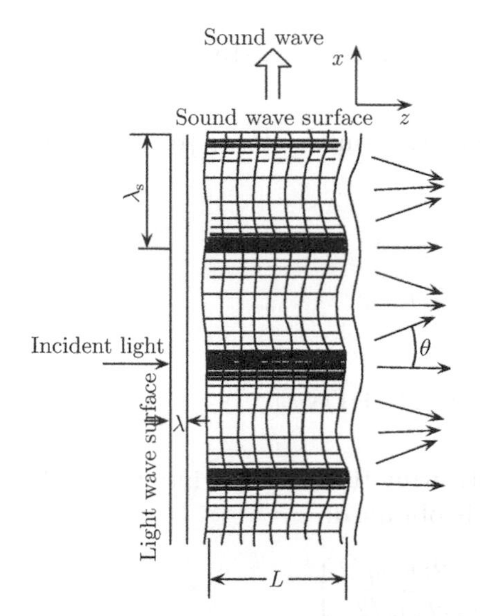

Fig. 1.3-3 The Raman-Nath diffraction

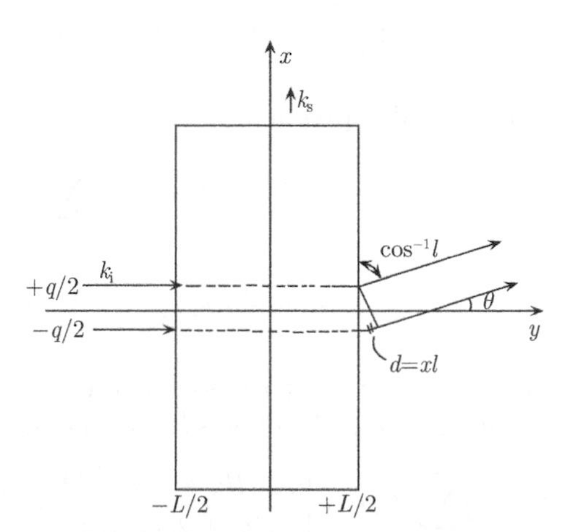

Fig. 1.3-4 The case of normal incidence

When the acoustic traveling wave is approximately regarded as an ultrasonic field not varying with time, the relation of dependence on time can be neglected. If so, the distribution of the refractive index along the x direction can be reduced to

$$n(x) = n_0 + \Delta n \sin(k_s x) \tag{1.3-8}$$

where n_0 is the average refractive index, Δn the variation of sound-induced refractive index. As the medium's refractive index has undergone periodic variation, it is capable of modulating the phase of the incident light wave. If what is investigated is a case of normal incidence by a plane light wave where it is injected into the acousto-optic medium's front surface at $y = -L/2$ and the incident light wave is

$$E_{in} = A \exp(i\omega_c t) \tag{1.3-9}$$

then the light wave that goes out from $y = L/2$ will no longer be a monochromatic plane wave but a modulated light wave, whose equiphase surface is a creased curved surface determined by the function $n(x)$, whose light field can be written as

$$E_{out} = A \exp\{i[\omega_c(t - n(x)L/c)]\} \tag{1.3-10}$$

This outgoing wave front surface can be divided into a number of subwave sources. So at the point P very far away, the total intensity of the diffracted light field is a summation of the contribution of all subwave sources, which is determined by the following integral:

$$E_p = \int_{-q/2}^{q/2} \exp\{ik_i[lx + L\Delta n \sin(k_s x)]\}dx \tag{1.3-11}$$

where $l = \sin\theta$, representing sine of the direction of diffraction, q is the width of the incident beam of light. Substitute $v = 2\pi\Delta nL/\lambda = \Delta n k_i L$ into the above equation and expand to the following form using Euler's formula:

$$E_p = \int_{-q/2}^{q/2} \{\cos[k_i lx + v\sin(k_s x)] + i\ \sin[k_i lx + v\sin(k_s x)]\}dx$$

$$= \int_{-q/2}^{q/2} \{\cos(k_i lx)\cos[\upsilon\sin(k_s x)] - \sin(k_i lx)\sin[\upsilon\sin(k_s x)]\}dx$$
$$+ j\int_{-q/2}^{q/2} \{\sin(k_i lx)\cos[\upsilon\sin(k_s x)] - \cos(k_i lx)\sin[\upsilon\sin(k_s x)]\}dx \quad (1.3\text{-}12)$$

Use the relational expressions

$$\cos[\upsilon\sin(k_s x)] = 2\sum_{r=0}^{\infty} J_{2r}(\upsilon)\cos(2rk_s x)$$

$$\sin[\upsilon\sin(k_s x)] = 2\sum_{r=0}^{\infty} J_{2r+1}(\upsilon)\sin[(2r+1)k_s x]$$

where $J_r(\upsilon)$ is the rth-order Bessel function. By substituting this equation into Eq. (1.3-12) and after integration, the expression of the real part is obtained:

$$E_p = q\sum_{r=0}^{\infty} J_{2r}(\upsilon)\left\{\frac{\sin(lk_i+2rk_s)q/2}{(lk_i+2rk_s)q/2} + \frac{\sin(lk_i-2rk_s)q/2}{(lk_i-2rk_s)q/2}\right\}$$
$$+ q\sum_{r=0}^{\infty} J_{2r+1}(\upsilon)\left\{\frac{\sin[lk_i+(2r+1)k_s]q/2}{[lk_i+(2r+1)k_s]q/2} - \frac{\sin[lk_i-(2r+1)k_s]q/2}{[lk_i-(2r+1)k_s]q/2}\right\} \quad (1.3\text{-}13)$$

while the integral of the imaginary part of Eq. (1.3-12) is zero. It can be seen from Eq. (1.3-13) that the condition for all the terms of the diffracted light field intensity to take the maximum value is

$$lk_i \pm mk_s = 0 \quad (m = \text{integer} \geqslant 0) \quad (1.3\text{-}14)$$

After angle θ and the sound wave vector are determined with one of the terms in them being maximum, the contributions of the other terms are almost equal to zero. Therefore, when m takes different values, the diffracted light of different θ directions takes the maximum value. The azimuth of all levels of diffraction is determined by Eq. (1.3-14):

$$\sin\theta = \pm m\frac{k_s}{k_i} = \pm m\frac{\lambda}{\lambda_s} \quad (m = 0, \pm 1, \pm 2, \cdots) \quad (1.3\text{-}15)$$

where m represents the level of the diffracted light, and the intensity of all levels of diffracted light is

$$I_m \propto J_m^2(\upsilon), \quad \upsilon = \Delta n k_i L = \frac{2\pi}{\lambda}\Delta n L \quad (1.3\text{-}16)$$

To sum up, as a result of Raman-Nath acousto-optic diffraction, the light wave is resolved into a group of diffracted rays of light corresponding to definite diffraction angle θ_m (i.e., the direction of propagation) and diffraction intensity, of which the diffraction angle is determined by Eq. (1.3-15) while the diffraction intensity is determined by Eq. (1.3-16), showing that the group of diffracted rays of light is discrete. Since $J_m^2(\upsilon) = J_{-m}^2(\upsilon)$, the diffracted light of all levels is symmetrically distributed on the two sides of the zero-level diffracted light and the intensities of the same level are equal, which is one of the main features of Raman-Nath diffraction.

In addition, as

$$J_0^2(\upsilon) + 2\sum_{1}^{\infty} J_m^2(\upsilon) = 1$$

it can be seen that the sum of the maximum intensities of the diffracted light at all levels without absorption should be equal to the incident light intensity, that is, the optical power is conservative. In the above analysis, the factor of time is neglected and the physical image of the Raman-Nath acousto-optic action is obtained using a fairly simple processing method. But, owing to the action of the light wave and sound wave field, every level of diffracted light wave will generate Doppler frequency shift. According to the principle of conservation of energy, there should be

$$\omega = \omega_{\mathrm{i}} \pm m\omega_{\mathrm{s}} \tag{1.3-17}$$

and all levels of diffracted light intensities will be modulated by an angular frequency of $\sqrt{2}$ ω_{s}. But as the ultrasonic wave frequency is 10^9 Hz, while the frequency of the light wave is as high as an order of magnitude of 10^{14} Hz, the effect of frequency shift can be neglected.

2. The Bragg diffraction

(1) The normal Bragg diffraction in isotropic media

When the frequency of the sound wave is rather high, the length L of the acousto-optic action is great, and the beam of light is slantingly injected into the sound wave surface at a definite angle, the light wave will penetrate multiple sound wave surfaces in the medium. So the medium possesses the properties of "volume grating". When the included angle between the incident light and sound wave surface satisfies definite conditions, the different levels of diffracted light in the medium will interfere with one another while the higher-level diffracted light will offset one another, with only the level 0 or level +1 (or level −1) (depending on the direction of the incident light) diffracted light appearing; that is, Bragg diffraction is generated, as shown in Fig. 1.3-5. For this reason, if the parameters can be rationally chosen, the ultrasonic field is sufficiently powerful, and it will be possible to have the energy of the incident light almost completely transferred to the level +1 (or level −1) diffraction extremum, thus enabling the energy of the light beam to be fully utilized. Therefore, acousto-optic devices manufactured using the Bragg diffraction effect are fairly efficient.

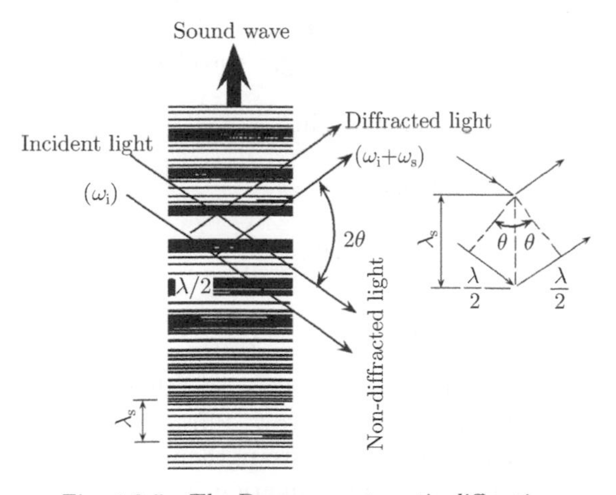

Fig. 1.3-5 The Bragg acousto-optic diffraction

Below let's derive the Bragg equation from the condition for strengthening the interference of the wave. For this purpose, the medium the sound wave passes through can be approximately regarded as many partially reflective partially transmissive mirror surfaces λ_{s} apart. For the traveling wave ultrasonic field, these mirror surfaces will shift along the x direction with velocity v_{s} (as $\omega_{\mathrm{m}} \leqslant \omega_{\mathrm{c}}$, in a certain instant, the ultrasonic field can be regarded as stationary and so having no effect on the distribution of the diffracted light intensity). For the standing wave ultrasonic field, it is completely stationary, as shown in

Fig. 1.3-6. When plane waves 1, 2 are injected into the acoustic wave field at angle θ_i, they are reflected in part at points B, C, E, generating diffracted light $1'$, $2'$, $3'$. The condition for the strengthening of coherence of the diffracted light is that the difference in the light path between them should be integral multiples of the wavelength, or it can be said that they must be in phase. Fig. 1.3-6(a) shows a case of diffraction on one and the same mirror surface, where the condition for the incident light 1, 2 to be in phase with $1'$, $2'$ reflected by them at points B, C is that the difference in the optical path AC–BD be equal to the integral multiples of the wavelength of the light wave. That is,

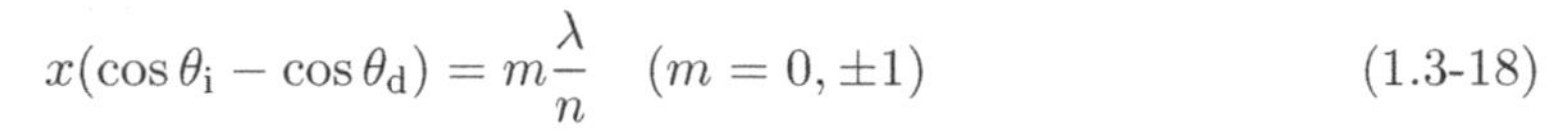

$$x(\cos\theta_i - \cos\theta_d) = m\frac{\lambda}{n} \quad (m = 0, \pm 1) \tag{1.3-18}$$

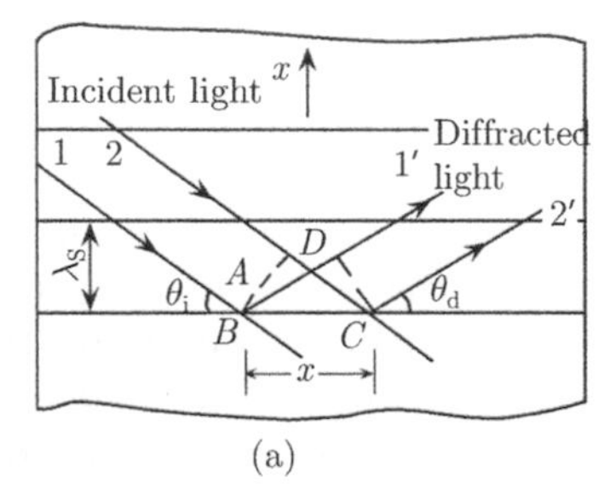

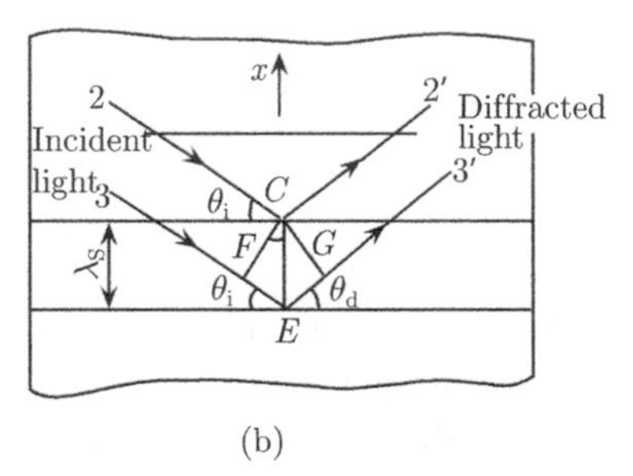

Fig. 1.3-6 A model for condition for generation of Bragg diffraction

If it is required that all the points on the acoustic wave surface satisfy this condition at the same time, there is no other choice but to make

$$\theta_i = \theta_d \tag{1.3-19}$$

That is, only when the incident angle is equal to the diffraction angle will this be realized. The diffraction on two different mirror surfaces λ_s apart is as shown in Fig. 1.3-6(b). The condition for $2'$, $3'$ reflected from points C, E to be in phase is that their optical path difference be equal to integral multiples of the light wavelength, i.e.,

$$\lambda_s(\sin\theta_i + \sin\theta_d) = \frac{\lambda}{n} \tag{1.3-20}$$

By taking into account $\theta_i = \theta_d$, it follows that

$$2\lambda_s \sin\theta_B = \frac{\lambda}{n}$$

or

$$\sin\theta_B = \frac{\lambda}{2n\lambda_s} = \frac{\lambda}{2nv_s} f_s \tag{1.3-21}$$

where $\theta_i = \theta_d = \theta_B$; θ_B is called the Bragg angle. It is clear that, only when the incident angle θ_i is equal to the Bragg angle θ_B, will the light wave diffracted on the acoustic wave surface be in phase so that the condition for coherence strengthening is satisfied and the maximum value of diffraction is obtained. The above equation is called the Bragg equation. For instance, for the acousto-optic diffraction in water, suppose the light wavelength $\lambda = 0.5\ \mu\text{m}$, $n = 1.33$, the acoustic wave frequency $f_s = 500$ MHz, and acoustic velocity $v_s = 1.5 \times 10^3$ m/s. Then $\lambda_s = \dfrac{v_s}{f_s} = 3 \times 10^{-6}$ m. From Eq. (1.3-21) we obtain the Bragg angle $\theta_B \cong 6 \times 10^{-2}$ rad $= 3.4°$.

Below we make a brief analysis of the relationship between the Bragg diffracted light intensity and the characteristics of acousto-optic material and acoustic field intensity. According to inference and proof, when the intensity of the incident light is I_i, the expressions of the zero-th and first level diffraction light intensities can be written as

$$I_0 = I_i \cos^2\left(\frac{v}{2}\right), \quad I_1 = I_i \sin^2\left(\frac{v}{2}\right) \tag{1.3-22}$$

It is known that v is the additional phase retardation generated when the light wave penetrates an ultrasonic field of length L. v can be represented by the variation Δn of the sound-induced refractive index, i.e.,

$$v = \frac{2\pi}{\lambda}\Delta n L$$

Thus

$$I_1/I_i = \sin^2\left[\frac{1}{2}\left(\frac{2\pi}{\lambda}\Delta n L\right)\right] \tag{1.3-23}$$

Suppose the medium is isotropic. It is known from crystal optics that when light wave and sound wave propagate along certain symmetrical directions, Δn is determined by the elasto-optic coefficient P of the medium and the elastic strain amplitude S of the medium under the action of the sound field, i.e.,

$$\Delta n = -\frac{1}{2}n^3 PS \tag{1.3-24}$$

where S is related to the ultrasonic driving power P_s while the ultrasonic power is related to the area of the transducer (H is the width of the transducer, L the length of the transducer), the sound velocity v_s, and energy density $\frac{1}{2}\rho v_s^2 S^2$($\rho$ is the density of the medium), i.e.,

$$P_s = (HL)v_s\left(\frac{1}{2}\rho v_s^2 S^2\right) = \frac{1}{2}\rho v_s^3 S^2 HL$$

Therefore

$$S = \sqrt{2P_s/HL\rho v_s^3}$$

Thus

$$\Delta n = -\frac{1}{2}n^3 P\sqrt{\frac{2P_s}{HL\rho v_s^3}} = -\frac{1}{2}n^3 P\sqrt{\frac{2I_S}{\rho v_s^3}} \tag{1.3-25}$$

where $I_s = P_s/HL$ is called the ultrasonic intensity. Substitution of Eq. (1.3-25) into Eq. (1.3-23) yields

$$\eta_s = \frac{I_1}{I_i} = \sin^2\left[\frac{\pi L}{\sqrt{2}\lambda}\sqrt{\left(\frac{n^6P^2}{\rho v_s^3}\right)I_s}\right] = \sin^2\left(\frac{\pi L}{\sqrt{2}\lambda}\sqrt{M_2 I_s}\right) \tag{1.3-26}$$

or

$$\eta_s = \frac{I_1}{I_i} = \sin^2\left[\frac{\pi}{\sqrt{2}\lambda}\sqrt{\left(\frac{L}{H}\right)M_2 P_s}\right] \tag{1.3-27}$$

where $M_2 = n^6P^2/\rho v_s^3$ is a combination of the physical parameters of the acousto-optic medium, a quantity determined by the properties of the medium itself and which is called the quality factor (or the acousto-optic high quality index) of the acousto-optic material. It is one of the principal indices of selecting the acousto-optic medium. It can be seen from Eq. (1.3-27) that: ① If under the condition that the ultrasonic power P_s is definite, it is desired that the intensity of the diffracted light should be as high as possible, then it is required that a material of great M_2 be selected and the transducer be made long and narrow (i.e., great in L, small in H); ② When the ultrasonic power P_s is sufficiently high, such that $\left[\frac{\pi}{\sqrt{2}\lambda}\sqrt{\left(\frac{L}{H}\right)M_2 P_s}\right]$ is enabled to reach $\frac{\pi}{2}$, $I_1/I_i = 100\%$; and ③ When P_s varies, I_1/I_i will also vary with it, hence by controlling P_s (i.e., controlling the electric power applied on the transducer), the goal of controlling the intensity of the diffracted light can be attained, thus realizing acousto-optic modulation.

(2) The particle model for Bragg acousto-optic diffraction

What is dealt with above is the principle of Bragg interaction between acoustic and light from the point of view of coherence and superposition of the light wave, but the condition for acousto-optic Bragg diffraction can also be obtained from the quantum characteristics of light and acoustic. A light beam can be regarded as a photon (particle) flow of energy $\hbar\omega_i$ and momentum $\hbar k_i$, of which ω_i and k_i are the angular frequency and wave vector of the light wave. Similarly, an acoustic wave can also be regarded as a phonon flow of energy $\hbar\omega_s$ and momentum $\hbar k_s$. The interaction between acoustic and light can be regarded as a series of collisions between photons and phonons. Every time a collision occurs there will be the annihilation of an incident photon (ω_i) and a phonon (ω_s) while a new (diffracted) photon of frequency $\omega_d = \omega_i + \omega_s$ will be generated at the same time. The flow of these new diffracted photons propagates along the direction of diffraction. According to the principle of conservation of momentum before and after collision, there should be

$$\hbar\boldsymbol{k}_i \pm \hbar\boldsymbol{k}_s = \hbar\boldsymbol{k}_d$$

that is,

$$\boldsymbol{k}_i \pm \boldsymbol{k}_s = \boldsymbol{k}_d \tag{1.3-28}$$

Similarly, according to the conservation of energy, there should be

$$\hbar\omega_i \pm \hbar\omega_s = \hbar\omega_d$$

That is,

$$\omega_i \pm \omega_s = \omega_d \tag{1.3-29}$$

where "+" represents absorbing the phonon, "−" releasing the phonon, depending on the relative direction of $\boldsymbol{k}_i$ and $\boldsymbol{k}_s$ during the collision between photon and phonon. That is, the diffracted photon is generated by the photon that has disappeared and the phonon that has been absorbed during the collision, which takes the sign "+" in the formulae and the frequency is $\omega_d = \omega_i + \omega_s$. If in the collision the disappearance of an incident photon leads to the generation of phonon and diffracted photon at the same time, then the "−" sign will be taken, the frequency being $\omega_d = \omega_i - \omega_s$. As the light wave frequency (ω_i) is far higher than the acoustic wave frequency (ω_s), according to Eq. (1.3-29), it can be approximately considered that

$$\omega_d = \omega_i \pm \omega_s \cong \omega_i \tag{1.3-30}$$

Therefore

$$\boldsymbol{k}_d = \boldsymbol{k}_i \tag{1.3-31}$$

So the vector graph of Bragg diffraction is an isosceles triangle, as shown in Fig. 1.3-7. From the figure, $k_i \sin\theta_i + k_d \sin\theta_d = 2k_i \sin\theta_B = k_s$ can be directly derived. Then there is

$$\sin\theta_B = \frac{k_s}{2k_i} = \frac{\lambda}{2n\lambda_s} \tag{1.3-32}$$

$$\theta_{\rm i} = \theta_{\rm d} = \theta_{\rm B}$$

This is the Bragg equation obtained previously.

(3) Abnormal Bragg diffraction

The above discussion is made under the condition that the wave vectors of the incident light and diffracted light are equal (i.e., $|\boldsymbol{k}_{\rm i}| = |\boldsymbol{k}_{\rm d}|$); that is, it is assumed that the direction of polarization of the incident light and that of the diffracted light are the same. Therefore, the corresponding refractive indices are equal ($m_{\rm i} = m_{\rm d}$). These properties can be fulfilled only in isotropic media (glass, liquid, and crystals of the cubic system). If the acousto-optic media are anisotropic crystals, the refractive index of the light beam is in general related to the direction of propagation. As the diffracted light propagates along a direction different from that of the incident light, the polarized state of the incident light is different from that of the diffracted light, and the indices of refraction corresponding to them are not equal ($n_{\rm i} \neq n_{\rm d}$), so $|\boldsymbol{k}_{\rm i}| \neq |\boldsymbol{k}_{\rm d}|$). Such Bragg diffraction that occurs in anisotropic media is called abnormal Bragg diffraction.

The abnormal Bragg diffraction no longer has the condition for $n_{\rm i} \neq n_{\rm d}$, and the corresponding geometric relationship is more complicated than that of the normal Bragg diffraction. Its momentum triangular closure condition $\boldsymbol{k}_{\rm d} = \boldsymbol{k}_{\rm i} \pm \boldsymbol{k}_{\rm s}$ is shown in Fig. 1.3-8 (corresponding to the case of taking "+" in the formulae), where the modes of $k_{\rm i}$, $k_{\rm d}$, and $k_{\rm s}$ are

$$\begin{aligned} k_{\rm i} &= 2\pi n_{\rm i}(\theta_{\rm i})/\lambda \\ k_{\rm d} &= 2\pi n_{\rm d}(\theta_{\rm d})/\lambda \\ k_{\rm s} &= 2\pi/\lambda_{\rm s} = 2\pi f_{\rm s}(\theta) \end{aligned} \tag{1.3-33}$$

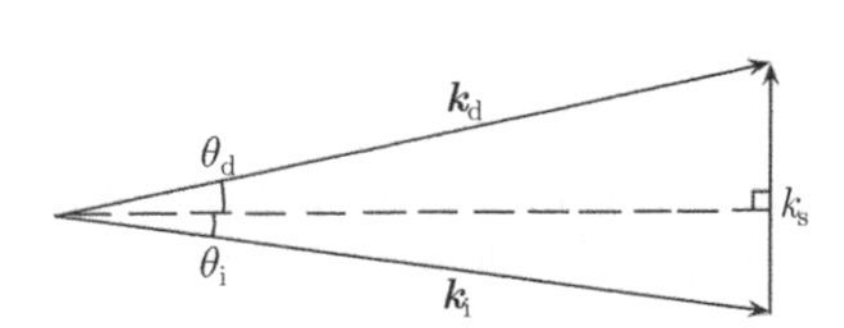

Fig. 1.3-7 The wave vector diagram of normal Bragg diffraction

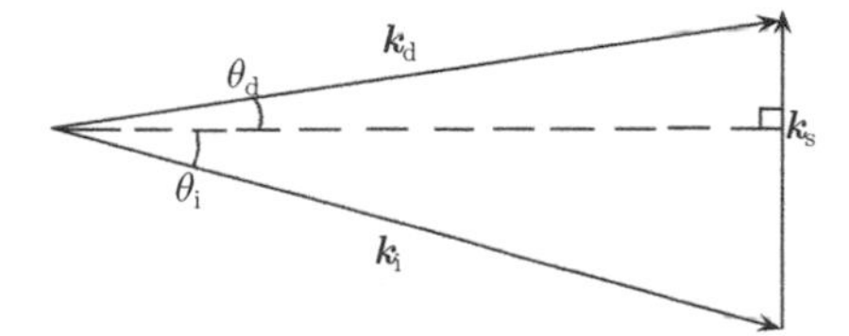

Fig. 1.3-8 The wave vector diagram of abnormal Bragg diffraction

respectively. From Fig. 1.3-8, according to the law of cosine, we have

$$k_{\rm d}^2 = k_{\rm s}^2 + k_{\rm i}^2 - 2k_{\rm s}k_{\rm i}\cos\left(\frac{\pi}{2} - \theta_{\rm i}\right) = k_{\rm s}^2 + k_{\rm i}^2 - 2k_{\rm s}k_{\rm i}\sin\theta_{\rm i}$$

$$k_{\rm i}^2 = k_{\rm s}^2 + k_{\rm d}^2 - 2k_{\rm s}k_{\rm d}\sin\theta_{\rm d}$$

from which $\sin\theta_{\rm i}$ and $\sin\theta_{\rm d}$ can be solved. Then substitution of Eq. (1.3-33) into the above equation yields

$$\left.\begin{aligned} \sin\theta_{\rm i} &= \frac{\lambda}{2n_{\rm i}(\theta_{\rm i})v_{\rm s}}\left\{f_{\rm s} + \frac{v_{\rm s}^2}{\lambda^2 f_{\rm s}}[n_{\rm i}^2(\theta_{\rm i}) - n_{\rm d}^2(\theta_{\rm d})]\right\} \\ \sin\theta_{\rm d} &= \frac{\lambda}{2n_{\rm d}(\theta_{\rm d})v_{\rm s}}\left\{f_{\rm s} - \frac{v_{\rm s}^2}{\lambda^2 f_{\rm s}}[n_{\rm i}^2(\theta_{\rm i}) - n_{\rm d}^2(\theta_{\rm d})]\right\} \end{aligned}\right\} \tag{1.3-34}$$

The above equation is called Dixon's equation, in which $n_{\rm i}$ and $n_{\rm d}$ are the functions of angles $\theta_{\rm i}$ and $\theta_{\rm d}$. Therefore, only after the functional relationship of variation of $n_{\rm i}$ and $n_{\rm d}$ with the angle is determined for a definite medium will it be possible to solve the relationship between $\theta_{\rm i} - f_{\rm s}$ and $\theta_d - f_{\rm s}$ using Dixon's equation, thereby determining the geometric relationship of abnormal Bragg diffraction. An analysis of Eq. (1.3-34) provides us with the following characteristics:

1) The first term $\frac{\lambda}{2nv_s}f_s$ in the equation is exactly the condition for normal Bragg diffraction while the second term exists only in an anisotropic medium, which differs from crystal to crystal. When $f_s = f_0 = \frac{v_s}{\lambda}\sqrt{n_{i0}^2 - n_{d0}^2}$, θ_i reaches maximum, and $\theta_d = 0$, where f_0 is called the maximum frequency of abnormal Bragg diffraction, $n_{i0} = n_i\ (\theta_{i0})$, $n_{d0} = n_d\ (\theta_{d0})$, while θ_{i0} and θ_{d0} are the θ_i and θ_d values corresponding to f_0. When $f_s = f_0$, the two terms on the right side of the above equation are equal but when $f_s \geqslant f_0$, the second term on the right side is small and can be neglected. Now Dixon's equation can be reduced to

$$\sin\theta_i = \sin\theta_d = \frac{\lambda}{2nv_s}f_s \tag{1.3-35}$$

That is, it has become a normal Bragg diffraction equation. But when f_s approaches or is smaller than f_0, then it will have a geometric relationship totally different from that of the normal Bragg diffraction.

2) If the two sub-equations of Eq. (1.3-34) are added up, then the second term on the right side of the equation can be eliminated (the difference between n_i and n_d in the denominator neglected). Then we have

$$\sin\theta_i + \sin\theta_d \cong \frac{\lambda}{2nv_s}f_s$$

If it is further considered that if both θ_i and θ_d are very small, then there is

$$\alpha = \theta_i + \theta_d \cong \frac{\lambda}{2nv_s}f_s \tag{1.3-36}$$

This shows that, although the $\theta_i - f_s$ relationship is different from the normal Bragg diffraction relationship, the relationship between α and f_s is completely identical to that of normal Bragg diffraction.

3) As the abnormal Bragg diffraction $|k_i| \neq |k_d|$, there can exist interaction in the same direction, that is, k_i, k_d, and k_s are all in the same direction. Obviously, the closure condition of the momentum triangle can be reduced to the scalar form, i.e., $k_d = k_i \pm k_s$. If we substitute into it $k_i = \frac{2\pi}{\lambda}n_i$, $k_d = \frac{2\pi}{\lambda}n_d$, and $k_s = \frac{2\pi}{\lambda_s} = \frac{\pi}{v_s}f_s$, then we can obtain

$$\lambda = \pm v_s(n_d - n_i)/f_s \tag{1.3-37}$$

This shows that for a definite acousto-optic medium and direction of propagation, the numerator part to the right of the equation is a constant. Now, when white light (or light with complex spectral components) is injected, for a certain definite acoustic frequency f_s, only the definite wavelength λ that satisfies Eq. (1.3-37) can be diffracted. If f_s is changed, so should the corresponding diffracted wavelength. By making use of this characteristic, tunable acousto-optic filters can be made.

3. The quantitative standard for distinguishing between Raman-Nath diffraction and Bragg diffraction

In theory, the Raman-Nath diffraction and Bragg diffraction are two extreme cases that appear when varying the acousto-optic diffractive parameters. The main parameters that influence the appearance of the two cases of diffraction are the wavelength of acoustic λ_s, the incident angle θ_i of the light beam, and the length of action L of acoustic and light. In order to give the quantitative standard for distinguishing between the two kinds of diffraction, the parameter G is introduced for characterization.

$$G = k_s^2 L/k_i \cos\theta_i = 2\pi\lambda L/\lambda_s^2 \cos\theta_i \tag{1.3-38}$$

When L is small and k_s is great ($G \ll 1$), the diffraction is Raman-Nath diffraction whereas if L is great and λ_s is small ($G \gg 1$), we have Bragg diffraction. In order to seek a practical standard, that is, when the parameter G is so great that, with the exception of the intensities of level 0 and level 1 diffraction light, all the remaining diffracted light intensities are very low and can be neglected. If such is the case, then it can be considered that the region of Bragg diffraction is entered. After years of practice, the following quantitative standard is now universally adopted:

$$\begin{cases} G \geqslant 4\pi & \text{(region of Bragg diffraction)} \\ G > \pi & \text{(region of Raman-Nath diffraction)} \end{cases} \tag{1.3-39}$$

To facilitate application, another quantity is introduced:

$$\begin{aligned} L_0 &= \lambda_s^2 \cos\theta_i/\lambda \cong \lambda_s^2/\lambda \\ G &= 2\pi L/L_0 \end{aligned} \tag{1.3-40}$$

Therefore, the above quantitative standard can be written as

$$\begin{cases} L \geqslant 2L_0 & \text{(region of Bragg diffraction)} \\ L \leqslant \dfrac{1}{2}L_0 & \text{(region of Raman-Nath diffraction)} \end{cases} \tag{1.3-41}$$

where L_0 is called the characteristic length of acousto-optic devices. The introduction of parameter L_0 will greatly simplify the design of devices. As $\lambda_s = v_s/f_s$, L_0 is related not only to the properties (v_s and n) of the medium, but also to the operating conditions (f_s and λ_0). In fact, L_0 is a representation of the main characteristics of the interaction between sound and light.

4. A theoretical analysis of the Bragg diffraction—the theory of the coupled wave

The interaction between sound and light can be regarded as a process of interaction between parameters, that is, first, because of the perturbation by the acoustic field, the electric susceptibility in the medium is made to vary periodically in time and space, making the incident light wave and the acoustic wave in the medium coupled to generate a series of polarized wave with composite frequency, whose angular frequency and wave vector component are, respectively,

$$\begin{aligned} \omega_d &= \omega_i \pm m\omega_s \quad (m = \pm 1, \pm 2, \cdots) \\ \boldsymbol{k}_d &= \boldsymbol{k}_i \pm m\boldsymbol{k}_s \quad (m = \pm 1, \pm 2, \cdots) \end{aligned} \tag{1.3-42}$$

The re-radiation by these polarized waves of the secondary waves with the above-mentioned composite frequency causes the strengthening of the radiation coherence of the corresponding secondary waves to form the diffracted light at different levels. The variation of the refractive index induced by the elastic wave in the medium is

$$\Delta n(x,t) = \Delta n \cos(\omega_s t - k_s x) \tag{1.3-43}$$

The interaction of such a variation with the light field of ω_i and ω_d will cause additional electric polarization in the medium:

$$\Delta P(x,t) = 2\sqrt{\varepsilon\varepsilon_0}\Delta n(x,t)E(x,t) \tag{1.3-44}$$

where $E(x,t)$ is the sum of the electric field with frequency $\omega_{\rm i}$ and $\omega_{\rm d}$; the medium polarization formed by ΔnE causes the energy exchange between the two light waves $\omega_{\rm i}$ and $\omega_{\rm d}$. In order to set up the coupling equation of the interaction between acousto and light, we proceed from the fluctuation equation

$$\nabla^2 E = \mu\sigma\frac{\partial E}{\partial t} + \mu\varepsilon\frac{\partial^2 E}{\partial t^2} + \mu\frac{\partial^2}{\partial t^2}P^{(NL)} \tag{1.3-45}$$

Suppose the medium is loss-free. Then $\sigma = 0$. Equation (1.3-45) is the basic equation of the parametric interaction, that is, the equation of fluctuation of the light wave propagating in a medium in which the refractive index varies. It satisfies both the light field with frequency $\omega_{\rm i}$ and that with $\omega_{\rm d}$. Suppose what is involved in the incident light field and diffracted light field is both linearly polarized light. For the incident light field with frequency $\omega_{\rm i}$, Eq. (1.3-45) can be written as

$$\nabla^2 E_{\rm i} = \mu\varepsilon\frac{\partial^2 E_{\rm i}}{\partial t^2} + \mu\frac{\partial^2}{\partial t^2}(\Delta P)_{\rm i} \tag{1.3-46}$$

where $(\Delta P)_{\rm i}$ is the component of the electric polarization in the direction parallel to $E_{\rm i}$, whose frequency is $\omega_{\rm i}$. The components of the other frequencies are not in step with $E_{\rm i}$, so the average value of their contributions to $E_{\rm i}$ is zero. Thus the general electric field of medium polarization is the sum of the following two traveling waves:

$$\begin{aligned} E_{\rm i}(x,t) &= \frac{1}{2}E_{\rm i}(x_{\rm i})\exp[{\rm j}(\omega_{\rm i}t - k_{\rm i}x)] + c.c \\ E_{\rm d}(x,t) &= \frac{1}{2}E_{\rm d}(x_{\rm d})\exp[{\rm j}(\omega_{\rm d}t - k_{\rm d}x)] + c.c \end{aligned} \tag{1.3-47}$$

where $k_{\rm i}$ and $k_{\rm d}$ are parallel to the propagation direction of the incident light and that of the diffracted light, respectively. Performing differentiation twice with respect to Eq. (1.3-47) gives

$$\nabla^2 E_{\rm i}(x,t) = -\frac{1}{2}\left[k_{\rm i}^2 E_{\rm i} + 2{\rm j}k_{\rm i}\frac{{\rm d}E_{\rm i}}{{\rm d}x_{\rm i}} + \nabla^2 E_{\rm i}\right]\exp[{\rm j}(\omega_{\rm i}t - k_{\rm i}x)]$$

Suppose $E_{\rm i}(x_{\rm i})$ varies so slowly that $\nabla^2 E_{\rm i} \ll k_{\rm i}{\rm d}E_{\rm i}/{\rm d}x_{\rm i}$. Combining the above equation with Eq. (1.3-46) and using the relation $k_{\rm i} = \omega_{\rm i}\sqrt{\mu\varepsilon_{\rm i}}$ gives

$$k_{\rm i}\frac{{\rm d}E_{\rm i}}{{\rm d}x_{\rm i}} = {\rm j}\mu\left[\frac{\partial^2}{\partial t^2}(\Delta P_{\rm i})\right]\exp[{\rm j}(\omega_{\rm i}t - k_{\rm i}x)] \tag{1.3-48}$$

Using the relation $\Delta P = 2\sqrt{\varepsilon\varepsilon_0}\Delta n(x,t)[E_{\rm i}(x,t) + E_{\rm d}(x,t)]$ gives

$$(\Delta P)_{\rm i} = \frac{1}{2}\sqrt{\varepsilon\varepsilon_0}\Delta n E_{\rm d}\exp\{{\rm j}[(\omega_{\rm s} + \omega_d)t - (k_{\rm s} + k_{\rm d})x]\} + c.c \tag{1.3-49}$$

Substitution of Eq. (1.3-49) into Eq. (1.3-48) yields

$$\frac{{\rm d}E_{\rm j}}{{\rm d}x_{\rm i}} = -{\rm j}\eta_{\rm i}E_{\rm d}\exp[-{\rm j}(k_{\rm i} - k_d - k_{\rm s})x]$$

Similarly, there is

$$\frac{{\rm d}E_{\rm d}}{{\rm d}x_{\rm d}} = -{\rm j}\eta_{\rm d}E_{\rm i}\exp[-{\rm j}(k_{\rm i} - k_{\rm d} - k_{\rm s})x] \tag{1.3-50}$$

where $\eta_{\rm i,d} = \frac{1}{2}\omega_{\rm i,d}\sqrt{\mu\varepsilon_0}\Delta n = \frac{\omega_{\rm i,d}\Delta n}{2c} = \frac{\pi n^3}{2\lambda}PS$, which is referred to as the coupling coefficient. It characterizes the intensity of the interaction between acousto and light.

Equation (1.3-50) shows that, under the action of a definite acoustic field, the condition for obtaining maximum energy coupling between the incident light field E_{i} and diffracted light field E_{d} is

$$k_{\mathrm{i}} = k_{\mathrm{s}} + k_{\mathrm{d}} \tag{1.3-51}$$

If this Bragg condition is satisfied and, since the acoustic frequency ω_{s} is much smaller than the light frequency $\omega_{\mathrm{i,d}}$, $\omega_{\mathrm{i}} \cong \omega_{\mathrm{d}}$, by taking $\eta_{\mathrm{i}} = \eta_{\mathrm{d}} = \eta$, then Eq. (1.3-50) becomes

$$\begin{aligned}\frac{\mathrm{d}E_{\mathrm{i}}}{\mathrm{d}x_{\mathrm{i}}} &= -\mathrm{j}\eta E_{\mathrm{d}}\\ \frac{\mathrm{d}E_{\mathrm{d}}}{\mathrm{d}x_{\mathrm{d}}} &= -\mathrm{j}\eta E_{\mathrm{i}}\end{aligned} \tag{1.3-52}$$

In order to solve the above equation, transform the two coordinates x_{i} and x_{d} onto a new coordinate ξ, as shown in Fig. 1.3-9. As it is the angle resolving line along directions k_{i} and k_d, we have

$$\begin{aligned}x_{\mathrm{i}} &= \xi\cos\theta\\ x_{\mathrm{d}} &= \xi\cos\theta\end{aligned}$$

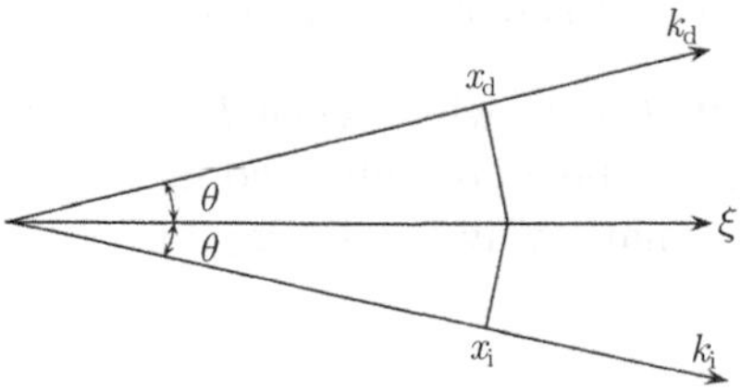

Fig. 1.3-9 The relationship between $x_{\mathrm{i,d}}$ and coordinate ξ

So Eq. (1.3-52) becomes

$$\begin{aligned}\frac{\mathrm{d}E_{\mathrm{i}}}{\mathrm{d}\xi} &= \frac{\mathrm{d}E_{\mathrm{i}}}{\mathrm{d}x_{\mathrm{i}}}\cos\theta = -\mathrm{j}\eta E_d\cos\theta\\ \frac{\mathrm{d}E_{\mathrm{d}}}{\mathrm{d}\xi} &= \frac{\mathrm{d}E_{\mathrm{d}}}{\mathrm{d}x_{\mathrm{d}}}\cos\theta = -\mathrm{j}\eta E_{\mathrm{i}}\cos\theta\end{aligned} \tag{1.3-53}$$

The solution for the above equation is

$$\begin{aligned}E_{\mathrm{i}}(\xi) &= E_{\mathrm{i}}(0)\cos(\eta\xi\cos\theta) - \mathrm{j}E_{\mathrm{d}}(0)\sin(\eta\xi\cos\theta)\\ E_{\mathrm{d}}(\xi) &= E_{\mathrm{d}}(0)\cos(\eta\xi\cos\theta) - \mathrm{j}E_{\mathrm{i}}(0)\sin(\eta\xi\cos\theta)\end{aligned}$$

By using the relationship between ξ and x_{i}, x_{d}, the above equation can be written as

$$\begin{aligned}E_{\mathrm{i}}(x_{\mathrm{i}}) &= E_{\mathrm{i}}(0)\cos(\eta x_{\mathrm{i}}) - \mathrm{j}E_d(0)\sin(\eta x_{\mathrm{i}})\\ E_{\mathrm{d}}(x_{\mathrm{d}}) &= E_{\mathrm{d}}(0)\cos(\eta x_{\mathrm{d}}) - \mathrm{j}E_{\mathrm{i}}(0)\sin(\eta x_{\mathrm{d}})\end{aligned} \tag{1.3-54}$$

Equation (1.3-54) is exactly the equation that describes the two beams of coupled wave in an isotropic medium that satisfy the Bragg condition. When the amplitude of the incident light is $E_{\mathrm{i}}(0)$, the frequency is ω_{i}, and $E_{\mathrm{d}}\ (0) = 0$, Eq. (1.3-54) becomes

$$\begin{aligned}E_{\mathrm{i}}(x_{\mathrm{i}}) &= E_{\mathrm{i}}(0)\cos(\eta x_{\mathrm{i}})\\ E_{\mathrm{d}}(x_{\mathrm{d}}) &= -\mathrm{j}E_{\mathrm{i}}(0)\sin(\eta x_{\mathrm{d}})\end{aligned} \tag{1.3-55}$$

and there is

$$|E_{\mathrm{i}}(x_{\mathrm{i}})|^2 + |E_{\mathrm{d}}(x_{\mathrm{d}})|^2 = |E_{\mathrm{i}}(0)|^2 \tag{1.3-56}$$

The above equation shows that, for the two kinds of light wave, the optical power is conservative, in the process of interaction between acoustic and light. From the afore-mentioned coupling coefficient $\eta_{\mathrm{i,d}} = \dfrac{\omega_{\mathrm{i,d}}\Delta n}{2c} = \dfrac{1}{2}\left(\dfrac{2\pi}{\lambda}\Delta n\right)$ and considering $x_{\mathrm{i}} = x_{\mathrm{d}} = L/\cos\theta_{\mathrm{B}}$, we define a new quantity $v = \frac{2\pi}{\lambda}\Delta n\frac{L}{\cos\theta_B}$. Therefore, represented by the light intensity, Eq. (1.3-55) can be written as

$$I_i = I_i(0)\cos^2\left(\frac{v}{2}\right)$$
$$I_d = I_i(0)\sin^2\left(\frac{v}{2}\right) \tag{1.3-57}$$

The variation of the optical power with the distance of action is as shown in Fig. 1.3-10. The acousto-optic diffraction efficiency is defined as the ratio of the output diffracted light intensity to the input light intensity, i.e.,

$$\eta_s = \frac{I_d(L)}{I_i(0)} = \sin^2\left(\frac{v}{2}\right) \tag{1.3-58}$$

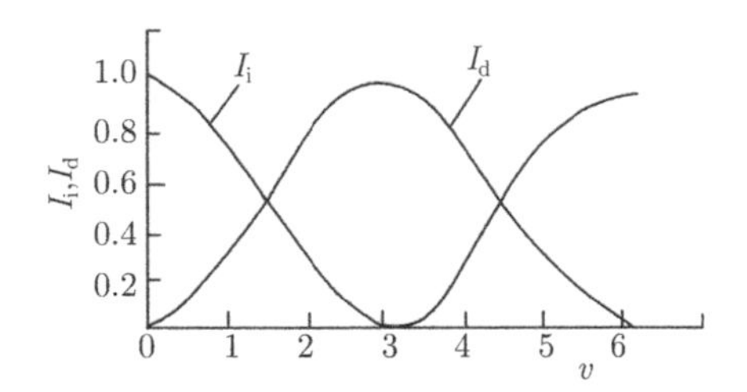

Fig. 1.3-10 Curves of variation of $I_{i,d}$ with v

It can be seen from Eq. (1.3-57) that, when $v/2 = \pi/2$, $I_i = 0$, while when $I_d = I_i\ (0)$, all the energy of the incident light will be transferred into the beam of diffracted light; that is, the ideal Bragg diffraction efficiency can reach 100%. That's why the Bragg diffraction effect is adopted so often in acousto-optic devices.

1.3.3 The acousto-optic volume modulator

1. The composition of the acousto-optic modulator

The acousto-optic modulator is composed of the acousto-optic medium, the electro-acoustic transducer, the acousto absorbing (or reflection) device, and the driving power source, as shown in Fig. 1.3-11.

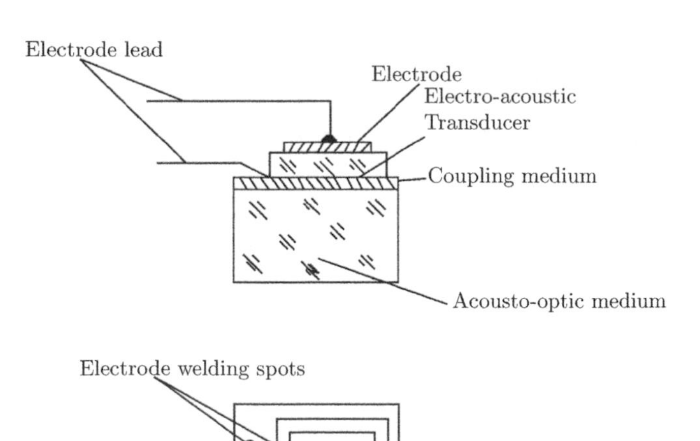

Fig. 1.3-11 The structure of the acousto-optic modulator

(1) The acousto-optic medium. The acousto-optic medium is the site of interaction between sound and light. When a beam of light passes through a variable ultrasonic field, because of the interaction between light and ultrasonic field, its outgoing light possesses all

levels of diffracted light that varies with time. Making use of the property of the variation of the intensity of diffracted light with the variation of the intensity of the ultrasonic wave, we shall be able to make the light intensity modulator.

(2) The electro-acoustic transducer (also referred to as the ultrasonic generator). By the use of the anti-piezoelectric effect of certain piezoelectric crystals (quartz, $LiNbO_3$, etc.) or piezoelectric semiconductors (CdS, ZnO, etc.) and under the action of the externally applied electric field, it generates mechanical vibration while forming the ultrasonic wave. So it plays the role of transforming the modulated electric power into sound power.

(3) The acoustic absorbing (or reflection) device. It is placed opposite the source of ultrasonic to absorb the acoustic wave (operating in the traveling wave state) that has already passed through the medium to prevent the generation of disturbance due to a return to the medium. But in order to make the ultrasonic field operate in the standing wave state, it is necessary to replace the acoustic absorbing device with a acoustic reflecting device.

(4) The driving power source. It is used to generate the modulating electric signal to be applied onto the electrodes at the two ends of the electro-acoustic transducer to drive the acousto-optic modulator (transducer) into operation.

2. The operation principle of acousto-optic modulation

Acousto-optic modulation is a physical process of loading information onto the optical frequency carrier wave by means of the acousto-optic effect. The modulating signal is transformed from acting on the electro-acoustic transducer in the form of an (amplitude modulating) electric signal into an ultrasonic field varying in the form of an electric signal. When the light wave passes through the acousto-optic medium, because of the acousto-optic action, the light carrier wave becomes an intensity-modulated wave upon getting modulated.

It is known from the afore-going analysis that, no matter whether it is Raman-Nath diffraction or Bragg diffraction, their diffraction efficiency is related to the additional phase retardation factor $v = \frac{2\pi}{\lambda}\Delta n L$, where the refractive index difference Δn of the sound wave is proportional to the elastic strain amplitude S, while $S \propto$ the sound power P_s. So, when the sound wave field following the modulation by the signal makes the sound wave amplitude vary accordingly, the diffracted light intensity, too, will experience corresponding variation with it. The Bragg acousto-optic modulation characteristic curve is similar to that of the electro-optic intensity modulation, as shown in Fig. 1.3-12. It can be seen from the figure that the diffraction efficiency η and the ultrasonic power P_s are in the form of nonlinear modulation curve. In order to prevent the modulation from getting distorted, it is necessary to employ an additional ultrasonic offset to enable it to operate in a region of fairly good linearity.

For the Raman-Nath-type diffraction, the operational acoustic frequency is lower than 10 MHz. The operating principle of this kind of modulator is shown in Fig. 1.3-13(a). All levels of its diffracted light intensity are proportional to $J_n^2(v)$. If a certain level of diffracted light is taken as the output, the other levels of diffracted light can be shielded with a diaphragm. Then the light beam going out of the aperture is a modulating light varying with v. As the efficiency of the Raman-Nath diffraction is low, the availability of light energy is also low. The length L of mutual action as specified in the criterion Eq. (1.3-38) is small. When the operating frequency is rather high, its maximum allowable length is too small and the sound power required is very high. Hence the Raman-Nath-type acousto-optic modulator is only limited to operation at low frequencies and has only limited bandwidths.

For the Bragg diffraction, its diffraction efficiency is given by Eq. (1.3-58) as

$$\eta_s = \frac{I_d}{I_i} = \sin^2\left(\frac{v}{2}\right)$$

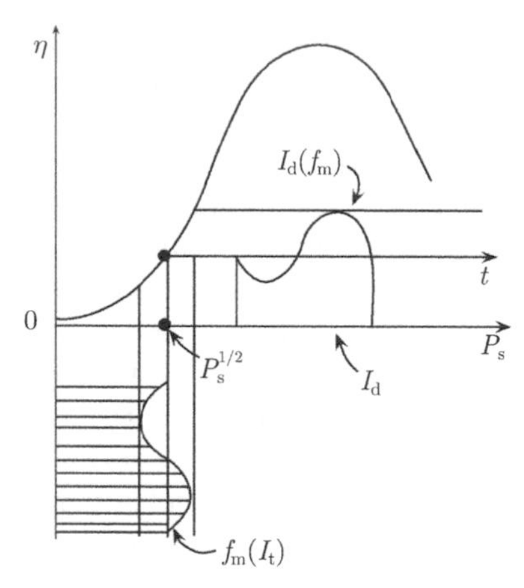

Fig. 1.3-12 The modulation characteristic curve

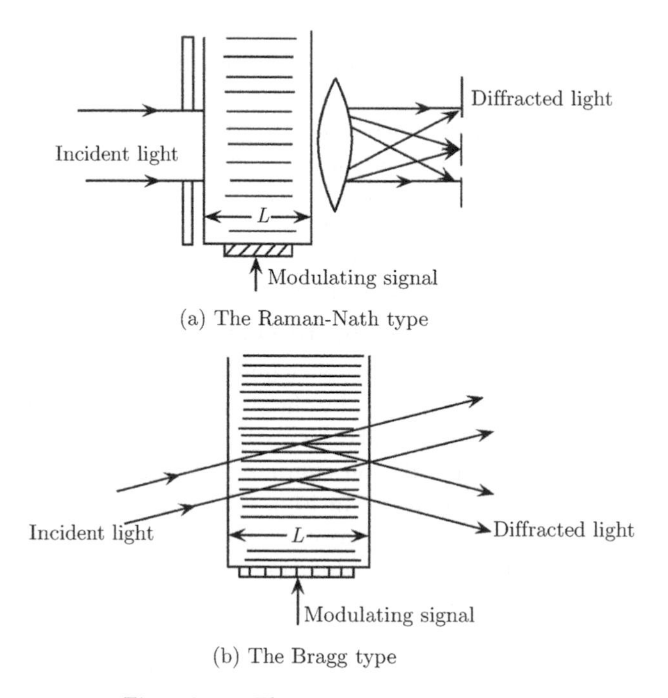

Fig. 1.3-13 The acousto-optic modulator

The principle of operation of the Bragg-type acousto-optic modulator is as shown in Fig. 1.3-13(b). Under the condition of rather low sound power P_s (or sound intensity I_s), the diffraction efficiency η_s monotonically increases with I_s (appearing to be a linear relation), i.e.,

$$\eta_s \approx \frac{\pi^2 L^2}{2\lambda^2 \cos^2 \theta_B} M_2 I_s \tag{1.3-59}$$

where the factor $\cos\theta_B$ has taken into account the influence of the Bragg angle on the acousto-optic action. It can be seen from this equation that, if the acousto intensity is modulated, the intensity of the diffracted light will also get modulated. The Bragg diffraction must make the diffracted light beam injected at the Bragg angle θ_B. Meanwhile, only when receiving the diffracted light beam along the symmetric direction relative to the sound wave surface, will a satisfactory result be obtained. For its high efficiency and fairly broad modulating bandwidth, the Bragg diffraction is frequently adopted.

3. The modulating bandwidth

As an important parameter of the acousto-optic modulator, the modulating bandwidth is a technical index for judging whether information can be transmitted distortion-free. It is subject to the restriction of the Bragg bandwidth. For the Bragg acousto-optic modulator, under the ideal plane light wave and acoustic wave condition, the wave vector is deterministic. Hence for light wave with a given incident angle and wavelength, only a sound wave with a definite frequency and wave vector can fulfill the Bragg condition. When the finite divergent light beam and acoustic wave field are adopted, the finite angle of the wave beam will expand. Therefore, only within a finite acoustic frequency range is the Bragg diffraction allowed to be generated. According to the Bragg diffraction eq. (1.3-21), the relation between the permissible acoustic frequency bandwidth Δf_{s} and the likely quantity of variation of the Bragg angle is obtained:

$$\Delta f_{\mathrm{s}} = \frac{2nv_{\mathrm{s}}\cos\theta_{\mathrm{B}}}{\lambda}\Delta\theta_{\mathrm{B}} \tag{1.3-60}$$

where $\Delta\theta_{\mathrm{B}}$ is the quantity of variation of the incident angle and diffraction angle induced by the divergence of the light beam and acoustic beam, or the quantity of variation allowed by the Bragg angle. Suppose the angle of divergence of the incident light beam is $\delta\theta_{\mathrm{i}}$, and that of acoustic wave beam is $\delta\phi$, For a diffraction-restrained wave beam, the relationship between these beam divergence angles with the wavelength and that with the beam width are approximated respectively as

$$\delta\theta_{\mathrm{i}} \approx \frac{2\lambda}{\pi n w_0}, \quad \delta\phi \approx \frac{\lambda_{\mathrm{s}}}{L} \tag{1.3-61}$$

where w_0 is the beam waist radius of the incident light beam, n the refractive index of the medium, and L the width of the acoustic beam. Obviously, the range covered by the incidence angle (the included angle between the light wave vector $\boldsymbol{k}_{\mathrm{i}}$ and the acoustic wave vector $\boldsymbol{k}_{\mathrm{s}}$) is

$$\Delta\theta = \delta\theta_{\mathrm{i}} + \delta\phi \tag{1.3-62}$$

If the incident (divergent) light beam propagating inside angle $\delta\theta_{\mathrm{i}}$ is resolved into a number of plane waves along different directions (i.e., different wave vectors $\boldsymbol{k}_{\mathrm{i}}$), for the component of the light beam in a particular direction, there is an acoustic wave with a suitable frequency and wave vector that satisfies the Bragg condition, while the acoustic wave beam contains many Fourier frequency spectral components of an acoustic carrier wave of the central frequency because of the modulation by the signal. Therefore, for every acoustic frequency, the acoustic wave component with many wave vectors in different directions can induce the diffraction of the light wave. Thus, the incident light corresponding to every definite angle has a beam of diffracted light with a divergence angle $2\delta\phi$, as shown in Fig. 1.3-14, while each direction of diffraction corresponds to a different frequency shift. So, in order to resume the modulation of the intensity of the diffracted light beam, it is necessary to make the component of the diffracted light with different frequency shifts mixing in the square-law detector. So it is required that two beams of diffracted light on the farthest boundary (e.g., OA′ and OB′ in the figure) overlap to a certain extent. This requires that $\delta\phi \approx \delta\theta_{\mathrm{i}}$. If we take $\delta\phi \approx \delta\theta_{\mathrm{i}} = \dfrac{\lambda}{\pi n w_0}$, then the modulated bandwidth is obtained from Eq. (1.3-60) and Eq. (1.3-62):

$$(\Delta f)_{\mathrm{m}} = \frac{1}{2}\Delta f_{\mathrm{s}} = \frac{2nv_{\mathrm{s}}}{\pi w_0}\cos\theta_{\mathrm{B}} \tag{1.3-63}$$

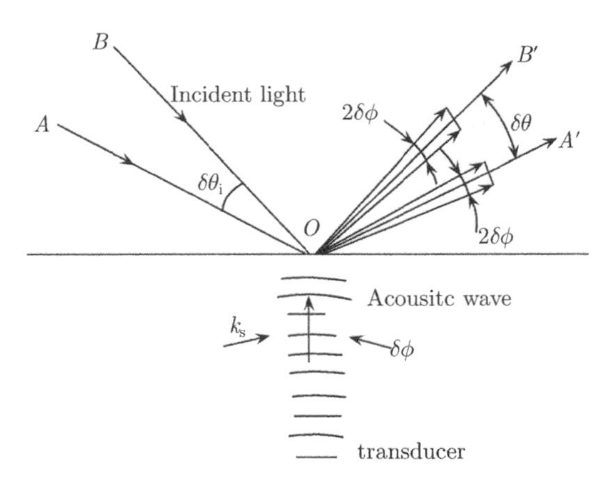

Fig. 1.3-14 The Bragg diffraction with wave beam divergence

The above equation shows that the bandwidth of the acousto-optic modulator is inversely proportional to the transition time $\left(\frac{w_0}{v_{\mathrm{s}}}\right)$ at which the sound wave passes through the light beam and that a broad modulated bandwidth can be obtained with a light beam of small width. But the divergence angle of the light beam should not be too big, or the level-0 and level-1 diffracted light beams will partially overlap and the modulating effect will be reduced. So it is required that $\delta\theta_{\mathrm{i}} < \theta_{\mathrm{B}}$. Thus, we can find from Eq. (1.3-21), Eq. (1.3-60), and $\Delta\theta_{\mathrm{i}} = \theta_{\mathrm{B}}$ that

$$\frac{\Delta f_{\mathrm{m}}}{f_{\mathrm{s}}} \approx \frac{\Delta f}{2f} < \frac{1}{2} \tag{1.3-64}$$

That is, the maximum modulated bandwidth Δf_{m} is approximately equal to half of the sound frequency f_{s}. Therefore, a large modulated bandwidth can be obtained only by adopting high frequency Bragg diffraction.

4. The diffraction efficiency of the acousto-optic modulator

Another important parameter of the acousto-optic modulator is the diffraction efficiency. According to Eq. (1.3-26), the acoustic intensity needed for obtaining 100% modulation is

$$I_{\mathrm{s}} = \frac{\lambda^2\cos^2\theta_{\mathrm{B}}}{2M_2L^2} \tag{1.3-65}$$

If represented as the sound power needed, it is

$$P_{\mathrm{s}} = HLI_{\mathrm{s}} = \frac{\lambda^2\cos^2\theta_{\mathrm{B}}}{2M_2}\left(\frac{H}{L}\right) \tag{1.3-66}$$

It can be seen that, the greater the quality factor M_2 of the acousto-optic material, the lower the sound power needed for obtaining a 100% diffraction efficiency. Moreover, the cross section of the electro-acoustic transducer should be made long (L long) and narrow (H small), but although an increase in the action length L is helpful to raising the efficiency of diffraction, it will lead to a decrease in the modulated bandwidth (as the acoustic beam divergence angle $\delta\phi$ is inversely proportional to L. A small $\delta\phi$ implies a small modulated bandwidth).

Letting $\delta\phi = \dfrac{\lambda_s}{2L}$ and using Eqs. (1.3-60)~(1.3-62), we can write the bandwidth as

$$\Delta f = \frac{2nv_s\lambda_s}{\lambda L}\cos\theta_B \qquad (1.3\text{-}67)$$

Solving L using the above formula and substituting it into the previously given Eq. (1.3-26), we have

$$2\eta_s \Delta f f_0 = \left(\frac{n^7 P^2}{\rho v_s}\right)\frac{2\pi^2}{\lambda^3 \cos\theta_B}\left(\frac{P_s}{H}\right) \qquad (1.3\text{-}68)$$

where f_0 is the acoustic center frequency ($f_o = v_s/\lambda_s$). We introduce the factor

$$M_1 = \frac{n^7 P^2}{\rho v_s} = (nv_s^2)M_2 \qquad (1.3\text{-}69)$$

M_1 is the quality factor that characterizes the modulated bandwidth characteristics of the acousto-optic material. The greater the M_1 value of an acousto-optic material of which a modulator is made, the greater the modulated bandwidth allowed will be.

In addition, for certain acousto-optic devices (such as the acousto-optic deflector to be discussed below), another factor should be considered; that is, during acousto-optic action, as the incident light beam has a definite width, and the acoustic wave is propagating in the medium with a limited velocity, the acoustic wave needs definite transition time to pass through the light beam, that is, $\tau = \dfrac{w_0}{v_s}$. For the acousto-optic deflector, its number of resolvable spots is proportional to the transition time ($N = \Delta f \cdot \tau$). When choosing the material for the deflector, the sound velocity v_s is required to be low. So another quality factor M_3 is introduced:

$$M_3 = \left(\frac{1}{v_s}\right)M_1 = \frac{1}{v_s}(nv_s^2)M_2 = \frac{n^7 P^2}{\rho v_s^2} \qquad (1.3\text{-}70)$$

Table 1.3-1 lists a number of acousto-optic materials and their physical properties.

1.3.4 The acousto-optic waveguide modulator

The structure schematic diagram of the acousto-optic Bragg diffraction-type waveguide modulator is as shown in Fig. 1.3-15. It is composed of a plane waveguide and a transducer with crossed electrodes. To effectively stimulate the surface elastic wave in the waveguide, piezoelectric materials (such as ZnO, etc.) are adopted for the waveguide and the underlay can be either piezoelectric material or non-piezoelectric material. For example, the underlay shown in Fig. 1.3-15 is y cut up $LiNbO_3$ piezoelectric crystal material, the diffusive Ti waveguide, and the electro-acoustic transducer with crossed electrodes on the surface made by the photoetching method. The whole device can rotate round the y-axis to enable the included angle between the guided-wave light and the electrode strip to be regulated to a Bragg angle. The incident light now passes the waveguide via the input prism, while the

Tab. 1.3-1 The properties of acousto-optic materials

Material	λ/μm	n	$\rho/(\mathrm{g/cm^3})$	Acoustic wave polarized direction	$v_s/(10^3\mathrm{m/s})$	Light wave polarized direction	Quality factor		
							$M_1 = n^7\rho^2/\rho v_s$	$^*M_2$	$^{**}M_3$
Vitreosil	0.63	1.46	2.2	Longitudinal	5.95	$\perp$	7.89×10^{-8}	1.51×10^{-15}	1.29×10^{-11}
				Transverse	3.76	$\perp$ or $\parallel$	0.963	0.467	0.0256
GaP	0.63	3.31	4.13	Longitudinal, [110]	6.32	$\parallel$	590.00	44.60	93.50
GaP				Longitudinal, [100]	4.13	$\perp$ or $\parallel$ to [010]	137.00	24.10	33.10
GaAs	1.15	3.37	5.43	Longitudinal, [110]	5.15	$\parallel$	925.00	104.00	179.00
GaAs				Longitudinal, [100]	3.32	$\parallel$ or $\perp$ to [010]	155.00	46.30	49.20
TiO_2	0.63	2.58	4.60	Longitudinal, [11 20]	7.86	[010]	62.50	3.93	7.97
$LiNbO_3$	0.63	2.20	4.70	Longitudinal, [11 20]	6.57	$\perp$ to [001]	66.50	6.99	10.10
YAG	0.63	1.83	4.20	Longitudinal, [100]	8.53		0.16	0.012	0.019
				Longitudinal, [110]	8.60	$\parallel$	0.98	0.073	0.114
YIG	1.15	2.22	5.17	Longitudinal, [100]	7.21	$\perp$+	3.94	0.33	0.53
$LiTaO_3$	0.63	2.18	7.45	Longitudinal, [001]	6.91	$\perp$	11.40	1.37	1.84
As_2S_3	0.63	2.61	3.20	Longitudinal,	2.60	$\parallel$	762.00	433.00	293.00
	1.15	2.46		Longitudinal,		$\perp$	619.00	347.00	236.00
SF-4	0.63	1.61	3.59	Longitudinal,	3.63	$\parallel$	1.83	1.51	3.97
β-ZnS	0.63	2.35	4.10	Longitudinal, [110]	5.51	$\perp$	24.30	3.41	4.41
				Transverse, [110]	2.165	$\parallel$ to [001]	10.60	0.57	4.90
α-Al_2O_3	0.63	1.76	4.00	Longitudinal, [001]	11.15	$\parallel$ or $\perp$[001]	7.32	0.34	0.66
CdS	0.63	2.44	4.82	Longitudinal, [11 20]	4.17	$\parallel$ to [11 20]	51.80	12.10	12.40
ADP	0.63	1.58	1.803	Longitudinal, [100]	6.15	$\parallel$	16.00	2.78	2.62
				Transverse, [100]	1.83	$\parallel$ to [010]	3.34	6.43	1.83
KDP	0.63	1.51	2.34	Longitudinal, [100]	5.50	$\parallel$ or $\perp$ to [001]	8.72	1.91	1.45
				Transverse, [100]		$\parallel$ to [010]	1.57	3.83	0.95
						$\parallel$ or $\perp$ to [001]			
H_2O	0.63	1.33	1.00	Longitudinal	1.50		4.36	160	29.10
Te	10.60	4.80	6.24	Longitudinal, [11 20]	2.20	$\parallel$ to [0001]	10.2000	4400.00	4640.00
TeO_2	0.63		5.99	Longitudinal, [100]	2.98	[100] or [010]	0.097	0.048	
				Longitudinal, [100]		[001] or [010]	22.90	10.60	
				Longitudinal, [100]	4.26	[001] or [010]	142.00	34.50	
				Longitudinal, [001]		[001] or [010]	113.00	25.60	
				Longtdn., [010] or [100]	3.04	arbitrary. [001]	3.70	1.76	
				Longitudinal, [110]	4.12	[110] or [110]	323.00	0.802	
				Longitudinal, [110]		[001] or [110]	16.20	3.77	
				Longitudinal, [101]	3.64	[010] or [101]	101.00	33.4	
				Longitudinal, [010]	2.98	[101] or [101]	42.6	20.40	
				[110] or [110]	0.617	arbitrary. [001]	68.60	793.60	

Continued

Material	$\lambda/\mu m$	n	$\rho/(g/cm^3)$	Acoustic wave polarized direction	$v_s/(10^3 m/s)$	Light wave polarized direction	Quality factor $M_1 = n^7\rho^2/\rho v_s$	$^*M_2$	$^{**}M_3$
				[101] or [101]	2.08	[100] or [010]	76.40	77.00	
$PbMoO_4$	0.63		6.95	Longitudinal, [100]	3.98	[001] or [010]		7.50	
				Longitudinal, [100]	3.98	[100] or [010]		24.00	
				Longitudinal, [100]	3.98	[010] or [001]		24.00	
				[100] or [010]	2.20	[100] or [001]		12.80	
				[100] or [001]	1.99				35.60
				Longitudinal, [001]	3.75	[100] along [010]			35.60
				Longitudinal, [001]	3.75	[001] along [010]		43.40	
	0.52			Longitudinal, [001]	3.75	[001] along [010]		56.10	
	0.49			Longitudinal, [001]	3.75	[100] along [010]			
$Bi_{12}GeO_{20}$ (BGO)	0.63	2.55	9.29	Longitudinal, [110]	3.42	arbitrary. polarized	29.50	9.91	8.64
$Bi_{12}SiO_{20}$ (BSO)	0.63			Longitudinal, [100]	3.83	arbitrary.polarized	33.80	9.02	8.83
$Sr_{0.75}Ba_{0.25}Nb_2O_3$	0.63	2.30		Longitudinal, [001]			26.80	38.60	48.80
						[001] polarized			
(SBN)		2.31		Longitudinal, [100]		[100] polarized	26.90	2.66	4.08
		2.27		Longitudinal, [001]		[001] polarized	59.30	8.62	10.80
		2.31		Longitudinal, [001]		[100] polarized	36.40	5.19	6.62
$Ge_{33}As_{12}Se_{55}$ glass	1.06	2.55	4.00	Longitudinal	2.50			246.00	
Transparent lead glass	0.63	1.72	4.80	Longitudinal	3.80	⊥		10.00	
						‖		8.00	
Yellow lead glass	0.63	1.96	6.30	longitudinal	3.10	⊥		16.00	
				Longitudinal		‖		20.00	
2-methacrylate (lucite)	0.63	1.55	1.18	Longitudinal	2.68	⊥		50.00	
				Longitudinal		‖		56.00	
polystyre	0.63	1.59	1.06	Longitudinal	2.35	⊥		127.00	
				Longitudinal		‖		113.00	
				Longitudinal	2.11	⊥	29.50	180.80	
neKRS-5	0.63	2.60	7.37	Longitudinal		‖		254.00	
				Longitudinal				26.00	
$PbNO_3$	0.63	1.78	4.70	Longitudinal	2.62	⊥		16.00	
				Longitudinal		‖			
dimethylb enzene	0.63	1.50	0.86	Longitudinal	1.30	⊥ or ‖		40.00	

$* M_2 = n^6p^2/\rho v_s^3; ** M_3 = n^7p^2/\rho v_s^3.$

transducer-generated ultrasonic wave will cause periodic variation of the refractive indices in the waveguide and underlay. Therefore, relative to the acoustic wave front, after the light wave injected into the waveguide at an angle θ_B penetrates the output prism, level-1 diffracted light along the direction at an angle $2\theta_B$with the primary light beam is obtained, whose intensity is

$$I_1 = I_i \sin^2\left(\frac{\Delta\varphi}{2}\right) = I_i \sin(BV)$$

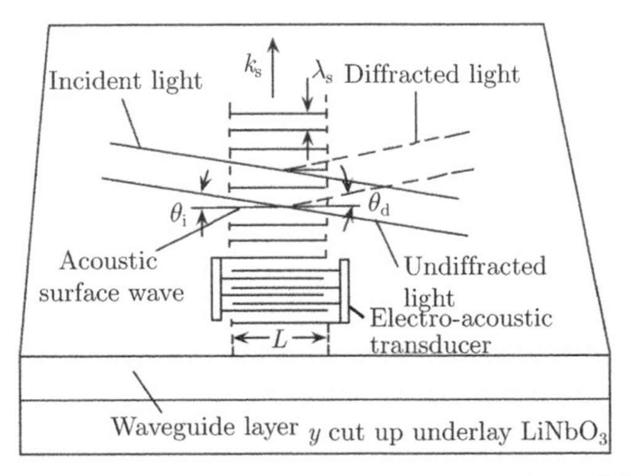

Fig. 1.3-15 The acousto-optic Bragg waveguide modulator

where $\Delta\varphi$ is the phase retardation experienced by the guided-wave light passing through the length L under the action of the electric field, and B is a proportionality constant dependent on such factors as the waveguide effective refractive index n_{eff}, etc. The above formula shows that the refracted light intensity I_d varies with the variation of voltage V, realizing the modulation of the intensity of the guided-wave light. For instance, when $\lambda = 0.6328$ μm, $V = 9$ V, 100% intensity modulation is obtained. The capacitance $C = 20$ pf, from which the driving power needed by unit bandwidth is 27 mW/MHz. When the frequency band is sufficiently broad, as the spatial locations of the output diffracted light are differently distributed, it can be used for optical deflection or for optical switching.

The Bragg modulated bandwidth is approximately inversely proportional to the acoustic wave aperture L. As L cannot be taken too small, which will reduce the diffraction efficiency, the bandwidth of an actual device is restrained by this factor. So there should be a compromise when considering this factor. In addition, to obtain definite diffraction efficiency while increasing the bandwidth of the modulator, a change in the structure of the transducer is often made in actual application. That is, the equal period interdigital transducer can be changed to variable period interdigital in form (i.e., the spacing gradually changing along the direction of acoustic wave propagation). Owing to the fact that the efficiency of the electro-acoustic transducer is highest when the interdigital spacing is equal to the half-wave length of the acoustic wave, the variable period transducer generates acoustic waves of different wavelengths at different positions of digital strips, thus broadening the bandwidth of the transducer.

1.3.5 Acousto-optic deflection

Another important use of the acousto-optic effect is using it to make a light beam deflect. The structure of the acousto-optic deflector is basically the same as that of the Bragg light modulator, the difference being in that the modulator is intended to change the intensity of the diffracted light while the deflector is intended to change the direction of the diffracted light by changing the acoustic wave frequency to make it deflect so that not only is the light beam made to continuously deflect, but also the separate light spots can be made to deflect by scanning.

1. The principle of acousto-optic deflection

It is known from an analysis of the previously mentioned theory of acousto-optic diffraction that the generation of the maximum of diffraction by a light beam injected into a medium at angle θ_i should satisfy the Bragg condition

$$\sin\theta_B = \frac{\lambda}{2n\lambda_s}, \quad \theta_i = \theta_d = \theta_B$$

The Bragg angle is in general very small and can be written as

$$\theta_B \cong \frac{\lambda}{2n\lambda_s} = \frac{\lambda}{2nv_s} f_s \tag{1.3-71}$$

Hence the included angle (deflection angle) between the diffracted light and incident light is equal to twice the Bragg angle θ_B, i.e.,

$$\theta = \theta_i + \theta_d = 2\theta_B = \frac{\lambda}{nv_s} f_s \tag{1.3-72}$$

It can be seen from the above equation that, by changing the frequency f_s of the ultrasonic wave, we can change its deflection angle θ, thereby attaining the goal of controlling the direction of light beam propagation. That is, the variation of the deflection angle of the light beam due to the change in the ultrasonic frequency Δf_s is

$$\Delta\theta = \frac{\lambda}{nv_s}\Delta f_s \tag{1.3-73}$$

This can be explained by Fig. 1.3-16 and the acousto-optic wave vector relationship. Suppose the acousto-optic diffraction satisfies the Bragg condition when the acoustic wave frequency is f_s. Then the acousto-optic wave vector plot is a closed isosceles triangle and the diffraction maximum is along the direction at angle θ_d with the ultrasonic wave face. If the acoustic wave frequency changes to $f_s + \Delta f_s$, according to the relation $k_s = \frac{2\pi}{v_s} f_s$, the acoustic wave vector quantity will have a change of $\Delta k_s = \frac{2\pi}{v_s}\Delta f_s$. Since the incident angle θ_i remains unchanged, as does the magnitude of the wave vector of the diffracted light, the acousto-optic plot will no longer be closed. The light beam will be diffracted along the OB direction, the corresponding light beam deflection is $\Delta\theta_d$, and both the angles θ and $\Delta\theta$ are very small. Hence it can be approximately considered that

$$\Delta\theta = \frac{\Delta k_s}{k_s} = \frac{\lambda}{nv_s}\Delta f_s$$

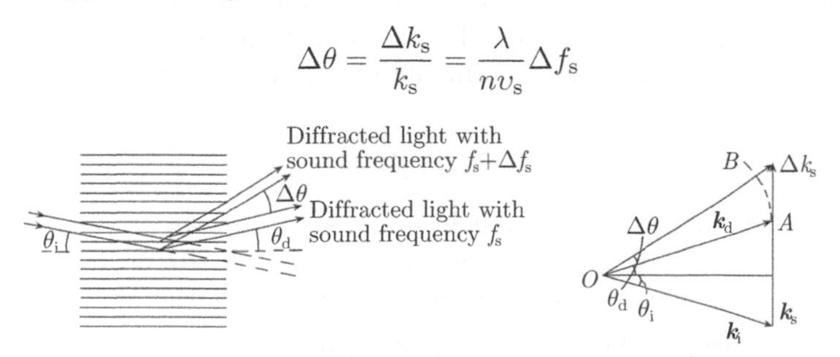

Fig. 1.3-16 The plot of acousto-optic deflection principle

So the deflection angle is proportional to the change of the sound frequency.

2. The main performance parameters of the acousto-optic deflector

There are three main performance parameters of the deflector, namely, the number of resolvable spots (which determines the capacity of the deflector), the deflection time τ (whose reciprocal determines the velocity of the deflector), and the diffraction efficiency η_S (which determines the efficiency of the deflector). The diffraction efficiency has already been discussed earlier, so we shall mainly examine the number of resolvable spots, the deflection velocity, and the operating bandwidth.

For an optical deflector, not only should we study the magnitude of its deflection angle $\Delta\theta$, but, we should mainly consider the number of its resolvable spots N, which is defined as the ratio of the deflection angle $\Delta\theta$ to the diffusion angle $\Delta\phi$ of the incident light beam itself; that is,

$$N = \frac{\Delta\theta}{\Delta\phi} \quad (\Delta\phi = R\lambda/w) \tag{1.3-74}$$

where w is the width of the incident light beam and R is a constant, whose value depends on the property of the incident light beam (homogeneous light beam or Gaussian light beam) and the resolvability criterion (the Rayleigh criterion or resolvability criterion). For instance, for deflectors used in displaying or recording, the Rayleigh criterion is adopted, whose $R =$ 1.0~1.3, while deflectors for optical memories adopt the resolvability criterion, whose $R =$ 1.8~2.5. The number of resolvable spots by scanning is

$$N = \frac{\Delta\theta}{\Delta\phi} = \frac{w}{v_s}\frac{\Delta f_s}{R} \tag{1.3-75}$$

where $\dfrac{w}{v_s}$ is the transition time of the ultrasonic wave, denoted by τ, which is exactly the deflection time of the deflector. So Eq. (1.3-75) can be written as

$$N\frac{1}{\tau} = \frac{1}{R}\Delta f_s \tag{1.3-76}$$

$N\dfrac{1}{\tau}$ is called the capacity-velocity product of the acousto-optic deflector, which characterizes the number of resolvable positions the light beam can point to within unit time. The above equation shows that it only depends on the operation bandwidth Δf_s while unrelated to the properties of the medium. Therefore, once the light beam width and sound velocity are determined, parameter τ is determined. Only when the bandwidth is increased, will it be possible to enhance the resolving power of the deflector. For instance, if the diameter of the incident beam $w = 1$ cm and the sound velocity $v_s = 4 \times 10^5$ cm/s, then $\tau = 2.5$ μs. If it is required that $N = 200$, then Δf_s is 100~200 MHz.

The bandwidth of the acousto-optic deflector is restrained by two factors, namely, the transducer bandwidth and the Bragg bandwidth because, when the acoustic frequency varies, so will the corresponding Bragg angle, the amount of variation being

$$\Delta\theta_B = \frac{\lambda}{2nv_s}\Delta f_s \tag{1.3-77}$$

So it is required that the acoustic beam and light beam possess a matched angle of divergence. Generally, the collimated parallel light beams are adopted for the acousto-optic deflector and, as its angle of divergence is very small, it is required that the angle of divergence of the acoustic wave $\delta\phi \geqslant \Delta\theta_B$. Taking $\delta\phi = \dfrac{\lambda_s}{L}$ and considering Eq. (1.3-77), we have

$$\frac{\Delta f_s}{f_s} \leqslant \frac{2n\lambda_s^2}{\lambda L} \tag{1.3-78}$$

In reality, the selection of the operation bandwidth is determined by the given indices N and τ, as is the central frequency of the operation frequency band. As it is in general not easy to make the Q value of normal Bragg devices very big, there always exist some influences of surplus high-level diffraction and various nonlinear factors as well as the components of driving source harmonics. In order to prevent the appearance of false spots in the operating frequency band, according to inference and calculation, it is required that the central frequency of the operating bandwidth

$$f_{s0} \geqslant \frac{3}{2}\Delta f_s = 1.5\Delta f_s$$

or

$$\frac{\Delta f_s}{f_{s0}} \leqslant \frac{2}{3} = 0.667 \tag{1.3-79}$$

This equation is the basic relational expression for designing the bandwidth of the Bragg acousto-optic deflector.

In order to enable the Bragg acousto-optic diffraction deflector to have good bandwidth characteristics, that is, to generate Bragg diffraction within a rather wide range of frequency and reduce as much as possible deviation from the Bragg condition, it is necessary to provide ultrasonic wave along an appropriate direction in a rather wide range of angle while doing everything possible to make the ultrasonic wave surface do the corresponding inclined revolution with the variation of the frequency so as to make the main direction of propagation of the ultrasonic wave equally share the direction of the incident light and that of the diffracted light from beginning to end so that the ultrasonic direction will automatically track the Bragg angle (referred to as ultrasonic tracking). The method of realizing ultrasonic tracking is in general to adopt a so-called "array transducer"; that is, by dividing the transducer into several pieces, the ultrasonic wave that makes them enter the acousto-optic medium is the ultrasonic wave emitted by all the transducers synthesized by superposing one over another to form an inclined wave surface. The primary direction of the synthesized ultrasonic wave varies with the variation of the sound wave frequency. Such a structure can ensure the fulfillment of the Bragg condition in a rather wide range of frequency. The array transducer is divided into the step type and plane type in terms of form. The structure of the former is shown in Fig. 1.3-17(a), which consists in grinding the acousto-optic medium into a series of steps; the height difference between steps is $\lambda_S/2$, the width of a step is S. The pieces of transducer are each pasted onto a step. The phase difference between two adjacent pieces is π, so there is also a phase difference of π rad between the surfaces of ultrasonic wave generated by each transducer. This makes the sound wave equiphase surface propagating

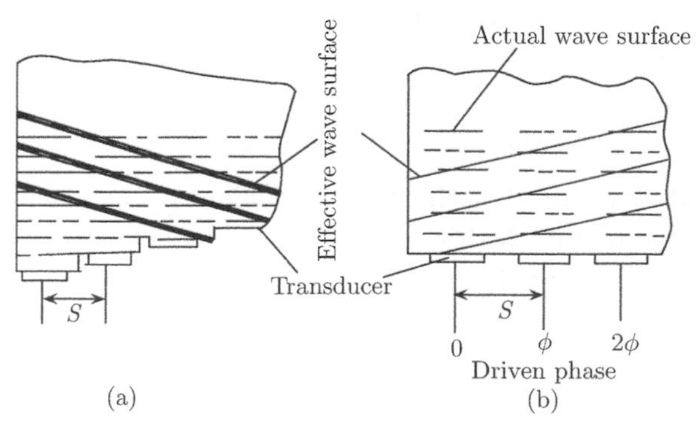

Fig. 1.3-17 The array transducer

in the medium do inclined revolution with it, the angle of revolution varying with frequency. Such a situation is equivalent to a change in the angle of the incident light beam to make it satisfy the Bragg condition. The latter is a plane-type structure as shown in Fig. 1.3-17(b). Its operation is basically the same as the former and is omitted.

1.3.6 Matters for consideration in designing the acousto-optic modulator

According to the process of operation of the acousto-optic modulator, first the electric oscillation is converted to ultrasonic vibration by the electro-acoustic transducer. Then, by means of the adhesive layer between the transducer and the acousto-optic medium, the vibration is transferred into the medium to form the ultrasonic wave. Therefore, it is necessary to consider how to effectively convert the electric power provided by the driving source into ultrasonic wave power in the acousto-optic medium. Through the interaction between sound and light, the ultrasonic wave will obtain diffracted light by inducing Bragg diffraction of the incident light beam. For this reason, we just have to consider how to enhance its diffracting efficiency and in what frequency range modulation can be performed distortion-free, that is, how to make the design so that the ultrasonic wave along an appropriate direction can be provided in a rather wide frequency range, enabling both the incident light direction and the included angle θ_i between the ultrasonic wave surfaces to meet the Bragg condition in this frequency range. That is, the method of design determines the possibility of enhancing the Bragg bandwidth. It is based on this requirement of the acousto-optic modulator that a definite analysis is made in this section on such issues as the selection of material for the acousto-optic medium, the design of the electro-acoustic transducer, etc.

1. The selection of material for the acousto-optic medium

The performance of the medium material exerts a direct influence on the quality of the modulator. Hence the great importance of rational selection of the medium material. During design, mainly the following factors should be considered:

(1) The modulating efficiency of the modulator should be high while the acoustic power needed should be as low as possible. The modulating efficiency of the modulator is characterized by the ratio of the post-modulation light intensity (i.e., diffracted light intensity) to the incident light intensity, or $\eta_s = \dfrac{I_1}{I_i} = \sin^2\left(\dfrac{v}{2}\right)$, where $v \propto M_2$; that is, the greater M_2 is, the greater v will be, hence the high modulating efficiency. When choosing the acousto-optic material, material with great M_2 value should be chosen while making an overall consideration of the physical and chemical performance of the material.

For instance, if water is used as the acousto-optic medium, its characteristic parameters will be $n = 1.33$, $P = 0.31$, $v_s = 1.5 \times 10^3$ m/s, $\rho = 1000$ kg/m^3, and the wavelength λ of the incident light is taken as $\lambda = 0.6328$ μm. Substituting these values into Eq. (1.3-26), we have

$$(\eta_s)_{\text{water}} = \sin^2(1.4L\sqrt{I_s}) \tag{1.3-80}$$

For other materials and other wavelengths, for convenience in calculation, the medium's M_2 can be represented as the quality factor relative to water, denoted by $M_\omega = M_{2(\text{material})}/M_{2(\text{water})}$. Then Eq. (1.3-26) is expressed in a form suitable for an arbitrary wavelength, i.e.,

$$\eta_s = \sin^2\left(1.4\frac{0.6328}{\lambda}L\sqrt{M_\omega I_s}\right) \tag{1.3-81}$$

Listed in Tab. 1.3-2 are the M_ω values of some frequently seen acousto-optic materials.

Tab. 1.3-2 The M_ω values of acousto-optic materials

Material	$\rho/(\mathrm{kg/m^3})$	$v_s/(\mathrm{km/s})$	n	P	M_ω
Water	1.00	1.50	1.33	0.31	1.00
Dense flint glass	6.30	3.10	1.92	0.25	0.12
Vitreosil (SiO_2)	2.20	5.97	1.46	0.20	0.006
Polystyrene	1.06	2.35	1.59	0.31	0.80
KRS-5	7.40	2.11	2.60	0.21	1.60
Lithium niobate ($LiNbO_3$)	4.70	7.40	2.25	0.15	0.012
Lithium fluoride (LiF)	2.60	6.00	1.39	0.13	0.001
Titanium dioxide (TiO_2)	4.26	10.30	2.60	0.05	0.001
Sapphire (Al_2O_3)	4.00	11.00	1.76	0.17	0.001
Lead molybdate ($PbMoO_4$)	6.95	3.75	2.30	0.28	0.22
α iodic acid (HIO_3)	4.63	2.44	1.90	0.41	0.50
Tellurium dioxide (TeO_2) (Slow shear wave)	5.99	0.617	2.35	0.09	5.00

If lead molybdate ($PbMoO_4$) is taken as an example, we find from the table that $M_\omega = 0.22$. Suppose the acoustic power $P_S = 1$ W ($I_S = 1\ \mathrm{W/mm^2}$), the cross-sectional area of the acoustic beam is 1 mm × 1 mm, and the acousto-optic action length $L = 1$ mm. Substituting the above-mentioned data into Eq. (1.3-81), we obtain

$$\eta_s = 40\%$$

(2) It is advisable to make the modulator possess a rather broad bandwidth. We already know that the Bragg condition is $\sin\theta_B = \dfrac{\lambda}{2\lambda_s}$. Obviously, when the optical and acoustic wavelength vary, variation of the Bragg angle will be induced. In fact, the optical wave possesses definite frequency spectral width. When the modulator operates within a wide range of frequency band, the deviation of the acoustic frequency relative to the central frequency will cause the diffraction angle to deviate from the Bragg angle. When a definite value is exceeded, the operation state of the modulator will be made to not satisfy the Bragg condition, so that the level 1 diffracted light intensity becomes lower. The frequency variation Δf_s to which half of the diffracted light intensity that the level-1 diffracted light intensity is lowered to relative to when it is at the central frequency is defined as the Bragg bandwidth. According to inference and proof, there is approximately $\Delta f_s = 1.8\dfrac{nv_s^2\cos\theta_B}{\lambda L f_s}$. So, the greater the nv_s^2 (i.e., $M_1 = nv_s^2 M_2$) is, the broader the Bragg bandwidth is. In order that the modulator will have a rather large bandwidth, a material with large quality factor M_1 should be chosen.

Of course, to obtain broadband modulation, apart from the requirements on the acousto-optic material, it is advisable to adopt the lens-focused fine Gaussian light beam when designing the acousto-optic modulator to make the transition time of the acoustic wave as short as possible. However, when the light beam emission angle $\Delta\theta_i$ is greater than the ultrasonic angle of divergence $\Delta\phi$, the rays at the edge will not be diffracted because of the absence of the light wave to satisfy the Bragg condition, affecting the performance of the modulator.

When assessing the performance of an acousto-optic medium, it is often necessary to consider the two indices, the bandwidth and the diffraction efficiency, at the same time. Hence the introduction of the parameter of efficiency bandwidth product ($\eta_s\Delta f_s$), i.e.,

$$\eta_s\Delta f_s \approx \frac{9nv_s^2M_2}{\lambda^3 f_s H}P_s = \frac{9M_1}{\lambda^3 f_s H}P_s \tag{1.3-82}$$

So when choosing the material, an overall consideration should be made of various factors prior to deciding on an appropriate material according to the specific requirements of the acousto-optic device.

2. The electro-acoustic transducer

The role of an electro-acoustic transducer is transform the electric power into acoustic power so as to set up the ultrasonic field in a medium. Usually the anti-piezoelectric effect of a certain material is made use of to generate mechanical vibration under the action of an externally applied electric field. So it is not only a mechanically vibrating system but also an electrically oscillating system associated with an externally applied modulating source. Here we shall mainly analyze such issues as the mechanism of the electro-acoustic transducer, the electric characteristics, and the matching of mechanical vibration with an electric tank.

(1) The vibration equation of the transducer crystal. The x-0° cut-up quartz crystal wafer is in general adopted for the transducer, with an alternating electric field applied onto the quartz plate. When the direction of the electric field is the same as that of the piezoelectric axis, the crystal plate will stimulate elastically mechanical vibration along the direction of the thickness, as shown in Fig. 1.3-18. When the frequency of the electric field is equal to that of the inherent mechanical vibration of the crystal, the amplitude of the elastic vibration will reach maximum.

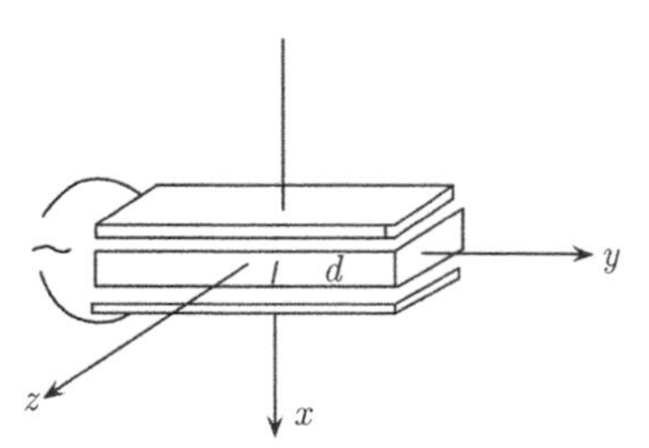

Fig. 1.3-18 The quartz crystal wafer transducer

If the transverse vibration induced by transverse compression is neglected, that is, if it is considered that the crystal wafer is infinitely large, then the natural resonance frequency of the quartz crystal wafer thickness vibration is

$$f_d = \frac{1}{2d}\sqrt{\frac{C_{11}}{\rho}} \tag{1.3-83}$$

where d is the thickness of the crystal wafer (cm), ρ the density (g/cm^3), and C_{11} the elastic modulus (N/m^2) of the vibrating form and orientation. For the quartz crystal, $\rho = 2.65$ g/cm^3, $C_{11} = 86.05 \times 10^3$ N/m^2. Thus, $f_d = \frac{285}{d}$kHz (d calculated in cm). It has been proved by experiment that, for the thickness-wise vibration of the quartz crystal wafer, it is suitable to adopt Eq.(1.3-84).

$$f_d = \frac{2580}{d}\text{kHz} \tag{1.3-84}$$

where d is in mm. Thus, in actual application, it will be possible to calculate and determine the thickness of the crystal wafer according to the frequency needed. For instance, if it is required that the resonator frequency of the quartz transducer be 30 MHz, then the thickness of the crystal wafer $d = 2880/(30 \times 10^3)$ mm $= 0.096$ mm.

The highest frequency that can be obtained by the piezoelectric quartz is about 50 MHz. Operating at such a high frequency and orientated to be normal to the x-axis, the quartz wafer has a thickness of merely 0.05 mm. So it's very difficult in terms of manufacturing technology. Furthermore, when strongly stimulated, the crystal wafer may break from electric breakdown. So, if the transducer is required to obtain a higher frequency, this can be realized by having it operate in a higher harmonic state, actually proving that the quartz crystal wafer operating at an odd resonant frequency can satisfy the condition for resonant oscillation. Although the oscillation output of the transducer is decreased, such a method of stimulation makes it possible to increase the input electric power without running the risk of electric breakdown. In particular, in certain applications where the power is not required to be very high, this method is especially suitable. The frequency of the kth-order harmonic vibration of the quartz crystal wafer is

$$f_k = kf$$

where f is the natural frequency of the basic vibration. Table 1.3-3 shows the quartz wafer with orientations as shown in Fig. 1.3-18 ($L = H = 19.96$ mm, $d = 9.99$ mm), the higher-order resonant frequency f_k measured, as well as the f_k values calculated by formulae. The difference between the two $\Delta f = (f_k - f_k')$ is very small.

Tab. 1.3-3 Comparison between measured and calculated resonant frequencies of quartz wafer

k	f_k/kHz	$f_{k'}$/kHz	Δf/‰
11	3170	3170	0.00
21	6061	6052	+1.43
31	8926	8934	−0.90
41	11816	11816	0.00
51	14713	14697	+1.09
61	20422	17579	+0.34

(2) The electric characteristics of the transducer crystal wafer. Although the vibration made by the electro-acoustic transducer is a mechanical one, it is driven by the electric oscillating energy as a load on the source. As it is equivalent to a subcircuit in the oscillating circuit, it is necessary to analyze its equivalent circuit and characteristics. The electrodes on the piezoelectric crystal wafer's two surfaces along the direction of the x-axis (Fig. 1.3-18) are equivalent to a capacitor under the action of the source, which, represented by C_1, is

$$C_1 = \frac{\varepsilon_{11} A}{d}$$

where ε_{11} is the dielectric tensor component of the piezoelectric crystal wafer, A its area, and d its thickness. When this capacitor is charged to voltage V, its electrical charge is equal to $q_0 = C_1 V$. In addition, under the action of the electric field, the piezoelectric crystal will deform, which will lead to positive piezoelectric effect and so cause the surface of crystal wafer to generate electrical charge q_1. So the total charge on the two plates of the crystal is

$$q = C_1 V + q_1 \tag{1.3-85}$$

where $q_1 = d_{11} P$, where d_{11} is the piezoelectric modulus and P is the force of the electric field acting on the quartz wafer. When the driving voltage is $V = V_0 \sin\omega_{\mathrm{m}} t$, the current flowing through the crystal wafer is

$$i = C_1 \frac{\mathrm{d}V}{\mathrm{d}t} + \frac{\mathrm{d}q_1}{\mathrm{d}t} = \mathrm{i}\omega_{\mathrm{m}} C_1 V_0 \cos\omega_{\mathrm{m}} t + i_1 \tag{1.3-86}$$

This equation shows that the piezoelectric crystal transducer in a circuit can be replaced by a capacitor C_1 and a parallel equivalent circuit, as shown in Fig. 1.3-19. According to inference and proof, the equivalent electric parameter of the crystal is obtained as follows:

$$L = \frac{\rho d^3}{8HLe_{11}^2}, \quad R = \frac{\pi^2}{8} \frac{\rho\eta d}{HLe_{11}^2}$$

$$C = \frac{8e_{11}^2 HL}{\pi^2 C_{11} d}, \quad C_1 = \frac{\varepsilon_{11} HL}{4\pi d}$$

where e_{11} is the piezoelectric constant, ε the dielectric constant of the crystal, η the mechanical loss constant of the medium, and C_{11} the elastic modulus in the x-axis direction.

When a voltage is applied on the crystal, definite electric energy will be stored in it. Because of the piezoelectric performance of the crystal, part of the electric energy is consumed in the crystal to generate elastic stress and is transformed into mechanical energy. The ratio between the two kinds of energy is exactly the measure of the efficiency of the transducer. This ratio is called the electromechanical coupling coefficient. Under the condition of vibration along the direction of the thickness, the mechanical energy of every unit volume is equal to $\frac{1}{2}C_{11}d_{11}^2E_x^2$ while the electric energy of every unit volume is $\frac{\varepsilon E_x^2}{8\pi}$. Therefore,

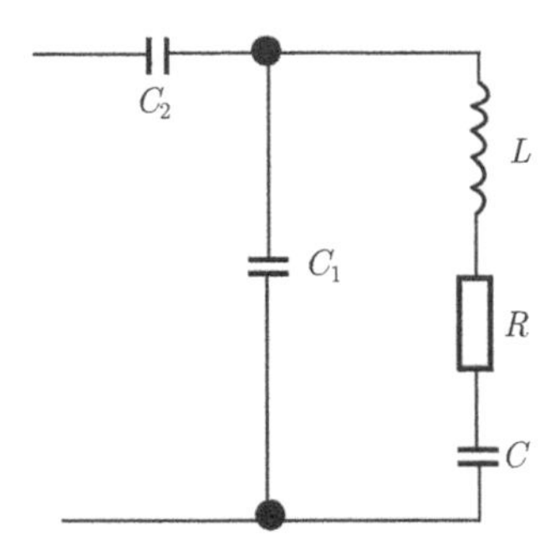

Fig. 1.3-19 The equivalent circuit of piezoelectric crystal

$$k^2 = \frac{4\pi C_{11}d_{11}^2}{\varepsilon} \quad \text{or} \quad k = d_{11}\sqrt{\frac{4\pi C_{11}}{\varepsilon}} \tag{1.3-87}$$

For the quartz crystal, $k = 10\%$. The electromechanical coupling coefficient is an important parameter for characterizing the characteristics of the transducer, the value of which differs from crystal to crystal. So in application one always hopes to adopt a crystal with big k values.

In order to be able to transfer the ultrasonic energy to the acousto-optic medium with no or fairly little loss, the sound impedance of the transducer should be as close to that of the medium as possible. This will reduce the reflection loss of the contact interface between the two. In fact, it is the rule for a modulator to introduce a transitional layer of coupled medium (metal or nonmetal) between the two. It can do three things, namely, it can transfer the ultrasonic energy into the medium, then it can reliably paste the transducer onto the medium, and, finally, it can play the role of the electrode of the transducer (if nonmetal is used as the coupling medium, an additional electrode must be used). If the requirement is that the sound impedance of the coupling medium match well the acousto-optic medium and the transducer, epoxy resin is in general adopted as the coupling adhering medium when the operation frequency is rather low. On the contrary, when the operation frequency is rather high, as it is possible that the thickness of epoxy resin may be close to the length of the sound wave, which will affect the response of the acousto-optic medium to the transducer-emitted ultrasonic wave, it is advisable to adopt a metallic material. Very often indium or an indium-tin alloy is adopted for experiments to obtain fairly good coupling effects. If indium and lead are alternately plated, not only can we enhance the emitting strength of the ultrasonic wave, the stability of the modulator can be improved as well.

3. Matching between the acoustic beam and light beam

As the incident light beam possesses a definite width, and the acoustic wave propagates in a medium with a limited velocity, the acoustic wave needs a definite transition time to penetrate the light beam. It is impossible for the response of the variation of the light beam intensity to that of the acoustic wave intensity to be instantaneous. In order to shorten its transition time to enhance its velocity of response, the modulator has the lens focus the light beam into the center of the acousto-optic medium when operating. The light beam becomes an extremely fine Gaussian light beam, thus reducing its transition time. As it is, to make full use of acoustic energy and light energy, it is considered that the relatively rational case of the acousto-optic modulator is one in which it operates at the ratio of the angle of divergence of the acoustic beam to that of light beam $\alpha \cong 1 \left(\alpha = \dfrac{\Delta\theta_{\rm i}(\text{light beam divergence angle})}{\Delta\phi(\text{acoustic beam divergence angle})}\right)$. This is because, when the angle of divergence of the acoustic beam is greater than that of the light beam, the ultrasonic energy on its edge will be wasted. Conversely, if the light divergence angle is greater than the acoustic divergence angle, then the rays on the edge cannot be diffracted since there is no longer any ultrasonic of appropriate direction (i.e., one that satisfies the Bragg condition). So, in designing the acousto-optic modulator, the ratio of one to the other should be accurately determined. In general, the angle of divergence of the light beam is taken as $\Delta\theta = 4\lambda/\pi d_0$, where d_0 represents the diameter at the waist of the Gaussian light beam focused in the acousto-optic medium. The angle of divergence of the ultrasonic wave beam is $\Delta\phi = \lambda_{\rm s}/L$, where L is the length of the transducer. Thus the ratio is found to be

$$\alpha = \frac{\Delta\theta_{\rm i}}{\Delta\phi} = (4/\pi)(\lambda L/d_0\lambda_{\rm s}) \tag{1.3-88}$$

It has been proved by experiment that the performance of the modulator is best when $\alpha = 1.5$.

In addition, for the acousto-optic modulator, in order to enhance the extinction ratio of the diffracted light, in the hope that the diffracted light will be separated from the level-0 light as much as possible, it is also necessary to adopt the rigorous separability condition, that is, requiring that the included angle between the center of the diffracted light and the level-0 light be greater than $2\Delta\phi$, that is, greater than $8\lambda\pi d$. As the included angle (i.e., the deflection angle) between the diffracted light and level-0 light is equal to $\dfrac{\lambda}{v_{\rm s}} f_{\rm s}$, the separability condition is

$$f_{\rm s} \geqslant \frac{8v_{\rm s}}{\pi d_0} = \frac{8}{\pi\tau} \cong \frac{2.55}{\tau} \tag{1.3-89}$$

or, because $f_{\rm s} = v_{\rm s}/\lambda_{\rm s}$, it can also be written as

$$\frac{1}{d_0} \leqslant \frac{\pi}{8\lambda_{\rm s}} \tag{1.3-90}$$

Substitution of Eq. (1.3-90) into Eq. (1.3-88) yields

$$\alpha = \frac{\lambda L}{2\lambda_{\rm s}^2} \cong \frac{L}{2L_0}$$

When the optimal performance condition for the modulator $\alpha = 1.5$, then

$$L = 3L_0 \tag{1.3-91}$$

from which the length L_0 of the transducer is determined. Then, by using Eq. (1.3-89), the diameter of the waist of the laser beam focused in the medium can be found:

$$d_0 = v_{\rm s}\tau = \frac{2.55v_{\rm s}}{f_{\rm s}} \tag{1.3-92}$$

from which the focal length of the lens can be chosen.

1.4 magneto-optic modulation

1.4.1 The magneto-optic effect

The magneto-optic effect is the physical basis of magneto-optic modulation. For some materials, such as the paramagnetic material, ferromagnetic material, and ferrimagnetic material, the atoms or ions in their internal composition all have definite magnetic moments. The compounds composed of such magnetic atoms or ions and possessing powerful magnetism are called magnetic materials. It has been found that there are many small regions in the magnetic material, in each of which the magnetic moments of all the atoms or ions are arranged parallel to one another. These small regions are referred to as magnetic domains. As the magnetic moments of the magnetic domains are different in direction, the action of one would offset that of the other. So, macroscopically, no magnetism is exhibited. If along a certain direction of the object an external magnetic field is applied, then the magnetic moments of all the magnetic domains of the object will turn from different directions to the direction of the magnetic field. Thus magnetism is manifested externally. When the light wave passes through such a magnetized material, changes in its propagation characteristics will take place. This phenomenon is called the magneto-optic effect.

The magneto-optic effect includes the Faraday rotation effect, Kerr effect, the Cotton Mouton effect, etc., the most important of which being the Faraday rotation effect, which makes the polarization direction of a beam of linearly polarized light propagating in a medium under the action of an externally applied magnetic field rotate and the size of its polarization angle θ proportional to the product of the magnetic field intensity H along the light beam direction and the length L of light propagating in the medium, i.e.,

$$\theta = VHL \tag{1.4-1}$$

where V is called the Verdet constant, which represents the angle the polarization direction has rotated after the linearly polarized light passes through the magneto-optic medium of unit length under unit magnetic field intensity. Table 1.4-1 lists the Verdet constants of some magneto-optic materials.

Tab. 1.4-1 The Verdet constants of different materials (in $(')/(\text{cm} \cdot \text{T}) \times 10^{-4}$)

Name of material	Coronal glass	Flint glass	Sodium chloride	Diamond	Water
V	$0.015 \sim 0.025$	$0.03 \sim 0.05$	0.036	0.012	0.013

With respect to the physical reasons for the phenomenon of optical rotation, it can be explained that the externally magnetic field makes the moment of the medium molecules arranged in a fixed direction. When passing through it, a beam of linearly polarized light is resolved into two streams of circularly polarized light of identical frequency and identical initial phase, of which one rotates clockwise and is called the right rotating circularly polarized light whereas the other rotates counterclockwise and is referred to as the left rotating circularly polarized light. The two streams of circularly polarized light propagate with different velocities without acting on each other. Their phase retardations generated after they have passed through a medium of thickness L are, respectively,

$$\varphi_1 = \frac{2\pi}{\lambda} n_{\mathrm{R}} L, \quad \varphi_2 = \frac{2\pi}{\lambda} n_{\mathrm{L}} L$$

So, there exists a phase difference between the two streams of circularly polarized light.

$$\Delta\varphi = \varphi_2 - \varphi_1 = \frac{2\pi}{\lambda}(n_{\mathrm{R}} - n_{\mathrm{L}})L \tag{1.4-2}$$

After passing through the medium, they are synthesized as a stream of linearly polarized light, whose polarization direction has rotated an angle relative to that of the incident light.

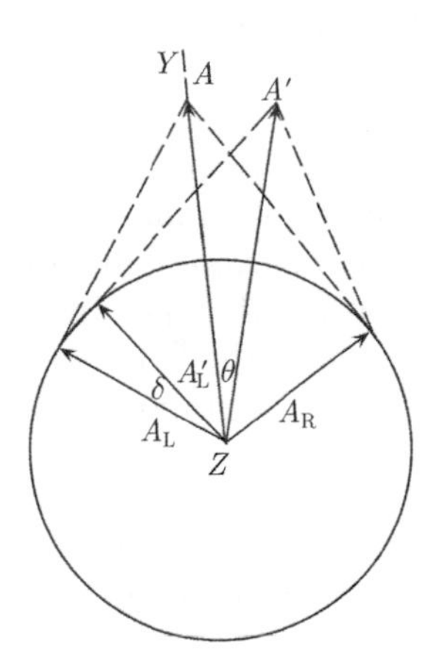

Fig. 1.4-1 When light passes through a medium, the polarizing direction rotates

In Fig. 1.4-1, YZ represents the vibration direction of the linearly polarized light injected into the medium, resolving the amplitude A into the left rotating and right rotating vectors A_L and A_R. Suppose the length L of the medium has made the right rotating vector A_R just rotate back to its original position. At this moment the left rotating light vector (as $v_L \neq v_R$) has rotated to A'_L. So the synthesized linearly polarized light A' has rotated an angle θ relative to the polarization direction of the incident light. This value is equal to half the angle δ, i.e.,

$$\theta = \frac{\delta}{2} = \frac{\pi}{\lambda}(n_R - n_L)L \tag{1.4-3}$$

It can be seen that the polarizing direction of A' will rotate rightward with the propagation of the light wave which is known as the rightward rotating light effect.

The rotating direction of the magnetically induced rotating light effect is only related to the direction of the magnetic field while independent of the direction of the rays propagating being forward or reverse, which is where the phenomenon of magnetically induced optical rotation is different from that of the natural optical rotation. During its round trip to and from a natural optically rotating material, the beam of light offsets itself since its rotating angles are equal but opposite in direction. But when passing through a magneto-optic medium, as long as the direction of the magnetic field remains unchanged, the rotating angle will always increase toward one direction. This phenomenon shows that the magnetically induced rotating light is an irreversible optical process. Hence its use in making such devices as the optical isolators and single-pass optical brakes.

Currently, the most frequently used magneto-optic material is mainly the yttrium iron garnet (YIG) crystal, whose absorption coefficient in the wavelength range $1.2 \sim 4.5$ μm is very low ($\alpha \leqslant 0.03\ \text{cm}^{-1}$) and has a rather big Faraday rotation angle. This wavelength range includes the optimum range ($1.1 \sim 1.5$ μm) in fiber-optic transmission and the frequency range of certain solid lasers. Therefore, it is possible to make magneto-optic devices such as modulators, isolators, ring-shaped devices, etc. using this material. As the variation of its physical performance with the temperature is not great, the magneto-optic crystal is not easily air-slaked and the modulating voltage is low. This is where it is superior to the electro-optic and acousto-optic devices, but when the operation wavelength has exceeded the above-mentioned range, the absorption coefficient will abruptly increase, even to the extent of making the device inoperable. This shows that it is in general not transparent in the region of visible light, while only usable in the near-infrared region and infrared region, which greatly limits its application.

1.4.2 The magneto-optic volume modulator

Similar to the electro-optic and acousto-optic modulation, the magneto-optic modulation also consists in having the information to be transmitted transformed into the variation of the intensity (amplitude) of the light carrier wave, the difference being that the magneto-optic modulation converts the electric signal into an alternating magnetic field corresponding to it first, with the polarized state of the light wave transmitted in the medium changed by the magneto-optic effect so as to attain the goal of changing the parameters like the light intensity, etc. The composition of the magneto-optic volume modulator is as shown in

Fig. 1.4-2. The operation material (YIG or Ga-doped YIG rod) is placed on the optical path along the axial direction, at whose two ends are placed the polarizing and analyzing devices, with the high frequency spiral-shaped coil wound around the YIG rod and controlled by the driving source. In order to obtain linear modulation, apply a constant magnetic field H_{dc} on the direction normal to light propagation, its intensity sufficient to make the crystal saturation magnetized. During operation, the high frequency signal current will induce and generate a magnetic field parallel to the direction of light propagation when passing through the coil. When the incident light passes through the crystal, because of the Faraday rotation effect, its polarized plane is seen to rotate, its rotation angle directly proportional to the magnetic field intensity H. Therefore, so long as the modulating signal is used to control the variation of the magnetic field intensity, the polarized plane of light will be made to undergo corresponding variation. But with the constant magnetic field H_{dc} here applied, which is normal to the light transmitting direction at that, the rotation angle is inversely proportional to H_{dc}. Thus

$$\theta = \theta_{\mathrm{s}} \frac{H_0 \sin(\omega_H t)}{H_{\mathrm{dc}}} L_0 \tag{1.4-4}$$

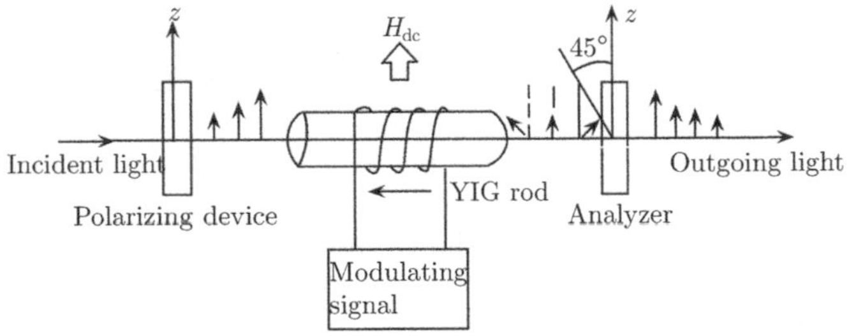

Fig. 1.4-2 Schematic diagram of magneto-optic modulation

where θ_{s} is the unit length saturated Faraday rotation angle and $H_0 \sin \omega_H t$ the modulating magnetic field. If the analyzer is also used, we can obtain the modulated light with definite intensity variation.

1.4.3 The magneto-optic waveguide modulator

Figure 1.4-3 shows the structure of a magneto-optic waveguide mode transforming modulator. On the disc-shaped yttrium-gallium garnet ($Gd_3Ga_5O_{12}$-GGG) substrate, the epitaxial growth Ga- and Se-doped YIG magnetic thin film as the waveguide layer (thickness $d =$ 3.5 μm, $n = 2.12$). On the surface of the magnetic thin film a metallic snake-shaped circuit is made using the photoetching technique. When an electric current flows through the snake-shaped circuit, the electric current in a certain channel of the snake-shaped circuit is along the y direction, the current in the adjacent channel is along the $-y$ direction, and this current is capable of generating a magnetic field with the $+z$ and $-z$ directions alternately changing. Then there may appear in the magnetic thin film the alternate saturation and magnetization along the $+z$ and $-z$ direction. Suppose the period of the magnetic field variation (i.e., the period of the snake-shaped structure) is

$$T = \frac{2\pi}{\Delta\beta}$$

where $\Delta\beta$ is the difference between the TE mode and TM mode propagation constants. Owing to the mismatch of the lattice constant with thermal expansion between the thin film and the substrate, the magnetization-prone direction is located within the plane of the thin

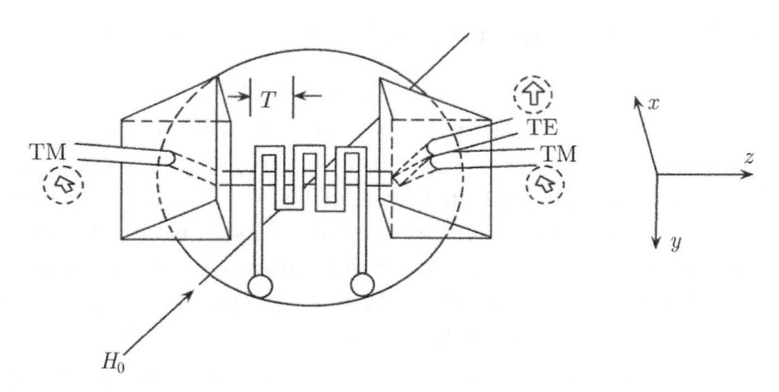

Fig. 1.4-3 The magneto-optic waveguide mode transforming modulator

film, the demagnetizing factor in which is zero. So the magnetization intensity M can be made to revolve freely in the thin film plane with a small magnetizing field. If laser ($\lambda = 1.152$ μm) is input and output by two prism couplers, what is injected in is the TM mode. Because of the Faraday magneto-optic rotation effect, with the transmission of the light wave along the z direction (the direction of magnetization) in the waveguide thin film, the electric field vector (along the x direction) originally located within the thin film will turn to the normal direction (the y direction). That is, the TM mode is transformed into the TE mode. Since the magneto-optic effect is proportional to the component M_z of the intensity of magnetization M along the direction of propagation z, changing the electric current in the snake-shaped circuit by applying a direct current magnetic field H_{dc} between the z-axis and y-axis along the 45° direction can change M_z, thus changing its transforming efficiency. When the input electric current is so large that it makes M saturate along the z direction, the transforming efficiency will reach maximum. If the device's $T = 2.5$ μm, a direct current of 0.5 A is input in the snake-shaped circuit, and the magneto-optic interaction length $L =$ 6 mm, then 52% of the power of the input TM mode ($\lambda = 1.152$ μm) transformed to the TE mode. The output coupler of the magneto-optic waveguide mode transforming modulator is a rutile prism with a high birefringence. No matter whether the output TE and TM modes are resolved into two light beams with a field angle of $20°11'$ each, or when the frequency of the electric current input in the snake-shaped circuit is from 0 to 80 MHz, the modulation of the light intensity in the two modes can be observed just the same.

1.5 The direct modulation

By the direct modulation is meant transforming the information to be transmitted into electric current signal that is injected into the semiconductor light source (the laser diode LD or semiconductor diode LED) so as to obtain the modulated signal. As it is carried on in the interior of the light source, it is also called the internal modulation. Such a direct modulation is both convenient and highly efficient besides being capable of high-speed modulation. It is the practical modulating method universally used in the fiber-optic communication system. In terms of the type of the modulating signal, the direct modulation can also be divided into analog modulation and digital modulation, the former carrying out direct modulation of the light intensity for the light source using continuous analog signals (e.g., TV, voice signals, etc.) and the latter carrying out the intensity modulation for the light source using the pulse code modulated (PCM) digital signals. Below we shall make a brief description of the two modulating methods.

1.5.1 The principle of direct modulation by the semiconductor laser (LD)

The semiconductor laser is a device by means of which electrons and photons act on each other and perform direct energy transformation. Figure 1.5-1 shows the curve of relationship between the output light power and the driving current of the arsenic-gallium-aluminum (AlGaAs) double heterojunction injection laser. The semiconductor laser has a threshold electric current I_t; when the driving current density is smaller than I_t, the laser basically does not shine or merely emits very faint fluorescent light of very broad spectral line width and rather poor directionality. When the density of the driving current is greater than I_t, however, it begins to emit laser. Now the width of the spectral line and the direction of radiation become evidently narrower while the intensity greatly increases. Furthermore, with an increase in the electric current, it appears to increase linearly, as shown in Fig. 1.5-2. It can be seen from Fig. 1.5-1 that the strength of laser emission is directly related to the magnitude of the driving current. If the modulating signal is applied on the laser (power source), it will be possible to directly change (modulate) the intensity of the signal of the laser's output light. By virtue of its simplicity and ability to operate at high frequency plus the guarantee of a good linear operation region and bandwidth, this modulating manner has found wide application in fiber optic communication, optical disk, and optical duplication, etc.

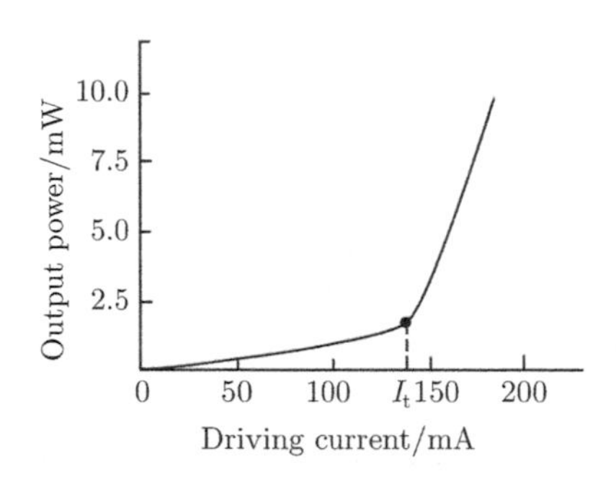

Fig. 1.5-1 The output characteristics of semiconductor laser

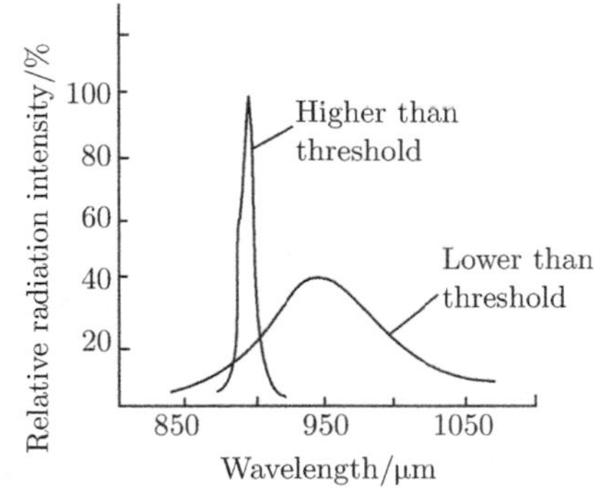

Fig. 1.5-2 The spectroscopic characteristics of semiconductor laser

What is shown in Fig. 1.5-3 is a schematic diagram of the modulating principle of the semiconductor laser, in which (a) shows the schematic of the electrical principle while (b) shows the curve of the relationship between the output optical power and the modulating signal. In order to obtain linear modulation so that the operating point will be located at rectilinear part of the output characteristic curve, it is necessary to apply a suitable biased current I_b while introducing the modulating signal current so that the optical signal output will not become distorted. But care must be taken that the modulating signal source be isolated from the direct current bias source to prevent the latter from affecting the former. When the frequency is rather low, this can be done by connecting the capacitance to the inductance coil in series. When the frequency is very high (>50 MHz), the high pass filter circuit must be adopted. In addition, the biased electric current directly affects the modulating performance of the LD. Usually, I_b should be chosen to be in the vicinity of the threshold current and slightly lower than I_t. Then the LD can obtain a fairly high modulating efficiency because, such being the case, the LD has no need for time to prepare for continuous emission of light signals (i.e., the delay time is very short), its modulating rate not limited by the average life of the carrier in the laser. At the same time, the relaxation oscillation, too, will be restrained to some extent. However, if the I_b is chosen too high, the extinction ratio of the laser will be deteriorated. So an all-around consideration should be made when choosing the bias current.

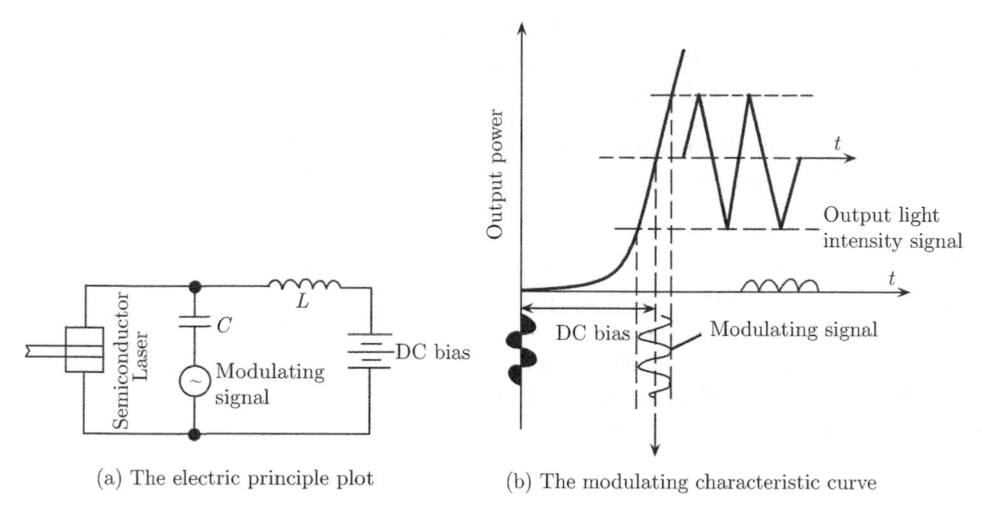

(a) The electric principle plot (b) The modulating characteristic curve

Fig. 1.5-3 Semiconductor laser modulation

When the semiconductor laser is in the state of continuous modulating operation, the power loss is rather large owing to the direct current bias, thus inducing a rise in temperature which will affect or damage the normal operation of the device. Now the emergence of the double heterojunction laser has made the threshold current density greatly decrease compared with that of the homojunction laser, making it possible to satisfactorily operate in the mode of continuous modulation at room temperature.

To prevent modulation distortion when the semiconductor laser is performing modulation at a high frequency, the fundamental requirement is that the output power maintain a good linear relation with an electric current above the threshold. In addition, in order to prevent the emergence of relaxation oscillation as much as we can, it is advisable to adopt a laser structure of rather narrow strip width. Furthermore, direct modulation will reduce the strength of the laser's principal mode while increasing that of the second mode, thus broadening the spectral line of the laser. Meanwhile, the pulse width Δt generated by modulation and the spectral line width $\Delta\nu$ constrain each other, constituting the so-called bandwidth limit to the Fourier transform. Therefore, the capability of the semiconductor laser for direct modulation is restricted by the $\Delta t \cdot \Delta\nu$ product. So, for modulation at a high frequency, one had better adopt a modulator of the quantum well structure or an external modulator.

1.5.2 The modulating characteristics of the semiconductor light-emitting diode (LED)

As the semiconductor light-emitting diode is not a threshold device, unlike the semiconductor laser, its output light power will not undergo an abrupt change with the variation of the injected electric current. Hence the linearity of the LED's *P-I* characteristic curve is fairly good. Figure 1.5-4 shows a comparison of the *P-I* characteristic curve between LED and LD. It can be seen from the figure that LED_1 and LED_2 are the *P-I* characteristic curves of the front luminous-type light-emitting diodes while LED_3 and LED_4 are those of the end face luminous-type light-emitting diodes. It can be seen from the figure that the *P-I* characteristic curve of the light-emitting diode is obviously better than that of the semiconductor laser. So it has found wide application in the analog fiber optic communication system. But in the digital fiber optic communication system, because of its incapability of obtaining very high modulating rate (the highest being merely 100 Mb/s), its application is limited.

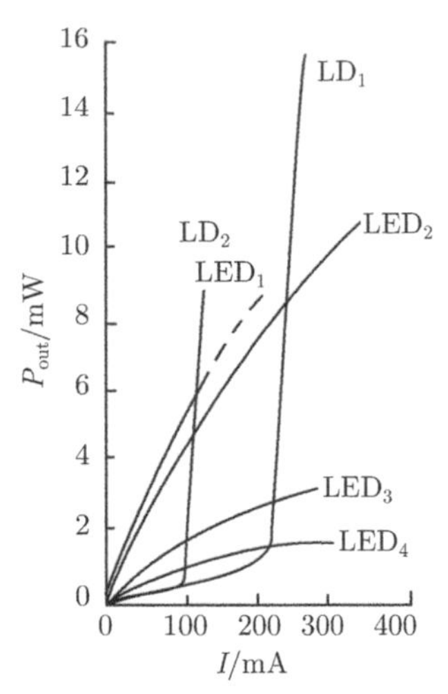

Fig. 1.5-4 The LED and LD P_{out}-I curves compared

1.5.3 The analog modulation of the semiconductor light source

No matter whether the LD or the LED is used as the light source, it is required that a bias current be applied to make the operating spot located in the straight line section of the P-I characteristic curve of the LD or LED, as shown in Fig. 1.5-3(b) and Fig. 1.5-5(b). The quality of the modulated linearity is related to the modulating depth m.

$$\text{LD}: \quad m = \frac{\text{modulating current amplitude}}{\text{bias current- threshold current}}$$

$$\text{LED}: \quad m = \frac{\text{modulating current amplitude}}{\text{bias current}}$$

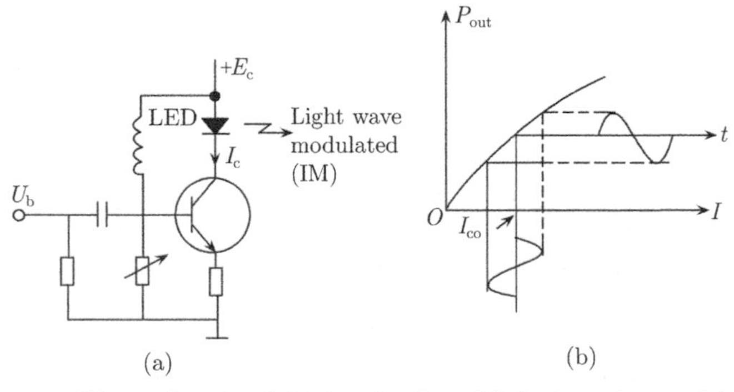

Fig. 1.5-5 The analog signal driving circuit and light intensity modulation

It can be seen from the figure that when m is great, the amplitude of the modulating signal will be great. So, the linearity will be poor. Conversely, if m is small, although the linearity is good, the amplitude of the modulating signal will be small. Therefore, a suitable m value should be chosen. Moreover, in the analog modulation, the linear characteristic of the light source device itself is the principal factor in determining the quality of the analog modulation. So, in applications requiring rather high linearity, it is necessary to make nonlinearity compensation; that is, the nonlinear distortion due to the light source should be corrected by the electronic technology.

1.5.4 The digital modulation of the semiconductor light source PCM

Digital modulation is the modulation of the optic carrier wave emitted by the light source using the binary signal code "1" and code "0", while digital signals very often adopt pulse code modulation (PCM), that is, first, the continuous analog signals are transformed into a group of amplitude modulated pulse sequences by "sampling". Then, through the processes of "quantization" and "coding", they form a group of rectangular pulses of equal amplitude and equal width as "code elements". For instance, the different combinations of "with pulse" and "without pulse" (with definite digit capacity of pulse code elements) represent the amplitude of the sampled value. This is pulse coding, which results in transforming the continuous analog signals into PCM digital signals, called "analog/digit" or A/D transformation (see books on relevant fiber optic communication for the specific process). Then, the PCM digital signal is used to perform intensity modulation for the light source, the characteristic curve of whose modulation is as shown in Fig. 1.5-6.

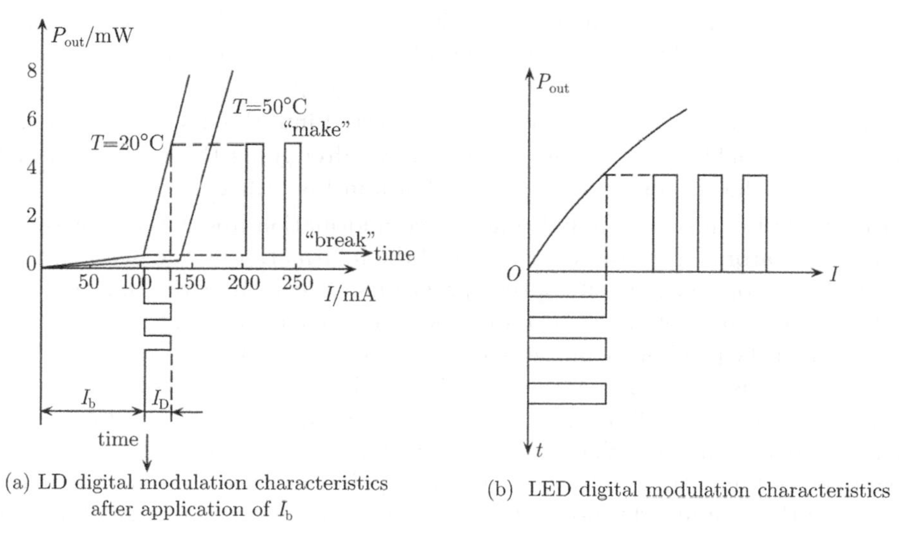

(a) LD digital modulation characteristics after application of I_b

(b) LED digital modulation characteristics

Fig. 1.5-6 The digital modulation characteristics

Because of its remarkable merits, digital optical communication has very good prospects for application. To begin with, the noise and distortion introduced in the course of transmission of the digital optical signals in the channel can be removed by adopting an indirect relay. Therefore it has strong anti-interference capability. Second, it does not make high requirements on the linearity of the digital fiber communication system, with the light-emitting power of the light source (LD) made full use of. Third, the digital optical communication equipment is conveniently connected to the PCM telephone terminal, the PCM digital color TV terminal, and the computer terminal, thereby making up a comprehensive communication system capable of transmitting not only telephone and color TV information but also computer data.

1.6 The spatial light modulator

1.6.1 The basic concept of the spatial light modulator

The different kinds of modulators described previously consist in exerting an action on a beam of light as a "whole" and, for each and every point on the x-y planes normal to the direction of light propagation, the effect is identical. The spatial light modulator (SLM) can form amplitude (or intensity) transmissivity varying with the xy coordinate.

$$A(x,y) = A_0T(x,y)$$

Or it may form phase distribution varying with the coordinate.

$$A(x,y) = A_0T\exp[\mathrm{i}\theta(x,y)]$$

Or it may form different scattering states varying with the coordinate. Judging by its name, this is a kind of device that carries out modulation of the spatial distribution of the light wave. It contains many independent elements (called pixels) that are arranged as one- or two-dimensional arrays, each capable of independently accepting the control by the light signal or electric signal and changing its own optical properties (transmissivity, reflectivity, refractive index, etc.) in accordance with the signal, thus performing modulation of the light wave passing through it. The signal that controls the optical properties of these pixels is called the "write in signal" (W), which can be either the optical signal or the electric signal; the light wave injected into the device and modulated is called the "read out light" (I_R); the output light wave after going through the spatial light modulator is called the "output light" (I_O), as shown in Fig. 1.6-1.

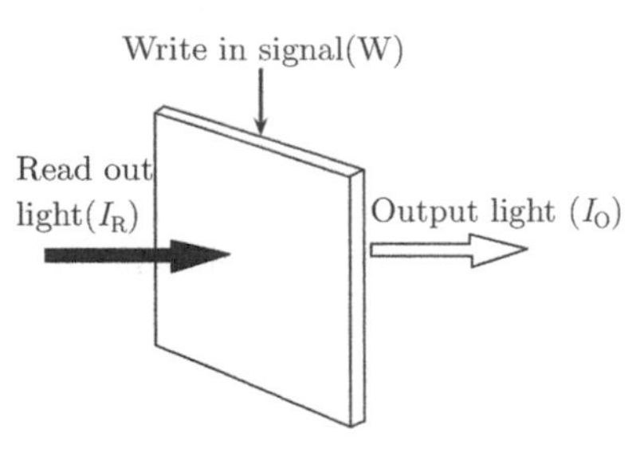

Fig. 1.6-1 Schematic diagram of the spatial light modulator

Obviously the write in signal should contain the information about the control over the pixels of the modulator and transmit each piece of the information to the modulator's corresponding pixel position to change its optical properties. When the write in signal is a light signal, it is usually represented as a two-dimensional image of light intensity distribution, which is imaged on the pixel plane of the spatial light modulator via an optical system. This process is spoken of as addressing, which, in this way, is called optical addressing. Since in terms of time addressing by all pixels is accomplished at the same time, optical addressing is a parallel addressing mode. When the read out light passes through the modulator, its optical parameters (amplitude, intensity, phase, or polarized state) are modulated by the various pixels of the spatial light modulator. As a result, it becomes a beam of output light with new optical parametric spatial distribution. This mode is mainly used in light-light transform devices, which can be applied in optical information processing and in optical computers for image transformation, display, storage, or filtering. In particular, when performing real-time two-dimensional parallel processing to exhibit the advantages of optical information processing, there is even a greater need for the real-time spatial light modulator. When the write in signal is an electric signal, it is necessary to adopt the "electric addressing" mode because the electric signal is a time sequence. In principle, the signals can only be output to the pixels of the modulator one by one. So the electric addressing is a mode of serial addressing, which is mainly used for the electro-optical real-time interfacing devices. Its merit is its capability of directly using the electric signal to control the amplitude or phase of the output light, getting easily connected to the computer as well as to electronic analog signals such as the TV camera signals. Under the action of the write in signal, the optical properties (transmissivity, refractive index, reflectivity, optical activity, and surface deformation, etc.) of the pixels of the spatial light modulator will vary accordingly. The mechanisms that cause the variation of the optical properties are mainly the electro-optic effect, acousto-optic effect, and magneto-optic effect as well as the electric absorption effect of such materials as all kinds of crystals, liquid crystals, and organic polymers (the principles of which are as treated previously).

1.6.2 The basic functions of the spatial light modulator

The basic function of the spatial light modulator is to provide real-time or quasi-real-time one-dimensional or two-dimensional optical sensing devices and operating devices. Different types of spatial light modulators each have their own characteristics, but they have some common or similar performance and functions which, summed up, are as follows:

1. The function of a transducer

In the electro-optic mixed processor, the write in electric signals can be transformed into output light signals, and such an output can be either one- or two-dimensional data groups arranged in the format needed or two-dimensional images. For instance, the to-be-processed information is from the analog signals of a camera or computer, which is often an electric signal varying with time. To input the signal into an optical processing system, we have to use a spatial light modulator. On the one hand, the serial electric signals arranged in the order of time are converted to control signals arranged in the form of one- or two-dimensional arrays and, on the other, the control signal on each pixel in the array is converted to one capable of modulating the read out light following the change in its optical properties.

In a real-time processing system, a write in non-coherent light signal can be converted to an output coherent light signal. As the object for processing by a real-time processing system is often an actual object, an ordinary optical system can only make it form a non-coherent image. But the processing system, on the contrary, requires a coherent image for frequency domain processing or light interference-based processing, etc. For example, in Fig. 1.6-2, the write in signal $I_{\rm W}$ is a two-dimensional image made up of a beam of non-coherent light while the read out light $I_{\rm R}$ is a beam of coherent light with a uniform amplitude. Then, when the spatial light modulator adopts the optical addressing mode to transform the write in light illuminance distribution into the light intensity transmission coefficients of the pixels, its output light $I_{\rm O}$ is a beam of coherent light carrying the write in image information.

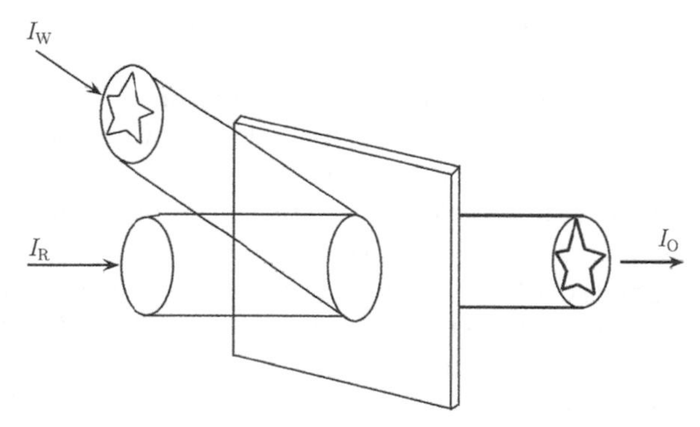

Fig. 1.6-2 The principle of optical addressing spatial light modulation

2. The amplification function

When the write in light intensity is rather weak, or when the image signal becomes weak in the process of information processing, it can be strengthened by adopting a spatial light modulator with spatially uniformly distributed read light of high intensity to obtain amplified output coherent light signals.

3. The operation function

For most spatial light modulators, signal multiplication is its intrinsic performance. As is shown in Fig. 1.6-1, the read light $I_{\rm R}$ carries the information on a two-dimensional image, with the write in signal $I_{\rm W}$ for controlling the transmittance of the spatial light modulator's pixels. Thus, the light intensity distribution of the output light signal $I_{\rm O}$ on the surface of the spatial light modulator is equal to the product of $I_{\rm O}$ and the signal $I_{\rm W}$. If the write in signal represents a matrix, and the read out light another, then multiplication between matrices of numbers can be realized using the spatial light modulator. In addition, operations related to the basic multiplying functions can also be performed, for example, programmable match filtering, the computer-controlled reconstructible optical interconnection, etc.

4. The threshold operation function

By means of the threshold characteristics of devices, a continuous write in signal can be converted to a number of separate "values" to output. The simplest operation is quantize the write in signal as 0, 1 to output, with a threshold given. When the write in signal exceeds the threshold, the output is "1", but when the write in signal is smaller than the threshold, the output is "0" (i.e., there is no output light). Such an operation is spoken of as the threshold operation. By utilizing this characteristic, it is possible to realize the binary logic operation and A/D transformation. A spatial light modulator that exhibits the threshold characteristic can be regarded as a two-dimensional array of the nonlinear optical switching.

Apart from the above-mentioned functions, the spatial light modulator has such functions as short-term storage (memory), optical amplitude-restriction, wave face recovery, etc., which are not dealt with here.

1.6.3 Several typical spatial light modulators

1. The Pockels readout optical modulator

An optical addressing-type spatial light modulator made by using the electro-optic effect, the Pockels readout optical modulator (PROM) is currently in application for its good performance.

(1) The structure of the Pockels readout optical modulator

To meet the requirement on real-time processing, there have emerged devices of a multitude of structural principles one after another. Some are manufactured by combining the photosensitive thin film with ferroelectric crystal, some by utilizing the photoconductive crystals that possess photosensitive performance themselves, of which the spatial light modulators made of $Bi_{12}SiO_{20}$ (BSO) for short) have been developing fairly quickly. BSO is a kind of noncentrosymmetric cubic crystal (of point group 23) that possesses not only the photoconductive effect, but also the linear electro-optic effect. It has a rather low half wave voltage and is fairly sensitive to the blue light of $\lambda = 400 \sim 450$ nm (as blue light has a very high photon energy) while its photoconductive effect on red light of 600 nm is very weak. As the characteristic of photosensitivity varies abruptly with the difference in wavelength, the material is sensitive to blue light but insensitive to red light. So blue light can be used as the write in light while red light can be used as the read out light, thereby reducing the interference with each other.

A schematic diagram of the structure of the BSO-PROM spatial light modulator is as shown in Fig. 1.6-3.

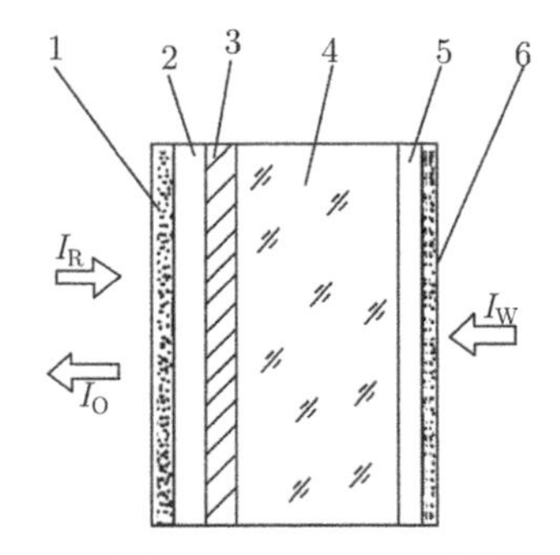

Fig. 1.6-3 Schematic diagram of the structure of reflective BSO spatial light modulator: 1,6-transparent electrodes; 2,5-insulating layers; 3-bichromatic reflective layer; 4-BSO crystal

The two sides of the BSO crystal are coated with a 3 μm thick insulating layer and the outermost layer is plated with transparent electrode to make it a transmissive device. If the write in side is plated with a two-color reflective layer to reflect the red light and let the blue light penetrate, then we have the reflective device. The reflective structure can not only reduce the half-wave voltage, but also it has removed the influence of optical activity of the crystal itself.

(2) The operation principle of the BSO-PROM spatial light modulator

The principle consists in converting the light intensity distribution of the image into spatial distribution of the voltage applied on the BSO crystal, thereby transmitting the image onto the readout light beam, the former utilizing the crystal's opto-electric conductivity and the latter, its Pockels electro-optic effect. The specific process of operation is: if there is no illumination after the operation voltage is applied on the transparent electrode, there is no variation of the optical properties of the crystal. Since the value of the photosensitive layer's resistance is very great, the greater part of the voltage having fallen on the photosensitive layer. If powerful blue light is used to illuminate the photosensitive layer, then the photons are excited and the electrons are enabled to obtain sufficient energy to surmount the forbidden zone and jump into the conductive zone. So there will be large quantities of free electrons and cavities taking part in conduction to make the resistance in the photosensitive layer decrease to scantiness (called the photoconductive effect). Thus the greater part of the voltage is applied on the BSO crystal. Since the resistance value of the photosensitive layer varies with the intensity of the incident light from without, the electro-optic effect of the crystal, too, undergoes corresponding variation with the intensity of the incident light. For instance, use a beam of laser to carry image information as the write in signal I_{W} for injection to the device from the right of the figure, to be illuminated onto the BSO crystal. As the electron-cavity pair is excited by the opto-electric effect in the crystal, the electrons are pulled toward the positive pole while the cavity causes the spatial variation of the potential according to the image shape distribution of the write in light. Thus, the illuminance distribution of the write in light is turned into electric field distribution in the BSO crystal via the opto-electric effect so that the image is stored. When fetching an image, use long wave light, such as the red light of 633 nm as readout light I_{R} to illumine the device from the left of the figure via the polarizer (in the x direction). The light becomes elliptically polarized light owing to the electro-optic effect, its ellipticity depending on the spatial variation of the voltage in the crystal. Therefore, the light intensity distribution from the analyzer (in the left of the figure, placed orthogonal to the polarizer) output light I_{O} will be proportional to the bright-and-dark distribution; that is, the spatial modulation of light is realized.

The process of operation of the above-mentioned electro-optic spatial modulator is as shown in Fig. 1.6-4. What is shown in Fig. 1.6-4(a), (b), and (c) represents the preparatory stage before write in. Shown in Fig. 1.6-4(a) is the application of voltage V_{o} between the two electrodes of the crystal; shown in Fig. 1.6-4(b) is illuminating the photosensitive layer with uniform lamp light to make it generate electron-cavity pairs and drift towards the electrode interface of the crystal under the action of the external electric field to make the electric field in the crystal zero, that is, eliminate the originally stored images (as the hidden resistance of the BSO is very great, the images stored can be kept for a very long period of time). What is shown in Fig. 1.6-4 is the reversal of the voltage to make the voltage on the crystal rise to $2V_0$ while Fig. 1.6-4(d) shows the write-in situation, where blue light of a rather short wavelength is used to carry the image information as the write in light I_{W} to be imaged on the surface of a BSO crystal wafer, which is converted into electric field distribution within the crystal via the opto-electric effect and then into the distribution of double refractive index via the electro-optic effect. Figure 1.6-4(e) shows the read out situation in which the linearly polarized red light of long wavelength is used as the readout light I_{R}. The reason for choosing red light of a long wavelength as the read out light is because it is basically incapable of generating the opto-electric effect on the BSO crystal, thus unable to destroy the originally written in image of the electric field. Because of birefringence, it is resolved into two mutually normal polarized components after it is injected into the crystal. As there is a phase difference between the two, the polarization state of their synthetic light varies accordingly. Therefore the light I_O output from the analyzer is the intensity modulated light. In the bright region of the recording screen, which is the unexposed region of the crystal, the crystal's birefringence effect is very weak. As the polarization state of the light beam in this region is almost unchanged, there is no display of image.

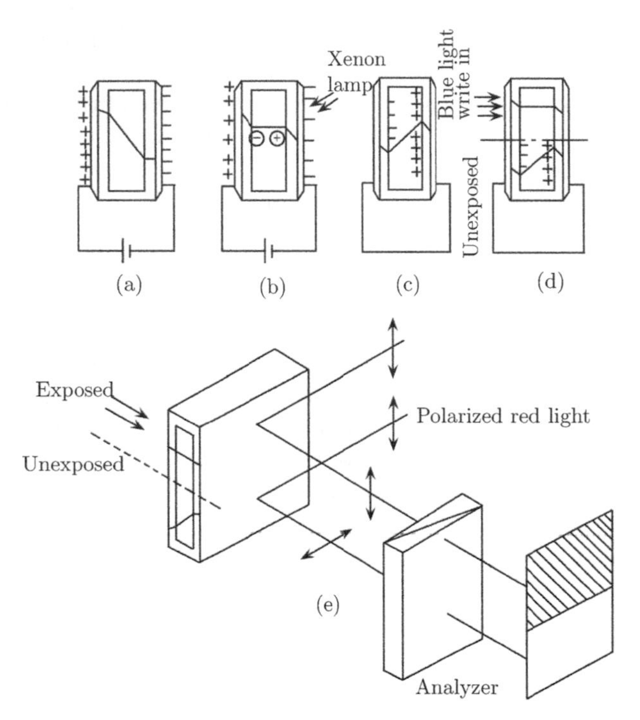

Fig. 1.6-4 The BSO-PROM spatial light modulator

2. The liquid crystal spatial light modulator

For an ordinary crystal, there is a definite melting point at which it changes from the solid state into the liquid state. Below the melting point it is in the solid state while above it, the liquid state, with the properties of a crystal lost at that. But some materials do not directly change from solid into liquid, but do so via a transitional phase. Now, on the one hand, it possesses the property of fluidity of liquids and, on the other, it has the characteristics of a crystal (e.g., optical, mechanical and thermal anisotropy). This transitional phase is spoken of as the "liquid crystal".

The liquid crystal is an organic compound composed in general of rod-shaped cylindrical symmetric molecules. It possesses a very strong electric dipole moment and an easily polarized chemical group. Owing to the anisotropic characteristics of the liquid crystal's molecules in terms of shape, dielectric constant, refractive index, and electric conductivity, when an external (electric, thermal, magnetic, etc.) field is applied on such a material, changes in the liquid crystal molecules' direction of arrangement and flowing position will take place, or the physical state of the liquid crystal will be changed. For instance, if an electric field is applied on it, its optical properties will change. This is the liquid crystal's electro-optic effect, which mainly includes the twisted effect, dynamic scattering effect, electro-controlled birefringence effect, phase change effect, guest-host effect, etc.

The fairly typical applied device of the liquid crystal modulator is the cadmium sulfide (CdS) optical valve. Its structural schematic is as shown in Fig. 1.6-5 along with and the plate glass that is used for the purpose of maintaining the fixed shape of the device. The transparent electrode material is a mixed material indium-tin oxide (ITO) of indium oxide (In_3O_3) and tin oxide (Sn_2O_3), the material for the liquid crystal molecule orientation film layer is SiO_2, which makes the liquid crystal molecule thin film in contact with it arrange itself along the plane; the reflectivity of the multi-layer medium membrane reflective mirror is about 90%, which is also used as the insulator between the two transparent electrodes to prevent the direct current from flowing through the liquid crystal layer; the material of the optically isolated layer is CdTe, which keeps the write in light I_W injected from the right side from injecting the left side of the optically isolated layer, while keeping the light leaking out from the reflective film from injecting the optical guide layer so as to isolate the write in light from the read out light. The material for the optical guide layer is CdS, the action of which will be described below.

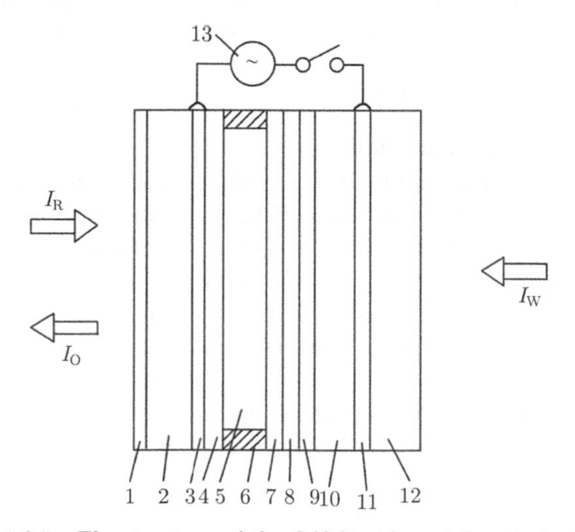

Fig. 1.6-5 The structure of the CdS liquid crystal optical valve;
1-medium film; 2,12-plate glass; 3,11-transparent electrodes; 4,7-liquid crystal molecular orientation film layer (SiO_2); 5-liquid crystal; 6-insulating ring; 8-multilayer medium reflective mirror; 9-light isolating layer; 10-photoconducting layer (CdS); 13-power source

The main function of the liquid optical valve under discussion is the realization of the non-coherent/coherent transformation of images, the operation process of which consists in making the to-be-transformed non-coherent image imaged onto the photoconductive layer from the right side of the device via an optical system (as the write in light I_W), with a beam of linearly polarized coherent light (as the read out light) injected toward the liquid crystal layer from the left side of the device, its polarizing direction consistent with the direction of the molecular long axis at the left end of the liquid crystal layer. Because of the action of the highly reflective film, this beam of light will twice pass through the liquid crystal layer to finally go out from the left and, by means of an analyzer, the direction of whose polarizing axis is normal to the direction of the I_R polarization to obtain the output light I_O.

When the power source applies the voltage on the series entity of the liquid crystal layer, the highly reflective film, the light isolating layer, and the photo conductive layer via the two transparent electrodes, as both the light isolating layer and the highly reflective film are very thin and the alternating current impedance is rather low, the voltage mainly falls on the liquid crystal layer and the photoconductive layer. Obviously, the proportion of voltage allocation on the two layers depends on the illumination of the photoconductive layer. For the dark portion in the incident light image, as the photoconductive layer is not illuminated and the electric conductivity is very low (i.e., the resistance is very great), the voltage is mainly allocated on the photoconductive layer while the voltage obtained by the liquid crystal is scanty, which is insufficient to generate appreciable electro-optic effect. Therefore, in the corresponding position, the liquid crystal is still in the original state (i.e., with a structure of arrangement of a 45° twist). After the read out light passes through this position, its output light intensity I_o is zero. For the pixel position of maximum illuminance in the incident light image, because of the internal opto-electric effect, the impedance of the photoconductive layer drastically diminishes, with the greater part of the voltage falling on the corresponding position of the liquid crystal layer, thus generating an apparent electro-optic effect. When the read out light passes through this position, its output light I_o is maximum. Then, for the pixel position of other densities of illuminance in the incident light image, the corresponding I_o values are between zero and maximum. Thus, the spatial distribution of the output light intensities is modulated by the spatial distribution written into the optical image, realizing the non-coherence/coherence transformation of images.

3. Other types of spatial light modulators

(1) The acousto-optic spatial light modulator

The acousto-optic spatial light modulator is a device for performing optical modulation using the acousto-optic effect. The operation of the acousto-optic modulator (whose structure is described in section 1.3 of this chapter) begins by transforming the electric write in signal into an ultrasonic wave carrying write in information via an electro-acoustic transducer. This acoustic wave acts on an acousto-optic medium to generate the distribution of an internal stress field which, in turn, is converted into the distribution of variation of the medium's refractive index via the opto-elastic effect to constitute a "phase grating". When passing through it, the read out light is modulated under the action of this "grating". It is known from the previously discussed principle of the interaction between acoustic and light that the intensity of its diffracted light can be controlled by the power of the ultrasonic wave or, otherwise, by the electric driving power of the electro-acoustic transducer. Therefore, by changing the ultrasonic power, it is possible to attain light intensity modulation. If the frequency modulating function of the acousto-optic device is utilized, it is also possible to realize the phase modulation for the read out light. This is because the rate of variation of the light wave phase with time is proportional to the angular frequency ω. Hence for light waves of different frequencies, after their propagation for identical periods of time, the amount of change in the phase is different.

However, compared with the previously described spatial light modulator, the acousto-optic spatial light modulator is different in two aspects; one is the spatial distribution of the write in information is not fixed but is slowly moving with the velocity of sound; second, the write in information is distributed only along the one-dimensional space (parallel to the propagation direction of the sound wave). So the acousto-optic modulator is most suitable for optically parallel processing of one-dimensional images (or information).

For instance, the acousto-optic modulator can be used to perform real-time frequency spectrum analysis of the broad frequency band radio frequency signal.

In radio astronomy, the composition of a celestial body can be learned by an analysis of the radio frequency electromagnetic wave radiated by it. A device for implementing frequency spectrum using the acousto-optic modulator is as shown in Fig. 1.6-6. The device is made up of two parts. The first part is the input circuit, including the receiving antenna A, the local oscillator (LO), the mixer M, and the power amplifier AMP. Following the mixing of the radio frequency signal RF with the local oscillation signal, the frequency falls from the radio frequency region to the ultrasonic region, the amplified ultrasonic signal being used to drive the electro-acoustic transducer T. The second part of the device is an integrated optical device D, which carries out Ti diffusion on the surface of the $LiNbO_3$ substrate to form a waveguide and the transducer T. The acoustic wave excited by the transducer in this waveguide forms the acousto-optic grating G. In the waveguide layer, the two lenses L_1 and L_2 are also fabricated. On one side of the waveguide is a laser diode LD and on the other there's an opto-electric detector DA.

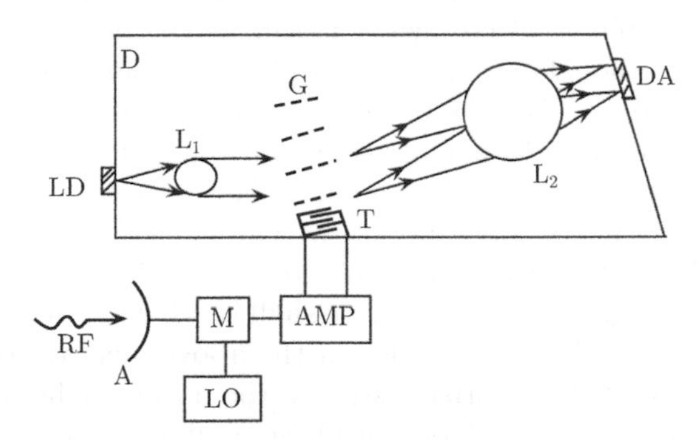

Fig. 1.6-6 The acousto-optic frequency spectrum analyzing device

The operation principle of the said frequency spectrum analyzing device is: the light beam emitted from LD, following collimation by L_1, serves as the incident light (read out light). This light beam experiences Bragg diffraction on the grating containing multiple frequency components while generating diffracted light beams in multiple directions at the same time. The intensity of all diffracted light beams depends on the average power of all the frequency components. After the detector array DA has detected the positions and intensities of the beams of diffracted light, it will be possible to find the frequency components and their relative intensities of the detected signals, completing the frequency spectrum analysis of the signals.

(2) The magneto-optic spatial light modulator

The magneto-optic spatial light modulator realizes modulation of the read out light using the magneto-optic effect after the write in information is recorded by inducing and magnetizing the ferromagnetic material.

1) The recording of the write in information. We know that some magnetic materials are magnetized when induced by an external magnetic field. When the external magnetic field is removed, the magnetic induction intensity is not restored as zero, but still has a "residual magnetic intensity". Now, even if there is an external magnetic field in the opposite direction, as long as its intensity does not exceed the critical value, the direction of the above-mentioned residual magnetic intensity will still not change. Only after the size of the external magnetic field in the opposite direction exceeds the critical value, will the direction of the residual magnetic intensity change with it. Therefore, the direction of the magnetic material's stable residual magnetic intensity can be utilized to "memorize" the direction of the original external magnetic field. If it is required to vary, then it is necessary to apply a sufficiently large magnetic field in the opposite direction. As there are two stable residual magnetic directions, the information recorded is binary. If a magnetic material is made into a thin film in shape, and divided into a large quantity of image elements independent of each other (etched into rectangular image elements arrays), by fabricating orthogonal addressing electrodes between one image element and the other, a two-dimensional data array represented with binary digits can be recorded.

The method of specific data recording is what is known as the matrix addressing, by which an electric current is applied on the electrode to generate a rather strong local magnetic field in the opposite direction at a certain unit that needs to change its residual magnetic direction to the effect that the designated image element will have its residual magnetic direction reversed. When the electric current flows through two electrodes in the orthogonal direction, the image element at the junction of the electrodes will be addressed (which one of the four image elements around the junction is addressed after all depending on the design of the magneto-optic thin film and the direction of the current in the electrodes). The state of magnetization of the thin film varies with the addressing magnetic field. Thus, by means of line by line write in, the binary electric write in signals can be transformed into an array of information as a 2-dimensional one characterized by the residual magnetic direction.

2) The information read out. For the magneto-optic modulator, the modulation of the read out light is realized through the magneto-optic effect; that is, when a beam of linearly polarized light passes the magneto optic medium, if there exists a magnetic field along the direction of light propagation, then, because of the Faraday effect, the polarizing direction of the incident light will rotate with the propagation of the light, the direction of rotation depending on that of the magnetic field. Thus, we shall be able to transform the information on the residual magnetic distribution recorded in the above-mentioned magnetic thin film into a different distribution of the polarizing state of the output light. If an analyzer is also used, then the binary amplitude modulation or phase modulation can be accomplished.

The specific modulating process can be illustrated with Fig.1.6-7. If the two image elements "1" and "2" have already been modulated into residual magnetic intensity of the opposite direction (denoted by direction of an arrow in the figure, in which "1" represents that thin film magnetization has the same direction with the light beam, "2" the converse), because of the Faraday effect, after the linearly polarized light polarizing along the y-axis direction has passed through the two units, the polarizing direction will rotate a θ and $-\theta$ angle, respectively, to yield two streams of outgoing light P_1 and P_2, (one rotating a θ angle clockwise, one rotating a θ angle counterclockwise). Then, have an analyzer A arranged in the rear of the device, whose light transmittance direction forms an angle φ with the y-axis. After P_1 passes through A, the light intensity is proportional to $\cos^2(\varphi - \theta)$ while after P_2 passes through A, the light intensity is proportional to $\cos^2(\varphi + \theta)$, realizing binary amplitude modulation. If the φ angle is appropriately chosen so that $\varphi - \theta = \pm 90°$, complete contrast output can be obtained; that is, one image element is in the "closed state" with no light passing through it while the other image element can pass through in part or in whole, or in the "open state". The magneto-optic spatial light modulator can reach a very high frame rate and has a stable characteristic of storage as well as a very high contrast ratio. It can be made into large array devices (e.g., those of 512 × 512 image elements). This kind of spatial light modulator has found wide application in optical pattern recognition, optical information processing, image coding, optical interconnection, etc.

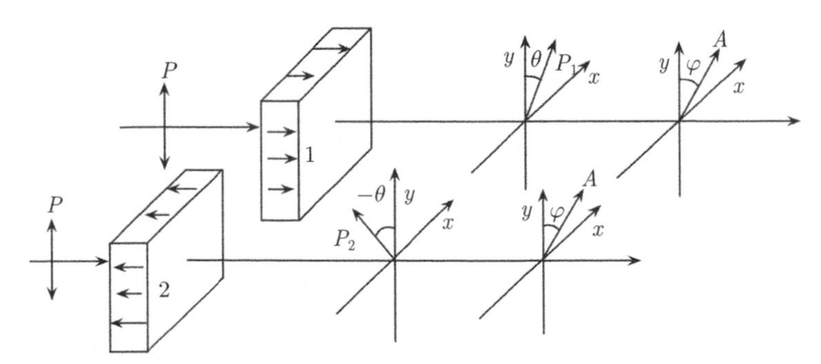

Fig. 1.6-7 The information readout of the magneto-optic modulator

What is described above is spatial light modulators based on the electro-optic, acousto-optic, and magneto-optic effect. Apart from these, there have emerged in recent years many other kinds of spatial light modulators such as ferroelectric ceramic (PLZI) modulators, microchannel plate (MSLM) modulators, multiquantum well modulators, etc., which will not be dealt with here.

Exercises and questions for consideration

1. A longitudinally applied KDP electro-optic modulator length is 2 cm and refractive index $n = 1.5$. If the operation frequency is 1 GHz (1000 kHz), try to find the transition time of light at this moment and the decay factor induced.

2. In order to obtain linear modulation from an electro-optic modulator, insert a 1/4 wave plate in the modulator. How would one best place its axial layer? If the 1/4 wave plate is rotated, what is the change in the dc bias it provides?

3. In order to reduce the half-wave voltage of the electro-optic modulator, adopt four pieces of z-cut KD*P crystal to connect (the optical path in series, the electric path in parallel) as a longitudinal series structure. (1) In order to make the electro-optic effect of the four pieces of crystal superimposed piece by piece, how should the x-, y-axes of the pieces of crystal be orientated? (2) If $\lambda = 0.628$ μm, $n_O = 1.51$, $\gamma_{63} = 23.6 \times 10^{-12}$ m/V, calculate its half-wave voltage and compare with the single crystals.

4. Try to design an experimental device, and explain how to examine the polarized state of the incident light (the linearly polarized light, the elliptically polarized light, and spontaneous light), and point out the phenomenon based on which your conclusion is made. If a longitudinal electro-optic modulator does not have a polarizer, can the incident spontaneous light be modulated? Why?

5. Suppose a lead molybdate (PbM_0O_4) acousto-optic modulator is used to modulate the He-Ne laser. It's known that the acoustic power $P_s = 1$ W, the acousto-optic interaction length $L = 1.8$ mm, the transducer width $H = 0.8$ mm, $M_2 = 36.3 \times 10^{-15}$ s^3/kg. Find the Bragg diffraction efficiency of the lead molybdate acousto-optic modulator.

6. What reaction will a standing wave ultrasonic field make to the Bragg diffracted light? Give the frequency shift it brings about and the direction of diffraction.

7. Make an acousto-optic deflector using a lead molybdate crystal. Take $n = 2.48$, $M_2 = 25$ (relative to fused quartz $M_{2\text{quartz}} = 1.51 \times 10^{-15}$ s^3/kg). The transducer length $L = 1$ cm, the width $H = 0.5$ cm. The acoustic wave propagates along the direction of the optical axis, the acoustic frequency $f_{s0} = 150$ MHz, $v_s = 3.66 \times 10^5$ cm/s, the light beam width $w = 0.85$ cm, $\lambda = 0.5$ μm.

(1) Prove the said deflector can only generate normal Bragg diffraction.

(2) To obtain 100% diffraction efficiency, find the sound power P_s.

(3) If we take the Bragg bandwidth $\Delta f_s = 125$ MHz, how much is the diffraction efficiency reduced?

(4) Find the resolvable number of points N.

8. A beam of linearly polarized light passes through a solid glass rod of $L = 25$ cm, diameter $D = 1$ cm, its exterior wound by $N = 250$ coils of conductor, energized with current $I = 5$ A. Take Verdet constant $V = 0.5'/\text{G} \cdot \text{cm}$. Try to calculate the rotation angle θ of light.

9. Give an account of the difference and association between the waveguide modulator and volume modulator using electro-optic modulation as an example.

References

[1] A. Yariv, An Introduction to Optical Electronics (2nd ed.), New York, 1976.

[2] Li Yinyuan,Yang Shunhua, Nonlinear Optics, Beijing, Science Press, 1974.

[3] A. Yariv and Yeh P. Optical Waves in Crystals. John Wiley Sons. Inc., 1984.

[4] A. Waksberg. Electro-Optic Modulators for Laser, *Opt. and Laser Techn.*, 1976.

[5] F. T. Arechi and E.O. Schulz-Dubois. Laser Handbook (Volume 1). North-Holland Publishing Company, 1972.

[6] B. A. Auld. Acoustic Fields and Waves in Solids. John Wiley, 1973.

[7] Xu Jieping, The Principles, Design and Application of Acousto-optic Deices, Science Press, Beijing, 1982 (in Chinese).

[8] Dong Xiaoy, Light Wave Electronics, Nankai University Press, Tianjin, 1985.

[9] Fang Junxin, Sheng Yuqin, The Physical Basis of Integrated Optics (lecture notes), Shanghai Jiaotong University, 1982.

[10] Ogawa Aiya, Applied Crystal Physics, translated from Japanese by Cui Chengjia, Science Press, Beijing, 1985.

[11] Sasakiakiwo et al., Fundamentals and Application of Liquid Electronics translated from Japanese by Zhao Jingan et al., Science Press, Beijing , 1985.

[12] Fundamentals of Information Optics (in Chinese), Teaching & Research Group of Optical Instrumentation, Tsinghua University, Engng Industry Press, Beijing, 1985.

[13] A. K. Ghatak and K. Thyagarajan. Optical Electroncs. Cambridge University Press. New York, 1989.

[14] D. Armitage et al. High-Speed Spatial Light Modulator. *IEEE Trans.*, 1985, QE-21. 1241-1247.

[15] D. Casaseut. Spatial Light Modulators. *Proc. IEEE*, 1977, 65(1), 143-157.

[16] Li Zemin, Fiber Optics (Principles and Technology) (in Chinese), Science and Technological Literature Press, Beijing, 1992.

[17] Zhao Dazun, Zhang Huaiyu, Spatial Light Modulators (in Chinese), Beijing Polytechnical University Press. Beijing, 1992.

[18] I. C. Chang. Acoustooptic Devices and Applications. *IEEE Trans.*, 1976, 2-22.

[19] Holler F. etal. A Spatial Light Modulator Using BSO Crystals. *Opt. Commum.*, 1986, 58(1), 20-24.

[20] W. E. Ross et al. ZD Magneto-optic Spatial Light Modulator for Signal Processing. *Proc. SPIE*. 1982, 341, 191-198.

[21] I. C. Chang. Sonics and Ultrasonics. *IEEE Trans. SU*-23, 1976.

CHAPTER 2

The Q Modulating (Q-switching) Technology

2.1 Overview

Laser is a kind of electromagnetic wave in the optical frequency range whose frequency is far higher than that of the ordinary electromagnetic wave. This enables the optical wave to be used in fields in which it could never be applied before, such as high-speed broad frequency band communication and cases of interaction between light and material requiring that the single photon energy be sufficiently great.

According to the uncertainty principle of quantum mechanics and the wave packet theory of electrodynamics, the pulse width of the electromagnetic wave is inversely proportional to the width of its spectral line:

$$\Delta\tau \propto \frac{1}{\Delta\nu}$$

from which it can be seen that the greater $\Delta\nu$ is, the narrower the pulse width likely to be obtained will be. In particular, as the spectrum width of the solid-state laser operating material is rather broad, a very narrow pulse width can be obtained. This kind of short pulse technique has already become one of the most important component parts of the laser technology and has formed a unique disciplinary field with the Q-switching technology, mode-locking technology, and chirp technology as its representative.

Q-switching is a kind of technology for compressing the generally output continuous or pulsed laser energy into pulses of an extremely narrow width for emission, making it possible for the peak power of the light source to increase by several orders of magnitude. It is now no longer difficult to obtain laser pulses of a peak power above the megawatt level (10^6 W) and pulse width of the nanosecond level (1 ns $= 10^{-9}$ s). The emergence of the Q-switching technology has tremendously promoted the application of the laser technology in two aspects. On the one hand, the interaction between the very powerful coherent light of the Q-switching laser pulses and material will generate a series of new phenomena and new technologies of great significance that will directly push forward the development of nonlinear optics and, on the other, the very short pulse width of the Q-switching laser pulse can promote the advance of such applied technologies as pulse laser ranging, laser radar, and high-speed holography. It can be said that the Q-switching laser has become an irreplaceable important technology in many application fields. The emergence and development of the Q-modulating technology is a breakthrough of importance in the history of the development of laser.

2.1.1 The output characteristics of the pulsed solid-state laser

When making an observation and recording of the pulse output by a common pulsed solid-state laser with an oscilloscope (OSC), we have discovered that its waveform is not a smooth optical pulse but is composed of many peak pulses whose amplitude, pulse width, and time interval vary randomly, as shown in Fig. 2.1-1(a). The width of each peak is $0.1 \sim 1$ μm, the interval several microseconds, and the length of the pulse sequence approximately equal to the time duration of the flashlight pump. What is shown in Fig. 2.1-1(b) is the peak of the ruby laser observed. This phenomenon is called the relaxation oscillation of lasers.

The principal reason for the generation of relaxed oscillation is when the operating material of the laser is being pumped, the upper energy level inverted population exceeds the threshold condition, and laser oscillation will be generated. The density of photon number in the cavity is increased and laser is emitted. With the emission of laser, large quantities of the upper energy level population are consumed, leading to a decrease of the inverted population. If the decrease is lower than the threshold, laser oscillation will stop. Now, owing to the continued pumping operation of the optical pump, the upper energy level inverted population re-accumulates. When it exceeds the threshold, the second pulse will be generated. The above-mentioned process will be repeated over and over again until the stop of the pump. It's clear that every spike pulse is generated in the vicinity of the threshold. So the peak power level of the pulse is rather low. Meanwhile, it can be seen from this process of action that increasing the energy of the pump cannot help enhance the peak power other than increase the number of small spikes.

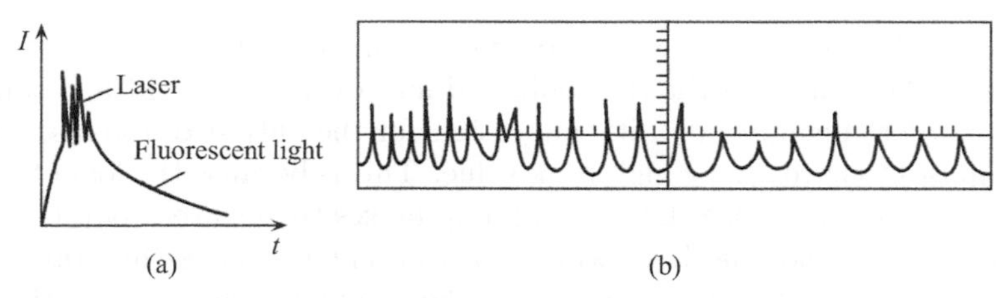

Fig. 2.1-1 The peak structure of pulsed laser output

The physical process of the generation of relaxation oscillation can be depicted with Fig. 2.1-2, which shows the variation of the inverted population Δn and the photon number ϕ in the cavity. Each spike can be divided into four stages (Before moment t_1, because of the action of the pump, the inverted population Δn increases, but has not reached the threshold Δn_{th} as yet. Hence it's impossible for laser oscillation to form.):

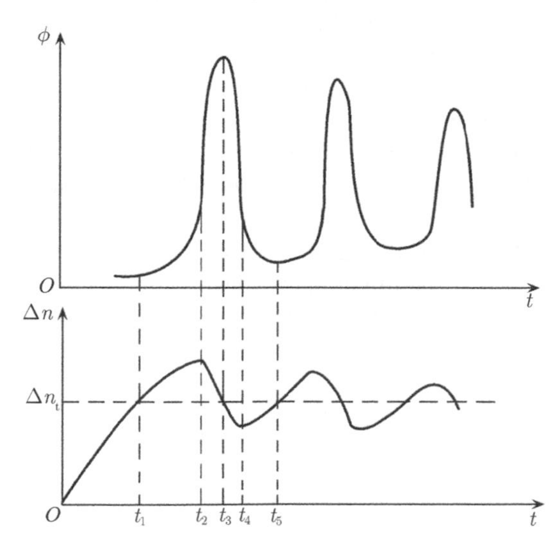

Fig. 2.1-2 Variation of photon number and inverted population in cavity with time

Stage 1: ($t_1 - t_2$): When laser oscillation has just begun, $\Delta n = \Delta n_{\text{t}}$, $\phi = 0$. Because of the action of the optical pump, Δn continues to increase. In the meantime, the density ϕ of photon number in the cavity, too, begins to increase. The rate at which Δn decreases because of the increase of ϕ is lower than that at which the pump makes Δn increase. So Δn has been increasing until it reaches maximum.

Stage 2 ($t_2 - t_3$): Δn begins to decrease after reaching maximum, but is still greater than Δn_{t}. Therefore ϕ continues to increase and very quickly at that until it reaches maximum.

Stage 3 ($t_3 - t_4$): $\Delta n < \Delta n_t$, the gain is smaller than the loss, and the density ϕ of the photon number decreases and drops abruptly.

Stage 4 ($t_4 - t_5$): The photon number decreases to a definite extent and the pump again plays the leading role. So Δn begins to rise again and by the moment t_5, Δn again reaches the threshold Δn_t and begins to generate the second spike pulse. As the duration of the pumping process of the pump is much greater than the width of every spike pulse, the above-mentioned process is repeated over and over again, generating a train of spike pulses. The greater the power of the pump, the quicker the formation of the spike pulse and hence the shorter the interval between spikes.

2.1.2 The basic principles of Q modulation

As the output of the pulsed laser is composed of a number of irregular spike pulses, each of which occurs in the vicinity of the threshold and has a very short width (being merely of the microsecond order of magnitude), it's impossible for the output of the laser scattered in such a train of pulses to have a very high peak value. This is because the loss of the ordinary laser resonator is invariable. Once the optical pump makes the inverted population reach or slightly exceed the threshold, the laser will begin to oscillate. Thus the population of the laser's upper energy level will decrease because of the stimulated radiation so that the upper energy level is unable to amass a very large inverted population but is only confined to near the inverted threshold population. That is the reason why the peak power of the ordinary laser cannot be raised.

Since the laser upper level maximum inverted population is restricted by the threshold of the laser, then, to enable the upper energy level to accumulate large quantities of particles, effort should be made to change the threshold of the laser. Specifically, when the laser is about to get pumped at the initial stage, measures should be taken to set the oscillation threshold of the laser very high to restrain laser oscillation from being generated. In so doing, the inverted population of the laser upper energy level will be able to accumulate a great deal. When the inverted population is accumulated to maximum, set the threshold to very low all of a sudden. Now, the large quantities of particles accumulated on the upper level will jump to the lower energy level as if in an avalanche. So, energy is released in an extremely short period of time, yielding the giant pulse laser output with an extremely high peak power.

It can be seen from the above that changing the threshold of a laser is an effective method of enhancing the accumulation of the upper energy level population. Then, what parameters should be changed to change the threshold? It is known from "the principle of laser" that the threshold condition for oscillation of a laser can be expressed by

$$\Delta n_{\text{th}} \geqslant \frac{g}{A_{21}} \cdot \frac{1}{\tau_{\text{c}}} \tag{2.1-1}$$

while

$$\tau_{\text{c}} = \frac{Q}{2\pi\nu}$$

So

$$\Delta n_{\text{th}} \geqslant \frac{g}{A_{21}} \cdot \frac{2\pi\nu}{Q} \tag{2.1-2}$$

where g is the number of modes, A_{21} the probability of spontaneous radiation, and τ_{c} the life of photons in the cavity. The Q value is called the quality factor, which is defined as

$$Q = 2\pi\nu_0 \left(\frac{\text{energy stored in cavity}}{\text{energy lost per second}} \right)$$

where ν_0 is the central frequency of laser, W is used to represent the energy stored in the cavity, and δ the energy loss rate for single-path propagation of light in the cavity. Thus, the energy loss of light in a single path is δW. Suppose L is the length of the resonator, n the refractive index of the medium, and c the light velocity. Then the time needed for light to travel a single path will be nL/c. Hence the energy lost per second by light in the cavity is $\dfrac{\delta W}{nL/c}$. So the Q value can be expressed by

$$Q = 2\pi\nu_0 \frac{W}{\delta W c/nL} = \frac{2\pi nL}{\delta\lambda_0} \tag{2.1-3}$$

where λ_0 is the laser's central wave length in the vacuum. It can be seen from Eqs. (2.1-2) and (2.1-3) that when λ and L are definite, the Q value will be inversely proportional to the resonator loss; that is, if the loss is great, the Q value will be low and the threshold high, making it difficult for oscillation to be initiated whereas if the loss is small, the Q value will be high and the threshold low, and oscillation will be easily initiated. So it's clear that, if the threshold of a laser is to be changed, the change can be effected by suddenly changing the Q value (or the loss δ) of the resonator, which is both effective and simple.

The Q-modulating technology is simply a technology that uses a certain method to make the Q value of the resonator vary with time in accordance with a definite procedure. When starting the pump, the resonator should be made to be in the low Q value state, that is, the oscillation threshold should be raised to prevent oscillation from forming so that the upper energy level inverted population will be accumulated in large quantities. The time in which energy can be stored depends on the lifetime of the laser upper energy level. When accumulation reaches maximum (saturated value), decrease the cavity loss all of a sudden. Then the Q value abruptly increases and laser oscillation is expeditiously set up. Within an extremely brief period of time, the inverted population of the upper energy level is consumed, turned into optical energy in the cavity, and released from the cavity output end in the form of single pulses. Thus giant pulses with very high peak power are obtained.

The process of building of the Q laser pulse and the variation of the parameters with time are shown in Fig. 2.1-3, in which (a) represents the variation of the pump rate W_{p} with time, (b) shows that the cavity's Q value is the step function of time, (c) shows the variation of the inverted population Δn, and (d) the variation of the photon number ϕ in the cavity with time.

During the greater part of the pumping process, the resonator is in the low Q value (Q_0) state. So the threshold is very high and oscillation just cannot be initiated, leading to the steady accumulation of the population of the laser upper energy level until the t_0 moment when the inverted population reaches maximum Δn_{i}. At this moment Q value rises abruptly (loss decreases), with the oscillation threshold value decreasing with it. Thus laser oscillation begins to be set up. As $\Delta n_{\mathrm{i}} \gg \Delta n_{\mathrm{t}}$ (threshold inverted population), the stimulated radiation is strengthened very rapidly. The energy stored by the laser medium is converted in an extremely short period of time into the energy of the stimulated radiation field, resulting in the generation of a narrow pulse with a very high peak power.

It can also be seen from Fig. 2.1-3 that there is a process for the building of the Q-modulating pulse. Oscillation begins when the Q value step rises. During a rather long time after oscillation is set up at $t = t_0$, the photon number ϕ increases very slowly as shown in Fig. 2.1-4. Its value is very small from beginning to end ($\phi \cong \phi_1$), as the probability of stimulated radiation is small. Now it is still the spontaneous radiation that is dominant, and it is only when oscillation is sustained to $t = t_{\mathrm{D}}$ and ϕ increased to ϕ_{D} that the avalanche process

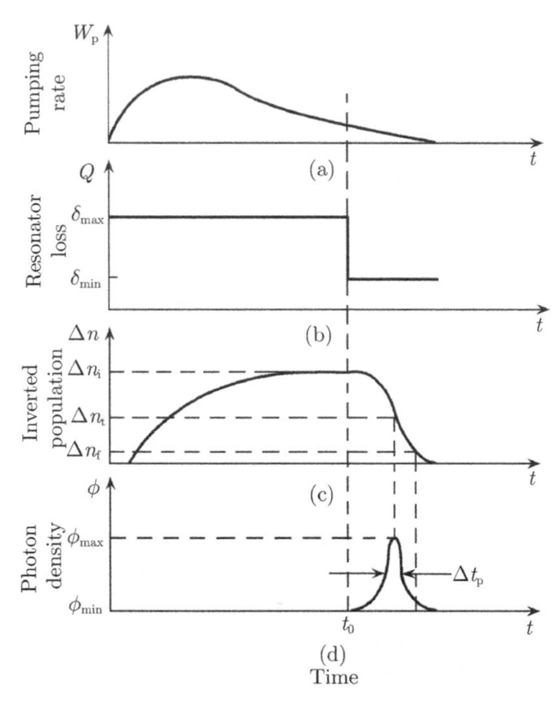

Fig. 2.1-3 The process of buildup of the Q-switching laser pulses

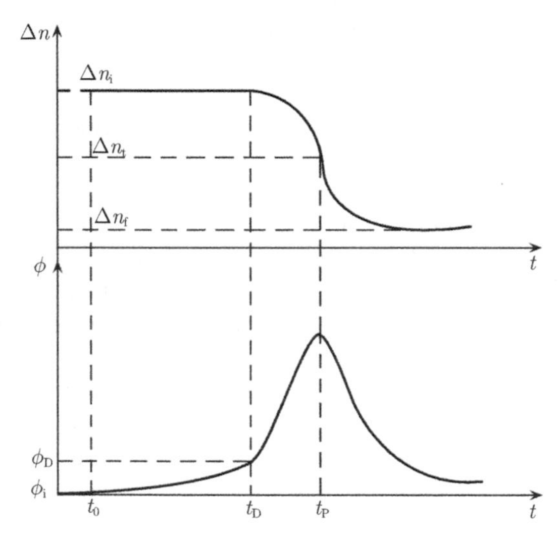

Fig. 2.1-4 The process from start of oscillation to formation of pulses

is formed, ϕ quickly increases, and stimulated radiation quickly surpasses spontaneous radiation to become dominant. Therefore, a definite retardation time Δt (i.e., the duration of the Q switch remaining on) is needed by the Q modulating pulse from the moment oscillation is set up to the formation of the giant pulsed laser. The expeditious growth of the photon number makes Δn quickly decrease. By the moment $t = t_p, \Delta n = \Delta n_t$, and the photon number has reached maximum ϕ_m. As $\Delta n < \Delta n_t$, ϕ quickly decreases. Now $\Delta n = \Delta n_t$ is the remaining population in the operation material after the termination of oscillation. It is clear that the peak of the Q pulse occurs at the moment when the inverted population is equal to the threshold inverted population ($\Delta n = \Delta n_t$). As the Q value is inversely proportional to δ, by changing the δ value of the resonator, we shall be able to make the Q value undergo the corresponding variation. The loss of the resonator in general includes the reflection loss, diffraction loss, absorption loss, etc. So, by controlling different types of the variation of loss, different types of Q modulating technologies can be developed. For instance, by controlling the variation of the reflection loss, there is Q modulation by mechanical rotating mirror, the electro-optic Q modulation technology, the acousto-optic Q modulation technology for controlling the variation of the diffraction loss, the dye Q modulation technology for controlling the variation of the absorption loss, etc. Of these, although the mechanical rotating mirror Q modulating technology was developed the earliest, it is at present seldom used. So we shall mainly discuss the latter three Q modulating technologies and their implementation while the mechanical rotating mirror Q modulation, as a type of typical Q modulating methods, will only be briefly dealt with.

2.1.3 Basic requirements on a laser for carrying out Q modulation

(1) As Q modulation consists in storing energy in the form of active particles in the high energy state of the operation material to be concentrated within a very short period of time for release, the operation material is required to work under powerful pumping, i.e., it should have a high anti-damage threshold. Secondly, it is also required to have a fairly long lifetime. If the upper energy level of the laser operation material is τ_2, then the velocity at which the inverted population n_2 on the upper energy level decreases owing to spontaneous radiation is $\dfrac{n_2}{\tau_2}$. Thus, when the pumping rate is W_p, under the condition balance is attained, satisfy

$$\frac{n_2}{\tau_2} = W_\mathrm{p}$$

and the upper energy level will reach the maximum inverted population

$$n_2 = W_\mathrm{p}\tau_2$$

Therefore, in order to make the upper energy level of the laser operation material accumulate into a population as large as possible, the $W_\mathrm{p}\tau_2$ value is required to be rather large, but τ_2 should not be too large, or it may affect the velocity of energy release. According to the above-mentioned requirements, the requirements on the operation materials for all solid-state lasers can be satisfied, as can those of liquid lasers. However, for some gaseous lasers, such as He-Ne lasers, as they can only operate under the low-ionized condition, the pumping rate must not be too high, or there will be no way to implement Q modulating technology.

(2) The pumping rate must be faster than the spontaneous radiation rate of the laser upper energy level. That is, the duration of pumping (the half-width of the waveform) must be shorter than the lifetime of the laser medium's upper energy level, or it will be impossible to realize sufficiently frequent population inversion.

(3) The change of the resonator's Q value should be quick, which should in general be comparable to the time for the resonator to set up oscillation. If the Q-switching time is too slow, the pulse will become broader and there may even be the phenomenon of multiple pulses.

2.2 The basic theory of Q modulating lasers

For the process of the formation of Q modulating pulses and the influence of various parameters on the laser pulse, the rate equation can be adopted for an analysis. The set of rate equations is a set of equations for depicting the laws of variation of the photon number in the cavity and the inverted population of the operation material with time. Based on these laws, it is possible to infer the relationships between the peak power of the Q-modulating pulse, the pulse width, etc. and the population inversion.

2.2.1 The rate equation for Q modulation

The laser forming rate equation is established based on the inherent relationship between the variation of the population of the operation material and that of the photon number in the cavity. In laser physics, the rate equations of the three energy level system and four energy level system of ordinary lasers are already given, from which we can directly write the equations for the variation of the inverted population and that of the photon number in the cavity with time:

$$\left.\begin{aligned}\frac{\mathrm{d}\Delta n}{\mathrm{d}t} &= 2n_1W_{13} - \Delta n\frac{A}{g}\phi - 2n_2A \\ \frac{\mathrm{d}\phi}{\mathrm{d}t} &= \Delta n\frac{A}{g}\phi - \delta\phi\end{aligned}\right\} \text{three-level system} \tag{2.2-1}$$

$$\left.\begin{aligned}\frac{\mathrm{d}\Delta n}{\mathrm{d}t} &= n_1W_{14} - \Delta n\frac{A}{g}\phi - \Delta nA \\ \frac{\mathrm{d}\phi}{\mathrm{d}t} &= \Delta n\frac{A}{g}\phi - \delta\phi\end{aligned}\right\} \text{four-level system} \tag{2.2-2}$$

where Δn is the density of the inverted population, ϕ is that of the photon number in the cavity, g is the number of the spontaneous radiation wave shape in the cavity, W_{13} and W_{14} are the probabilities of stimulated transition, and A is the probability of spontaneous radiation.

It can be seen from the above-mentioned two sets of rate equations that they are equivalent because the particles that are in the process of transition due to the stimulation from without are mainly concentrated between two energy levels in realizing population inversion. Therefore, in order to facilitate analysis, we use a model for the two energy level system to replace the actual three energy level and four energy level systems.

The rate equation of the Q modulating laser is a special case of the laser oscillator. In the process of sudden change of Q, as the laser is in the transient process of abrupt change, the effects of the two processes of pumping stimulation and spontaneous radiation can be neglected. Meanwhile, for simplicity, it is considered in the following analysis that the Q value is step variable. Thus Eqs. (2.2-1) and (2.2-2) can be reduced to

$$\begin{aligned}\frac{\mathrm{d}\Delta n}{\mathrm{d}t} &= -2\Delta n\frac{A}{g}\phi \\ \frac{\mathrm{d}\phi}{\mathrm{d}t} &= \left(\Delta n\frac{A}{g} - \delta\right)\phi\end{aligned} \tag{2.2-3}$$

In the above equation, let $\frac{\mathrm{d}\phi}{\mathrm{d}t} = 0$ (cavity gain equal to the threshold condition for loss). Then the threshold inverted population Δn_{t} during steady-state oscillation can be found and there is

$$\Delta n_{\mathrm{t}} = \frac{\delta}{A}g$$

Substitution of the above equation into Eq. (2.2-3) gives

$$\begin{aligned}\frac{\mathrm{d}\Delta n}{\mathrm{d}t} &= -2\Delta n\frac{\Delta n}{\Delta n_{\mathrm{t}}}\delta\phi \\ \frac{\mathrm{d}\phi}{\mathrm{d}t} &= \left(\frac{\Delta n}{\Delta n_{\mathrm{t}}} - 1\right)\delta\phi\end{aligned} \tag{2.2-4}$$

The coupled equations (2.2-4) are the rate equations of Q modulating laser oscillation.

2.2.2 Solving the rate equation

For the above-mentioned first-order differential equation set, the numerical method is in general employed to find the parameters of the Q modulating pulse. In order to solve the Q modulating rate equation, it is necessary to give the specific form of the Q-switching function. Generally, to facilitate solution, it is customary to assume in advance several typical Q-switching functions (the step switching function, linear switching function, and parabolic switching function), while the actual Q-switching function is often rather complicated, and even very hard to express in a simple form. Here, we shall focus on the ideal step switching function.

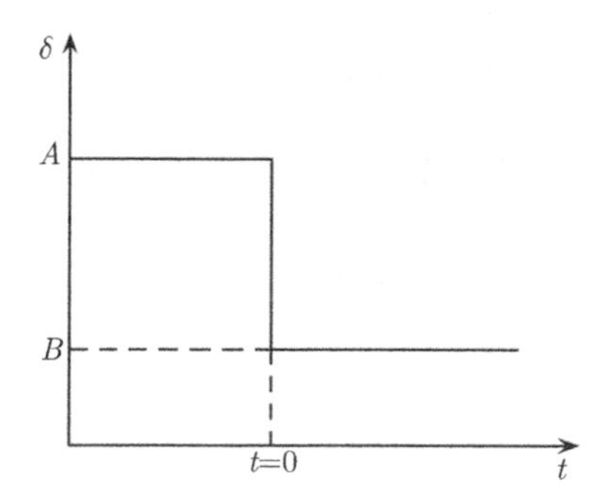

Fig. 2.2-1 The step variation of loss δ

Suppose the loss δ in the cavity experiences a sudden change as the step function shown in Fig. 2.2-1, i.e.,

$$\delta = \begin{cases} A & (t<0) \\ B & (t>0) \end{cases}$$

The process before $t = 0$ is only a preparation for the initial condition of the density Δn_{i} of the initial inverted population. So the process of accumulation of Δn can be forgone while simply considering the process of variation after $t = 0$.

In Eqs. (2.2-4), the upper formula divided by the lower formula and elimination of time t gives

$$\frac{\mathrm{d}\phi}{\mathrm{d}\Delta n} = \frac{1}{2}\left(\frac{\Delta n_{\mathrm{t}}}{\Delta n} - 1\right) \tag{2.2-5}$$

At moment $t = 0$, Δn reaches maximum Δn_{i}, while the stimulated radiation photon number is zero; that is, $\phi = \phi_{\mathrm{i}} = 0$. Later, ϕ begins to increase, with the process of avalanche formed when reaching t_0 (see Fig. 2.1-4). ϕ increases abruptly whereas Δn begins to decrease. This process is sustained until moment t_{p} when $\Delta n = \Delta n_{\mathrm{t}}$ and the photon number in the cavity reaches maximum ϕ_{m}. Integrating over Eq. (2.2-5) and taking into account that the integrating limit to Δn is $\Delta n_{\mathrm{i}} \to \Delta n_{\mathrm{t}}$, we have

$$\int_0^{\phi_{\mathrm{m}}} \mathrm{d}\phi = \frac{1}{2}\int_{\Delta n_{\mathrm{i}}}^{\Delta n_{\mathrm{t}}} \left(\frac{\Delta n_{\mathrm{t}}}{\Delta n} - 1\right)\mathrm{d}\Delta n \tag{2.2-6}$$

After integration there is

$$\phi_{\mathrm{m}} = \frac{1}{2}\left[\Delta n_{\mathrm{i}} - \Delta n_{\mathrm{t}} + \Delta n_{\mathrm{t}}\ln\left(\frac{\Delta n_{\mathrm{t}}}{\Delta n_{\mathrm{i}}}\right)\right] \tag{2.2-7}$$

Using Taylor's series and expanding, we have the approximate expression

$$\phi_{\mathrm{m}} = \frac{\Delta n_{\mathrm{t}}}{4}\left(\frac{\Delta n_{\mathrm{i}}}{\Delta n_{\mathrm{t}}} - 1\right)^2 \tag{2.2-8}$$

It is clear that there exists the quadratic relation between ϕ_m and the parameter $(\Delta n_i/\Delta n_t)$, whose curve of variation is as shown in Fig. 2.2-2. Therefore, enhancing the ratio of the initial inverted population Δn_i to the threshold value inverted population is helpful to the increase of the maximum photon number ϕ_m in the cavity.

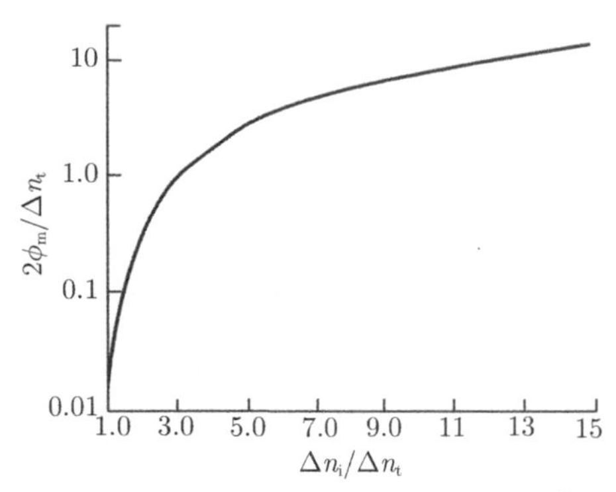

Fig. 2.2-2 The relation between ϕ_m and $\dfrac{\Delta n_i}{\Delta n_t}$

1. The peak power of the Q modulating pulse

It can be approximately considered that these photons escape within the lifetime t_c in the cavity while the energy of each photon is $h\nu$. Then the transient power of laser $P = \phi h\nu/t_c$. Using Eq. (2.2-7), we have

$$P = \frac{h\nu}{2t_c}\left(\Delta n_i - \Delta n + \Delta n_t \ln\frac{\Delta n}{\Delta n_i}\right) \tag{2.2-9}$$

When $\Delta n = \Delta n_t$, the output power reaches maximum, i.e., the peak power is

$$P_m = \frac{h\nu}{2t_c}\left(\Delta n_i - \Delta n_t + \Delta n_t \ln\frac{\Delta n_t}{\Delta n_i}\right) \tag{2.2-10}$$

If the initial inverted population Δn_i is well in excess of the threshold inverted population Δn_t (the case of high Q value), then we have

$$P \cong \frac{1}{2}\frac{\Delta n_i}{t_c}h\nu \tag{2.2-11}$$

2. The energy and energy availability of the Q modulating pulse

The energy of the laser pulse is provided by the process of stimulated radiation that consumes the inverted population. If the time for the photon number to fall from its maximum value ϕ_m to ϕ_f is taken as the ending of the pulse, then the corresponding inverted population of ϕ_f is Δn_f. So, the total energy of the Q modulating pulse can be determined by the following equation:

$$E = \frac{1}{2}(\Delta n_i - \Delta n_f)h\nu V \tag{2.2-12}$$

where V is the volume of the active medium in the cavity and Δn_f is the density of the inverted population when laser oscillation ends and can be found using the integral equation (2.2-6) as:

$$\Delta n_f = \Delta n_i \exp\left[\frac{\Delta n_i}{\Delta n_t}\left(\frac{\Delta n_f}{\Delta n_i} - 1\right)\right] \tag{2.2-13}$$

$$\frac{\Delta n_f}{\Delta n_i} = \exp\left(\frac{\Delta n_f - \Delta n_i}{\Delta n_t}\right) \tag{2.2-14}$$

Usually, $\Delta n_i \gg \Delta n_f$, so it can be seen from Eq. (2.2-12) that the energy of the Q modulating pulse linearly increases with the growth of the parameter $\Delta n_i/\Delta n_t$.

Through the above analysis, we can find the proportion of energy a Q modulating pulse can extract from the energy stored in an active medium. No contribution is made by Δn_f. These remaining inverted particles dissipate in the form of fluorescent light after the giant pulses come to an end. Therefore, $(\Delta n_i - \Delta n_f)/\Delta n_i$ is used to represent the energy the Q

modulating pulse can extract from the medium, which is spoken of as the energy availability of the single pulses and denoted by η. In what follows an analysis will be made of the relation between η and $\Delta n_{\rm i}/\Delta n_{\rm t}$ and that between $\Delta n_{\rm f}/\Delta n_{\rm i}$ and $\Delta n_{\rm i}/\Delta n_{\rm t}$. Figure 2.2-3 shows the relation of η with $\Delta n_{\rm i}/\Delta n_{\rm f}$ and $\Delta n_{\rm i}/\Delta n_{\rm t}$. It can be seen from the figure that η grows with the increase of $\Delta n_i/\Delta n_{\rm t}$, showing that the energy availability is high while $\Delta n_{\rm f}/\Delta n_{\rm i}$ decreases. When $\Delta n_{\rm i}/\Delta n_{\rm t} > 3$, about over 90% of the energy will be extracted by the pulse while when $\Delta n_{\rm i}/\Delta n_{\rm t} = 1.5$, the energy availability is only 60% with $\Delta n_{\rm i}/\Delta n_{\rm t}$ increased. Therefore, for the Q modulating laser, the step variation of the Q-switching function should be adequate so that $\Delta n_{\rm i}/\Delta n_{\rm t} > 3$ and above since only then will a fairly high operation efficiency be ensured.

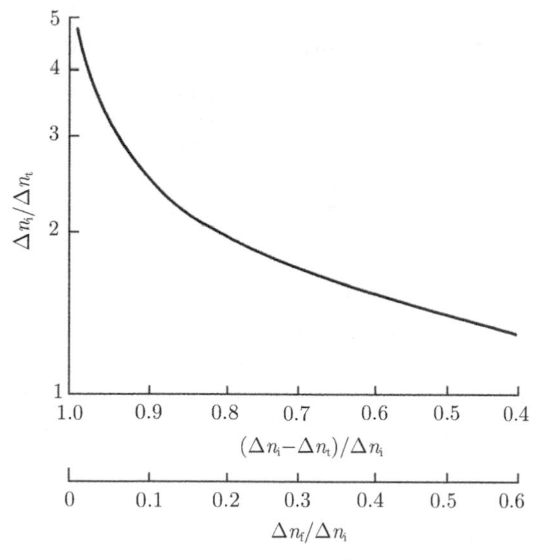

Fig. 2.2-3 The relation of η with $\dfrac{\Delta n_{\rm f}}{\Delta n_{\rm i}}$ and $\dfrac{\Delta n_{\rm i}}{\Delta n_{\rm t}}$

3. The time characteristics of the Q modulating pulse

Let's have a discussion on the pulse width and waveform of the Q modulating pulse below. From Eqs. (2.2-4) we find

$$dt = -\frac{\Delta n_{\rm t}}{2\Delta n \delta \phi} \mathrm{d}\Delta n \tag{2.2-15}$$

Substituting ϕ into the above equation and performing integration, we have

$$\Delta t = -\int_{\Delta n_{\rm i}}^{\Delta n} \frac{\mathrm{d}\Delta n'}{2\Delta n' \delta \left[\dfrac{\phi_0}{\Delta n_{\rm t}} + \dfrac{1}{2}\left(\dfrac{\Delta n_{\rm i}}{\Delta n_{\rm t}} - \dfrac{\Delta n'}{\Delta n_{\rm t}} + \ln \dfrac{\Delta n'}{\Delta n_{\rm i}}\right)\right]} \tag{2.2-16}$$

If the time Δt under discussion only refers to an interval of the width of the laser pulse, then, in that interval, the initial photon number density ϕ_0 can be neglected and Eq. (2.2-16) can be written as

$$\Delta t\ \delta = -\int_{\Delta n_{\rm i}}^{\Delta n} \frac{\mathrm{d}\Delta n}{\Delta n \left(\dfrac{\Delta n_{\rm i}}{\Delta n_{\rm t}} - \dfrac{\Delta n}{\Delta n_{\rm t}} + \ln \dfrac{\Delta n}{\Delta n_{\rm i}}\right)} \tag{2.2-17}$$

It's not easy to directly find the analytic solution for this integral equation but, based on the given initial value $\frac{\Delta n_\mathrm{i}}{\Delta n_\mathrm{t}}$, the numerical solution for Δt can be found using numerical integration, the results being listed in Tab. 2.2-1, in which Δt_1 is the time needed (time for the pulse to rise) for the photon number to rise from semi-maximum to the peak, Δt_2 the time needed for the photon number to fall from the peak to semi-maximum value (the time for the pulse to fall), while $\Delta t_1 + \Delta t_2$ is the pulse width Δt.

Tab. 2.2-1 The relation of pulse width Δt with $\frac{\Delta n_\mathrm{i}}{\Delta n_\mathrm{t}}$

$\frac{\Delta n_\mathrm{i}}{\Delta n_\mathrm{t}}$	$\frac{2\phi_\mathrm{m}}{\Delta n_\mathrm{t}}$	Δt_1	Δt_2	$\frac{\Delta n_\mathrm{i}}{\Delta n_\mathrm{t}}$	$\frac{2\phi_\mathrm{m}}{\Delta n_\mathrm{t}}$	Δt_1	Δt_2
1.105	0.0052	12.291	12.632	4.055	1.655	0.782	1.263
1.221	0.0214	7.960	8.437	4.482	1.982	0.702	1.186
1.350	0.0499	5.335	5.803	4.953	2.353	0.633	1.120
1.492	0.0918	3.892	4.356	5.474	2.774	0.572	1.064
1.649	0.149	3.016	3.480	6.050	3.250	0.518	1.015
1.822	0.222	2.432	2.896	6.686	3.786	0.471	0.973
2.014	0.314	2.016	2.481	7.389	4.389	0.429	0.936
2.226	0.426	1.704	2.171	8.166	5.066	0.391	0.905
2.460	0.560	1.463	1.931	9.025	5.825	0.357	0.877
2.718	0.718	1.271	1.741	9.974	6.674	0.327	0.854
3.004	0.904	1.114	1.586	11.023	7.623	0.300	0.833
3.320	1.120	0.984	1.459	12.182	8.683	0.275	0.816
3.669	1.396	0.875	1.352				

Figure 2.2-4 gives the calculation results at several different initial values. The longitudinal coordinate in the figure is the normalized photon number density $2\phi_\mathrm{m}/\Delta n_\mathrm{t}$ while the horizontal coordinate represents the time parameter t/t_c with the lifetime of the photons in the cavity (t_c) as the unit. It can be seen from the solution to the above-mentioned rate equation that, in the Q modulating laser, $\frac{\Delta n_\mathrm{i}}{\Delta n_\mathrm{t}}$ is an extremely important parameter, which directly influences the output power and pulse width, or the overall efficiency. When the $\frac{\Delta n_\mathrm{i}}{\Delta n_\mathrm{t}}$ value increases, the peak photon number increases; the pulse rising time (front edge) and falling time (rear edge) are shortened at the same time. The pulse becomes narrow, only the rear edge changes rather slowly. This is because the process of stimulated radiation is basically completed at the peak of the pulse, the phenomenon being simply a result of the free decay of photons in the cavity.

When designing a Q modulated laser, it is thus very important to raise the pump rate of the optical pump to increase Δn_i. At the same time, it is advisable to choose the laser operation material of fairly high efficiency and a suitable resonator structure to reduce Δn_t and other losses.

Example Consider a ruby step Q modulating laser, the rod length is equal to the length of the cavity $L = 10$ cm, the sectional area $S = 0.75$ cm^2, Cr^+concentration $N = 1.58\times 10^{19}$ cm^{-3}, $n = 1.76$, the output mirror reflectivity $R = 0.5$, and internal scattering loss $\delta_\mathrm{i} = 0.01$ cm^{-1}. It is known that the initial inverted population $\Delta n_\mathrm{i} = 0.22N$ and the threshold inverted population $\Delta n_\mathrm{t} = 0.1N$. Find the peak power P_m, pulse energy E, and pulse width Δt of the Q-modulated giant pulse.

Solution As required, we give the following results:

$$\frac{\Delta n_\mathrm{i}}{\Delta n_\mathrm{t}} = \frac{0.22}{0.1} = 2.2$$

$$\Delta n_\mathrm{i} = 0.22N = 0.22 \times 1.58 \times 10^{19}\mathrm{cm}^{-3} = 3.45 \times 10^{18}\mathrm{cm}^{-3}$$

$$\Delta n_\mathrm{t} = 0.1N = 0.1 \times 1.58 \times 10^{19}\mathrm{cm}^{-3} = 1.58 \times 10^{18}\mathrm{cm}^{-3}$$

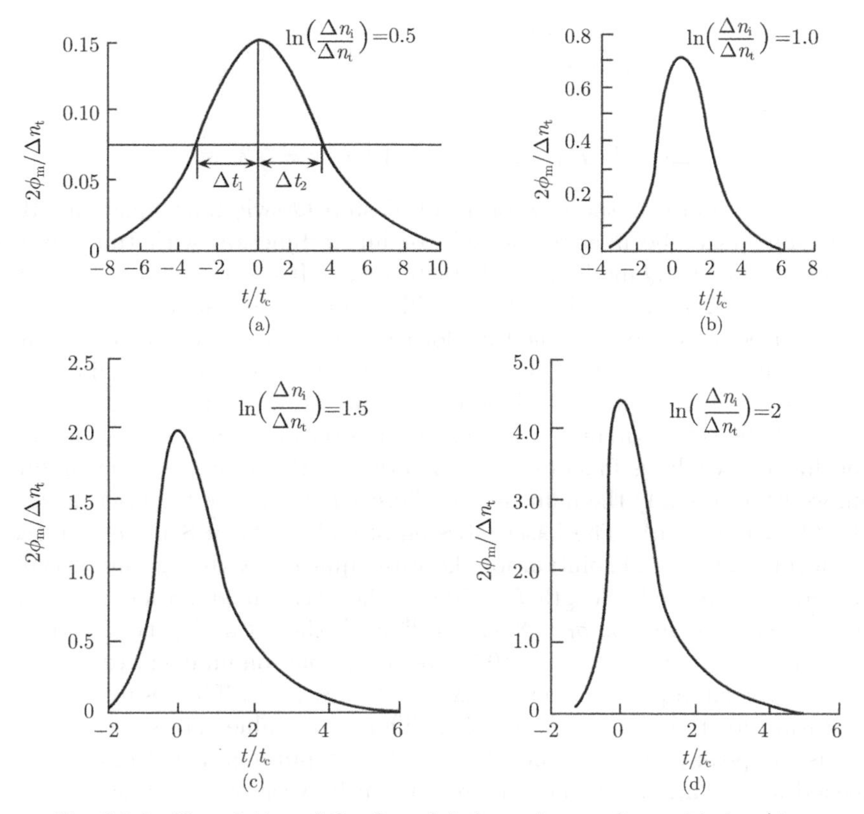

Fig. 2.2-4 The relation of the Q modulating pulse waveform with $\Delta n_{\rm i}/\Delta n_{\rm t}$

from which the peak photon number density can be found:

$$\begin{aligned}\phi_{\rm m} &= \frac{1}{2}\left[\Delta n_{\rm i} - \Delta n_{\rm t} + \Delta n_{\rm t}\ln\left(\frac{\Delta n_{\rm i}}{\Delta n_{\rm t}}\right)\right]\\ &= \frac{1}{2}\Big[(3.45-1.58)\times 10^{18} + 1.58\times 10^{18}\times \ln 2.2\Big]\mathrm{cm}^{-3}\\ &= 3.25\times 10^{17}\mathrm{cm}^{-3}\end{aligned}$$

The peak power of the giant pulse:

$$\begin{aligned}P_{\rm m} &= h\nu\phi_{\rm m}V\delta = h\nu\phi_{\rm m}VSL\left(\frac{1}{2}\ln\frac{1}{R}\right)c/(Ln)\\ &= \frac{1}{2}h\nu\phi_{\rm m}VS\ln\left(\frac{1}{R}\right)\frac{c}{n}\\ &= \frac{1}{2}\times 6.626\times 10^{-34}\times 3.25\times 10^{17}\times 0.75\times \ln\left(\frac{1}{0.5}\right)\\ &\quad\times\frac{(3\times 10^{10})^2}{0.6943\times 10^{-4}\times 1.76}(\mathrm{MW}) = 412(\mathrm{MW})\end{aligned}$$

The energy of the giant pulse:

$$\begin{aligned}E &= \frac{1}{2}(\Delta n_{\rm i} - \Delta n_{\rm t})h\nu V\\ &= \frac{1}{2}(3.45-0.54)\times 10^{18}\times 6.626\times 10^{-34}\end{aligned}$$

$$\times \frac{3 \times 10^8}{0.6943 \times 10^{-6}} \times 0.75 \times 10 = 2.2(\mathrm{J})$$

The width of the giant pulse:

$$\Delta t \approx E/P_\mathrm{m} = 2.2/(412 \times 10^6)\ \mathrm{ns} = 5.3\ \mathrm{ns}$$

What is discussed above is the case of an ideal step Q-switching function. If the loss in the cavity appears to be in a relationship of linear function with time, as shown in Fig. 2.2-5, what will be its Q modulation characteristics? It is shown by theoretical analysis and experimental research that, if the critical oscillating point is assumed to be a point at $t = 0$, then $t = t_\mathrm{s}$ can be understood as the time for the cavity loss to decrease from maximum (point A) to minimum (point B). Obviously, the difference between the linear switching function and the ideal step switching function lies in the difference in t_s. When $t_\mathrm{s} = 0$, we have a case of the step switching function. Therefore, in order to study the association between the linear switching function Q modulation and the step switching function Q modulation, we have to study the influence of different t_s values on the laser pulse (t_s being the time the Q switch is on). The Laser Division of University of Science and Technology of China performed numerical solution for the rate equation with a group of typical laser parameters, namely, the cavity length $L = 50$ cm, the active medium length $l = 6$ cm, the stimulated emissive cross section $\sigma_0 = 8.8 \times 10^{-19}\ \mathrm{cm}^2$, the total density of active particles $n_0 = 6 \times 10^{19}\ \mathrm{cm}^{-3}$, $R_1 = 8\ \%$, $R_2 = 5 \times 10^{-3}$, the initial photon number density $\phi_\mathrm{i} = 10^{-14}$, and the initial inverted population density $\Delta n_\mathrm{i} = 2.48 \times 10^{-2}$. The results concerning the numerical solution for the rate equation using different t_s values are shown in Fig. 2.2-6, in which ϕ_m is the peak photon number density, Δt the pulse half-width, and t_D the time of delay needed for setting up the laser pulse. It can be seen from the figure that the law of variation of ϕ_m and Δt with the Q-switching time t_s has two characteristics: ① There exists a characteristic point $t_\mathrm{s} = t_\mathrm{D}$. When t_s of the Q-switching is smaller than or equal to this characteristic time ($t_\mathrm{s} \leqslant t_\mathrm{D}$), ϕ_m and Δt of the giant pulse have nothing to do with the Q-switching time t_s. ② When $t_\mathrm{s} \leqslant t_\mathrm{D}$, the solution for the linear switching function is not different from that for the step switching function in essence, the difference being only in the time of delay for setting up the pulse as the former has an additional switching time t_s than the latter.

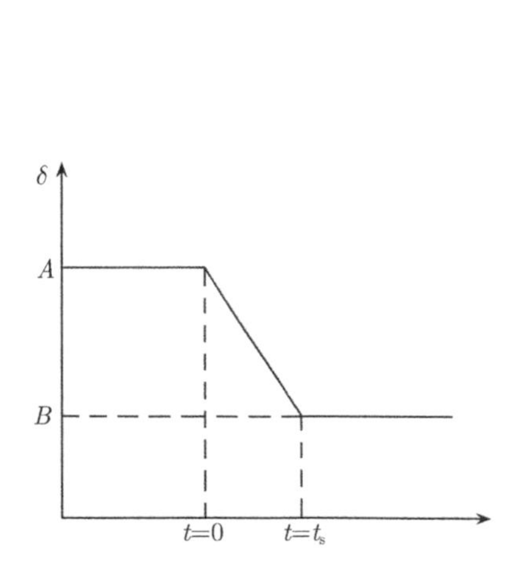

Fig. 2.2-5 The linear Q-switching function

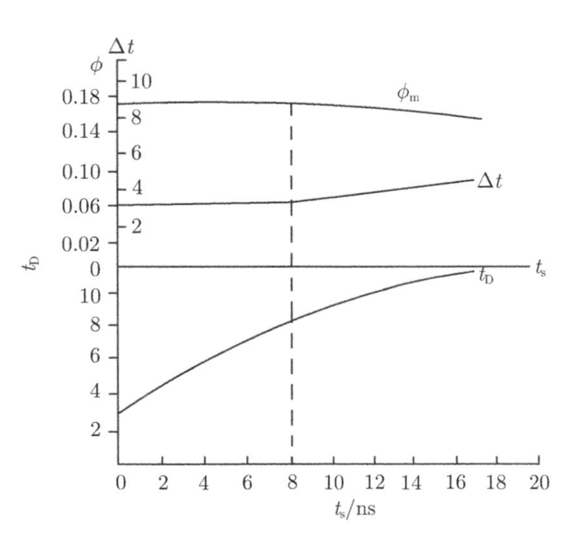

Fig. 2.2-6 The relationship of ϕ_m and Δt with t_s

In addition, there is a steadily variable nonlinear (parabolic) switching function as shown in Fig. 2.2-7. The results concerning the calculation of the characteristics of the steadily variable parabolic switching and the linear switching function Q modulating pulse using the above-mentioned typical laser parameters and taking $t_s = 5$ ns are shown in Fig. 2.2-8. It can be seen from the figure that, under the $t_s \leqslant t_D$ condition, parabolic switching Q modulation is completely identical to linear switching Q modulation.

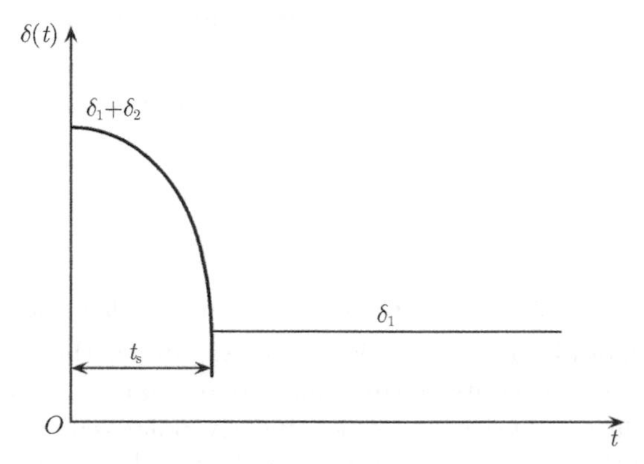

Fig. 2.2-7 The parabolic Q switching function

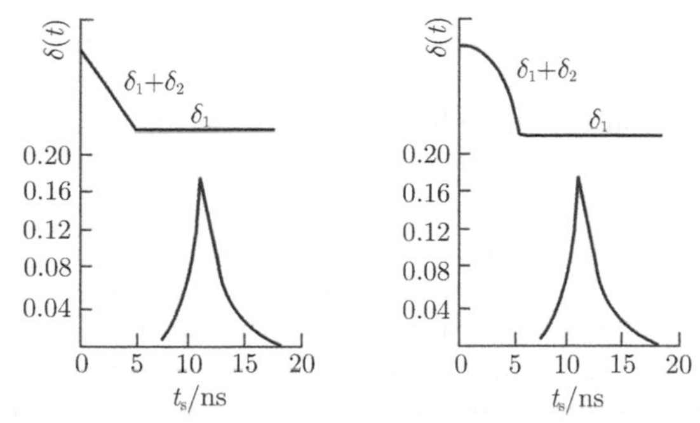

Fig. 2.2-8 Comparison between parabolic switching and linear switching Q modulation

In a word, according to the results obtained by the Laser Division of University of Science and Technology of China in their research, Q-switching can be divided into two classes: ① When $t_s \leqslant t_D$, the performance of the Q modulating pulse has nothing to do with the form of the Q-switching function, nor is it related to the Q-switching time t_s, while the results are the same as those with the ideal step switching Q modulation, which is called "quick switching"; ② When Q-switching $t_s > t_D$, it's just like transient switching function Q modulation, that is, the Q-switching is one with which the loss of the cavity is varying all the time in the course of Q modulation, whose Q modulating pulse performance is related to the Q-switching function and Q-switching time t_s, which is called "slow switching".

2.3 Electro-optic Q modulation

By using the electro-optic effect of certain crystals, electro-optic Q-switching devices can be fabricated. As electro-optic Q modulation possesses such merits as short switching time (about 10^{-9} s), high efficiency, Q modulating moment capable of accurate control, narrow output pulse width (10~20 ns), and high peak power (above dozens of MW), it is at present a fairly widely applied Q modulating technology.

2.3.1 Electro-optic Q modulating devices with a polarizer

Shown in Fig. 2.3-1 is a diagram of the operation principle of electro-optic crystals Q modulating devices. The laser operation material is Nd:YAG crystal, the polarizer adopts the calcite air-gap Glan-Foucault prism, and KD*P (potassium ditritium phosphate)crystal is used as the modulating crystal, which is z-0° cut (to make the light transmitting surface perpendicular to the z-axis). The longitudinal electro-optic effect of its γ_{63} is used to connect the ring-shaped electrodes at the two ends of the modulating crystal to the Q modulating power source.

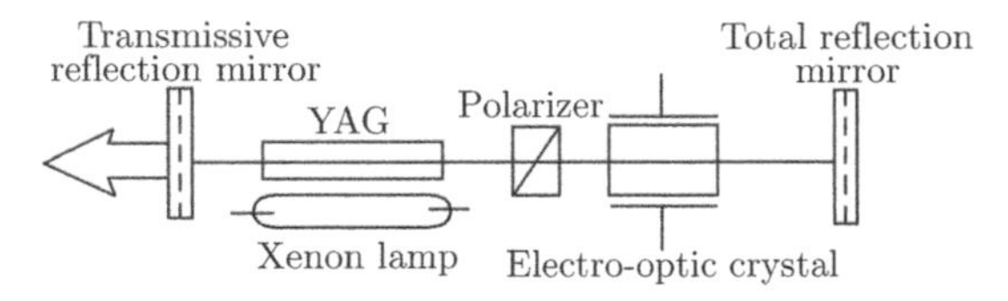

Fig. 2.3-1 Electro-optic Q modulating device with a polarizer

The electro-optic Q modulating process is as follows: under the optical pumping by the xenon lamp, the YAG crystal emits natural light (irregular polarized light) which, after passing through the polarizing prism, becomes linearly polarized light along the x-axis. If no voltage is applied on the modulating crystal, the light will pass through the crystal along the axial direction (optical axis), with no change ever taking place in its polarized state. After being reflected by the total reflective mirror, it once again (invariably) passes through the modulating crystal and the polarizing prism. The electro-optic switch is in the "on" state.

If $\lambda/4$ voltage is applied on the modulating crystal, because of the longitudinal electro-optic effect, after the linearly polarized light along the x direction passes through the crystal, a phase of $\pi/2$ will be generated between the two components which, after going out of the crystal, are synthesized into a circularly polarized light. Following being reflected back by the total reflection mirror, it once again passes through the crystal and once again a $\pi/2$ phase difference will be generated, a π phase difference in all being accumulated in a round trip. After synthesis, the linearly polarized light vibrating along the y direction will be obtained, which is equivalent to the polarized face having rotated 90° relative to the incident light. So it can no longer pass through the polarizing prism. Now, the electro-optic Q switch is in the "off" state. If just as the xenon lamp is ignited, a $\lambda/4$ voltage is applied on the modulating crystal in advance, to enable the resonator to be in a low Q value "off" state, to block the formation of laser oscillation and, when the inverted population of the laser upper energy level is accumulated to maximum, remove the $\lambda/4$ voltage on the crystal all of a sudden to enable the laser to be in a high Q value state instantaneously, generating an avalanche-like laser oscillation to output a giant pulse.

It is known from the basic principle of electro-optic Q modulation that one of the keys to obtaining highly efficient Q modulation is to accurately control the time delay of "on" of the Q switch; that is, the delay begins as the xenon lamp is ignited. The effect is best when the switch is immediately turned on as the inverted population of the operation material's upper energy level reaches maximum. If the Q switch is turned on unduly early, the oscillation is initiated when the inverted population at the upper energy level has not yet reached maximum, it is obvious the giant pulse power output will be reduced. Worse still, multi-pulse may occur. Conversely, if the delay is too long, that is, the Q switch is turned on late, because of the losses such as spontaneous radiation, the power of the giant pulse will be affected.

Shown in Fig. 2.3-2 is a diagram of the operation procedure for Q modulation, the process being: ① First turn on the main source to charge capacitor C and connect it to the xenon lamp electrode but do not energize and or ignite it; ② Start the crystal power source to apply a voltage on the KD*P crystal to put the cavity in the closed state; ③ Have a pulse time mark signal generated by a single junction transistor oscillator input to the control circuit, by which the signal should be sent to the laser main source to make it stop charging the capacitor and to the flip-flop to ignite the xenon lamp so as to give the operation material energy, making the inverted population accumulate in large quantities, respectively. But, as the KD*P crystal now has a voltage $V_{\lambda/4}$ applied on it, the resonator loss is greatest such that no laser oscillation can form. When the population is inverted to maximum, it is applied on the grid of the thyratron via the signal of the delay circuit (to make it) to instantaneously remove the voltage on the KD*P crystal so that the resonator Q value is increased abruptly, forming laser oscillation and outputting giant pulses. The delay circuit can be accurately regulated by experiment until the output laser is most intense.

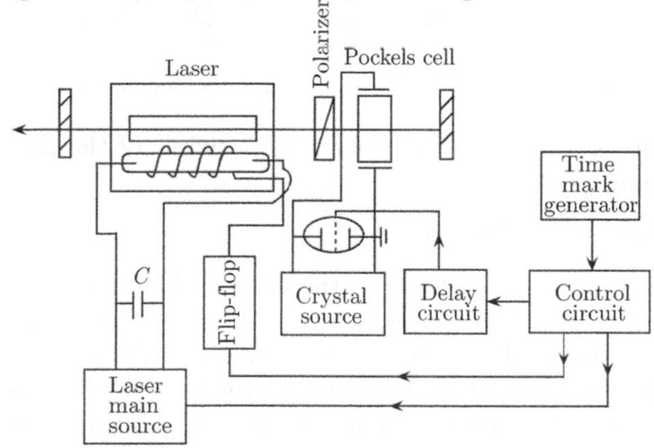

Fig. 2.3-2 Schematic diagram of the electro-optic Q modulating procedure

One of the crucial factors for the electro-optic Q modulating devices with a polarizer to obtain an ideal switching effect is to strictly ensure that the polarizing direction of the prism is kept consistent with the direction of the x-axis (or y-axis) of the modulating crystal to ensure that the polarizing direction will form an angle of 45°with the inductive main axes x', y' of the modulating crystal. A simple modulating method is, with the modulating crystal with a voltage applied on it, turn the relative bearings of the Glan-prism and the crystal until laser is unable to oscillate any longer.

The requirements on the modulating crystal are that the $\lambda/4$ voltage ($V_{\lambda/4}$) should be low, the extinction ratio high, the absorption coefficient at the wavelength of laser small, and the density of power that can be borne high. The currently used crystal is mainly the KD*P crystal, its $\lambda/4$ voltage for a laser of 1.06 μm being 3000~4000 V, which is lower than that of the KDP crystal. The lithium niobate ($LiNbO_3$) crystal is another kind of frequently used crystal, which is characterized by the absence of the longitudinal electro-optic effect while only capable of horizontal operation. The optimum mode of its application is to have the electric field applied on the crystal along the direction of the x-axis (or y-axis) while the light beam passes through along the direction of the z-axis (i.e., the optical axis). This mode of application has not only bypassed the bad influence caused by natural birefringence but also the half-wave voltage is low, its $V_{\lambda/4}$ being 2000~3000 V. It does not deliquesce, but its performance of bearing high-power laser is poor, which limits its application to a certain extent.

2.3.2 Monolithic double 45° electro-optic Q modulating devices

This is a kind of Q-switching that can forgo the polarizer. What is shown in Fig. 2.3-3 is a schematic diagram of this kind of Q-switching laser. The $LiNbO_3$ crystal is machined into a cuboid with two slopes of 45°, the optical axis (z-axis) along the cuboid's axial direction, and the voltage applied on the crystal along the direction of the x-axis. This way, not only is the transmittance of light not affected, but also the electric field is very homogeneous. It can be seen from the figure that such a structure does not ask for the insertion of a polarizer, thus reducing the cavity's insertion loss. So this is a fairly ideal electro-optic Q switch of simple structure.

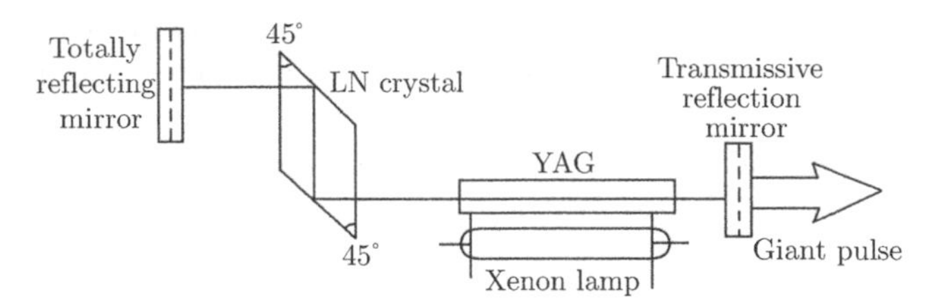

Fig. 2.3-3 Monolithic double electro-optic Q modulating laser

Below we shall make an analysis of the operation principle of the monolithic double 45° electro-optic Q-switching.

1. The case of no voltage applied ($V_x = 0$)

As shown in Fig. 2.3-4, when injected into the crystal along the direction of the y-axis, a beam of irregularly polarized light is resolved into the o light normal to the major cross-section and vibrating along the direction of x and e light parallel to the major cross-section and vibrating along the direction of z. According to the property of birefringence, the two beams of light propagate in the same direction, not separated, but $n_o > n_e$. As the reflective slope forms an angle of 45° with the optical axis z, the two linearly polarized lights will be totally reflected on the 45° reflective surfaces. The o light obeys the reflection law of the homogeneous medium, its reflective angle is equal to the incident angle, or equal to 45°. So o light propagates along the direction of the crystal's optical axis after getting reflected. But, for the e light, it's another story. Before being reflected, its direction of vibration is z while after reflection, it propagates approximately along the z direction. Though still parallel to the major cross-section, its vibrating direction changes from z to y and its refractive index changes to n_e' ($\cong n_o$). So the situation of the e light before and after its reflection is equivalent to its propagation in media with different refractive indices. According to the formula of reflection of anisotropic media, the angle of reflection θ_1' can be found, $n_o \sin 45° = n_e \sin\theta_1'$. For the LN crystal, for the 1.06 μm light wave, $n_o = 2.233$ and $n_e = 2.154$, which substituted into the equation gives $\theta_1' = 42°54'$, which is smaller than the angle of reflection of the o light, the included angle between the two being $\Delta\theta$ (2°6′). After the two beams of light go through the second 45°reflective surface, the o light is still reflected at 45°, so the outgoing o′ light is parallel to the incident light. The refractive index of the e light changes from n_e' ($\cong n_o$) before reflection to n_e, and its angle of reflection becomes 45° once again, so the reflected beam of the e light is also parallel to the direction of the incident light. The distance between the two outgoing beams of the o′ and e′ light is approximated to be $\delta = l \tan\Delta\theta$ (in which l is the geometric length of the crystal in the direction of the optical axis). As the δ value is very small, the two beams of light nearly overlap.

2. The case of a voltage applied ($V_x = V_{\lambda/2}$)

As shown in Fig. 2.3-5, after a voltage is applied on the crystal along the direction of the x-axis, for the incident light in section AB of the crystal, the propagation of the polarized light is basically the same as that when $V_x = 0$, the difference lying in section BC of the optical axis before and after the application of voltage.

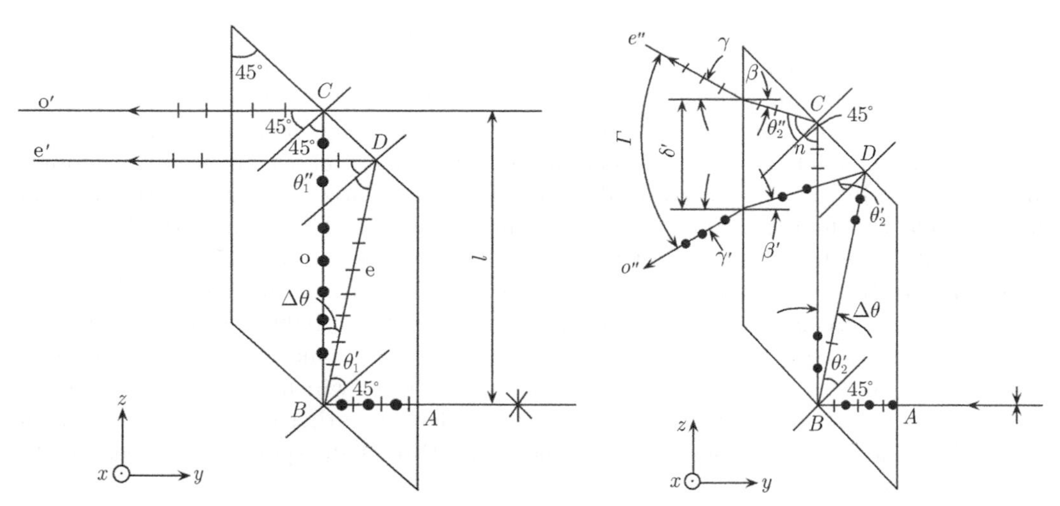

Fig. 2.3-4 The plot of the optical path at $V_x = 0$ Fig. 2.3-5 Plot of the optical path when $V_x \neq 0$

When $V_x = V_{\lambda/2}$ is applied on the crystal, its section BC is equivalent to a half-wave plate. For the propagation of the o light over this distance, its polarized surface has rotated 90°, that is, the original o light has become the e′ light. Also, for the propagation of the e light over this distance, its polarized surface has also rotated 90°, that is, the e light has become the o′ light. When they are reflected as they reach the second 45° reflective surface, there will be the following situation:

For the e′ light, because of the variation of the direction of polarization before and after reflection, according to $n_e' \sin 45° = n_e \sin \theta_2''$, its angle of reflection θ_2' is found by calculation to be 47°12′. As the e′ light is not emitted from the crystal perpendicular to the crystal outgoing surface, refraction will be generated on the outgoing surface. Suppose the angle of incidence of the ray on the outgoing surface is β and the refractive angle is γ. Then $\beta = \theta_2'' - 45° = 2°12'$. According to the law of refraction, it is found that $\gamma = 4°28'$.

Similarly, when the o′ light is reflected by the second 45° reflective surface, its angle of reflection is found by calculation to be 42°55′ according to $n_o \sin \theta_2 = n_e \sin \theta_2'$, which can be refracted on the outgoing surface just as well. In accordance with the law of refraction, it is found that $\gamma' = 4°39'$. Therefore, after being emitted from the crystal, the angle separating two beams of ray $\Gamma = \gamma + \gamma' = 9°7'$. The distance δ' between the outgoing points is approximated to be $\delta' = l \tan \Delta\theta + (\tan\beta + \tan\beta')d/2$ (d being the dimensions of the crystal along the y direction). If the dimensions of the crystal are $l = 30$ mm, $d = 10$ mm, then $\delta' = 1.48$ mm.

It can be seen from the above analysis that:

(1) For the monolithic double 45° $LiNbO_3$ crystal, the front section of the first 45°reflective surface is equivalent to an initiating polarizer, which is capable of generating the two beams of linearly polarized light, the o light and the e light, while the rear section of the second 45° reflective surface is equivalent to an analyzer. So the double 45°$LiNbO_3$ crystal is equivalent to sandwiching a modulating crystal between two polarizers.

(2) When a half–wave voltage $V_x = V_{\lambda/2}$ is applied on the crystal, both the o and e light that have passed through the crystal deviate from the propagating direction of the original incident light. Now, the optical path in the cavity is blocking, which is tantamount to being in the "closed" state, or the resonator is in the low Q value state that makes it impossible for laser oscillation to form. When the optical pump stimulates the operation material and the upper energy level inverted population accumulation reaches maximum, the half-wave voltage is instantaneously removed. Then the outgoing light from the o and e light after passing through the crystal will be parallel to the incident light which, after reflection by the

cavity mirror, still returns by the original route. A clear optical passage is made in the cavity, which is equivalent to being in the "open" state. The Q value increases sharply, and laser oscillation is able to form. Therefore, if whether to apply a voltage or not is under control, then we can change the Q value of the resonator thereby playing the role of Q-switching.

What should be pointed out here is that all the results of the above analysis are obtained from ideal cases. In reality, as crystals all have definite depolarizing character, the incident light beam has a definite angle of divergence, plus the errors of the geometric dimensions of crystals, all these factors will affect the light reflected from the first 45° slope so that it is impossible for it to be completely polarized light. Rather, it is elliptically polarized light with a very great eccentricity, that is, a small portion of the e light is contained in the o light and vice versa. So after reflection from the second 45° plane, there will appear four beams of light, o′, e′, o″, and e″, as shown in Fig. 2.3-6, of which the two light spots in the middle are fairly bright while those on the two sides are rather dark. As the two light spots in the middle are very close to each other, only three light spots can be seen on the observing screen.

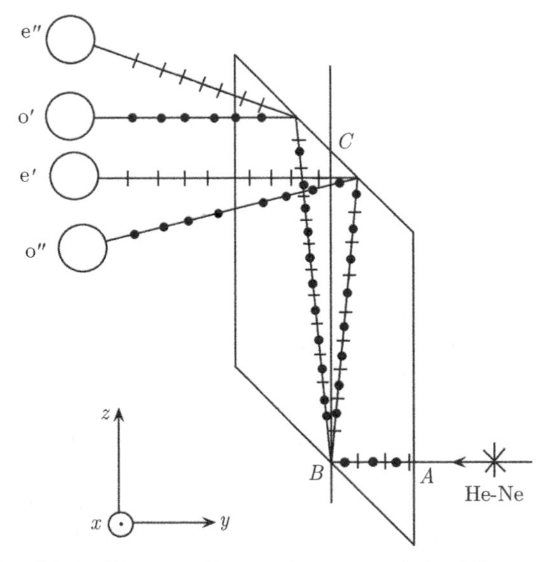

Fig. 2.3-6 The actually outgoing spots of the 45° crystal

(3) Because of the transverse application of the $LiNbO_3$ crystal with voltage applied along the x direction, the half-wave voltage is related to the dimensions of the crystal. It is found from the afore-going analysis that, for the o light and e light in section BC of the crystal, one beam propagates along the optical axis (z-axis) while the other does so near the optical axis. Hence the routes passed through are different and so is the half-wave voltage required by them. But since only one voltage can be applied on a crystal, the requirement that both beams of linearly polarized light rotate 90° just cannot be met. Under the circumstances, there's definitely no way to obtain optimum Q modulating effect. In order to solve this problem, a so-called "optical pre-biasing technique" is often adopted in practice. By optical "pre-biasing" we mean in the monolithic double 45° crystal Q-switching, before a voltage is applied, make the incident light beam incline a prescribed angle in advance so that the o light and e light will propagate in the plane made up of the optical axis and the main inductive axis, respectively, and have identical small included angles with the optical axis. The distances they travel between the two 45° reflective planes are equal. For this reason, after a half-wave voltage is applied, they receive the same modulation. For the principle and specific steps of operation of the optical pre-biasing technique, see Ref. [13] at the end

of this chapter.

The monolithic double 45° electro-optic Q modulating laser is already rather mature, the single pulse laser energy generated by the typical double 45° $LiNbO_3$ crystal Q-switching laser can be as high as 200 mJ, the pulse width is 6~10 ns, and the repetition rate can be as high as dozens of times, the peak value power is greater than 10 MW, and the divergence angle of the light beam is 2~5 mrad.

There are also the orthogonal incident mode, parallel incident mode, etc. of monolithic double 45° electro-optic Q-switching. But they are seldom adopted for not being optimum operating states. In addition, there is the monolithic single 45° Q switching electro-optic Q modulating method (see Fig. 2.3-7), the operating process of which is similar to that of the double 45° Q-switching, for which polarization is initiated mainly only for one kind of linearly polarized light (o or e light), while the other linearly polarized light (e or o light)

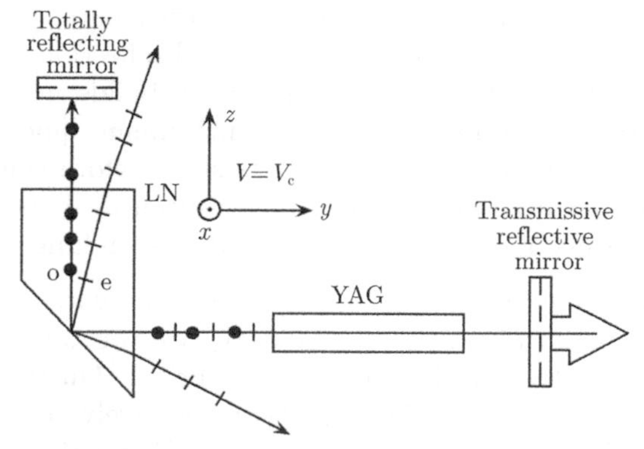

Fig. 2.3-7 The single 45°crystal Q switch

escapes out the cavity, so that polarized light pulse output is generated. Of low efficiency, the performance not as good as that of the double 45° counterpart, it is rarely adopted.

2.3.3 Pulse transmission mode (PTM) Q modulation

The electro-optic Q switches described above all belong in the Q modulation by means of the stored energy in the operation material, that is, energy is stored in the form of active particles in the high energy state of the operation material. When it reaches maximum, the Q switch should be "turned on" and extremely powerful laser oscillation will very quickly be set up in the cavity, while outputting laser from the output mirror end. Therefore, the intensity of the output light is proportional to that of the light field within the cavity. This mode of Q modulation by outputting laser pulses from the output reflective mirror is called the pulse reflection mode (PRM) Q-switching. Additionally, there is another resonator stored energy Q modulation switch, that is, energy is stored in the resonator in the form of photons (the light radiation field). This outputting mode of Q modulation is different from that of the PRM in that it has the coupled output mirror of the PRM Q modulating laser resonator replaced with a totally reflecting mirror. After the Q switch is turned on, photons simply oscillate in the resonator back and forth without output until the inverted particles' stored energy of the operation material is completely converted into the energy of photons in the cavity and placed in a designated optical device in the cavity (usually a polarizing prism) that the maximum oscillating energy stored in the cavity is instantaneously completely output via transmission. This Q modulating mode is spoken of as the pulse transmission mode (PTM) Q-switching. Further, as it is not oscillating cum outputting at the same time, but oscillating first and is not released in an instant until the maximum is reached, it is also called "cavity dumping".

Below we shall briefly treat several kinds of PTM Q modulating lasers.

Shown in Fig. 2.3-8 is a PTM Q modulating laser with an initiating polarizer P_1 and an analyzer P_2. $P_1//P_2$, M_1, are M_2 are total reflective mirrors and M_2 is placed on the optical path of the interface of the polarizing prism P_2 where the polarized light is reflected. When no voltage is applied on the electro-optic crystal, with the laser operation material stimulated by the optical pump, the density of the upper energy level inverted population gradually increases. The spontaneous radiated light started by the operation material can pass through P_1 and P_2 without a hitch. But as there is no reflection mirror at the output end, the cavity Q value is very low. Therefore no laser oscillation can be formed. When the stored energy in the operation material reaches maximum, apply half-wave voltage ($V_{\lambda/2}$). Now, after passing through the crystal, the linearly polarized light that has passed through P_1 will have its polarized surface rotate 90°. Hence it cannot pass through the polarizing prism P_2. But it can be reflected by the interface of the prism onto the total reflection mirror M_2. Thus, the loss of the resonator made up of two total reflection mirrors is very low, the Q value increases abruptly, and laser oscillation quickly builds up. When the density of photons of laser oscillation in the resonator reaches maximum, quickly remove the voltage on the crystal and the optical path will return to its state before voltage is applied. Thus, the maximum light energy stored in the cavity is instantaneously output coupled through prism P_2. This is the process of operation of the PTM Q-switching.

There is another mode of applying the PTM electro-optic switch with polarizing prisms, whose devices are as shown in Fig. 2.3-9. M_1 and M_2 are two total reflective mirrors, PC is a KD*P electro-optic crystal, and P is a polarizing prism. On the two electrodes of the KD*P crystal are applied voltage V_1 and voltage V_2, respectively, of which V_1 is a commonly applied voltage $V_{\lambda/4}$ while V_2 is a square-wave voltage. When the square-wave voltage V_2 is not on, the Q switch is in the completely "off" state (the resonator in the low Q value state). Now as the xenon lamp is ignited, the operation material is in the energy storing stage. The instant (moment t_0) the upper energy level inverted population of the operation material reaches maximum, apply the square-wave $V_2 = (V_{\lambda/4})$. Thus the combined voltage $V_{12} = 0$ (see Fig. 2.3-10) and the Q switch is completely in the "on" state. Furthermore, as the reflectivity of the two reflective mirrors of the resonator M_1 and M_2 is 100%, the resonator suddenly changes to a state of high Q value, with laser oscillation expeditiously built up in the resonator (but with no output). When the photon number density in the cavity reaches maximum, the square-wave V_2 quickly changes from $V_{\lambda/4}$ to 0 and the voltage V_{12} on the crystal jumps back to $V_{\lambda/4}$. Then, the intense light field formed in the resonator passes through the crystal in two round trips to make the polarized surface rotate 90° to be finally output out the resonator when reflected by the side of the polarizing prism P. It can be seen that the optimum efficiency of such PTM mode Q modulation depends on the time and width at which the square-wave V_2 is applied.

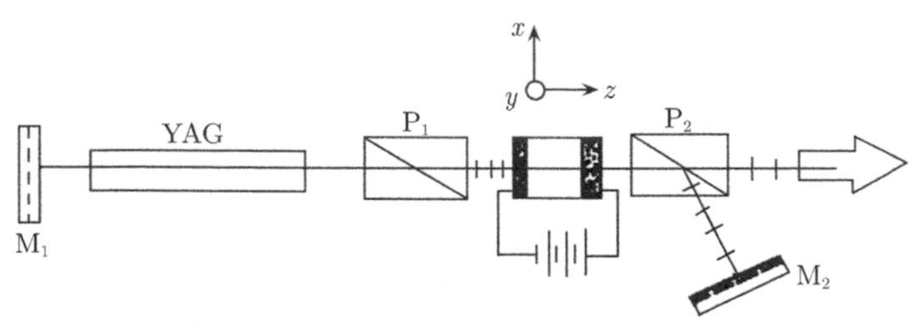

Fig. 2.3-8 The PTM Q modulating laser

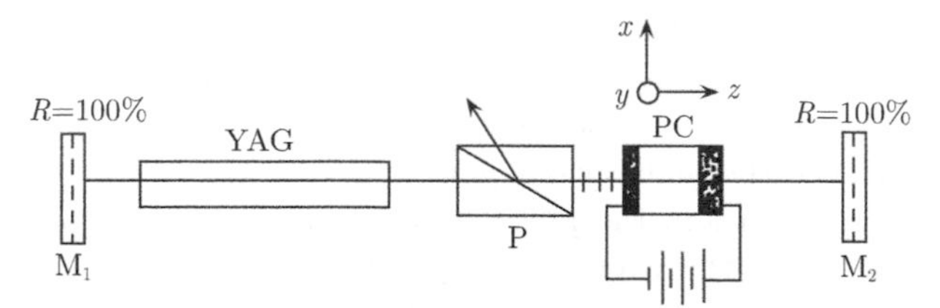

Fig. 2.3-9 The PTM electro-optic Q switch

In addition, the monolithic double 45° electro-optic Q modulating devices described above can be used not only for the PRM Q modulation, but also for the PTM operation, the operating principle is as shown in Fig. 2.3-11. Its difference from the former lies in replacing the original cavity mirror M_2 with two total reflective mirrors M_2 and M_3 placed at designated positions. When operation begins, no voltage is applied on the LN crystal ($V_x = 0$). But the pump lamp is ignited and particles are beginning to accumulate in large quantities on the upper energy level of the operation material. Now, although the optical path is in the straight-through state, as the two total reflective mirrors have already been moved from the center of the optical axial line to the two sides, the cavity loss is very great so that no laser oscillation can be formed. When the operation material's inverted population is amassed to the maximum, with voltage $V_x = V_{\lambda/2}$ applied on the LN crystal, as M_1, M_2, and M_3 are all totally reflecting mirrors, powerful laser oscillation is formed in the cavity, but without output. When the photon number density in the resonator reaches maximum, quickly remove the voltage, and the optical path in the resonator restores its through state so that giant pulses will directly be transmitted out by the LN crystal. For this kind of monolithic double 45° Q-switching it will also do by placing two totally reflecting mirrors at the two ends of the optical axial line in the cavity to construct the PTM Q modulation, its mode of application being simply applying and removing the voltage, which is just the opposite of the former.

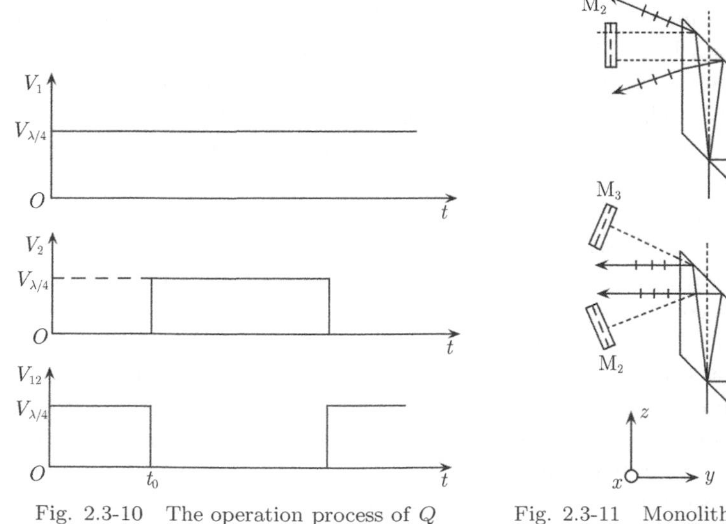

Fig. 2.3-10 The operation process of Q switch

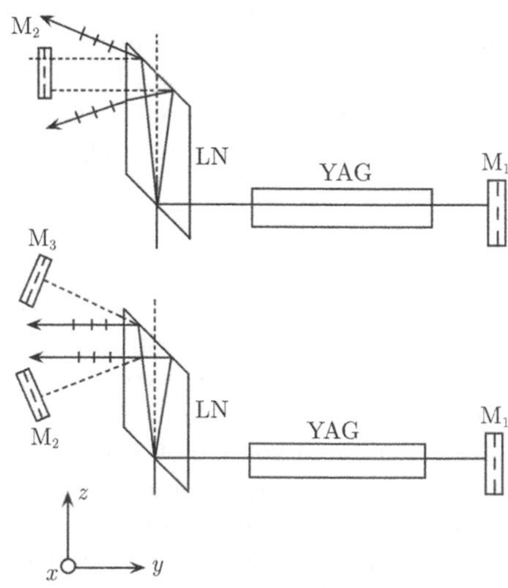

Fig. 2.3-11 Monolithic double 45° PTM electro-optic Q modulation

The PTM Q modulation has several remarkable advantages:

(1) Its efficiency is higher than that of the PRM Q modulation because, for the latter, oscillation and output proceed at the same time. It is after the Q switch is turned on that

laser oscillation begins to set up. Moreover, every time a round trip is made, there is laser output while for the PTM Q modulation, oscillation happens before outputting. When the density of the photon population in the cavity reaches maximum, all the light energy is output in an instant. Hence the high output power.

(2) For the PRM Q modulation, the pulse width mainly depends on the time at which laser oscillation is built up in the cavity. The process of formation cannot be completed until several round trips have been made by light in the cavity. Hence the rather broad pulse width. For the PTM Q modulation, however, the light energy is output within time $2L/c$ (completed once only). So the pulse width will be greatly compressed. But, to obtain the above-mentioned ideal results, it is necessary to accurately connect the resonator within the time pulse is formed. Furthermore, the trailing edge time of the electrical pulse should also be extremely accurate. That's why high requirements are made on the driving circuit system.

2.3.4 Other functions of Q modulating technology

With respect to the Q modulating technology described previously, be it the PRM operation with energy stored in the operation material or the PTM operation with energy stored in the cavity, the ultimate result is invariably the acquisition of the output of high peak power giant pulses of a compressed pulse width. But it has been found by research that, as the Q modulator can effectively control the loss and gain (gain Q switch) of the laser, that is, it can effectively control the net gain of the laser, a Q modulator can not only effectively control the energy (or power) characteristics of laser but also it can control the spatial (transverse mode) characteristics and frequency (longitudinal mode) characteristics of laser as well as the stability of output, etc.

1. The function of selecting the transverse mode

In view of the difference in the loss existing among different transverse modes, the cavity loss is controlled by the Q modulator inserted in the cavity. At the beginning, the laser is made to operate in the high threshold and low gain critical oscillating state (called "prelaser" technique). Under definite pumping power intensity, oscillation is built up only for the transverse mode with minimum loss (TEM_{00} mode); the remaining transverse modes with rather great losses are also restrained from oscillating (for details of the basic principles of mode selection see Chapter 5). Thus the single transverse mode "seed" is generated. In order to obtain great energy output, the Q switch is then completely turned on to enable seed laser to get fully amplified. So what is finally output will be the fundamental transverse mode laser of sufficiently high power. An experimental study is reported in Ref. [4] on the experimenting devices adopted as shown in Fig. 2.3-12, in which M_1 and M_2 are cavity mirrors, PC is the Pockels cell, P is the polarizing mirror, and on the two electrodes of the KD*P crystal are applied voltages V_1 and V_2, of which V_1 is commonly applied voltage and V_2 is square-wave voltage. The operation process is as shown in Fig. 2.3-13, in which V_{12} is the combined voltage on the KD*P crystal, and V_{os} is the commonly applied voltage value needed by the crystal under the condition of critical oscillation corresponding to the transverse mode with minimum cavity loss under definite pumping power. The V_1 value is slightly lower than the V_{os} value, barely enough to make the transverse mode with minimum loss build up oscillation. The selection of the square-wave V_2 should be made on condition that the laser of the transverse mode with minimum loss be fully amplified, while none of the remaining transverse modes can. So, as long as the width of the square-wave V_2 is suitably chosen so that the fundamental transverse mode with minimum loss is fully amplified while the Q switch is timely "turned off" before laser is formed for the other transverse modes, what is output out the cavity is simply the laser of the transverse mode with minimum loss.

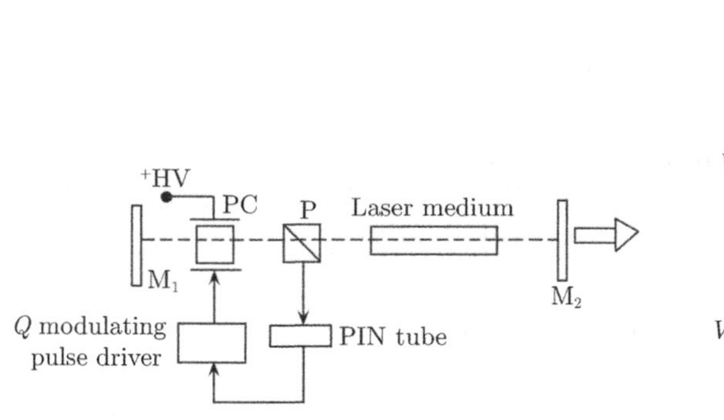

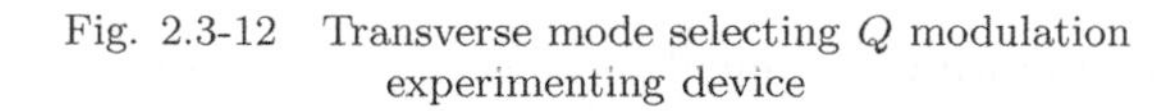
Fig. 2.3-12 Transverse mode selecting Q modulation experimenting device

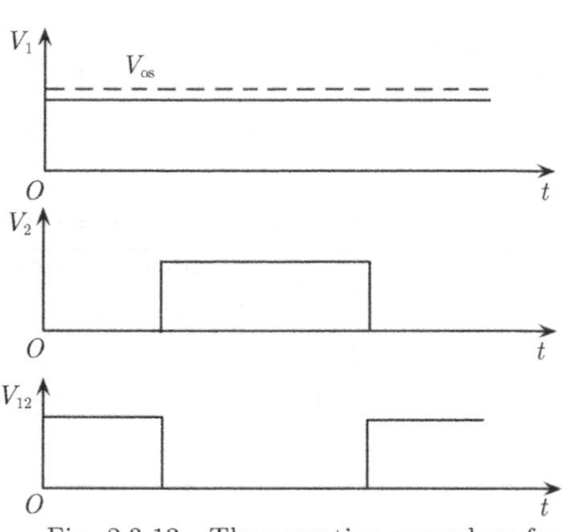

Fig. 2.3-13 The operation procedure for Q modulated selection of transverse mode

2. The function of selecting the single longitudinal mode

The Q modulating technology for the selection of the single longitudinal mode is implemented on the basis of the single transverse mode by making use of the difference in gain existing among different longitudinal modes. At the beginning, the Q switch should be put in a not completely turned off state. The loss of the cavity is controlled by the Q switch, so that, under definite pumping power, only the few longitudinal modes with rather big gains in the vicinity of the central frequency ν_0 can build up oscillation. In addition, as the few longitudinal modes are oscillating near the threshold, the time needed for laser forming is rather long, and the mode competition among different longitudinal modes is fairly sufficient. So, what is finally formed and sufficiently amplified is merely the single longitudinal mode with maximum gain. After the single longitudinal mode laser is formed, turn on the Q switch completely and the single longitudinal pulse laser output can be obtained.

The operation procedure for selecting the single longitudinal mode by Q modulation is basically the same as that shown in Fig. 2.3-13, the difference being that, under definite pumping power, corresponding to the critical oscillating condition of the longitudinal mode (central frequency ν_0) with maximum gain, voltage V_{os} is the voltage that should be applied on the crystal. As the longitudinal modes are in the vicinity of the central frequency, there is little difference in their gains. So insert an F-P etalon plate (reflectivity $R = 8\%$) so that there will exist loss difference in the longitudinal modes near the central frequency at the same time. Their net gain difference will thus increase, making the selection of longitudinal modes by the Q modulator even more efficient.

The Q modulating technology also has the mode locking and clipping functions. It is clear that the development of multifunction Q modulating technology will be a breakthrough in laser technology that will greatly enhance the practicality of Q modulating lasers.

2.4 Problems for consideration in designing electro-optic Q modulating lasers

In contrast to ordinary pulsed lasers, the Q modulating laser is characterized by supercritical oscillation. Therefore, there are some new requirements on the Q-switching devices, the laser operation material, the optical pump lamp, and the condition for coupled output.

2.4.1 Selection of the modulating crystal material

The quality of the electro-optic crystal plays a very important role in the Q modulating performance. At present, linear electro-optic crystal materials capable of yielding rather high optical quality are limited in number. Those already in wide application are mainly the KDP type crystals (KD*P, KDP, etc.) and the ABO_3 type crystals ($LiNbO_3$, $LiTaO_3$, etc.), whose general properties are as shown in Tab. 2.4-1.

Tab. 2.4-1 The performance of some electro-optic crystals

Name of crystal	Refractive index n				Electro-optic coef.	Half-wave voltage
	0.6328 μm		1.06 μm		$\gamma_{63}/(10^{-19}\text{cm/V})$	$V_{\lambda/2}/\text{V}$
	n_o	n_e	n_o	n_e		
KDP	1.508	1.467	1.494	1.460	10.5	~15000
KD*P	1.508	1.468	1.494	1.461	26.4	~6000
$LiNbO_3$	2.286	2.200	2.233	2.154	6.80	$\sim 9250\left(\frac{d}{l}\right)$

When choosing electro-optic materials, the following technical indices should be considered:

(1) The extinction ratio should be high. This is the main index for judging the performance of electro-optic Q-switching. The magnitude of the extinction ratio depends on the homogeneity of the crystal's refractive index. The extinction ratio of the KDP class crystals is in general as high as 10^4 above while that of $LiNbO_3$ crystals is rather low. It can be as high as 10^3, but is in general only 250 or thereabouts and can be used in Q-switching.

(2) The transmissivity should be high. The spectral transmission range of the KDP class crystals is 0.2~2.0 μm; from visible light to 1.4 μm, the transmissivity is greater than 85%. The insertion loss of Q-switching 10%~12%. The range of light transmittance of $LiNbO_3$ crystals is 0.4~5.0 μm, the highest can reach as high as 98%.

(3) The half-wave voltage should be low. The $V_{\lambda/2}$ of KD*P crystals is 6000 V; that of $LiNbO_3$ is 9000 (d/l)V. As it is applied transversely, the half-wave voltage is lower than that of the KDP class crystals.

(4) The anti-damage threshold value should be high. The high power density borne by the crystal should be high. The density borne by the KDP class crystals can reach 500 MW/cm^2 while the anti-damage threshold of the $LiNbO_3$ is rather low. So high-power Q-modulated lasers are optical damage prone.

(5) The crystal should be moisture-proof. The KDP class crystals are apt to deliquesce so that the transparent surface gets fluffy, causing an increase in transmittance loss. Therefore it should be sealed. By injecting a kind of matching liquid [17] into the Pockels cell (Fig. 2.4-1), the moisture-proof ability can be enhanced and the surface surplus reflection can be reduced. After the ZP matching liquid is injected into the Pockels cell, have the transmittance ratio of the 1.06 μm wavelength laser measured. It is found that the moisture-proof ability is increased 3%~5% higher than without the ZP matching liquid, making the total transmittance ratio close to the numerical value (90% or so) when the two ends of the Pockels cell crystal are plated with antireflective film. As the LN crystal does not deliquesce, no sealing device is needed.

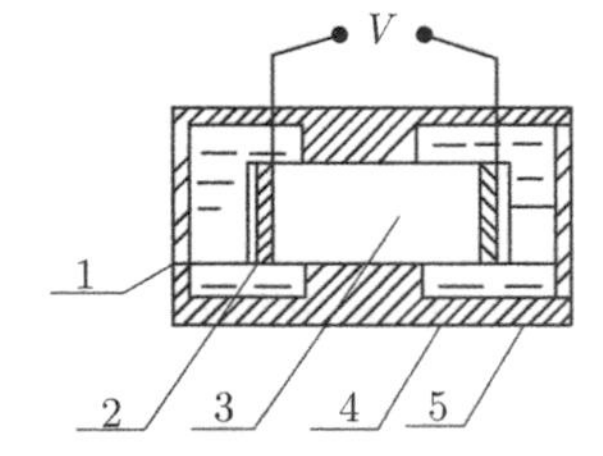

Fig. 2.4-1 The devices in the Pockels cell:
1-1.06 μm antireflective film plated on two sides of K9-glass window
2-Copper-ring electrode 3-KD*P crystal
4-Organic glass box 5-ZP liquid

2.4.2 The electrode structure of the modulating crystal

The structural form of the electrode and the quality of crystal contact directly influence the homogeneity of the electric field in the crystal. An extremely inhomogeneous electric field may lead to the loss of Q modulating effect of the device. Therefore, the presence of a homogeneous electric field in the crystal is the basic point of departure for designing the structure of the electrode.

The vast majority of KDP crystals are applied longitudinally, with the direction of the externally applied electric field consistent with the light transmitting direction. Such being the case, it is very difficult to work out a homogeneous electrode structure. As it is, very often the approximately homogeneous electric field mode is adopted. The ring-shaped electrode structure is the optimum mode in longitudinal application, the advantages of which include the electrode is not in contact with the light transmitting surface. When the clear aperture is equal to the structure of the clear hole opened in the center, the structural dimensions can be made somewhat small, or the effective light transmitting cross-section can be made bigger. But there still exists the problem that the electric field is not homogeneous. When designing such an electrode structure, on the premise that no high-voltage sparking will be caused, the widths of the two ring-shaped electrodes should preferably be broad.

For the $LiNbO_3$ class crystals, as they are all applied transversely and the direction of the electric field and that of light transmission are normal to each other, so long as plate type electrodes are made, homogeneous electric field distribution can be obtained.

The electrode is made of soft metals (aluminum foil, copper foil, silver foil, etc.) and glued together with the two ends of a crystal with adhesive. The contact surface of the electrode is required to possess fairly high finish to ensure reliable contact with the crystal. In addition, it is also feasible to have such metals as Au or metal oxides SnO plated on the two lateral surfaces of the two ends of the crystal. The homogeneity of the electric field of steam plated electrodes is fairly good.

2.4.3 Requirements on the laser operation material

If good Q modulation results are desired, not only is a high quality electro-optic crystal needed, but also, apart from the general requirements on the laser operation material, there are some new requirements. To begin with, it should possess the performance of a high energy storing density, that is, on the laser upper energy level large quantities of particles can be accumulated. So it is required that the stimulated radiation cross-section be small; that is, the lifetime of the upper energy level should be long and the spectral line be rather broad. If so, the occurrence of super-radiation can be prevented or weakened. In addition, a rather high anti-intense light damage threshold value is also required so that high laser power density can be withstood. The operation materials Nd:YAG, ruby, and neodymium glass can all meet the above-mentioned requirements in the main.

2.4.4 Requirements on the optical pump lamp

In order to reduce the loss of the inverted population due to spontaneous radiation, it is required that the pump lamp shining time duration (half-width of the pulse waveform) be shorter than the operation material's fluorescence lifetime (lifetime of the laser upper energy). It is shown by practice that different requirements are made on different operation materials and their dimensions. For instance, for the Q-modulated YAG laser, the half-width of its pump lamp pulse waveform is 200~300 μm while that of ruby is 1 ms or so. But, if the half-width of the lamp light wave is too narrow, the lamp efficiency will be reduced. So, it is advisable to choose a pump lamp for which the two are fairly well matched.

2.4.5 Requirements on the Q switch control circuit

It can be seen from the afore-going analysis that to obtain an optimum Q-modulating effect, the Q switch speed is required to be high, that is, the resonator optical path can be expeditiously and accurately turned on and closed. This is controlled through the electro-optic modulated source, usually including the crystal high voltage source, the control circuit, the delay circuit, the switching devices, the flip-flop circuit, etc. If it is desired that the Q-modulated laser will be able to operate highly efficiently, it is necessary to accurately design all the circuits so that they will be well coordinated in operation.

2.5 Acousto-optic Q modulation

2.5.1 The basic principles of acousto-optic Q modulation

The structure of the acousto-optic Q-switching device is basically the same as that of the acousto-optic modulator described in Chapter 1, being composed of the acousto-optic medium, the electro-acoustic transducer, sound absorbing material, and the driving source. The schematic diagram of its devices is as shown in Fig. 2.5-1. Mainly fused quartz, glass, lead molybdate ($PbMoO_3$), etc. are adopted as the acousto-optic medium. The transducer is often made of quartz, lithium niobate ($LiNbO_3$) crystal, etc. Lead rubber and glass cotton are often used as sound absorbing materials. Insert the Q-switching device in the resonator. When the high frequency oscillating signal generated by the acousto-optic source is applied on the transducer of the acousto-optic Q modulating device, the ultrasonic wave vibration formed makes the refractive index vary in the acousto-optic medium forming the equivalent "phase grating". When the light beam passes through the acousto-optic medium, Bragg diffraction will be generated, the diffracted light deviating a 2θ angle relative to 0-level light (e.g., when the ultrasonic frequency is in the range of 20~50 MHz, the angle of diffraction of quartz with respect to a light wave of 1.06 μm is $0.3^\circ \sim 0.5^\circ$). Such an angle is definitely capable of making the light wave deviate out of the resonator so that it will be in a high loss low Q-value state, which makes it impossible for oscillation to be generated, or, put another way, the Q switch has "turned off" laser. When the action of the high frequency signal comes to a sudden stop, then the ultrasonic field in the acousto-optic medium will vanish. Thus the resonator will abruptly change to the high Q-value state, which is equivalent to the Q switch "turned on". Every time the Q value changes alternately, the laser will be made to output a Q modulating pulse.

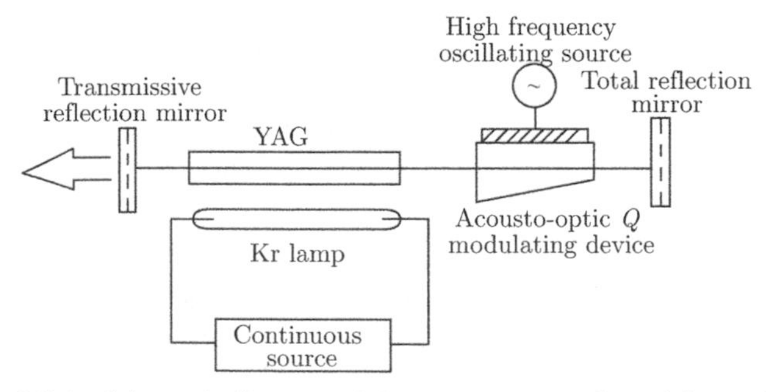

Fig. 2.5-1 Schematic diagram of the acousto-optic Q modulating laser

For the previously described electro-optic Q modulating laser, in order to make the stored energy of the operation material emitted as a single pulse in a very short period, the Q switch must complete the change from low to high Q value (stepped change) within a time shorter than that for building up the laser pulse. For the acousto-optic Q switch, the off time mainly depends on the transition time in which the acoustic wave passes through the light beam. (The electronic switch on or off time is not the main thing.) Take fused quartz as an example. The time needed for the acoustic wave to pass through a length of 1 mm is about 200ns (sound velocity is 5 mm/μs). This time duration appears too long for certain high gain pulsed lasers. Therefore, the acousto-optic Q switch is generally used in continuous lasers of a rather low gain. Also, the driving modulating voltage needed by the acousto-optic Q switching is very low (<200 V), so it is easy to realize Q modulation for continuous lasers to obtain a pulse output of high repetition frequency; the repetition rate in general can reach as high as 1~20 kHz. But as the switching capability of the acousto-optic Q switch for high energy lasers is rather poor, it should not be used in high gain Q modulating lasers.

When the acousto-optic Q switch is used in the continuous laser, a pulse modulator should be used to generate rectangular pulses of frequency f to modulate the signal of the high-frequency oscillator. Therefore, the frequency appearing in the ultrasonic field of the acousto-optic medium is the frequency of the pulse modulating signal. Thus the laser outputs a Q modulating pulse sequence of repetition rate f. In order that as many particles as possible can be accumulated on the laser upper energy level of the operation material, while avoiding too great losses of spontaneous radiation so that the laser can have maximum inverted population availability while ensuring definite peak value, the time interval between two adjacent pulses $1/f$ should be approximately equal to the lifetime of the upper energy level of the laser operation material. For instance, for the Nd: YAG laser, its upper energy level lifetime is about 230 ns. So, it is advisable for the Q modulating repetition rate f to be chosen as 4~5 kHz. In this case, the availability of the inverted population is highest, which makes it possible to obtain a Q modulating pulse sequence with a peak power of 20~30 kW. Too high or too low a repetition frequency will affect the Q modulating effect just the same. The switching time of acousto-optic Q modulation is in general shorter than the time for building up a pulse and is therefore of the fast switching type.

The mode of acousto-optic Q modulating operation is used by the continuous laser, as shown in Fig. 2.5-2. In this case, the pump rate W_p remains unchanged (see Fig. 2.5-2(a)),

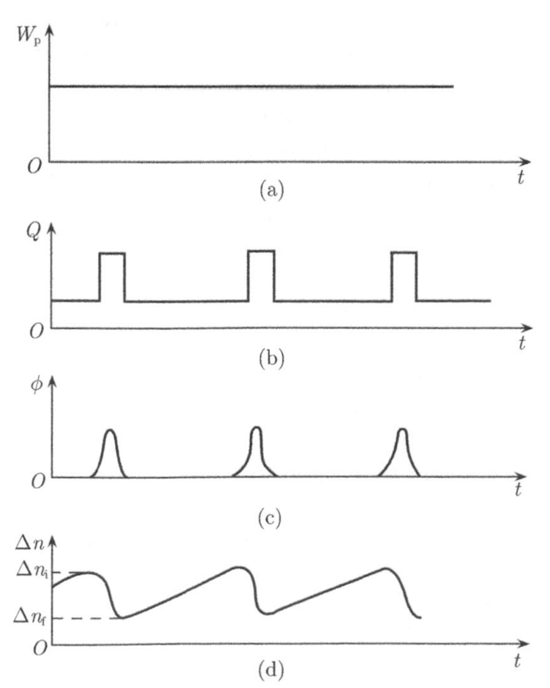

Fig. 2.5-2 The high repetition rate Q modulating process of the continuous laser
(a) pump rate; (b) Q value; (c) photon number; (d) inverted population

but the Q value of the resonator varies periodically (see Fig. 2.5-2(b)), whose period of variation is dependent on the pulse modulating signal frequency f, with a series of Q modulating pulses with a high repetition rate (See Fig. 2.5-2(c)) output. As the pumping is continuous, the Q value of the resonator (i.e., the loss of the cavity) varies periodically from the high Q state to the low Q state at frequency f. Hence the corresponding variation of the inverted population of the laser operation material (see Fig. 2.5-2(d)).

2.5.2 The structure and design of acousto-optic Q modulating devices

After the 1970s, the appearance of several acousto-optic materials of excellent performance and the development of the theory and technology of electro-acoustic transducers operating in the microwave frequency range brought signal breakthroughs to the performance of acousto-optic devices that found increasingly wide application. Therefore, when designing acousto-optic Q modulating devices, how to select suitable materials, rational structural dimensions, and manufacturing technologies and methods—all these are crucial links in ensuring good performance of the devices.

To construct a specific ultrasonic field, acousto-optic Q modulating devices generally adopt the traveling wave operation mode. For this reason, in the direction of the ultrasonic wave's travel forward sound absorbing material or a sound absorber must be mounted on the surface of the medium to eliminate the reflection of the ultrasonic wave. As the traveling wave field is quickly eliminated and the switching time is short, Q modulation can be conveniently carried out. On the contrary, the standing wave ultrasonic field is not quickly eliminated in the acousto-optic medium (definite time of attenuation is needed) and the switching time is so long that the acousto-optic device will lose its role as a switch. This is why it is not adopted.

1. The selection of material

For the electro-acoustic transducer, a material with a large electromechanical coupling coefficient should be chosen to enhance the efficiency of conversion from electric power to acoustic power. The frequently used material is the x-0° cut quartz crystal plate or y-36° cut $LiNbO_3$ crystal plate. The electromechanical coefficient of the latter is twenty-five times greater than that of the former and is an ideal material. However, at the present time, the large area LN crystal thin plate is very difficult to machine. What with the poor stability of its piezoelectric ceramic frequency, it is seldom, if ever, adopted.

For the selection of the acousto-optic medium material, an overall consideration should be given to the following requirements: the medium's quality factor M_2 should be large; the absorption of light should be small (i.e., the light transmission rate should be high); the absorption of ultrasonic wave should be small, possessing good thermal stability; and the medium should be optically homogeneous and have sufficiently large dimensions. For Q modulating sentence incomplete devices of large power, the adoption of materials with a high anti-damage-to-laser threshold. It is known by putting the aforementioned requirements together and looking up Tab. 1.3-2 that TeO_2 and $PbMoO_4$ are fairly ideal acousto-optic materials. But their light transmittance performance for light waves of wavelength 1.06 μm is rather poor, often seriously affecting the output efficiency of the Q modulating laser. Except for its rather low M_2 value, fused quartz can satisfy all the other requirements and is cheap and easily machined optically. So it is chosen as the acousto-optic medium for large power acousto-optic Q modulating lasers.

2. The design of the devices

The cruz of acousto-optic devices is to rationally determine the dimensions of the ultrasonic column, as shown in Fig. 2.5-3, in which the acousto-optic action distance L

can be determined by the Bragg criterion, that is, $L \geqslant 2L_0$, the characteristic length $L_0 = \lambda_s^2/\lambda = v_s^2/\lambda f_s$.

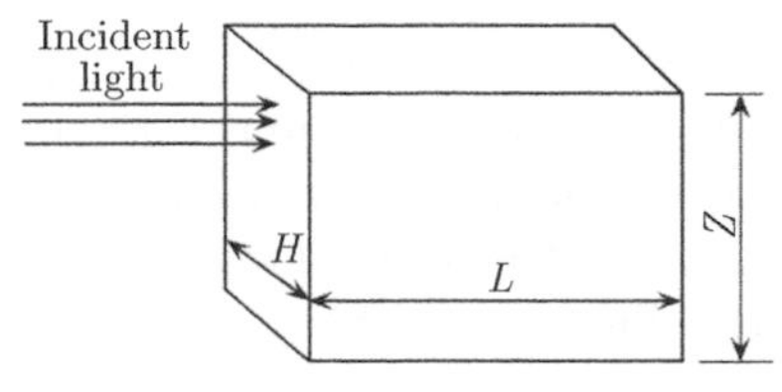

Fig. 2.5-3 The ultrasonic field dimensions

Calculate the efficiency of Bragg diffraction.

$$\eta_1 = I_1/I_i = \sin^2\left(\frac{v}{2}\right)$$

$$v = \frac{\pi}{\lambda}\sqrt{\left(\frac{2L}{H}\right)M_2 P_s}$$

It can be seen that, under definite acoustic power P_s, the greater the ratio L/H value, the higher the diffraction efficiency. So L can be taken great as much as possible according to the actual conditions of the material while the acoustic column width H should be as small as possible. In general, it should be equal to or slightly longer than the diameter of the laser beam. The length and width of the transducer will be okay if only they are a little greater than the above-mentioned L and H values. (This is for the purpose of ensuring sufficient insulation distance.) As the high frequency electric field is applied on the transducer along the thickness direction, the thickness of the transducer is the half wavelength of the ultrasonic wave, which is found by calculation, i.e., $d = \lambda_s/2 = v_s/2f_s$. It will do if the dimensions of the acousto-optic medium are slightly greater than those of the acoustic column. The surface facing the transducer is preferably ground into a compound angle, as shown in Fig. 2.5-4. This way, when collaborating with the sound-absorbing material, the influence of reflection of the ultrasonic wave can be minimized. In the meantime, it is customary to have the included angle between the acousto-optic medium's light transmitting surface and ultrasonic wave surface (i.e., the transducer contact surface) ground into a $(90° - \theta_B)$ angle so that while meeting the Bragg incidence condition $\sin\theta_B = \lambda/2\lambda_s$, it can be ensured that the beam of light will be injected perpendicular to the light transmitting surface (the moment at which the reflection loss on the medium's surface is minimum), which can be seen from Fig. 2.5-5.

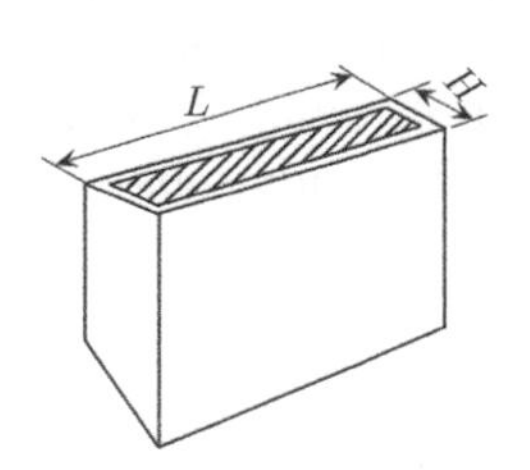

Fig. 2.5-4 The structural form of acousto-optic devices

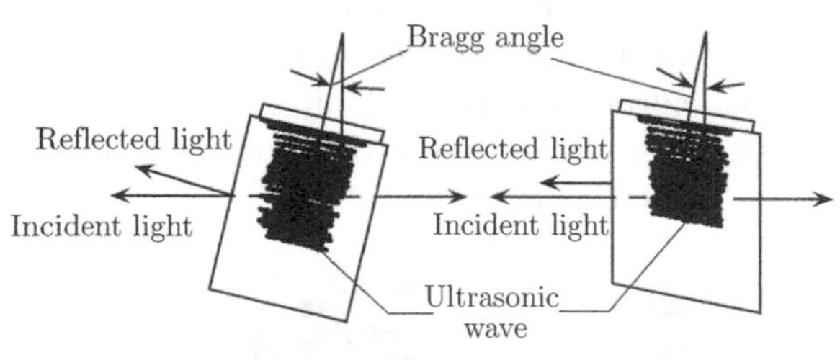

Fig. 2.5-5 The operation mode of acousto-optic devices

The adhesive technology of the transducer with the acousto-optic medium is also a problem of great importance. As it is through this combined layer that the ultrasonic power of the transducer enters the acousto-optic medium, the adhered layer has to be of low loss. It is known from the theory of ultrasonic propagation that, when the adhered layer is completely surrounded by one and the same medium, its acoustic wave transmission coefficient

$$D = \frac{1}{1 + \dfrac{1}{4}\left(m - \dfrac{1}{m}\right)^2 \sin^2\left(\dfrac{2\pi d'}{\lambda_s}\right)} \tag{2.5-1}$$

where the acoustic impedance ratio $m = Z_1/Z_2$, and d' is the thickness of the adhered layer.

It has been proved by experiment that there are two situations that can make the transmission coefficient close to maximum. One is when $m \approx 1$, that is, when the acoustic impedance Z_2 of the adhered layer is close to the acoustic impedance Z_1 of the medium, $D \approx 1$; or, when the thickness of the adhered layer is very thin (≈ 0), $D \approx 1$. So, as long as the adhered layer material and the acousto-optic material are well matched in acoustic impedance, or when the thickness of the adhered layer is smaller than 1 μm, satisfactory effects can be obtained just the same. Currently adopted technologies include the thermal compression welding and supersonic welding with indium as the transitional layer, etc while epoxy resin and glue 502, etc. can also be adopted as adhesives. Table 2.5-1 gives the acoustic impedance of a number of materials.

Tab. 2.5-1 gives the acoustic impedance of a number of materials

Name of material	Al	Au	Ag	In	Epoxy resin	Quartz	Glue 502
Density/(g/cm^3)	2.7	19.3	10.5	7.2	1.18	2.6	1.36
Acoustic imp/[$\times 10^5$/(s·cm^2)]	16.9	62.6	38	19.6	3.2	14.5	3.59

3. The structural form of devices

Both the transducer electro-acoustic energy transforming process and the situation where the ultrasonic wave has been absorbed will generate heat which, if not dissipated in time, will form a temperature gradient field in the acousto-optic medium, thereby disturbing the "phase grating" action of the ultrasonic field. Serious disturbance will cause the device to lose its Q modulating action. Therefore, it is necessary to consider the problem of heat dissipation for the device. Figure 2.5-6 shows three typical structures of an acousto-optic Q modulating device, in which (a) shows the fully-water-cooled type structure, in which the electrode briquetting on the transducer and the clamps on the acousto-optic medium and sound absorbing material should both be cooled through water; (b) shows the half-water- and air-cooled type, which only ensures that the electrode briquetting on the transducer will be cooled through water, while the medium clamp can be made in the form of the radiator to be spontaneously cooled by air or forcibly cooled by appropriately blown-in air; and (c) shows the half-water-cooled multiple reflection absorptive type, which is different from the second type in that it causes the acoustic wave to be absorbed through multiple reflections by the reflective surface on the medium clamp to attain the goal of sound absorbing and cooling. When designing devices with rather high acoustic power, it is advisable to adopt the fully water-cooled structure.

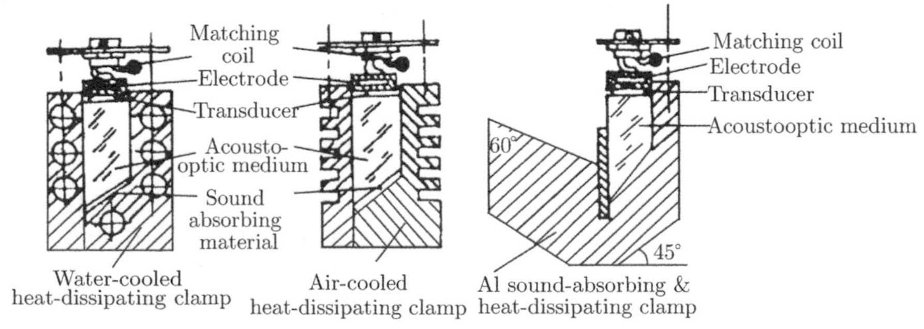

Fig. 2.5-6 Typical structure of acousto-optic Q modulating devices

4. Determination of the driving acoustic power

In order to shut off the laser oscillation with a definite gain, there should be sufficient acoustic power, which can be obtained by referring to the Bragg diffraction relational equation as follows:

$$\eta_1 = \frac{I_1}{I_i} = \sin^2\left(\frac{\upsilon}{2}\right) = \frac{1}{2}(1 - \cos\upsilon) \tag{2.5-2}$$

If it is desired that $\eta_1 = 100\%$, that is, the incident light can be completely diffracted into level 1 diffracted light, then $\upsilon = \pi$ is obtained from the above equation. $\upsilon = \sqrt{\left(\frac{2L}{H}\right) M_2 P_s} \cdot \pi/\lambda$, which, substituted, gives the acoustic power

$$P_s = \frac{1}{2}\left(\frac{H}{L}\right)\frac{1}{M_2}\lambda^2$$

or

$$P_s = 1.26\left(\frac{H}{L}\right)\frac{1}{M_w}\left(\frac{\lambda}{\lambda_r}\right)^2 \tag{2.5-3}$$

where M_w is the quality factor ratio of material to water, and λ_r is the He-Ne laser wavelength.

For instance, fused quartz is chosen as the acousto-optic medium, $M_w = 151/160 = 1/1.06$. If the dimensions of the ultrasonic field in the acousto-optic medium are $H = 5$ mm, $L = 50$ mm, substituted into Eq. (2.5-3), we have the acoustic power $P_s = 37.4$ W needed during 100% diffraction efficiency. So the per watt diffraction efficiency is $(100/37.4)\% = 2.7\%$. Therefore, to overcome the 20%~30% gain in the continuous YAG laser, only a power higher than 10 W can make the laser stop oscillating. In view of the fact that the electricity-acoustic transforming efficiency is in general lower than 50%, it is advisable to set the power of the high frequency driving source at 20 W or thereabouts.

In addition, in the process of high repetition rate operation, the acousto-optic Q modulating device keeps varying alternately between the state of the presence of an ultrasonic field (called state 1) and that of its absence (called state 2). It is impossible for the variation of its diffraction effect to be stepped but one with a definite rise and fall time (assumed to be t_r and t_f). For a particular laser, there is a definite time interval t_D between the moment of the sudden change of Q and the moment the giant pulse is output. Then when designing an acousto-optic Q modulating device, it is important that $t_f < t_D$ since only then can scattered interference be avoided, ensuring that Q modulated pulses of high peak power can be output. The t_f of acousto-optic devices has two sources. One is the resonator made up of the transducer as the driving field needs definite fall time, the other is the acoustic wave transition light beam that needs definite time.

2.5.3 The acousto-optic Q modulating dynamic experiment and output characteristics

The dynamic experimental devices of the acousto-optic Q continuous pump Nd:YAG laser are as shown in Fig. 2.5-7. First put the Nd:YAG laser into operation to output laser and have its power measured with a power meter. Then high frequency electric signals should be injected into the acousto-optic device while gradually increasing the high frequency voltage (i.e., augmenting the acoustic power) till the laser is turned off. This shows that the device has already met the requirement of the Q switch. Then, put in the pulse modulating signals of the kHz order of magnitude to make the high frequency electric signals injected modulated, thereby making the ultrasonic field in the acousto-optic medium now on and now off and the Q value in the resonator, correspondingly, alternately vary between high and low. It is then that there will be pulses of a high repetition rate that are output which, after getting attenuated by the attenuation plate, will be received by the photodiode and input into the dual line oscilloscope (pulse oscilloscope with a response frequency greater than 10 MHz) after amplification. Then the waveform shown in Fig. 2.5-8 can be observed on the fluorescent screen. The upper part of the plot is the Q modulated pulse waveform while the lower part is the voltage oscillation envelope that drives the transducer. It can be seen that the optical pulses appear in the vicinity of the middle part of the time within which the acoustic wave is interrupted.

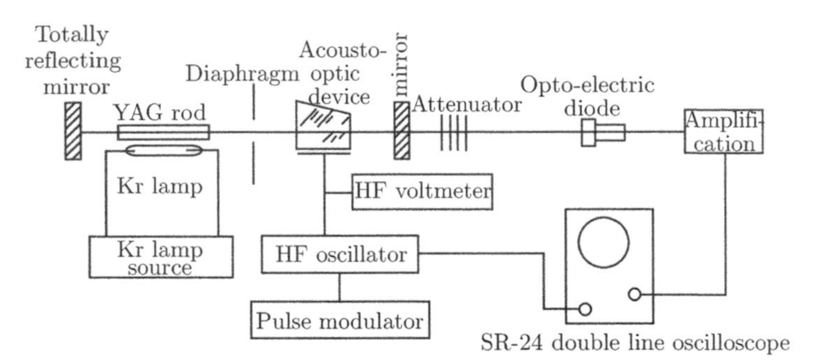

Fig. 2.5-7 Dynamic test device of acousto-optic Q modulating laser

It can be seen from Fig. 2.5-8 that there is a time interval from the moment the ultrasonic field vanishes to the formation of the giant pulses. This is the time t_D that the pulse is to be set up (which is in general 1~8 μs). So it is required that the switching time t_s of the Q switch should be shorter than t_D (which is also the condition for the fast switch), otherwise there may occur the frequently seen fall of the peak power, broadening of the pulse width, unstable output, or the generation of multiple pulses in the slow switch. The switching time of acousto-optic Q modulation is of the microsecond order of magnitude, which is much longer than the switching time of electro-optic Q modulation (more than a dozen ns), which will affect the peak power of the pulse and pulse width.

The dynamic characteristic curve of the acousto-optic Q modulating laser can be measured by experiment. What is shown in Fig. 2.5-9 is the curve of variation of the output peak power P_M and pulse width Δt with the variation of the input electric power when the repetition frequency is given (1 kHz). It can be seen from the figure that with an increase of the input power, the output peak power will correspondingly increase while the pulse width

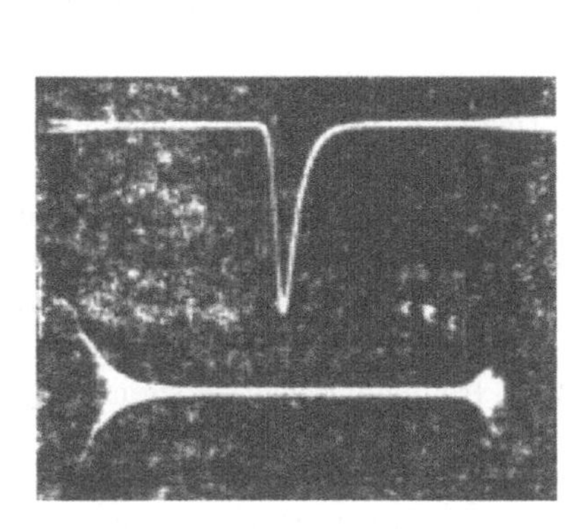

Fig. 2.5-8 The acousto-optic Q modulated waveform

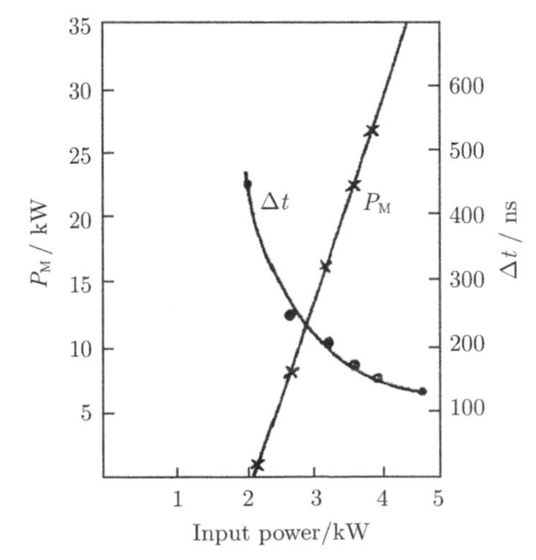

Fig. 2.5-9 The relationship of P_M, Δt with input power

becomes narrower. Shown in Fig. 2.5-10 is the curve of variation of P_M and the average power P with different repetition rates when the input power is not changed (3.5 kW). It can be seen that when the repetition rate is increased, P_M falls whereas P rises. When the repetition rate is in excess of 10 kHz, the average power varies very slowly and is close to the continuous power. Now the ratio of the average power to the Q not modulated continuous power is nearly equal to 1.

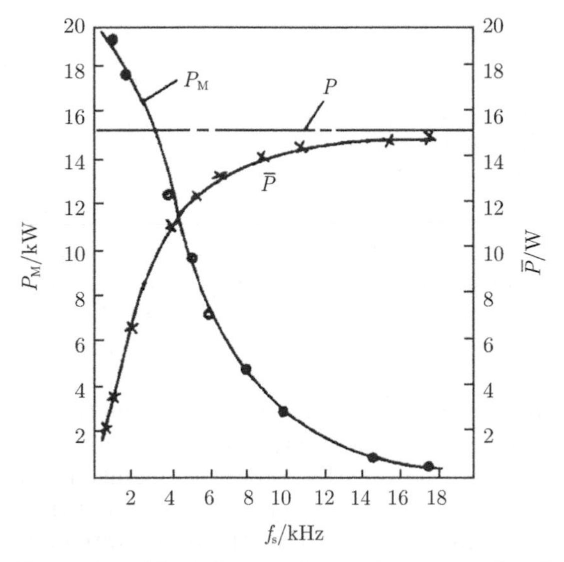

Fig. 2.5-10 The influence of repetition rate on P_M, P

Below we shall analyze the relationships of these dynamic characteristics with various parameters of the laser.

1. The influence of the repetition rate on the output performance

When the laser operates with the Q switch repeatedly varying, the operation material's inverted population is alternately being pumped and consumed. When the resonator is in the low Q state, stimulated by the pumping, the inverted population steadily accumulates on the high energy state. When Δn is increased to a definite quantity, the resonator suddenly changes to high Q state. Now the stimulated radiation light field is intensified, causing the Δn value to quickly decay, falling at an extremely steep gradient. The repetition of every time is shown by the curves in Fig. 2.5-11. Different repetition rates have different variation curves. When the repetition rate is smaller than the spontaneous radiation transition probability A_s, it is possible that the abrupt change of Q will not begin until the density of the inverted population has reached saturation. Now the initial population density Δn_i will have a rather large value. If the repetition frequency is greater than the spontaneous radiation transition probability A_s, then the inverted population will quickly attenuate for entering the high Q value state before its density has yet reached saturation. Now Δn_i will be rather small.

This characteristic of the repetition rate Q switching directly influences the output characteristic of the Q modulating laser. When the repetition frequency is rather high as time is insufficient between pulses for the inverted population of the laser upper energy state to reach maximum, that is, the Δn_i value is rather low, the output laser pulse peak power is bound to fall. And owing to the decrease of gain, both the pulse width and the time for pulse formation will increase. In the event the repetition frequency is too low, because of the spontaneous radiation transition, part of the inverted population will get lost, thus affecting the efficiency of the device. If the ratio of the peak power to the continuous output power is represented by H, then the variation of H with the repetition frequency is as shown in Fig. 2.5-12. The fluorescence lifetime of the Nd:YAG laser is about 230 μs, so if it is desired that the acousto-optic Q modulating device operate at a rather high efficiency, the repetition rate is always chosen to be within the range of 4~5 kHz.

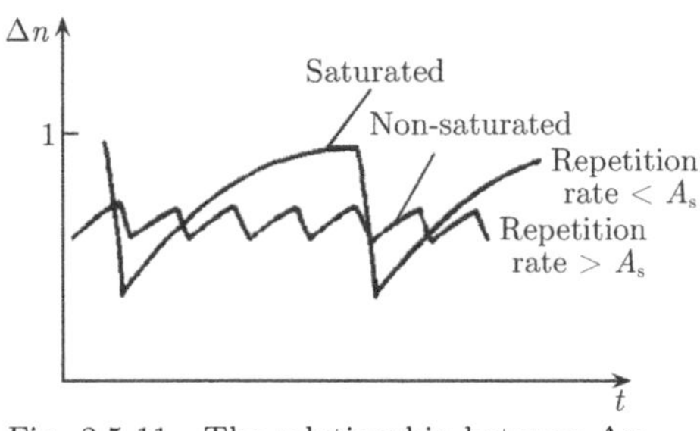

Fig. 2.5-11 The relationship between Δn and time at different repetition rates

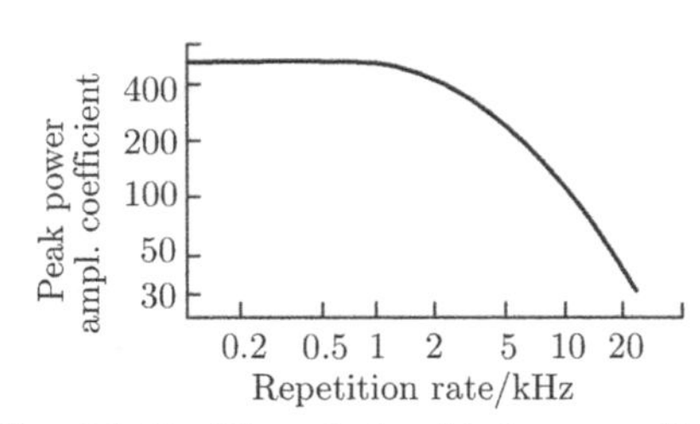

Fig. 2.5-12 The relationship between H and the repetition rate

2. The influence of parameter $\Delta n_\text{i}/\Delta n_\text{t}$

The pulse peak power output by the acousto-optic Q modulating YAG (continuous) laser is of the kW order of magnitude while that output by the electro-optic Q modulating YAG (pulsed) laser can reach as high as the MW order of magnitude, the difference being as great as three orders of magnitude. Moreover, the pulse width of the former is over a dozen times that of the latter. The main reason is that the pump rate of the continuous krypton lamp is two to three orders of magnitude lower than that of the pulse xenon lamp. So their $\Delta n_\text{i}/\Delta n_\text{t}$ values, too, are greatly different from each other. Although it is impossible for the $\Delta n_\text{i}/\Delta n_\text{t}$ values of acousto-optic operation to be very high, to obtain good enough output performance, it is still necessary to raise the $\Delta n_\text{i}/\Delta n_\text{t}$ value as best we can. For this reason, in the course of design and operation of the laser, on the one hand, the collecting performance of the collecting cavity should be enhanced while improving the electric power to increase the optical pump rate so as to obtain the highest possible density of the initial inverted population Δn_i when the switch is in the "off" state and, on the other, to cause the threshold inverted population Δn_t to decrease when the switch is in the "on" state, it is necessary to choose highly efficient operation material and suitably designed resonator structure (e.g., shorten the resonator length as much as possible and choose the optimum transmissivity of the output mirror).

It should be noted that the pump power of the acousto-optic Q modulating laser must not be increased without a limit since if the input power is too high, the Q switch would not be turned off but cause the generation of static laser deteriorating the output characteristics of the giant pulses. So, only under the condition of enhancing the acousto-optic diffraction efficiency to increase the diffraction loss shall we be able to further increase the pump power to obtain fairly high peak power.

For example, a silicon scriber requires that the laser peak power be > 5 kW, and the repetition frequency be 4~5 kHz. The parameters of the acousto-optic Q modulating laser designed to meet these requirements are as follows:

(1) Adopt the continuous Nd:YAG laser. The operation material is a YAG rod of ϕ 5 mm×70 mm. A 7 mm × 70 mm krypton (Kr) lamp pump is used, the collecting cavity is a single elliptical one of $2a = 16.8$ mm, $2b = 19$ mm, with the lamp, rod, and collecting cavity all cooled through water, respectively. When the input electric power is 3 kW, the laser has a >10 W continuous laser output (including the insertion loss after the Q modulating device is inserted in the cavity).

(2) The acousto-optic Q modulating device. As the laser power is rather high, and the light transmission coefficient is required to be great, it is advisable to choose fused quartz as the acousto-optic medium. Quarts crystal wafers can be chosen as the transducer. Since the transducer's thickness d is related to the ultrasonic frequency, $d = \lambda_s/2 = v_s/2f_s$, the higher f_s is, the smaller the thickness of the transducer will be. So it should be pointed out that an appropriate ultrasonic frequency f_s should be chosen according to the machining conditions available. For instance, if $f_s = 40$ MHz is chosen, then $d = 0.075$ mm, which is the small dimension range that general machining can attain. Therefore, the ultrasonic frequency can be set at 40 MHz.

The dimensions of the ultrasonic column are: the characteristic length $L_0 = \lambda_s^2/\lambda = v_s^2/\lambda f_s^2 = 18.3$ mm. From the Bragg diffraction condition $L \geqslant 2L_0 = 36.6$ mm. Take $L = 48$ mm (this length is fairly suitable for the machining of both the acousto-optic medium or the transducer crystal wafer), while $H = 5$ mm, the Bragg angle $\theta_B = \lambda/2\lambda_s = 0.14°$, $90° - \theta_B = 89.86°$. The dimensions of the acousto-optic medium are specified as shown in Fig. 2.5-13. A and B are two light transmitting planes, the roughness of which is required to be $<\lambda/5$, and the nonparallelism $<10''$.

The dimensions of the transducer crystal wafer are 50 mm × 5 mm × 0.075 mm. To satisfy the need for making plate electrodes and adhesion, a Cr-Au layer is steam plated each on the two sides of the quartz crystal wafer and side C of the acousto-optic medium. The vacuum thermal pressure indium welding is adopted as the technology for adhering the transducer and acousto-optic medium.

The acousto-optic device adopts the fully water-cooled structure. And the ultrasonic high frequency driving source power is 20 W.

(3) The output of the Q modulating laser. When the input electric power is 3 kW, the average power of the laser is 12 W, the repetition frequency is 4 kHz, the output Q modulating pulse peak power is 12 kW, the pulse width is 280 ns.

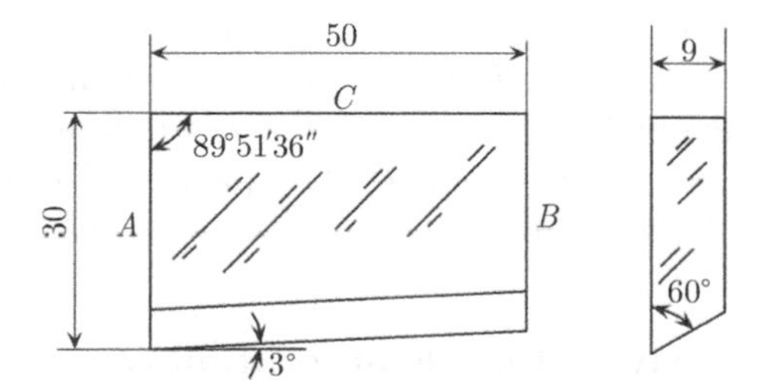

Fig. 2.5-13 The dimensions of the acousto-optic medium

2.5.4 The acousto-optic cavity dumping laser

Using the acousto-optic device as the switching component to realize "cavity dumping" is a kind of operating mode for storing (photon) energy in the cavity, the structural form of which is as shown in Fig. 2.5-14, in which M_1, M_2, M_3, and M_4 are all total reflection mirrors. The curvatures of M_2 and M_3 as well as the distance between the two mirrors should be so chosen that the centers of curvature of the two are exactly made to be focused into a very small diameter. The acousto-optic device will be placed in the beam waist portion of the light beam.

When no voltage is applied on the acousto-optic device, the resonator is in the high Q value state and extremely strong laser oscillation (but without output) can be set up in the cavity. After the density of the photon number in the cavity has reached maximum, apply a voltage on the acousto-optic device all of a sudden to form an ultrasonic field to make almost the entire laser beam undergo deflection and the photon energy in the resonator almost completely output coupled from the plane reflective mirror M_4. Hence the term "cavity dumping". Obviously, the output efficiency is rather high and the optical pulse width is very narrow, being equivalent to the time $2L/c$ of the nanosecond order of magnitude needed by the photons for a round trip in the resonator. The repetition frequency of the optical pulse can reach up to MHz order of magnitude and above, as shown in the figure. When no voltage is applied on the acousto-optic device, the optical path is as shown by the solid line in the figure. And Bragg diffraction will occur when a voltage is on; the optical path is as shown by the dashed line. After a round trip within the resonator, light is output when reflected by M_4.

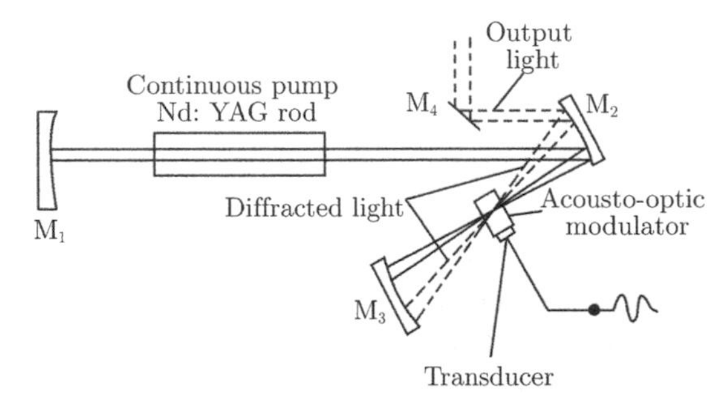

Fig. 2.5-14 The acousto-optic cavity dumping laser

Compared with the previously described Q-switching devices, acousto-optic cavity dumping devices make higher requirements on the devices. One is that in order to realize cavity dumping as best one can, the acousto-optic device used has to have only one level of diffracted light and the diffraction efficiency should be as close to 100% as possible, so it should be rigorously specified that the Bragg diffraction device be used. Second, the cavity dumping mode requires that the switching speed be much higher than the Q switching speed. Its rise time is about 5 ns and the light beam must be focused on a diameter of about 50 μm. Third, in order to enhance the Bragg diffraction efficiency, the modulating frequency of the cavity dumping device is much higher than that of the Q-switching device. So the ultrasonic frequency can be directly taken as the modulating frequency. Therefore, the repetition frequency of the output optical pulse can reach as high as the MHz order of magnitude and above.

2.6 The passive saturable absorption Q modulation

What have been described in the previous few sections are all active Q modulating methods, that is, methods of man-made utilization of certain physical effects in controlling the loss of the laser resonator to achieve an abrupt change of the Q value. This section will deal with a kind of passive Q-switching, or a method of utilizing the characteristics of certain saturable absorbers themselves to automatically change the Q value.

2.6.1 The Q modulating principle of saturable absorption dyes

Certain organic dyes are a kind of nonlinear absorbing media, that is, their absorption coefficients are not constant. Under the action of rather strong laser, their absorption coefficient will decrease with an increase in the light intensity until saturation exhibiting the transparency characteristic with respect to light. Such dyes are spoken of as saturable absorption dyes. The absorption coefficient is

$$\alpha = \frac{\alpha_0}{1 + \dfrac{I}{I_s}} \tag{2.6-1}$$

where α_0 is the absorption coefficient when the light intensity is very low ($I \to 0$) and I_s is the saturable absorption light intensity of the dye, whose magnitude is related to the kind and concentration of the dye. Generally speaking, when the concentration of a dye increases, so will the I_s value, where I is the incident light intensity. It can be seen from Eq. (2.6-1) that, when $I \gg I_s$, the absorption coefficient tends to zero, thus the dye becomes transparent with respect to the light beam passing through it. Then, when a dye (solution or solid-state slice) possessing such a performance is placed in a resonator, in the initial stage, the spontaneous fluorescence in the resonator is very faint, the dye's absorption coefficient is very great, the transmissivity of light is very low, and the resonator is in the low Q value (high loss) state. So it's impossible for laser oscillation to form. With the continued operation and action of the optical pump and the accumulation of the inverted population, the fluorescence in the cavity gradually becomes stronger. When the light intensity is comparable to I_s, the absorption coefficient of the dye becomes smaller while the transmissivity gradually becomes greater. When it reaches a definite value, the dye absorption reaches saturation value and becomes transparent upon being suddenly "bleached". Now, the Q value in the resonator increases abruptly, generating laser oscillation that outputs Q modulating laser pulses. As the pump is of the pulsed type, the light field in the cavity quickly weakens ($I \to 0$), making the dye restore its absorbing characteristic that plays the role of shutting the resonator. Then repeat the above process again.

The dye used for Q-switching must have an absorption peak with respect to the laser wavelength. For instance, dye BDN has an absorption peak near 1.06 μm and is therefore suitable for Nd:YAG laser Q modulation. The methyl solvent of cryptocyanine and chlorophyll has an absorption peak near 704 nm and so is suitable for ruby laser Q modulation. In addition, a rather large absorption cross-section is also important, requiring one much larger than that of laser materials (the absorption cross-section is in general 10^{-20} m^2 or thereabouts, while the absorption cross-section of laser operation materials is of the 10^{-24} m^2 order of magnitude) to make the saturated light intensity I_s rather low. Thus it can be "bleached" under the action of power density of the MW/cm^3 order of magnitude.

The time and spectral characteristics of the dye Q modulating laser output pulses mainly depend on the relaxation time of the dye. If the relaxation time has reached the optical cavity transition time order of magnitude, the output spectral line can be made to become narrow. Owing to the fact that the light intensity density in the continuous laser cavity usually cannot saturate the bleaching dye, the dye Q-switching can only be used in pulsed lasers.

2.6.2 The rate equation of saturation absorption

The principle of dye saturated absorption can be explained with the rate equation. By approximately regarding the dye as a two-energy level system, the resonant absorption cross-section of the dye molecules with respect to laser radiation of frequency ν can be expressed by

$$\sigma(\nu) = \frac{h\nu}{nc}[B_{12}(\nu)n_1 - B_{21}(\nu)n_2] \tag{2.6-2}$$

where $n = n_1 + n_2$ is the total dye molecule number in unit volume, in which n_1 and n_2 represent the population densities on the two energy levels, respectively, B_{12} (ν) and B_{21} (ν) represent the corresponding frequency Einstein absorption coefficient and emission coefficient, respectively; the resonant absorption cross-section is a nonlinear parameter. When illumination begins, there is $n_2 \approx 0$ and $n_1 \approx n$, which, substituted into Eq. (2.6-2), gives the initial resonant absorption cross-section

$$\sigma_0(\nu) = \frac{h\nu}{c}B_{12}(\nu) \tag{2.6-3}$$

Suppose the thickness of the dye material is l. Then the initial transmissivity of dye Q switching is

$$T_0(\nu) = \exp[-n\sigma_0(\nu) \cdot l] \tag{2.6-4}$$

Place such a dye possessing nonlinear absorption in the resonator. As the initial transmissivity is very low, laser oscillation cannot be formed. With the action of the optical pump, part of the dye molecules absorb fluorescence to transit from the ground state E_1 to high energy state E_2, causing a reduction of the absorption cross-section. If the fluorescence is further strengthened, and the absorption cross-section further reduced so that finally transition is in dynamic equilibrium, then there is $B_{12}(\nu)n_1 = B_{21}(\nu)n_2$. Now $\sigma(\nu) = 0$, and with absorption having reached saturation, the dye becomes transparent, realizing abrupt change of Q and generating laser oscillation with giant pulses output. Thereafter, because of the abrupt fall of the photon number in the cavity and the decrease in the absorption transition of the dye molecules, those molecules that have already jumped to high energy state can relax and return to the ground state, augmenting the absorption cross-section so that the resonator is again in the closed state. This is an action cycle of the dye Q-switching.

Below we shall derive the law of the variation of the absorption cross-section with time according to the rate equation of the dye molecule energy level transition. The rate equation of the high energy state dye molecule variation is

$$\frac{\mathrm{d}n_2}{\mathrm{d}t} = B_{12}(\nu)un - \{[B_{12}(\nu) + B_{21}(\nu)]u + A_{21}\}n_2 \tag{2.6-5}$$

This is a first-order constant coefficient differential equation, in which u is the density of laser radiation energy. In the equation, letting $[B_{12}(\nu) + B_{21}(\nu)]u + A_{21} = p$, $B_{12}(\nu)un = q$ and solving the above equation, we have

$$n_2 = q[1 - \exp(-p)t]/p \tag{2.6-6}$$

Substituting Eq. (2.6-6) into Eq. (2.6-2), after collation, we have

$$\sigma(\nu) = \frac{h\nu}{c} B_{12}(\nu) \frac{A_{21} + [B_{12}(\nu) + B_{21}(\nu)]u\exp(-p)t}{p} \tag{2.6-7}$$

This is the equation of variation of the dye Q-switching absorption cross-section with time and the incident laser radiation density. It can be seen from Eq. (2.6-7) that when the incident light intensity is very strong (i.e., u is very great), the absorption cross-section $\sigma(\nu)$ is very small. Now the maximum transmissivity $T_{\max} \approx 1$, showing that the dye is completely transparent when saturated. What is shown in Fig. 2.6-1 is the relationship of ferrovanadium cyanine dye transmissivity with the laser power density with respect to the ruby dye Q modulating laser. With an increase of the laser power density, the transmissivity of dye steadily increases; at $P=$ 5 MW/cm^2, $T \approx 100\%$. But it is shown by experiment that, even under very strong laser illumination, the maximum transmissivity of the dye during saturated absorption can only be close to 1.

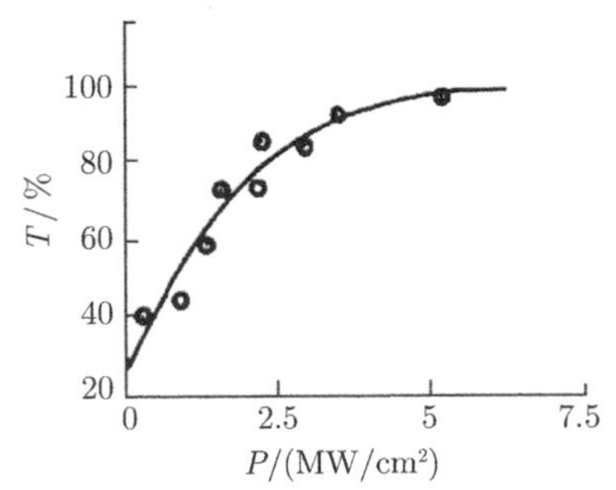

Fig. 2.6-1 The relationship between dye transmissivity and laser power density

2.6.3 The dye Q modulating laser and its output characteristics

The dye Q modulating laser is constructed just by inserting a dye cell in the resonator of the pulsed laser. It has two forms. One uses the rear wall of the dye cell as the total reflection mirror as shown in Fig. 2.6-2(a) while the other consists in inserting the dye cell alone in the cavity, as shown in Fig. 2.6-2(b). As the dye cell has many medium surfaces, to prevent the reflection from the various light transmitting surfaces from generating parasitic oscillation, the dye cell is preferably so placed that it forms a roll angle with the totally reflecting mirror.

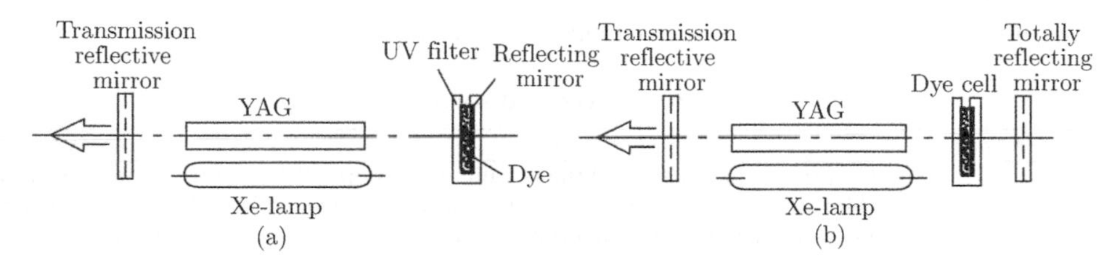

Fig. 2.6-2 The dye Q modulating laser

1. Requirements for consideration in choosing dyes

(1) The dye absorption peak should basically be in agreement with the laser wavelength. Figure 2.6-3 shows the absorption spectral lines of several saturable absorption dyes, of which the absorption peak central wavelength of the pentamethenyl dye is 1.06 μm (as shown in Fig. 2.6-3 (a)); the relative absorption spectra of BDN dye of definite concentration is as shown in Fig. 2.6-2(b); and the central wavelength of BDN's absorption peak is 700 nm, which is suitable for use in the ruby laser of a wavelength of 694.4nm.

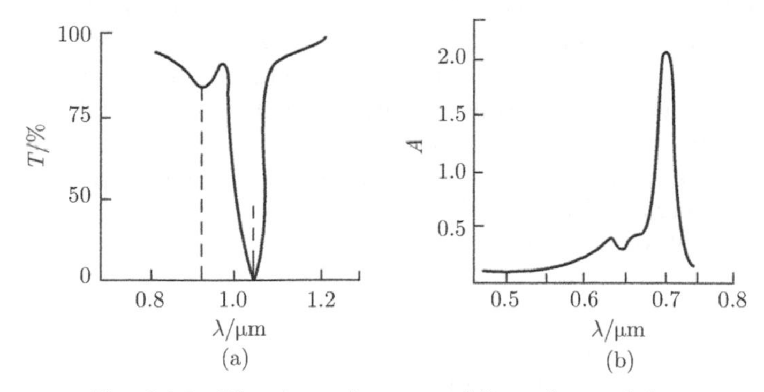

Fig. 2.6-3 The absorption spectral lines of several dyes

(2) Dyes should have suitable saturated light intensity values I_s. The saturation values needed can be obtained by changing the concentration of a dye. If I_s is too small, even a very weak light intensity will be able to make the dye "bleached" and transparent. If the accumulation of the inverted population of the laser operation material is not sufficient, then the role of optical switching cannot be effectively played. But, if I_s is too great, and it's extremely difficult for the dye to reach the saturated absorption state, then the switching speed will be too slow, thus seriously affecting the Q modulating efficiency.

(3) The dye solution should have definite stability. The preservation period of all dye solutions is rather short. For instance, the pentamethenyl chlorobenzene solution can only be preserved for 15 days when shielded from light while the hendecamethenyl acetone solution can only be preserved a single day in summer, causing definite difficulty in application. At

present, some dye solutions are made into solid-state tablets to improve their stability to a certain extent. For instance, BDN is known to have been doped in organic glass and made as thin film (of a thickness of about 0.15 mm) to make its lifetime more than a million times longer. Its transmissivity for laser depends on the concentration of the BDN dye contained in the film. In addition, there are some external factors that also have some influence on the stability of dye Q-switches. For example, strong laser radiation and illumination by the UV rays of the xenon lamp will damage the chain molecule junction and make the solution oxidize from heating, causing the dye to decompose. As a countermeasure, filters can be used to shield UV rays. To eliminate the effects of oxidization, deoxidizer can be put in the solution and the dye cell vacuum sealed or filled with inert gases. Table 2.6-1 lists several dye Q modulating materials and the corresponding solvents.

Tab. 2.6-1 Dyes and solvents

Laser operation materials	Dyes	Solvents
Ruby	Cryptocyanine, metallophthalocyanine, chlorophyll D, chlor-aluminiumthalocyanine, zircophthalocyanine	Acetone, nitrobenzene, methanol chlorobenzene
YAG, Nd:glass	Pentamethenyl, hendecamethenyl, BDN	Acetone, chlorobenzene, dichloroethane
CO_2	SF_6, BF_3	

2. The factors affecting dye Q modulating output characteristics

(1) The dye concentration. The higher the concentration of a dye, the lower its light transmissivity will be, which is equivalent to having raised the laser threshold value. This will make the operation material's initial inverted population density increase, and so the peak value power of the Q modulating pulse will be high. But, at the same time, the absorption loss of the dye in the cavity will increase, which prolongs the time for "bleaching" the dye, which is tantamount to an increase of the saturation light intensity I_s, which will, however, affect the Q modulating effect. For this reason, the Q modulating dye should have an optimum concentration (or optimal transmissivity). At this concentration, the ratio of the Q modulating laser's output energy to threshold energy has the maximum value. It has been found by measurement in experiment that, for medium-sized and small power lasers with an output of 1.06 μm wavelength, the optimum transmissivity is 50%~60%, whose corresponding dye solution concentration is the optimum concentration.

(2) The input energy. Put the Q modulating dye cell of definite concentration in the resonator, whose characteristic curves of input and output are as shown in Fig. 2.6-4. With an increase in the input energy, the energy of the output laser increases step by step, with the height of the steps approximately equal. The time characteristic of the dye Q modulating laser output can be observed with an oscilloscope. When the input energy is between E_{i1} and E_{i2} (the first step of the staircase), single pulses are output; when between E_{i2} and E_{i3} (the second step of the staircase), there appear double pulses; when between E_{i3} and E_{i4} (the third step of the staircase), there appear three optical pulses. This is because when the input energy is rather great, after the dye Q modulating laser outputs the first optical pulse, the surplus optical pump energy continues to stimulate and once again make the inverted

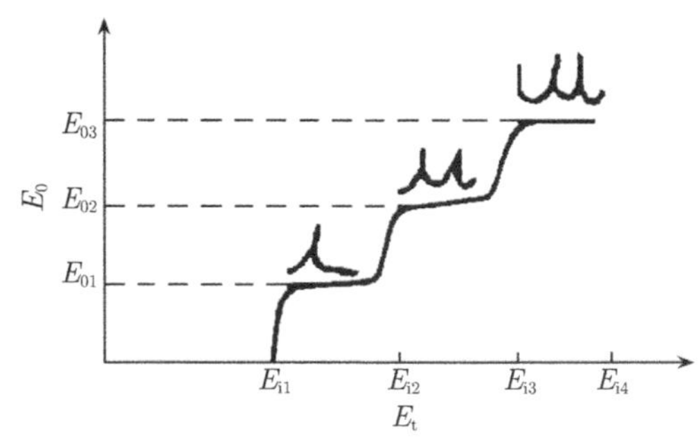

Fig. 2.6-4 Characteristic curves of dye Q modulating output

population reach Δn_t value and once again make the dye "bleached", so that the second pulse will be formed, whose energy is approximately equal to that of the first pulse. So the step heights on the output characteristic curve are approximately equal. If the optical pumping is very strong and the dye concentration is not high, a train of multiple pulses of approximately equal energy can be output. If we want to obtain single pulse output, then it is necessary that the input energy must not be sufficient to make the dye "bleached" once again. The corresponding minimum input energy that makes laser output single pulses is called the threshold energy of the dye Q modulating laser.

(3) The dye cell. This is mainly the influence of the thickness and volume of the liquid layer on the use performance. For a definite transmissivity, the thinner the dye layer is, the higher its concentration will be, while the thicker the dye layer, the lower the concentration. But the higher the percentage the solvent molecules account for will be, thus increasing such laser losses as scattering, absorption, etc. Shown in Fig. 2.6-5 are the output characteristic curves measured of two dye cells containing the same solvent (chlorobenzene) but of different thicknesses (being 3.5 mm and 8.5 mm, respectively), placed in one and the same laser one after the other. The curves show the quantity of the solvent's insertion loss. It can be seen that, under identical input energy condition, the output energy of the dye cell that is thin is high. At the same time, under the condition of maintaining identical optimum transmissivity, if the thickness is low, concentration will be high, which is conducive to the stable output of single pulses. But, the dye cell should not be made too thin, or it will hinder the flow and diffusion of the dye affecting the useful lifetime. This is especially true of lasers operating at high repetition frequencies. It is generally desirable for the dye cell to be about 1 mm thick. The dye cell shown in Fig. 2.6-6 is of a structural form that can not only reduce the dye's light transmitting thickness as best it can, but also it can help increase the flowability of the dye.

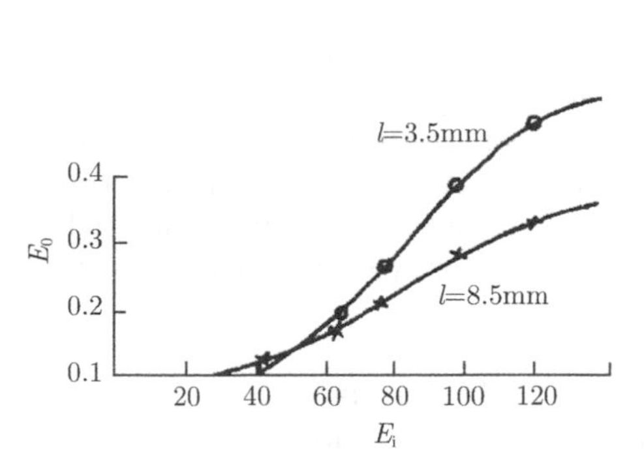

Fig. 2.6-5 Influence of liquid layer thickness on the output characteristics

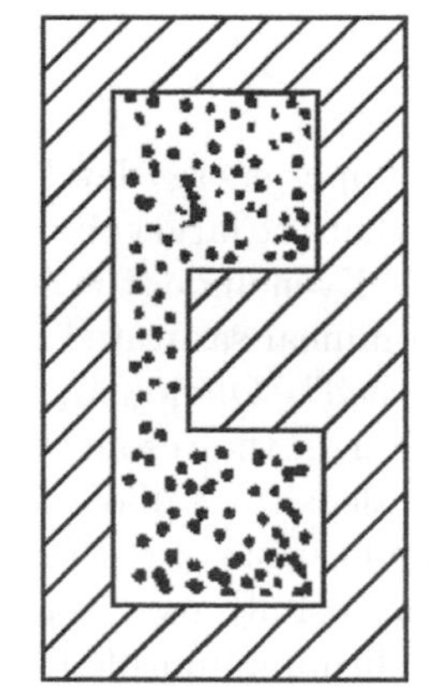

Fig. 2.6-6 The structure of the dye cell

(4) The resonator output mirror reflectivity. The output mirror reflectivity of commonly used lasers is chosen according to the gain and loss parameters in the resonator. But for the dye Q modulating lasers the selection of the output mirror reflectivity should consider that the selection should be helpful to the accumulation of the inverted population. If the output mirror reflectivity is raised, so too is the density of the initial optical power in the resonator so that the "bleaching" time of the dye is shortened. Now the inverted population can only accumulate to a rather low value and the natural output pulse peak power, too, will decrease. So the output mirror reflectivity of dye Q modulating lasers should in general take values lower than the optimum values of static-state lasers.

For instance, consider the experimenting devices of a BDN dye Q modulating laser, whose BDN dye film transamissivity $T = 30\%$. Insert the dye film in the resonator and giant pulse laser output of a single pulse energy of 51 mJ can be obtained. With the increase of the input energy, two, three, and more Q modulating pulse outputs can be obtained, the data on which are listed in Tab. 2.6-2. For BDN dye film of different concentrations, their output characteristics are listed in Tab. 2.6-3.

Tab. 2.6-2 Q modulating pulse output

Input, output	Single pulse threshold	Double pulse threshold	Three pulse threshold
Input/J	11	15	21.4
Output/mJ	51	101.5	150.5

Tab. 2.6-3 Output characteristic of BDN dye

Transmissivity	Threshold energy/J	Single pulse output/mJ
18%	16.87	59.35
36%	11.9	36.5
52%	9.2	15.2

2.6.4 LiF:F_2^- color-center crystal (saturable absorption) Q modulation

With respect to the saturable absorber, apart from certain organic dyes, it has been found that certain active color-centers also exercise very strong absorption of the incident light within their absorption zone and the oscillator strength of absorption transition is rather great. Or it can be said that the absorption cross-section σ is rather large. Hence its saturation light intensity I_s is low. According to the model for the particle transition of the two-energy level system, the medium's saturation light intensity

$$I_s = \frac{h\nu}{2\sigma\tau_a} \tag{2.6-8}$$

where τ_a is the upper energy level lifetime of the saturation absorber. It is known from the above equation that the larger the absorption cross-section σ is, the lower the saturation light intensity will be. Even under the action of a rather low light intensity, the active color-center can enter the nonlinear saturated state, making the light wave of that frequency transparent and then "bleached". Obviously, by putting such a saturable absorption material with an active color-center in the resonator of a laser, its transmissivity will quickly vary with the variation of the density of photons in the resonator, playing the role of the passive switch for changing the Q value of the resonator.

The fairly successfully used color-center saturable absorber is the F_2^- center of the LiF crystal (i.e., the lithium fluoride F_2^- color-center crystal), which not only possesses the easy-to-use merit of the dye Q switch but also it is characterized by its long-term stable optical quality like that of $LiNbO_3$ and some other crystals. What's more, the LiF crystal has a fairly high heat conductivity of 0.1 W/(cm · °C) and a high anti-optical damage threshold. So such color-center crystal passive Q modulation is very suitable for use by high repetition rate, high power lasers.

Under the illumination by the high energy γ ray, a piece of LiF crystal will have a color-center formed in the crystal and become dark brown. All kinds of color-centers formed under different illumination conditions correspond to different absorption peaks. The F_2^- center is a four-energy level system. The absorption by the F_2^- center in the LiF crystal has a strong absorption peak at 0.96 μm, with a bandwidth of 0.2 eV. This is in fairly good agreement with the emission wavelength of the Nd ions (the absorption spectrum of F_2^- is as shown in Fig. 2.6-7). So it can serve as the Q switch for YAG and neodymium glass lasers.

The absorption cross-section of the LiF:F_2^- center with respect to the 1.06 μm wavelength $\sigma \approx 2 \times 10^{-17}$ cm^2, the relaxation time $\tau_a \approx 10^{-7}$ s, and the saturated light intensity is merely 0.1 MW/cm^2. Compared with the gain medium of the Nd laser, the gain cross-section of Nd:YAG $\sigma \approx 5 \times 10^{-19}$ cm^2, and the stimulated state lifetime $\tau_g \approx 10^{-4}$ s. LiF:F_2^- is just the Q switch needed. From the point of view of application, the physical performance of the color-center crystal in question is also very good. Its optical property is homogeneous, it does not deliquesce, and its anti-light damage threshold is as high as 40 GW/cm^2. It is highly optically stable and does not change after experiencing 10^6 times of action by Nd:YAG Q modulating laser pulses and is equally thermally stable (with a thermal dissociation temperature of 130~160°C) and can be used for a long period of time at room temperature. It is also easy to machine and use. What is shown in Fig. 2.6-8 is the experimental curve of the LiF:F_2^- color-center crystal's transmissivity T under the action of a 1.06 μm laser. The static transmissivity T_0 of the LiF:F_2^- color-center crystal is related to the concentration and fabrication technology of F_2^- and the surplus transmissivity after saturation is related to the impurities in the LiF:F_2^- crystal, the alkali metal particles, etc., generated as a side effect in the process of γ-ray irradiation. They are all non-activated absorption. When the 1.06 μm effective absorption coefficient of the LiF:F_2^- crystal irradiated by C_0^{60}-γ-ray of 10^{-6}~10^{-8} rad is 0.6 cm^{-1}, the highest light density ratio of the initial static-state to the illumination saturated state can reach 14~18 fold, a ratio not inferior to the active electro-optic Q-switching. The crystal's initial transmissivity T_0 and the saturated transmissivity T during illumination directly affect the Q modulating performance. A LiF:F_2^- crystal of length 40 mm, initial transmissivity $T_0 = 15\%$, and saturated transmissivity $T = 90\%$ under illumination is used as the Q switch of the 1.06 μm Nd:YAG laser. When the laser's pump energy E_p is 37.5 J, single pulses of a giant pulse width of 25 ns are output, whose energy is 100 mJ. Compared with the free oscillation output pulse energy without the Q switch, the ratio value $\eta = 50\%$. If a LiF:F_2^- crystal of a rather high initial transmissivity ($T_0 = 64\%$) is adopted instead, then the operation of single pulses can be maintained only under a rather low optical pump energy ($E_p = 20$ J). Raising the optical pump input energy will generate multiple pulses.

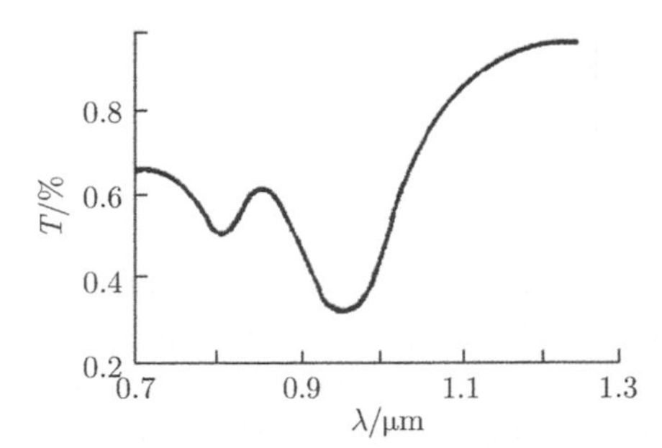

Fig. 2.6-7 Absorption spectral line of F_2^- center

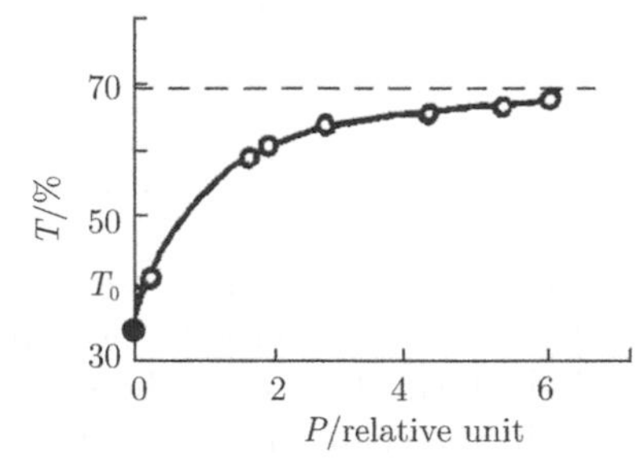

Fig. 2.6-8 Transmissivity of F_2^- center under the action of 1.06 μm laser

When the passive type Q modulating saturable absorber serves as the Q switch in a laser, the width of the spectral line becomes narrow while generating giant pulses of narrow pulse width. When the LiF:F_2^- serves as the Q switch in YAG and neodymium glass lasers, the same phenomenon can also be observed. That is, not only are giant pulses of extremely high peak power obtained, but also the role of selecting the longitudinal mode is played at the same time.

2.6.5 The diode pump passive Q modulating laser

The LD pumped completely solid laser is arousing increasing public recognition for its high efficiency, compact structure, good light beam quality, stable performance, and long lifetime. In particular, as far as the Q modulating mode operation is concerned, it has such advantages as narrow pulse width, high peak power, high repetition frequency, etc., for which it has found wide application in many fields including microprocessing, ranging, remote sensing, lidar, optical data storage, micromedicine, etc.

In contrast to the active Q modulating mode, the completely solid passive Q modulating laser does not need any external driving devices, its structure is simple and compact, the cost low, and it is easy to use. In recent years, the emergence of Cr:YAG crystals with good physicochemical performance has aroused extremely great interest of scholars both at home and abroad. The Cr:YAG passive Q modulating mode under the flashlight pump or LD pump has already been applied in Nd:YAG, Nd:YLF, Nd:S-FAP and other IR lasers, having output high repetition frequency pulsed IR laser.

It has been discovered by researchers of Changchun Optical Machinery Institute under the Chinese Academy of Sciences after in-depth investigations that, if only appropriate design parameters are chosen, even the Nd:YVO_4/Cr:YAG structure can attain highly efficient and stable Q modulating IR output. Further, if the frequency doubling crystal KTP is incorporated in the cavity, pulsed green light output of high repetition rate is obtained. This scheme has successfully been applied in the Nd:YAG/Cr:YAG/LBO structure yielding high peak power Q modulating green laser.

1. The Nd:YVO_4/Cr:YAG structured passive Q modulating IR laser

Using the continuous laser diode pump Nd:YVO_4 crystal can attain 1064 nm IR laser output. Then by inserting in the resonator a Cr:YAG crystal and we shall obtain Q modulating pulse IR laser output, the experimenting devices of which are as shown in Fig. 2.6-9.

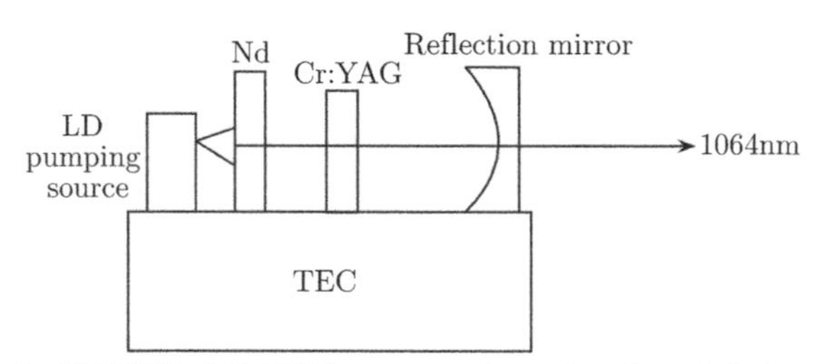

Fig. 2.6-9 Nd:YVO_4/ Cr:YAG structured passive Q modulating IR laser

The close-to-the-pump mode is adopted for experiment; that is, the light emitted by the LD is directly injected onto the immediately adjacent Nd:YVO_4 crystal. Specifically, as the oscillation light beam has the highest intensity near the central axial line, it first bleaches the positions in the center and vicinity of the axial line of the passive Q-switching Cr:YAG, while for the position farther from the axial line, as its incident light density is rather low, it's not easy to bleach and so the loss is big. Therefore, the total effect of inserting a Cr:YAG is equivalent to the insertion of a "dynamic diaphragm" in the resonator, leading to the higher-order transverse mode that makes the formation of oscillation difficult, thus improving the spatial distribution of the output light beam. In addition, the close-to-the-pump mode can avoid the loss of energy in the process of light beam transmission and reshaping, the device structure is simple and compact, and the cost is low.

The pump light threshold value measured is about 142 mW. With the increase of the injected pump power, the average power output by the Q modulating IR pulsed laser and the repetition frequency are appreciably increased, and the pulse energy and peak power, too, are increased correspondingly, while the pulse width tends to decrease. When the injected pump power is 600 mW, we can obtain a Q modulating pulsed laser output of an average power 138 mW, pulse width 19.8 ns, a repetition frequency as high as 170.1 kHz, and peak power 40.96 W. Now the system's light-light transforming efficiency is as high as 23%.

2. The Nd:YVO_4/Cr:YAG/KTP structured passive Q modulating green light laser

Using the continuous laser diode to pump the Nd:YVO_4 crystal, we obtain 1064 nm near-IR laser. Incorporate the frequency doubling crystal KTP in the cavity and we obtain 532 nm continuous green light output. Further, insert the saturable absorber Cr:YAG and we can obtain the passive Q modulating pulse green light output. The experimenting devices are as shown in Fig. 2.6-10.

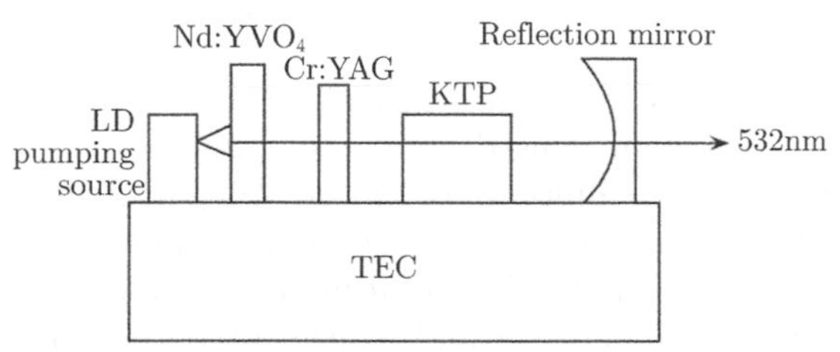

Fig. 2.6-10 The Nd:YVO_4/Cr:YAG/KTP structured passive Q modulating green light laser

The pump light threshold measured is about 240 mW. With the increase of the injected pump power, the average power and peak power of the Q modulating green light pulse output are appreciably increased, while the width of the pulse and the repetition frequency, in contrast, exhibit the tendency to increase first and decrease second. When the injected pump power is 750 mW, we obtain the passive Q modulating pulse green light output of an average power of 86 mW, pulse width of 26.4 ns, the repetition frequency as high as 79.2 kHz, the peak power 41.1 W, and the light-light transforming efficiency exceeding 11.4%. This laser continuously operates for an hour under an average power of 80 mW, the pulse peak and period stability both better than $\pm 5\%$.

What is especially worth noting is the tendency of the law of variation manifested in increase first and then decrease by pulse width and repetition frequency of the Nd:YVO_4/Cr:YAG/KTP structured passive Q modulating green light laser output with the increase of the injected pump power, which shows that under high power pumping, green light pulses of narrow pulse width and high peak, which is exactly the operating state we seek.

Fig. 2.6-11 A far field speckle photo of Q modulating pulse green laser

Figure 2.6-11 is a photograph of the far field light spot when the average power of the Q modulating pulse green light output is 80 mW, showing that it is a very good fundamental mode output and that, taking into account the "dynamic diaphragm" effect of Cr:YAG, good laser beam quality can be obtained even under the close-to-the-pump condition.

3. The Nd:YAG/Cr:YAG/LBO structured passive Q modulating green light laser

By using the continuous laser diode to pump the Nd:YAG crystal, we obtain 1064 nm continuous IR laser. Incorporating the frequency doubling crystal LBO in the cavity, we can obtain 532 nm continuous green light output. Then by inserting a saturable absorber Cr:YAG between Nd:YAG and LBO we can obtain passive Q modulating pulsed green light output.

What should be noted is that, under the action of the high repetition rate and high power density, the KTP crystal is apt to exhibit "gray line", which is unfavorable to long-term stable operation of lasers. For this reason, from the point of view of production, LBO of a rather small nonlinear coefficient but a very high anti-optical damage threshold value is used in experiment as the frequency doubling crystal. It has also been proved by experiment that for anisotropically emitted Nd:YAG laser medium, under the condition of absence of any polarization initiating element in the cavity, the class 1 phase matching with LBO is more likely to obtain high quality output light beam than the class II phase matching with the KTP crystal with intracavity frequency doubling. The experimenting devices are as shown in Fig. 2.6-12.

As the laser crystal Nd:YAG used is rather thick, it just cannot be stuck close to the pump. The coupled optical system is used, with the coupled optical part forming a high quality pumping spot, making the ellipticity of the pumping spot injected onto Nd:YAG as close to 1 as possible. At the same time, the size of the pumping spot is slightly smaller than that of the spot formed by the fundamental mode in the cavity on Nd:YAG so as to ensure full utilization of the pumping light and realize single transverse mode oscillation.

The pumping light threshold value measured is about 290 mW. With an increase of the pump power, the Q modulating green light pulse output average power and repetition frequency will appreciably increase, while the pulse width and peak power exhibit point fluctuation in small portions. When the injected pump power is 600 mW, we obtain the passive Q modulating pulse green light output of an average power of 27 mW, pulse width of 15.2 ns, a repetition frequency of 16.4 kHz, and a peak power as high as 108.1 W. The output light beam quality is of TEM_{00} mode, with no phenomenon of saturation ever appearing.

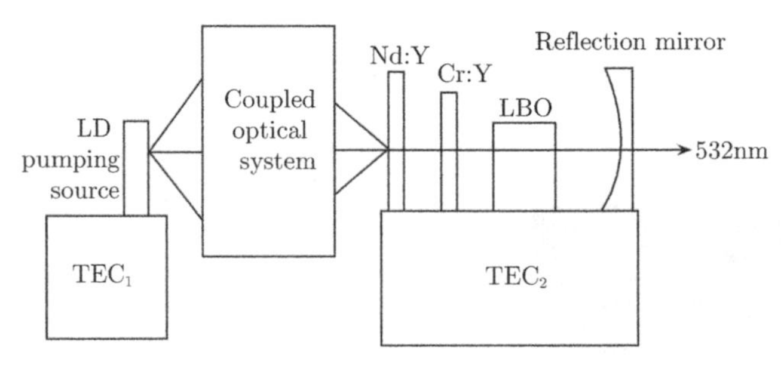

Fig. 2.6-12 The Nd:YAG/Cr:YAG/LBO structured passive Q modulating green light laser

If only the fluctuation point is not considered, the pulse width will slightly broaden with the increase in the pump power while the peak power will basically remain unchanged. Of course, in actual application, it is undesirable for fluctuation to occur. It has been found by experiment that, if only appropriate conditions are available (e.g., vary the injected pump power or change the temperature control condition, etc.), it is still easy to make the said laser output stable single peak Q modulating pulses without difficulty. Figure 2.6-13 shows the waveform of a stable green light pulse.

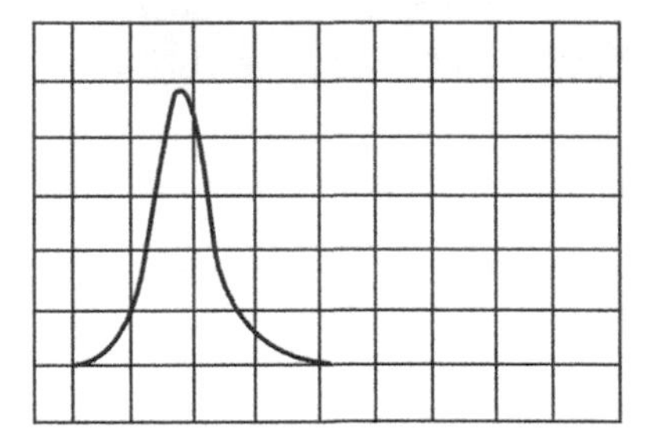

Fig. 2.6-13 Stable green light pulse waveform

2.7 A brief introduction to rotating mirror Q modulation

In the resonator of the laser, the parallelism between the two reflective mirrors directly affects the Q value of the resonator. Rotating mirror Q modulation is exactly the method for controlling the Q value by changing the parallelism of the reflective mirrors.

What is shown in Fig. 2.7-1 is a schematic diagram of the rotating mirror Q modulating laser. The total reflective mirror in the resonator of the pulsed laser is replaced with a right angle prism mounted on a rotor of a motor rotating at high speed. As it makes repeated rotations around the axial line perpendicular to the resonator, a resonator whose Q value periodicallyvaries is constituted. After the pump xenon lamp is ignited, as the prism surface is not perpendicular to the resonator axis, the reflection loss of the resonator is very great. Now the Q value of the resonator is very low, so no laser oscillation can be formed. Within this period, with the operation material stimulated by the optical pump, the inverted population at the laser upper energy level will be accumulated in large quantities. Meanwhile, the prism face is gradually turning to get close to the position perpendicular to the resonator axis. The Q value of the resonator also gradually rises and will form laser oscillation in due time and output giant pulses. This is the operation principle of rotating mirror Q modulation.

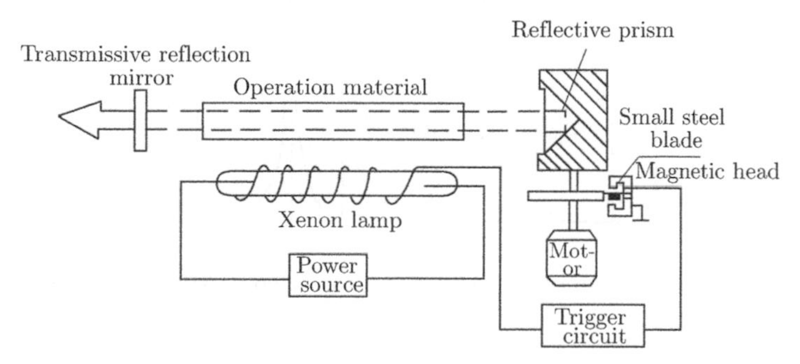

Fig. 2.7-1 Schematic diagram of the rotating mirror Q modulating laser

To enable the rotating mirror Q modulating laser to obtain stable maximum power output, a very crucial problem is accurately controlling the delay time. That is, it is required that, after the xenon lamp is ignited, there be a definite period of delay time to ensure that the inverted population will reach maximum (saturated value). This moment is just equal to the time needed by the prism to rotate to the cavity forming position (the position where the two reflection mirrors are parallel to each other) and only by making it form laser oscillation can the maximum laser power output be obtained. Therefore, neither too early nor too late a generation of laser oscillation is ideal. It has been discovered by experiment that there exists an optimum delay time. Figure 2.7-2 shows the process of operation of rotating mirror Q modulation.

To accurately control the delay time, usually the delay device shown in Fig. 2.7-3 is adopted, which consists in mounting a piece of magnetic steel on the prism support to rotate at a high speed together with the prism. When the magnetic steel rotates to where it becomes tangent to the head, the magnetic head coil will induce pulse signals, which after amplification will trigger on the xenon lamp via the trigger circuit. The magnetic head position can be adjusted and is determined according to when the normal direction of the prism surface forms a φ angle with the axial line direction of the resonator; the magnetic steel happens to pass through the magnetic head to trigger on the xenon lamp. The included angle φ is called the delay angle, and what corresponds to it is the delay time t. If the rotational speed of the motor is n(r/min), then there is the relationship of correspondence

$$t = 60\varphi/2n\pi \tag{2.7-1}$$

where t is in s, φ is in rad, and n is in r/min.

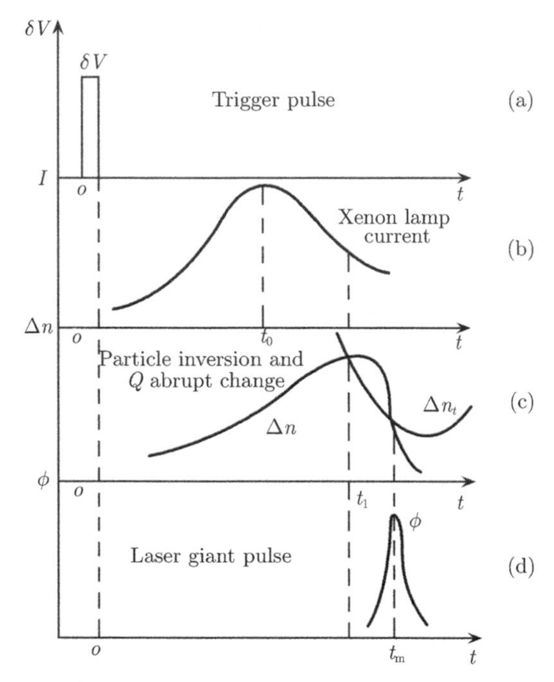

Fig. 2.7-2 The process of rotating mirror Q modulating operation

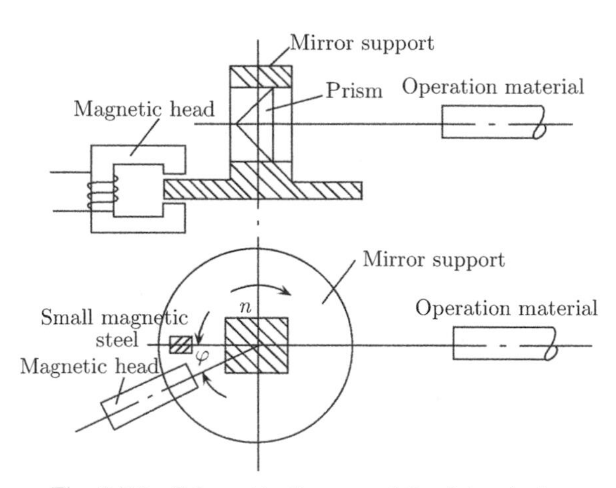

Fig. 2.7-3 Schematic diagram of the delay device

The delay time is related to the upper energy level particle lifetime of the operation material, the xenon lamp's discharge waveform, and the structure of the resonator. For different operation materials, their delay time also differs. For instance, the optimum delay time of ruby is about 1.5 ms, that of neodymium glass about 260 μs, that of YAG about 120 μs. In practice, the delay time can be estimated based on the xenon lamp waveform and the operation material upper energy level lifetime. Then calculate the φ angle and have the normal direction of the prism surface placed in a position deviating from the axial line φ angle. At the same time, place the magnetic head in the position facing the small magnetic steel and tightened. The position of the magnetic head can be adjusted later via experiment till the laser output is strongest.

Rotating mirror Q modulation is kind of slow switching. If the Q modulating parameters (e.g., the speed of motor, pump power, etc.) are not appropriately chosen, the phenomenon of multiple pulses is apt to occur. Shortening the switching time is a helpful measure to prevent the multiple pulses from emerging. So it is necessary to raise the rotational speed of the rotating mirror correspondingly. However, it has been shown by experiment and calculation that, for every particular situation, there exists an optimum rotational speed, that is, there is an optimum switching time. There is a time for forming Q modulating pulses. If it happens, the rotating mirror has rotated to the cavity forming position when giant pulses reach the peak; that is to say, at the moment giant pulses are formed, the Q value is maximum and the loss is minimum. Thus, the giant pulses output will be in the optimum state of high peak power and small pulse width. Now the corresponding speed of the rotating mirror is the optimum speed.

To sum up, there is a great variety of operating modes of Q modulating lasers, each with its own characteristics and which are optional for different applications. They are briefly described as follows:

(1) Rotating mirror Q modulation. This is a kind of Q-switching developed fairly early. As its switching time is approximately equal to the pulse time, it is of the slow switching type. When using this type of switch, special care should be taken to select the optimum rotational speed so as to eliminate the generation of multiple pulses. As this kind of switch has no insertion loss, there does not exist the problem of optical damage either. Therefore it can be used in pulsed lasers of rather great energy to obtain giant pulses with a pulse power of over dozens of MW and a pulse width of the ns order of magnitude. Its main shortcoming is that the mechanical wear at high revolving speed will affect the useful lifetime. Also, rather high requirements are made on the assembling technology. For these reasons, thistype of Q switch is basically no longer adopted at present.

(2) Electro-optic crystal Q modulation. As its switching time is mainly dependent on the rising and depressurizing time of the circuit high voltage pulses, it is in general possible to make it shorter than the time for setting up the pulses. It is thus of the fast switch type that is capable of generating narrow pulses and giant pulses of good synchronized performance, long lifetime, and stable output giant pulses of a peak power of over dozens of MW and a pulse width of longer than ten ns. Therefore it is the currently most widely applied Q-switching, whose main drawback is its rather high half-wave voltage requiring high voltage pulses of several thousand volts and it is apt to interfere with other electronic circuits.

(3) Acousto-optic Q modulation. Of the fast switching type, its switching time is shorter than the pulse setting up time. The switch needs a modulating voltage of only more than a hundred volts and is easily coordinated with the continuous laser for Q modulation to obtain giant pulses of a kHz high repetition rate. In addition, the pulse repeatability is good and giant pulses of a peak power of several hundred kW and a pulse width of dozens of ns can be obtained. But because of its poor switching capability for high energy lasers, it can only be used in low gain continuous lasers.

(4) Saturable absorber Q modulation. Of the passive type fast switching, such a Q switch has a simple structure, is convenient to use, and is free from electrical interference. Giant pulses of a peak power of several MW and a pulse width of over ten ns can be obtained using this modulation. Its main drawback is, as it is a passive type Q-switching, the time for generating the Q modulating pulses has definite randomness and is beyond man-made control. Furthermore, the dye is deterioration prone, should be replaced from time to time,

and has an unstable output.

Exercises and Questions for Consideration

1. Give an account of the principle of obtaining giant pulses of high peak power using Q modulating technology and briefly describe how the various parameters vary with time in the process of formation of the Q modulating pulses.

2. As the operation material of the Q modulating laser, what conditions should it possess? Why?

3. There is an electro-optic Q modulating YAG laser with a polarizing prism. Try to answer or calculate the following questions:

(1) Draw a schematic diagram of the Q modulating laser's structure and mark the directions of the polarizing mirror's polarizing axes relative to the various main axes of the electro-optic crystal.

(2) How should we adjust the polarization initiating direction of the polarizing prism relative to the position of the crystal to obtain ideal switching effect?

(3) Calculate 1/4 wavelength voltage $V_{\lambda/4}$ ($l = 25$ mm, $n_o = n_e = 1.05$, $\gamma_{63} = 23.6 \times 10^{-17}$ m/V).

4. Why does acousto-optic Q modulation operate in the traveling wave operating state, which is in general suitable only for the operation of the continuous laser at a high repetition frequency? Is it necessary to modulate the high frequency signals applied on an electro-acoustic transducer with a pulse voltage of frequency f?

5. When an ultrasonic wave of a frequency $f_s = 40$ MHz sets up an ultrasonic field ($v_s = 5.96 \times 10^5$ cm/s) in a fused quartz acousto-optic medium (n= 1.54), calculate the incidence angle θ of an incident light of wavelength $\lambda = 1.06$ μm for satisfying the Bragg condition.

6. An acousto-optic Q modulating device of ($L = 50$ mm and $H = 5$ mm) made of fused quartz is used in the continuous YAG laser for Q modulation. It is known that the laser's single path gain is 0.3 and the electro-acoustic efficiency of the acousto-optic device is 40%. (1) How high should the driving power P_s of the acousto-optic device be? (2) If the acousto-optic device is to operate in the Bragg diffraction region, how high should its acoustic field frequency be?

References

[1] Lan Xinju, Huang Guobiao et al., Laser Technology, Changsha, Science and Technology Press, 1988.

[2] W. Koechner, Solid-state Laser Engineering, Springer Verlag, 1976 (translated by Huaguang, Science Press, 1983).

[3] A.Yariv. Quantum Electronics. 2nd Ed. New York: John Wiley Sons, Inc., 1975.

[4] Editorial Group of Laser Physics, Laser Physics, Shanghai, Shanghai People's Press, 1976.

[5] A. E. Siegman. Lasers. Printed in the United States ofAmerica, 1986.

[6] Laser Technology, compiled and translated by Tianjin University, Beijing, Science Press, 1972.

[7] J. T. Verdegen. Laser Electronics. New Jersey:Prentice-Hall, Inc., 1981.

[8] Orazio Svelto (author). C. David, Hanna (translator).Principles of Lasers (3rd Ed). New York: Plenum Press, 1989.

[9] A. Szabo and R. A. Stein. Theory of Laser Giant Pulsingby a Saturable Absorber. *J. Appl. Phys.*, 1965, 36, pp.1562–1566.

[10] A. L. Egorov, V. V. Korobkin, and R. V. Serov.Single-frequency Q-Switched Neodymium laser. *Sov. J. Quant.Electr.*, 1975, 5, pp. 291–293.

[11] Wu Hongxing et al., Fast Switch Q Modulating Theory and Slow Switch QModulating Theory, Applied Laser, 1984, 4(6).

[12] Wu Hongxing et al., Laser Multiple Function QModulating Technology and Experimenting Research, Quantum Electronics, 1993, 10 (2).

[13] Crystal R&D Group, Shanghai Opto-mechanical Institute, Chinese Academy of Sciences, Near-optic Axis Electrooptic Modulation and Monolithic Crystal Laser Q Switch, Laser, 1975, 2 (2), pp.8–19.

[14] E. O. Ammann and J. M. Yarborough. Mode-SelectionTechnique for Continuously Pumped Repetitive Q-Switched Lasers.*Appl. Phys. Lett.*, 1972, 20, pp. 117–120.

[15] D. J. Kuizenga. Short-Pulse Oscillator Development forthe Nd:Gelass Laser-Fusion Systems. *IEEE J. Quant. Electr.*,1981, QE-17.

[16] R. B. Chesler, M. A. Kara, and J. E. Geusic. AnExperimental and Theoretical Study of High Repetition Rate$Q$-Switching in Nd:YAG Lasers. *Proc. IEEE*, 1970, 58, pp.1899–1914.

[17] Dong Huifang et al., A Study of the Dioptric Matching Liquid, Laser and IR, 1983, 12.

CHAPTER 3

Ultrashort Pulse Technology

3.1 Overview

Ultrashort pulse technology is an important means for such disciplines as physics, chemistry, biology, optoelectronics, and laser spectroscopy to study the microscopic world and reveal new ultrafast processes. The development of ultrashort pulse technology has experienced the stages of active mode-locking, passive mode-locking, synchronously pumped mode-locking, collision pulse mode locking (CPM), as well as additive pulse mode-locking (APM) that emerged in the 1990s or the coupled cavity mode-locking (CCM) and self-mode-locking. Since the realization of laser mode-locking in the 1960s, the mode-locked optical pulse width had reached nanosecond and subnanoseond 10^{-9} ~10^{-10} s order of magnitude by the mid- and late 1960s. By the mid- and late 1970s, the pulse width had reached the subpicosecond (10^{-13} s) order of magnitude and by the 1980s, there had appeared an upsurge, that is, definite breakthroughs had been made in both theory and practice, with the width of the ultrashort pulse entering the femtosecond (10^{-15} s) stage. In 1981, R.L. Fork and others with Bell Laboratories of the U.S. proposed the theory of collision mode-locking and realized collision mode-locking in a 6-mirror ring-shaped cavity by obtaining a stable optical pulse sequence of 90 fs. Following the adoption of the optical pulse compressing technique, 6 fs optical pulses were obtained. The emergence of the self-mode-locking technology in the 1990s made it possible to obtain the ultrashort typical pulse sequence of 8.5 fs in the titanium-doped sapphire self-mode-locked laser.

This chapter will be devoted to a discussion of the principle, characteristics, method of implementation, several typical mode-locked lasers, and the relevant ultrashort pulse techniques, such as the method of measuring the pulse width of the ultrashort pulses, the compression technique of ultrashort pulses, etc.

3.1.1 The output characteristics of multimode lasers

In order to better understand the principle of mode-locking, we shall first discuss the output characteristics of multiple longitudinal mode-free operating lasers that have not undergone mode-locking. For a laser of cavity length L, the frequency interval of the longitudinal mode is

$$\Delta\nu_q = \nu_{q+1} - \nu_q = \frac{c}{2L} \tag{3.1-1}$$

The output of the free operating laser in general contains a number of longitudinal modes above the threshold, as shown in Fig. 3.1-1. Neither the amplitude nor the phase of these modes is fixed. The variation of the laser output is a result of its irregular superposition and a time-averaged statistical value.

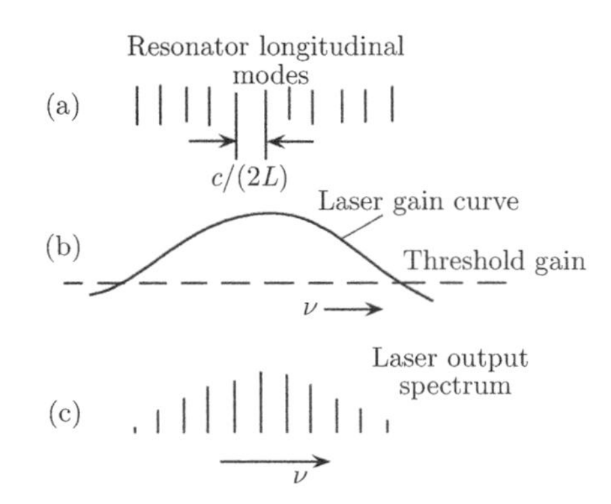

Fig. 3.1-1 The interaction between laser gain curve and resonator longitudinal modes

Suppose N longitudinal modes are contained within the net gain line width of the laser operation material. Then the electric field of the laser output is the sum of N longitudinal mode electric fields, i.e.,

$$E(t) = \sum_{q=0}^{N} E_q \cos(\omega_q t + \varphi_q) \tag{3.1-2}$$

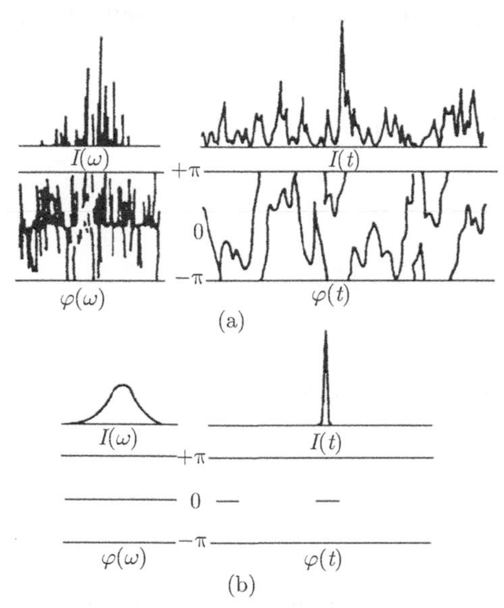

Fig. 3.1-2 The signal structure modes of the non-mode-locked and ideally mode-locked lasers: (a) non-mode-locked; (b) ideally mode-locked

where $q = 0, 1, 2, \cdots, N$ is the ordinal of the qth of the N longitudinal modes in the laser, ω_q and φ_q are the angular frequency and phase of the mode of longitudinal mode ordinal q, and E_q is the field intensity of longitudinal mode ordinal q. In general, the phases φ_q of the N longitudinal modes are unrelated to one another; that is, they are not associated with one another in time, but are completely independent and stochastic. This can be represented by $\varphi_{q+1} - \varphi_q \neq$ constant. On the other hand, influenced by various irregular perturbations such as the thermal deformation of the laser operation material and cavity length, the phases themselves of different longitudinal modes will be found to drift; that is, their respective phases on the temporal axis are unstable. φ_q itself is not a constant. Thus, the coherence condition among the longitudinal modes is destroyed. So the total light field of the laser output is a result of the irregular superposition of all the light fields of different frequencies. Their light field intensities, too, irregularly fluctuate with time. Figure 3.1-2 gives the time-depicted and frequency-depicted graphs of the non-mode-locked laser pulses and completely mode-locked laser pulses. Within the frequency domain, the optical pulse can be written as

$$\nu(\omega) = \alpha(\omega) \exp[-\mathrm{i}\varphi(\omega)] \tag{3.1-3}$$

where $\alpha(\omega)$ is the amplitude frequency spectrum and $\varphi(\omega)$ is the phase frequency spectrum. When the pulse bandwidth $\Delta\omega$ is narrower than the average optical frequency ω_0, the optical pulse in the time domain can be written as

$$V(t) = A(t) \exp\{\mathrm{i}[\varphi(t) - \omega_0 t]\} \tag{3.1-4}$$

where $A(t)$ is the amplitude of the pulse and $\varphi(t)$ is the phase. In Fig. 3.1-2, suppose the laser operates in 101 discrete longitudinal modes with equal frequency interval $\Delta\omega$. In the time domain, the repetition period of the field is $2\pi\ (\Delta\omega)^{-1}$, corresponding to twice the transition time in the laser resonator. It is known from Fig. 3.1-2(a) that, when the laser is in the free operating mode, within the range of the frequency spectrum, its laser frequency spectrum is made up of the equally spaced $(c/2L)$ separate spectral lines. Their amplitudes are irregular while the phases are randomly distributed between $-\pi$ and $+\pi$. Within the time domain, their phases, too, irregularly fluctuate from $-\pi$ to $+\pi$ while the intensity distribution is characterized by noise. When detecting the optical power of the non-mode-locked laser output with the receiving elements, the light intensity $I(t)$ received is the superimposition of all the longitudinal mode light intensities that satisfy the threshold condition. Now, the output light intensity in a certain instant is

$$I(t) = E^2(t) = \sum_q E_q^2 \cos^2(\omega_q t + \varphi_q) + 2\sum_{q \neq q'} E_q E_{q'} \cos(\omega_q t + \varphi_q) \cdot \cos(\omega_{q'} t + \varphi_{q'}) \quad (3.1\text{-}5)$$

The light intensity received is the average value in a period of time (t_1) better than $2\pi/\omega_q$; the average light intensity is

$$\overline{I(t)} = \overline{E^2(t)} = \frac{1}{t_1}\sum_q \int_0^{t_1} E^2(t)\mathrm{d}t$$

As

$$\frac{1}{t_1}\sum_q \int_0^{t_1} E_q^2 \cos^2(\omega_q t + \varphi_q)\mathrm{d}t = \sum_q \frac{1}{2}E_q^2$$

$$\frac{1}{t_1}\sum_{q \neq q'} \int_0^{t_1} E_q \cdot E_{q'} \cos(\omega_q t + \varphi_q) \cdot \cos(\omega_{q'} t + \varphi_{q'})\mathrm{d}t = 0$$

so

$$\overline{I(t)} = \sum_{q=0}^{N} \frac{1}{2}E_q^2 \quad (3.1\text{-}6)$$

This equation shows that the average light intensity is the sum of all the longitudinal mode light intensities.

If appropriate measures are taken to make the longitudinal modes independent of one another synchronous in time, that is, have their phases associated with one another so that they will be in a definite relation ($\varphi_{q+1} - \varphi_q$ = constant), then there will be an interesting phenomenon essentially different from the above-mentioned situation: what is output by the laser will be optical pulses of extremely narrow pulse width and very high peak power, as shown in Fig. 3.1-2(b). That is to say, the phases of all the modes of this laser are already locked in accordance with the $\varphi_{q+1} - \varphi_q$ = constant relationship. Such a laser is referred to as the mode-locked laser, the corresponding technique the "mode-locking technique".

3.1.2 The basic principle of mode-locking

To obtain ultrashort optical pulses of a narrow pulse width and high peak power, the only way is adopt the mode-locking method, that is, make the neighboring frequencies equally spaced and fixed at $\Delta\nu_q = \dfrac{c}{2L}$. This can be realized in single transverse mode lasers.

Below we shall analyze the relationship between the laser output and phase locking. For convenience in operation, suppose all the oscillation modes of a multi-mode laser possess equal amplitudes E_0, and there are $2N+1$ longitudinal modes in all that exceed the threshold value. For the modes located in the center of the medium gain curve, their angular frequency is ω_0, the initial phase is 0, and its mode ordinal $q = 0$; that is, with the central mode as the reference, the phase difference between all neighboring modes is α, and the frequency interval between the modes is $\Delta\omega$. Suppose the qth oscillating mode is

$$E_q(t) = E_0 \cos(\omega_q t + \varphi_q) = E_0[(\omega_0 + q\Delta\omega)t + q\alpha] \quad (3.1\text{-}7)$$

where q is the ordinal of the oscillation longitudinal modes in the cavity. The total light field of the laser output is a result of the coherence of the $2N+1$ longitudinal modes:

$$\begin{aligned} E(t) &= \sum_{q=-N}^{N} E_0 \cos[(\omega_0 + q\Delta\omega)t + q\alpha] \\ &= E_0 \cos\omega_0 t\{1 + 2\cos(\Delta\omega t + \alpha) \\ &\quad + 2\cos[2(\Delta\omega t + \alpha)] + \cdots + 2\cos[N(\Delta\omega t + \alpha)]\} \end{aligned}$$

Using the triangular function relationship

$$\cos\beta + \cos(2\beta) + \cdots + \cos(N\beta) = \frac{\sin\left(\dfrac{1}{2}N\beta\right)\cos\left[\dfrac{1}{2}(N+1)\beta\right]}{\sin\left(\dfrac{1}{2}\beta\right)} \tag{3.1-8}$$

we obtain

$$E(t) = E_0 \cos(\omega_0 t)\frac{\sin\left[\dfrac{1}{2}(2N+1)(\Delta\omega t + \alpha)\right]}{\sin\left[\dfrac{1}{2}(\Delta\omega t + \alpha)\right]} = A(t)\cos(\omega_0 t) \tag{3.1-9}$$

$$A(t) = E_0 \frac{\sin\left[\dfrac{1}{2}(2N+1)(\Delta\omega t + \alpha)\right]}{\sin\left[\dfrac{1}{2}(\Delta\omega t + \alpha)\right]} \tag{3.1-10}$$

It is known from Eqs. (3.1-8)~(3.1-10) that after the $2N+1$ modes of oscillation have undergone phase locking, the total light field becomes an amplitude modulated wave of frequency ω_0. $A(t)$ is a periodic function varying with time. The light intensity $I(t)$ is proportional to $A^2(t)$ and is also a function of time, with the light intensity modulated. According to Fourier analysis, the total light field is composed of $2N+1$ longitudinal mode frequencies. Therefore, the laser output pulse is a light wave including $2N+1$ longitudinal modes. Figure 3.1-3 gives the output light intensity curve of 7 oscillating modes.

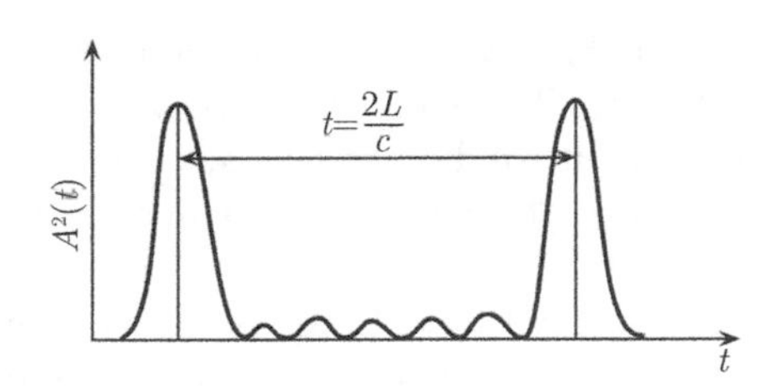

Fig. 3.1-3 The output light intensity of 7 oscillating modes

It is clear from the above analysis that, as long as the variation of the amplitude $A(t)$ is known, the characteristics of the output laser can be known. For convenience of discussion, we assume $\alpha = 0$. Then

$$A(t) = E_0 \frac{\sin\left[\dfrac{1}{2}(2N+1)\Delta\omega t\right]}{\sin\left(\dfrac{1}{2}\Delta\omega t\right)} \tag{3.1-11}$$

As the numerator and denominator of the above equation are both periodic functions, $A(t)$, too, is one. If only its period, extremum, and zero point are found, the law of variation of $A\ (t)$ can be obtained.

From Eq. (3.1-11) we can find the period of $A(t)$ as $2L/c$. Within a period there are $2N$ zero value points and $2N+1$ extremum points.

At $t = 0$ and $t = 2L/c$, $A(t)$ takes maximum value. As the numerator and denominator of $A(t)$ are zero at the same time, using the Robert rule, we can find the amplitude at this moment to be $(2N + 1)E_0$.

At $t = L/c$, $A(t)$ takes minimum value $\pm E_0$. When N is an even number, $A(t) = E_0$, when N is an odd number, $A(t) = -E_0$. Apart from the $t = 0$, L/c, $2L/c$ points, $A(t)$ possesses $(2N - 1)$ submaximum values.

As the light intensity is proportional to the $A^2(t)$, the maximum value at $t = 0$ and $t = 2L/c$ is called the main pulse. Between two adjacent main pulses. There are in all $2N$ zero points in addition to $(2N - 1)$ submaximum values; these pulses are called secondary pulses. Also, there are in general very many mode-locked longitudinal modes in a mode-locked laser, so the value of the secondary pulse is usually negligible while the interval between the two main pulses $\tau = 2L/c$ happens to be the time used for an optical pulse to make a round trip in the cavity. So the mode-locked oscillation can also be understood as there being only one optical pulse propagating back and forth in the cavity.

It is known from the above brief analysis that the mode-locking with a laser's multiple longitudinal modes has led to the following phenomena of significance.

1) The laser's output is a regular pulse sequence of interval $\tau = 2L/c$.

2) The width of every pulse $\Delta\tau = \dfrac{1}{(2N+1)}\dfrac{1}{\Delta\nu_q}$, or approximately equal to the reciprocal of the oscillating line width. The width $\Delta\tau$ of the main pulse is defined as the time interval for the pulse peak value to fall to the first zero value. As the oscillating line width cannot exceed the laser's net gain line width $\Delta\nu_{\mathrm{q}}$, under the extreme condition, $\Delta\tau_{\min} = 1/\Delta\nu_{\mathrm{q}}$. It is clear that the broader the gain line width is, the more likely will it be to obtain a narrow mode-locking pulse width. For instance, if a neodymium glass laser of $\Delta\nu_{\mathrm{q}} = 20{\sim}30$ nm is used for mode-locking, narrow pulses of $10^{-12}{\sim}10^{-13}$ s can be obtained. But in gas lasers, $\Delta\nu_{\mathrm{q}}$ is generally very small. For instance, for the Ne-He laser, $\Delta\nu_{\mathrm{q}} = 2 \times 10^{-3}$ nm, so no pulses narrower than 1 ns can be obtained.

3) The peak power of the output pulse is proportional to $E_0^2(2N + 1)^2$ while the average power of a free operating laser is proportional to $E_0^2(2N + 1)$. Therefore, because of mode-locking, the peak power has increased by $(2N + 1)$ times. In solid-state lasers, there can be as many as $10^3 \sim 10^4$ oscillating modes, so the peak power of a single pulse can be very high.

4) The result of phase-locking for the multi-mode $(\omega_0 + q\Delta\omega_{\mathrm{q}})$ laser is the realization of $\varphi_{\mathrm{q+1}} - \varphi_{\mathrm{q}} =$ constant, leading to the output of a pulse sequence of high peak power and narrow pulse width. Therefore, after a multiple longitudinal mode laser has undergone mode-locking, power coupling will take place among the oscillating modes, which will no longer be independent. The power of each of the modes should be regarded as being provided by all the oscillating modes.

3.1.3 The method of mode-locking

Mode-locking was first realized in the He-Ne laser using the acousto-optic modulator. Later, the same thing was done using the internal modulating method in other lasers of argon ion, carbon dioxide, ruby, yttrium aluminum garnet, etc. Still later, there appeared the saturable absorption dye mode-locking. With the development of the mode-locking techniques, the development of the ultrashort pulse measuring techniques, too, was pushed forward, the latter, in turn, promoted the advance of the mode-locking technique. Much research was done on the locking of the transverse mode and shortly after investigations were made on simultaneous locking of the longitudinal and transverse modes. After the 1970s, such technologies as the active plus passive mode-locking, double mode-locking (loss modulation plus phase modulation), mode-locking plus Q modulation, and synchronous mode-locking were followed by the realization of collision mode-locking self-mode-locking, etc. This section will mainly discuss the active mode-locking, passive mode-locking, self-mode-locking, synchronous pump mode-locking, etc.

1. The active mode-locking

What is adopted for the active mode-locking is the method of the periodic modulation of the resonator parameters, that is, the insertion of an externally controlled modulator in the laser resonator to periodically change the amplitude or phase of the oscillating mode in the cavity at a definite frequency. When the modulating frequency chosen is equal to the longitudinal mode interval, the modulation of the modes will generate marginal frequency, whose frequency is in agreement with those of two neighboring longitudinal modes. Because of the interaction between modes, all the modes will become synchronous when sufficiently modulated to form mode-locked pulse sequences.

2. The passive mode-locking

Another effective method of generating ultrashort pulses is the passive mode-locking. This method is implemented by placing a saturable absorber in a laser resonator. The saturable absorber is a kind of nonlinear absorptive medium, the absorption of the laser in the resonator varying with the intensity of the light field. When the light field is rather weak, the absorption of light is very strong, so the light transmissivity is very low. With an increase in the laser intensity, the absorption is reduced. When reaching a definite value, absorption is saturated, the light transmissivity reaching 100%, making the laser pulse of maximum strength suffer the minimum loss, thereby yielding very strong mode-locked pulses. This is similar to, yet different from, the passive Q switch. The passive mode-locking requires that the lifetime of the saturable absorber's upper energy level be exceptionally short and when placed in the cavity, the absorber be immediately close to the reflective mirror.

3. Self-mode-locking

By self-mode-locking we mean a method with which longitudinal mode-locking can be realized without having to insert any modulating element in the resonator when the nonlinear effect of the activated medium itself is capable of maintaining the equally spaced distribution of the longitudinal mode frequencies plus an ascertained initial phase relationship. The titanium-doped sapphire self-mode-locked laser is the currently most sought-after research topic as well as the most practical device in great demand in its field of application.

4. The synchronous pump mode-locking

Active mode-locking is realized by periodically modulating the loss or optical path of the resonator. If mode-locking is to be realized by periodically modulating the gain of the resonator, then this can be done by adopting the pulse sequence of an active mode-locked laser to pump another laser. This method is referred to as the synchronous pump mode-locking, whose advantage lies in that pulses much smaller than the pumped pulse width can be obtained during periodic pumping. In addition, in synchronous pumping dye lasers, the ultrashort pulse's frequency generated is continuously adjustable within a definite wavelength range.

3.2 Active mode-locking

Active mode-locking consists in inserting a modulator in the laser resonator, whose modulating frequency should be exactly equal to the longitudinal mode interval. If so, a mode-locked pulse sequence of a pulse repetition rate of $f = c/2L$ can be obtained.

According to the principle, modulation can be divided into phase modulation (PM) [(or frequency modulation (FM)] mode-locking and amplitude modulation (or loss modulation) mode-locking. Below we shall discuss their principles and method of realization.

3.2.1 The amplitude modulated mode-locking

Amplitude-modulated mode-locking can be realized by using either the acousto-optic or electro-optic modulator. As the frequency of loss modulation is $c/2L$, the period of modulation happens to be the time needed by the optical pulse to make a round trip in the resonator. Therefore, in its passage through the modulator, the laser beam that propagates back and forth in the resonator is always located in the same modulating period part. If the modulator is placed at one end of the cavity and at a certain moment t_1 the loss suffered by the optical signal passing through the modulator is $\alpha(t_1)$, then during the period the pulse makes a round trip transit $\left(t_1 + \frac{2L}{c}\right)$, this optical signal will suffer the same loss, $\alpha\left(t_1 + \frac{2L}{c}\right) = \alpha(t_1)$. If $\alpha(t_1) \neq 0$, then every time this part of the signal makes a round trip in the resonator the signal will suffer a loss once. If the loss is greater than the gain in the cavity, this part of the light wave will finally vanish while the light that passes through the modulator at moment $\alpha(t_1) = 0$ can always do so loss-free. Furthermore, this light wave will be steadily amplified in passing through the operation material back and forth in the resonator, making the amplitude increasingly great. If the loss and gain in the cavity can be suitably controlled, then a pulse sequence output of a very narrow pulse width and a period of $2L/c$ will be formed.

Taking the simplest case of sinusoidal modulation as an example, we can discuss the basic principle of amplitude modulation from the frequency characteristics. Suppose the modulating signal

$$a(t) = A_{\mathrm{m}} \sin\left(\frac{1}{2}\omega_{\mathrm{m}} t\right) \tag{3.2-1}$$

where A_{m} and $\frac{1}{2}\omega_{\mathrm{m}}$ are the amplitude and angular frequency of the modulating signal, respectively. When the modulating signal is of zero value, the loss in the cavity is minimum whereas when the modulating signal is equal to positive and negative maximum, the loss in the cavity is invariably of the maximum value. So the frequency of loss variation is twice that of the modulating signal, and the loss factor

$$\alpha(t) = \alpha_0 - \Delta\alpha_0 \cos(\omega_{\mathrm{m}} t) \tag{3.2-2}$$

where α_0 is the average loss of the modulator, $\Delta\alpha_0$ the amplitude of loss variation and ω_m the angular frequency of loss variation in the cavity, whose frequency is equal to the longitudinal mode frequency interval $\Delta\nu_{\mathrm{q}}$. The transmissivity of the modulator

$$T(t) = T_0 + \Delta T_0 \cos(\omega_{\mathrm{m}} t) \tag{3.2-3}$$

where T_0 is the average transmissivity and ΔT_0 the amplitude of transmissivity variation. With the modulator placed in the cavity, when no modulating signal is applied, the modulator loss

$$\alpha = \alpha_0 - \Delta\alpha_0 \tag{3.2-4}$$

where α is a constant, which represents such losses of the modulator as absorption, scattering, reflection, etc. The transmissivity

$$T = T_0 + \Delta T_0 \tag{3.2-5}$$

Furthermore,

$$\alpha + T = 1 \tag{3.2-6}$$

Suppose the light field in the cavity before modulation is

$$E(t) = E_c \sin(\omega_c t + \varphi_c) \tag{3.2-7}$$

After getting modulated, the light field in the cavity becomes

$$\begin{aligned} E(t) &= E_c T(t) \sin(\omega_c t + \varphi_c) = E_c[T_0 + \Delta T_0 \cos(\omega_m t)] \sin(\omega_c t + \varphi_c) \\ &= A_c[1 + m \cos(\omega_m t)] \sin(\omega_c t + \varphi_c) \end{aligned} \tag{3.2-8}$$

where $A_c = E_c T_0$ is the amplitude of the light wave field; $m = \dfrac{E_c \Delta T_0}{A_c}$ is the modulating coefficient of the modulator. To ensure distortion-free modulation, we should take $m < 1$. Shown in Fig. 3.2-1 are the waveform graphs of the principle of loss modulating mode-locking in the time domain. Figure 3.2-1(a) is the waveform of the modulating signal; (b) that of the loss in the cavity, the frequency being twice that of the modulating signal; (c) that of the modulator's transmissivity; (d) the unmodulated optical wave electric field in the cavity; (e) the modulated optical waves electric field in the cavity; and (f) the optical pulse of the mode-locked laser output.

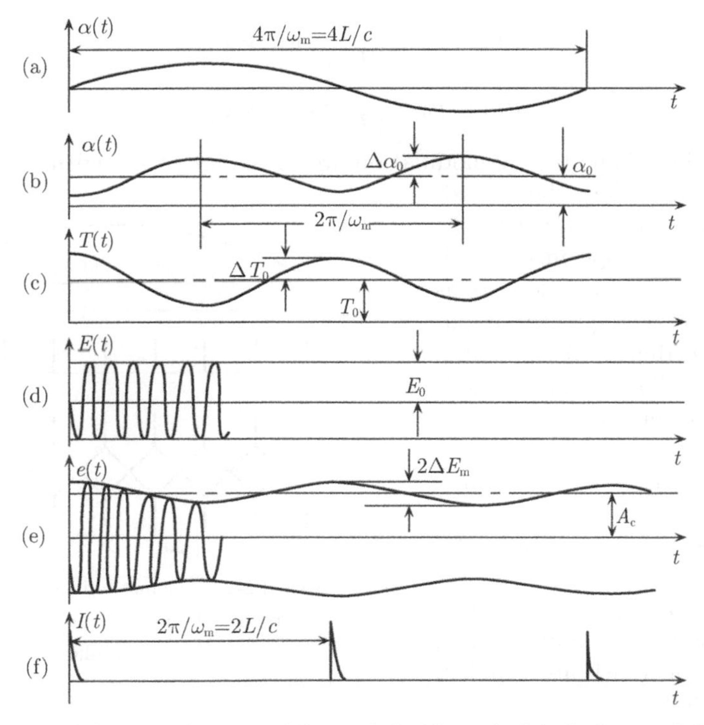

Fig. 3.2-1 Schematic diagram of the mode-locking principle for loss modulation

Below we shall discuss the mode-locking principle from the perspective of the frequency domain. Now expanding Eq. (3.2-8), we obtain

$$\begin{aligned} E(t) = A_c \sin(\omega_c t + \varphi_c) &+ \frac{1}{2} m A_c \sin[(\omega_c + \omega_m)t + \varphi_c] \\ &+ \frac{1}{2} m A_c \sin[(\omega_c - \omega_m)t + \varphi_c] \end{aligned} \tag{3.2-9}$$

The above equation shows that, for a light wave of frequency ω_c, after its modulation by the modulating signal of an externally applied frequency $\frac{1}{2}\omega_m$, its frequency spectrum includes three frequencies, namely, ω_c, the upper side frequency $(\omega_c + \omega_m)$, and the lower side frequency $(\omega_c - \omega_m)$. In addition, the light wave phases of the three frequencies are all identical. It is thus clear that the loss varies at $f_m = \omega_m/2\pi = \Delta\nu_q$. For this reason, in the qth oscillation mode there will appear oscillation of other modes. The result of loss modulation is the association of all the longitudinal modes. The mode-locking process is as follows:

Suppose the longitudinal mode frequency located at the center of the gain curve is ν_0. As its gain is maximum, it begins oscillation first, the electric field expression being

$$E(t) = E_0 \cos(\omega_0 t) \tag{3.2-10}$$

When passing through the modulator in the cavity, the light wave undergoes loss modulation, resulting in the generation of two side frequency components $\nu_0 \pm \nu_m$. When the frequency ν_m of the loss variation is equal to the frequency interval of the longitudinal modes in the cavity,

$$\nu_m = \frac{c}{2L} = \Delta\nu_q$$

So

$$\nu_1 = \nu_0 + \frac{c}{2L} = \nu_0 + \nu_m$$

$$\nu_{-1} = \nu_0 - \frac{c}{2L} = \nu_0 - \nu_m$$

The side frequencies stimulated by modulation are in reality two longitudinal mode frequencies adjacent to ν_0. Thus, the two adjacent longitudinal modes are made to oscillate, which possess specified amplitudes and the same phase relationship as ν_0. ν_1 and ν_{-1} are subsequently amplified via the gain medium and modulated via the modulator. The results of modulation, in turn, stimulate new side frequencies $\nu_2 = \nu_1 + \frac{c}{2L}$, $\nu_{-2} = \nu_{-1} - \frac{c}{2L}$, and $\nu_3 = \nu_2 + \frac{c}{2L}$, $\nu_{-3} = \nu_{-2} - \frac{c}{2L}$, and so on and so forth. This process proceeds on until all the longitudinal modes falling within the laser line width are coupled, as shown in Fig. 3.2-2.

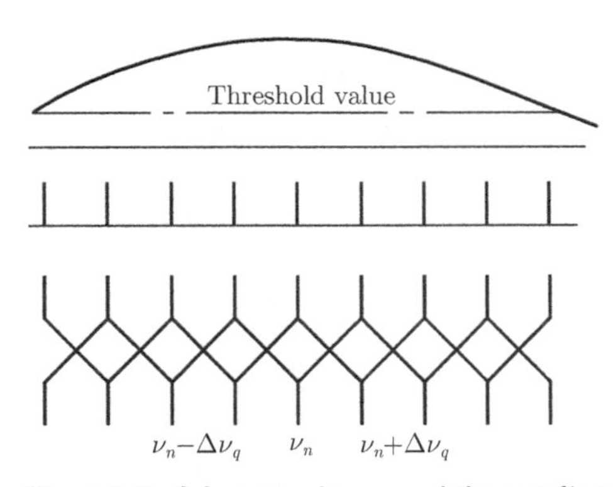

Fig. 3.2-2 Schematic diagram of the coupling process of longitudinal modes during loss modulation

Additionally, it is known from Eq. (3.2-9) that the amplitudes of the three components are all maximum at $t = 0$. At the same time, they possess identical phases as the original central longitudinal modes. When passing through the modulator, the light of the two side frequencies is once again modulated, with the oscillating mode $(\omega_0 \pm 2\omega_m)$ appearing. After many times of modulation, all the oscillating modes of frequency $(\omega_0 \pm N\omega_m)$ generated will possess identical phases and their frequencies, too, will be equally spaced, thereby attaining the goal of mode-locking. With the coherent superposition of these longitudinal modes, powerful coupling will take place, forming strong and narrow optical pulse sequences.

3.2.2 The phase modulated mode-locking

Phase modulation consists in inserting an electro-optic modulator in the laser cavity. When the refractive index of the modulator's medium periodically varies according to the externally applied signal, there will be different phase retardations when the light wave passes through the medium at different moments. This is the principle of phase modulation. Below we shall have a discussion with a lithium niobate (LN) crystal phase modulator as an example.

Suppose the light propagates along the x direction, with the modulating signal voltage applied along the z direction, that is, the mode of transverse application is adopted. Then the refractive indices of the crystal are

$$\begin{aligned} n'_x &= n_\mathrm{o} - \frac{1}{2} n_\mathrm{o}^3 \gamma_{13} E_z \\ n'_z &= n_\mathrm{e} - \frac{1}{2} n_\mathrm{e}^3 \gamma_{33} E_z \end{aligned} \tag{3.2-11}$$

where n_o is an ordinary light refractive index, n_e is an extraordinary light refractive index, γ_{13} and γ_{33} are electro-optic coefficients, are E_z is the electric field applied along the z direction.

$$E_z = \frac{V_0}{d}\cos(\omega_\mathrm{m} t) \tag{3.2-12}$$

where d is the length of the crystal in the z direction, V_0 is the amplitude of the externally applied voltage, and ω_m is the modulated angular frequency. If the crystal's length in the x direction is l, then the phase retardation generated after the light wave passes through the crystal is

$$\Delta\varphi(t) = \frac{2\pi}{\lambda} l \Delta n(t) = \frac{\pi}{\lambda}\frac{l}{d}\gamma_{33} n_\mathrm{e}^3 V_0 \cos(\omega_\mathrm{m} t) \tag{3.2-13}$$

As the variation of frequency is differentiation of phase variation with respect to time, we have

$$\Delta w(t) = \frac{\mathrm{d}\varphi(t)}{\mathrm{d}t} = \frac{-\pi}{\lambda}\frac{l}{d}\gamma_{33} n_\mathrm{e}^3 V_0 \omega_\mathrm{m} \sin(\omega_\mathrm{m} t) \tag{3.2-14}$$

Figure 3.2-3 shows the variation $\Delta n(t)$ of the crystal refractive index, the light wave phase retardation $\Delta\varphi(t)$, and frequency variation.

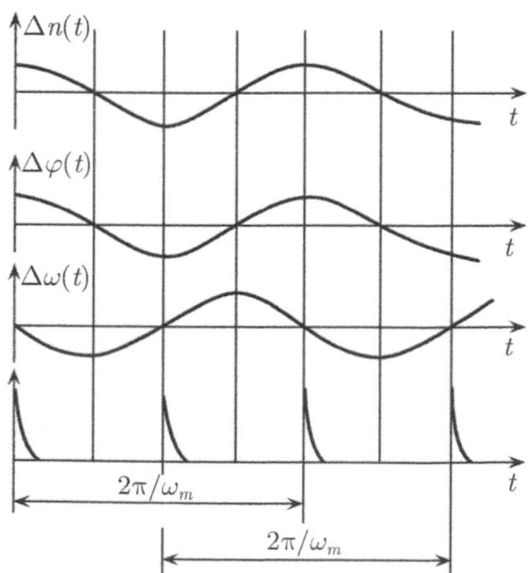

Fig. 3.2-3 The principle of phase modulated mode-locking

The role of the phase modulator can be understood as one of frequency shift that makes the frequency of the light wave move toward the large (or small) direction. Every time the pulse passes through the modulator, frequency shift will occur once, finally shifting beyond the gain curve. Similar to the loss modulator, this part of the light wave will vanish from within the cavity. Only those moments corresponding to the extremum points (maximum or minimum) of phase variation, through the optical signals of the modulator, their frequency not shifting, can survive in the cavity and are steadily amplified, forming a pulse sequence of a period of $\frac{2L}{c}$.

It is known from Fig. 3.2-3 that, in every period there exist two frequency invariant points, increasing the phase instability of the mode-locked pulse position. As the phase difference between two possible situations is π, this is also spoken of as 180° spontaneous phase switching. If no necessary measures are taken for the mode-locked laser, its output pulse may spontaneously jump from one column to another.

Similarly, an analysis can be made of the frequency characteristics first. Suppose the unmodulated light field is

$$E(t) = A_c \cos(\omega_c t) \tag{3.2-15}$$

The light field whose phase has been modulated becomes

$$E(t) = A_c \cos[\omega_c t + m_\phi \cos(\omega_m t)] \tag{3.2-16}$$

where $m_\phi = \frac{\pi}{\lambda} \frac{l}{d} \gamma_{33} n_e^3 V_0$. Expanding Eq. (3.2-16), when $m_\phi \ll 1$, we have

$$E(t) = A_c \cos(\omega_c t) + \frac{A_c}{2} m_\phi \cos[(\omega_c + \omega_m)t] - \frac{A_c}{2} m_\phi \cos[(\omega_c - \omega_m)t] \tag{3.2-17}$$

The frequency spectrum shown in the above equation is identical with that of the amplitude modulated oscillation, being composed of the carrier frequency ω_c and two side frequencies $(\omega_c \pm \omega_m)$. If the angular frequency ω_m of the modulating signal has the same interval as the adjacent longitudinal mode frequency, as the frequency of phase variation is also ω_m, the final result will be the same as amplitude modulation. When the modulating depth m_ϕ is rather great,

$$\begin{aligned} E(t) = A_c\{ & J_0(m_\phi)\cos(\omega_c t) + J_1(m_\phi)[\cos(\omega_c + \omega_m)t - \cos(\omega_c - \omega_m)t] \\ & + J_2(m_\phi)[\cos(\omega_c + 2\omega_m)t - \cos(\omega_c - 2\omega_m)t] \\ & + J_3(m_\phi)[\cos(\omega_c + 3\omega_m)t - \cos(\omega_c - 3\omega_m)t] + \cdots\} \end{aligned} \tag{3.2-18}$$

where $J_n\ (m_\phi)$ is the nth-order Bessell function of the first type. It is thus known that the frequency spectrum of the frequency modulated oscillation is composed of infinitely many side frequencies containing $\nu_q = n f_m (n = 0, 1, 2, 3, \cdots)$ frequency components. Moreover, these side frequency light all possess identical frequency intervals and phases and are in accord with the central longitudinal modes. When they have the corresponding longitudinal modes stimulated and coupled, the goal of mode-locking can be attained and an ultrashort pulse output with a period of $\frac{1}{f_m} = \frac{2L}{c}$ is obtained.

3.2.3 The structure of an active mode-locked laser and the essentials of its design

The simplest active mode-locked laser is made up by inserting a modulator in a free running laser. The modulator can be either an acousto-optic loss modulator or an electro-optic phase or loss modulator. The following points should be considered when designing an active mode-locked laser.

(1) The requirements on all the optical components in an active mode-locked laser should be more rigorous than those on general Q-switching devices. The end face reflectivity should of necessity be controlled to be minimum. Otherwise, as the etalon effect will reduce the number of longitudinal modes and destroy the effect of mode-locking, the reflective end faces of all the components should be cut into the Brewster angle, placed at an inclination or plated with an AR coating while the reflective mirror should be wedge-shaped, as shown in Fig. 3.2-4.

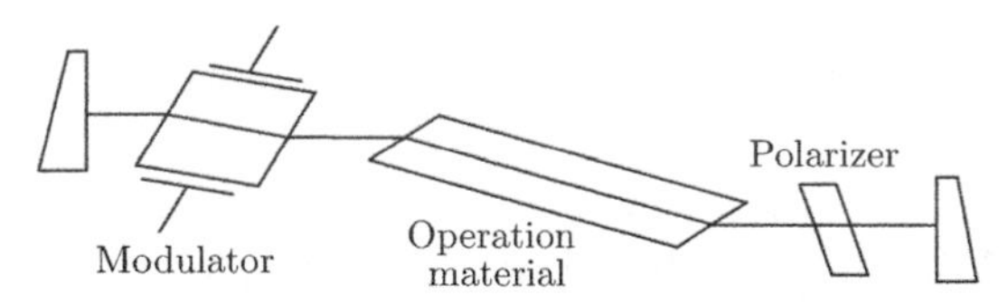

Fig. 3.2-4 Schematic diagram of the structure of an active mode-locked laser

(2) The modulator should be placed as close to the reflective mirror as possible in the cavity so as to obtain the maximum coupling effect among the longitudinal modes. If the modulator is far from the reflective mirror, then the mode-locking effect will deteriorate. Suppose the modulator is placed in the middle of the cavity as shown in Fig. 3.2-5. Then the interval between the two times the light beam passes through the modulator is $\frac{L}{c}$. If the loss variation frequency $\nu_m = \frac{c}{2L}$ in the cavity, suppose the loss is minimum when the light beam passes through the modulator the first time while it is maximum when the light beam passes through the modulator the second time. Then it cannot be ensured that, after it has passed through the modulator, the phase difference between adjacent longitudinal modes will possess the 0 or π condition. Nor can the mode-locked pulse output be obtained.

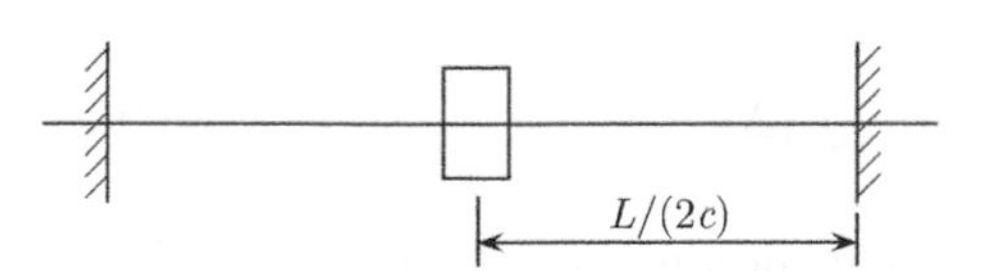

Fig. 3.2-5 Schematic diagram of the modulator's position

In addition, the dimensions of the modulator in the light transmitting direction should be as small as possible as this is when the mode-locking effect is best. If the dimensions of the modulator are rather large, as it takes the light wave a definite time to pass through the modulator, in which time not all the loss of the modulator is equal to zero. However, so long as the gain is greater than the loss, the light after modulation will be able to oscillate, thus broadening the pulse width. Of course, it is not possible to make the crystal very small, lest crystal machining would become difficult to some extent.

(3) The frequency of the mode-locked modulator should exactly be tuned to $f_{\rm m} = \Delta\nu_q = \frac{c}{2L}$ (phase modulated), or $f_{\rm m} = \frac{1}{2}\Delta\nu_q = \frac{c}{4L}$ (amplitude modulated). Otherwise, the laser will be made to operate beyond the mode-locked region and enter the quenching region or the frequency-modulated region, thereby destroying mode-locking.

3.2.4 Mismatch-free mode-locked pulse width and stabilization of the mode-locked system

It can be seen from the above analysis that the modulating period of the intracavity modulator should precisely be equal to the time of the optical pulse to make a round trip in the cavity; that is, it should be equal to the reciprocal of the longitudinal mode interval ($1/\Delta\nu_{\rm q} = 2L/c$). The situation in which the stimulating frequency of the modulator is strictly equal to the cavity's longitudinal mode interval is referred to as the mismatch-free situation; the contrary is the mismatch situation.

In the mismatch-free situation, through a self-consistent analysis of the mode-locking characteristics of a typical inner cavity phase modulated (PM) and amplitude modulated (AM) uniformly broadened laser, we can obtain a formula for calculating the mode-locked pulse width $\tau_{\rm p}$ and spectral width $\Delta f_{\rm p}$.

FM modulation:

$$\tau_{\mathrm{p}} = \left(\sqrt{2\sqrt{2}1n2/\pi}\right)(g_0/m_\phi)^{1/4}(1/f_{\mathrm{m}}\Delta\nu_{\mathrm{g}})^{1/2} \tag{3.2-19}$$

$$\Delta f_{\mathrm{p}} = \left(\sqrt{2\sqrt{2}\ln 2}\right)(m_\phi/g_0)^{1/4}(f_{\mathrm{m}}\Delta\nu_{\mathrm{g}})^{1/2} \tag{3.2-20}$$

where g_0 is the gain factor of the cavity, $\Delta\nu_{\mathrm{g}}$ is the laser gain line width, and m_ϕ is the phase modulating coefficient. The product of the pulse width and bandwidth is

$$\tau_{\mathrm{p}}\Delta f_{\mathrm{p}} = 2\sqrt{2}\ln 2/\pi = 0.624 \tag{3.2-21}$$

AM modulation:

$$\tau_{\mathrm{p}} = \left(\sqrt{\sqrt{2}\ln 2/\pi}\right)(g_0/m)^{1/4}[1/(f_{\mathrm{m}}\Delta\nu_{\mathrm{g}})]^{1/2} \tag{3.2-22}$$

$$\Delta f_{\mathrm{p}} = \left(\sqrt{\sqrt{2}\ln 2}\right)(m/g_0)^{1/4}(f_{\mathrm{m}}\Delta\nu_{\mathrm{g}})^{1/2} \tag{3.2-23}$$

where m is the amplitude modulating coefficient. The product of the pulse width and bandwidth is

$$\tau_{\mathrm{p}}\Delta f_{\mathrm{p}} = 2\ln 2/\pi = 0.441 \tag{3.2-24}$$

As m_ϕ and m are proportional to $\sqrt{P_{\mathrm{m}}}$ (acousto-optic modulator), where P_{m} is the electric power entering the modulator, $\tau_{\mathrm{p}} \propto \left(\frac{1}{P_{\mathrm{m}}}\right)^{\frac{1}{8}}$ while τ_{p}, in turn, is proportional to $(1/f_{\mathrm{m}})^{\frac{1}{2}}$, it is more effective to compress the pulse width by increasing the modulating frequency than increasing the modulator's power. On the other hand, we may obtain a rather narrow pulse width using a laser medium of a rather broad gain line width $\Delta\nu_{\mathrm{g}}$.

Many factors such as the thermal effect in the resonator, the fluctuation of the pump, and mechanical vibration will cause changes in the equivalent cavity length so that the longitudinal modes interval changes as well. For this reason, if now the frequency f_{m} of the modulator remains unchanged, then there will occur the so-called mismatch. Can we now perform mode-locking? How wide is the allowable range of mismatch? How accurate should the requirement on $f_{\mathrm{m}} = \Delta\nu_{\mathrm{q}}$ be? What measures should be taken to make the active mode-locked laser operate stably? What is shown in Fig. 3.2-6 can give qualitative answers. In the figure, the horizontal coordinate is $\Delta\Omega/\Delta\omega_{\mathrm{q}}$ ($\Delta\Omega$ is the amount of mismatch, or the difference between the longitudinal mode frequency interval and modulating frequency; $\Delta\,\omega_{\mathrm{q}} = \pi c/L$ is the longitudinal mode frequency interval); the longitudinal coordinate is the ratio of the output power to the free operating output power; $\Delta\varphi_{\mathrm{m}}$ is peak phase retardation after the light wave passes through the modulator. The operation conditions are: $g_0 = 0.075$, single pass power loss $\alpha_0 = 0.070$, and there are 5 modes that operate freely. It can be seen from the figure that the output power is maximum when $\Delta\Omega = 0$. In a small region (mode-locked region) near $\Delta\Omega = 0$, the output power rapidly decreases with the increase in $\Delta\Omega$ until it gradually decreases to zero. In a certain region (quenching region) of $\Delta\Omega$ that continues to increase, the no output state is maintained. When we continue to increase $\Delta\Omega$, the output power will, however, increase. Now the frequency modulating region is entered (analysis omitted). In a word, the influence of mismatch on the laser output power and frequency spectrum is very great.

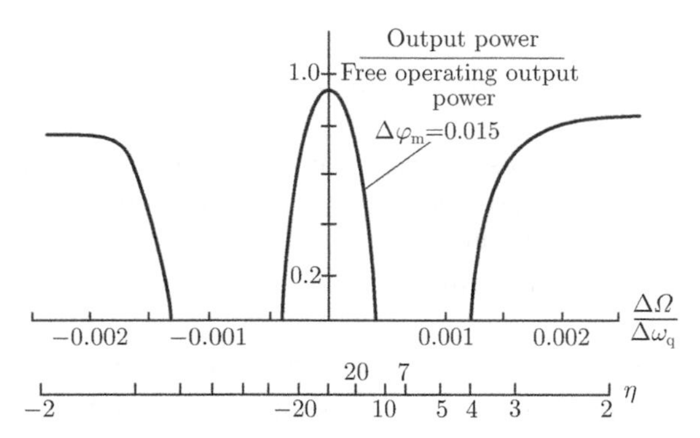

Fig. 3.2-6 The relationship of output power varying with mismatch

In addition, mismatch also affects the pulse width of the mode locked laser output. Figure 3.2-7 gives the curve of relationship between the pulse width and mismatch $\Delta\Omega = \omega_q - \omega_m$ during phase modulation. When the mismatch is negative and very small, the pulse width is narrowest whereas if the mismatch is positive, the pulse width rises monotonically. On the contrary, under the amplitude modulation condition, when $\Delta\Omega = 0$, the pulse width has the minimum value, and monotonically rises with the increase in positive and negative mismatch at that.

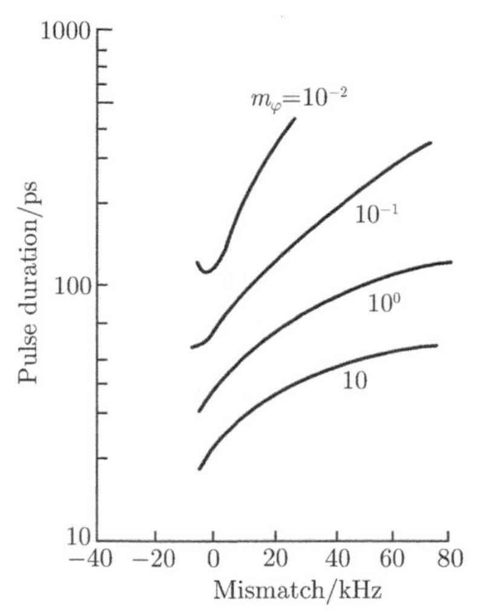

Fig. 3.2-7 Correlation between pulse width and mismatch during active phase modulated mode-locking

For lasers with definite gain and loss, the size of the mode-locked region is related to the peak phase retardation $\Delta\varphi_m$. If $\Delta\varphi_m$ is too great, the mode-locked region will be very small or even vanish so that the mode-locking device is rendered inoperable. So, when designing the mode-locked laser, care should be taken to choose a suitable $\Delta\varphi_m$. Also, it is necessary to choose the operation point according to the mismatch curve and the mode-locked output stability needed. For a specific laser and ascertained operation condition, the mismatch curve can be found by experiment to serve as a basis for adjusting devices and analyzing the device characteristics.

In order to make the laser output amplitude, pulse width, and pulse interval stable, it is necessary to make the mode-locked modulator's modulating frequency, amplitude, and its variation with time strictly match the laser's equivalent cavity length (a definite equivalent cavity length corresponds to a definite longitudinal mode frequency interval). Taking into account that the variation of the equivalent cavity length is mainly due to the thermal effect, etc., it is important to continuously compensate for the heat and sound variation in the cavity whenever necessary. In general the feedback circuit is adopted, that is, the variation of the cavity parameters are detected and, after amplification, used to control the modulator's frequency or cavity length, which, in essence, is compensation for the mismatch. Below we shall use a continuously operating YAG laser as an example to give an account of what measures to take to reduce mismatch and stabilize the cavity length:

(1) Enhance the stability of the pump source (e.g., require that the instability of the

current be better than 5%).

(2) Strengthen the cooling of the pump lamp and the operation material and ensure that the cooling water flow will be uniform and stable.

(3) When designing the thermo-stabilizing cavity, use invar or quartz of small linear expansion coefficient as much as possible to make the base for supporting the reflection mirror while taking thermostat measures.

(4) Reduce the effects of external vibration and impact by adopting vibration isolating equipment.

(5) Adopt electronic feedback circuit to compensate for mismatch whenever necessary.

Adopt the automatic compensating system as shown in Fig. 3.2-8. The phase modulator will be stimulated with 200 MHz frequency and light signals will be received with an opto-electric detector. The second harmonic of the beat frequency will be detected via the band-pass filter. The amplified 400 MHz signal and the signal whose frequency doubling with a 200 MHz oscillator are sent into the phase discriminator. The output signal is sent into the pressure controlled oscillator after amplification by DC to control the frequency of the oscillator. So this system is a negative feedback circuit that uses the error signal detected after mismatch to control the frequency of the oscillator, which is made to track the variation of the cavity length, thus realizing compensation for mismatch.

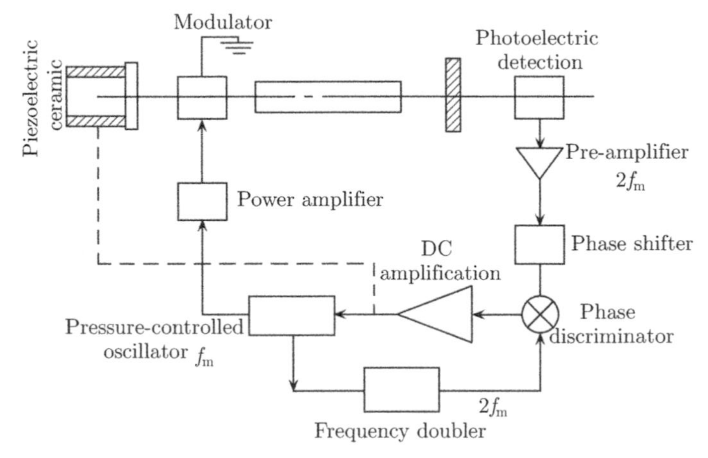

Fig. 3.2-8 The automatic compensating system

The circuit represented by the dotted line in the figure is different from the above-mentioned circuit in that the DC error signal obtained with the phase discriminator is not used to control the frequency of the pressure-controlled oscillator but to drive the piezoelectric ceramic of the cavity and then change the variation of the cavity length to compensate for the mismatch so as to finally obtain a stable mode-locked pulse sequence.

3.3 Passive mode-locking

By inserting a saturable absorption dye in the laser resonator to regulate the loss in the cavity, when the mode-locking condition is satisfied, a series of mode-locked pulses can be obtained. According to the mechanism and characteristics of the mode-locking forming process, passive mode-locking can be divided into two types, the solid laser passive mode-locking and the dye laser passive mode-locking.

3.3.1 The solid-state laser passive mode-locking

1. The operation principle

As a dye's saturable absorption coefficient decreases with an increase of the light intensity, the highly intensive laser generated by the high gain laser can make dye absorption saturated. Figure 3.3-1 shows the variation of laser with the laser intensity I through the dye's transmissivity T. The transmissivity of a strong signal is greater than that of a weak one, with only a small part absorbed by the dye. The strong and weak signals are approximately determined according to the dye's saturated light intensity I_s; one greater than I_s is a strong signal, otherwise, it is a weak one. Suppose before the occurrence of mode-locking, the distribution of photons in the cavity is basically uniform in spite of some fluctuation. As the dye possesses the characteristic of saturable absorption, the transmissivity of a weak signal is small and the loss inflicted great while the transmissivity of a strong signal is great and the loss small. Moreover, its loss can be compensated for by getting amplified through the operation material. So every time the optical pulse passes through the dye and operation material, the relative values of strength of its strong and weak signals will vary once. After cycle times in the cavity, the difference between the maximum and minimum values will become increasingly great, so that finally the leading edge of the strong pulse will steadily be cut steep while the spike part can effectively pass through to make the pulse narrower. From the point of view of the frequency domain, at the beginning, the fluorescent light of spontaneous radiation and the laser rise and fall pulses generated when the threshold is reached, through the selective action of the saturable absorption dye in the noise pulses, have only the high gain central wavelength ν_0 left with it and its side frequency. Subsequently, after several times of absorption by the dye and amplification by the operation material the side frequency signal, for its part, stimulates new side frequencies, the process going on until all the modes within the gain line width take part in the oscillation, yielding a series of pulse sequence outputs of period $\dfrac{2L}{c}$.

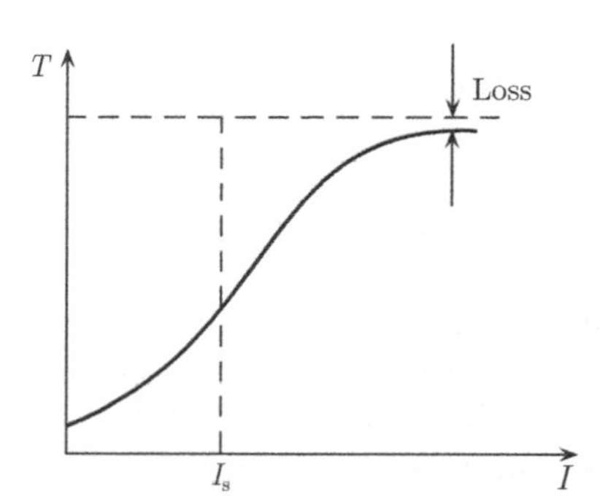

Fig. 3.3-1 The absorption characteristics of saturable dye

In a passive mode-locked laser, the physical process of evolution from irregular pulses to mode-locked pulses is approximately divided into three stages, as shown in Fig. 3.3-2. The essence of this process lies in the strongest pulse getting selectively strengthened while the background pulses are gradually restrained. The three stages can be briefly treated as follows:

(1) The linear amplification stage. The spontaneous radiation fluorescent light is generated when the flashlight pumping starts. When it exceeds the laser threshold, the initial laser pulse possesses the spectral content of the fluorescent light bandwidth as well as the interference among laser modes of a stochastic phase relationship, leading to the fluctuation of the light intensity of a very great pulse total amount, as shown in Fig. 3.3-2. Within the time of a period $\dfrac{2L}{c}$, the optical pulse passes through the organic dye and the laser medium once each. In the absorber dye, less strong pulses are absorbed than weak pulses while in the laser medium, linear amplification is generated, resulting in the action of natural mode

selection. The process of linear amplification makes the frequency spectrum narrower and the amplified signal fluctuation smoothened and broadened, as shown in Fig. 3.3-2(b) and (c). The time duration in the linear amplification stage is rather long. For instance, when the resonator length is 1 m and the effective gain is a few hundred, the linear amplification stage is about 2000 cycles.

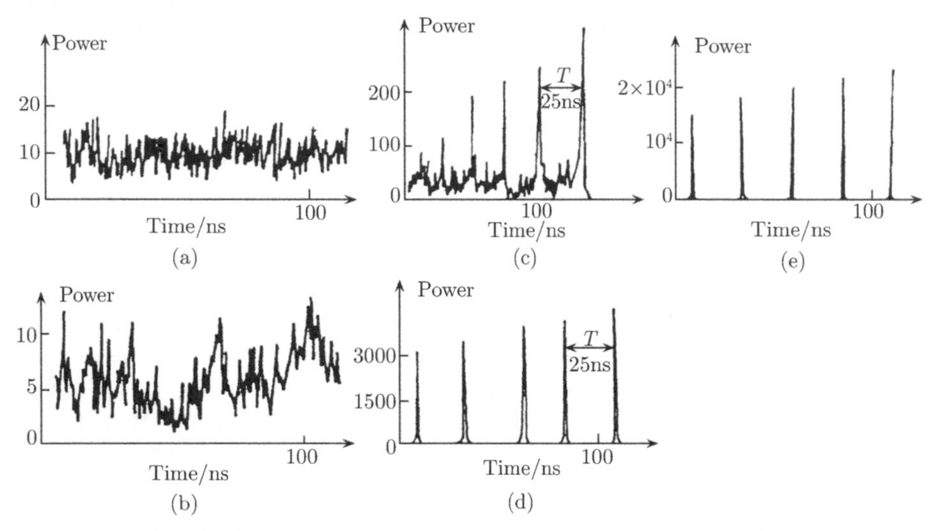

Fig. 3.3-2 The physical process of saturable dye mode-locking

(2) The nonlinear absorption stage. In this stage, although the gain of the laser medium is linear, the strongest pulses of the laser radiation field make the saturable absorption dye absorb nonlinearly. This stage is mainly characterized by the strong pulses making dye absorption saturated; the dye is "bleached", and the pulse strength is rapidly increased while large quantities of weak pulses are absorbed by the dye and restrained, with the emitted pulses narrower and the frequency spectrum broadened, as shown in Fig. 3.3-2(d).

(3) The nonlinear amplification stage. As the strong pulses selected can not only make the dye absorption saturated, but also can make the laser operation material's gain reach saturation, the amplification of the operation material enters the nonlinear stage. When the strong pulses pass through the active medium, much of the leading edge and the central portion is amplified. Owing to the consumption of the inverted population, the gain decreases, so that the rear edge of the pulse is less amplified or even not amplified at all, resulting in the front edge becoming steep and the pulse becoming narrower, with the small pulses almost completely restrained until finally a pulse sequence of high strength and narrow pulse width is output, as shown in Fig. 3.3-2(e). This stage is characterized by pulse compression and further broadening of the frequency spectrum.

2. The structure of the passive mode-locked solid laser

The structure of a typical mode-locked solid laser is shown in Fig. 3.3-3. Such mode-locked lasers mainly include the optical resonator, the laser rod, the dye cell, and the pinhole diaphragm. In order to obtain a high quality mode-locked pulse sequence of a high repetition rate, strict requirements are made on the concentration of the dye, the strength of the pump, and the design and adjustment of the resonator. Otherwise, the laser output would be extremely unstable. When designing a passive mode-locked laser, attention should be given to the following points:

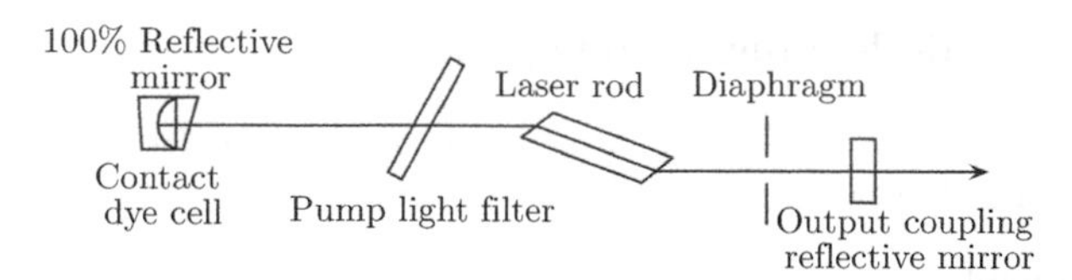

Fig. 3.3-3 Schematic diagram of the resonator structure of a passive mode-locked solid laser

(1) In order to eliminate the etalon effect, the surface of optical components should be cut into a Brewster angle (or a dip of 2° ~3°), plated with an AR coating and placed at an inclination so as to help eliminate the reflection from the non-operation surface. To prevent the reflected light from the end component from entering the cavity, the rear surface of the total reflective mirror should be ground into a wedge-shaped mirror.

(2) The saturable dye used for mode-locking should possess the following conditions: ① The dye's absorption spectral line should match the laser wavelength; ② its absorption line's width should be greater than or equal to the laser line width and; ③ its relaxation time should be shorter than the time for a pulse to make a round trip once in the cavity.

Table 3.3-1 lists the saturated light intensity (I_s) and relaxation time (T_{21}^b) of several dyes.

Tab. 3.3-1 The saturated light intensity (I_s) and relaxation time (T_{21}^b) of several dyes

Dye	Eastman Kodak		DDI	Cryptocyanine
	"9740"	"9860"		
I_s/W·cm^{-2}	4×10^7	5.6×10^7	$\approx 2\times10^7$	5×10^6
T_{21}^b /ps	8.3	9.3	14	22
Applicable lasers	Neodymium glass laser		Ruby laser	

For the above dyes, in terms of the absorption peak, the absorption of cryptocyanine methanol solution is 706 nm, those of DDI methanol and water solution are 760 nm and 703 nm, that of "9740" is 1.045 μm, and that of "9860" is 1.051 μm. For the ruby laser, the methanol or water solution of dicarbocyanine iodide (DDI) and the acetone solution of cryptocyanine should be used while for neodymium glass and yttrium-aluminum garnet (YAG) lasers, the pentamethenyl hendecamethenyl or Eastman Kodak "9740" and "9860" saturated absorption dyes. In terms of the relaxation time, it is 25~35 ps and 6~9 ps for "9740" and "9860", respectively. It is 25 ps for the acetone solution of cryptocyanine and 14 ps for DDI.

In terms of chemical stability, all organic saturated absorption dyes exhibit decomposition and color change when "bleached" by strong laser with none appearing to be stable. In order to reduce the destruction of the dye by the ultraviolet radiation from the pump lamp and background light, good care should be taken to the shield structure light, such as by adopting the liquid slot window made up of UV-ray absorbing glass or improving the pump lamp's spectral characteristics.

The dye cell should be placed as close to the reflective mirror as possible, usually 1~2 mm apart. Sometimes the dye cell and reflective mirror can be integrated, which is helpful in the pulse reflective front edge overlapping the incident trailing edge in the dye, thus helping the absorber reach saturation when the light intensity is great. So that there will be an appropriate initial transmissivity, the dye cell thickness and dye concentration should be rationally chosen. The static transmissivity during mode-locking should be greater than during Q modulation, which is 0.6~0.8 in general. The dye solution should be regularly replaced and the adoption of a cyclic structure can prolong the useful life of the dye.

(3) The pump power is preferably slightly higher than the laser threshold value. A high pump power may cause the generation of double pulses.

3.3.2 The passive mode-locking of the dye laser

The passive mode-locking of the dye laser can be realized by inserting a saturable absorption dye in the resonator of a dye laser.

1. The operation principle

Figure 3.3-4 shows the device of a passive mode-locked dye laser, with the rhodamine 6G dye as the laser gain medium and DODCI dye as the saturable absorber. The tuning elements are used to regulate the wavelength range of laser output.

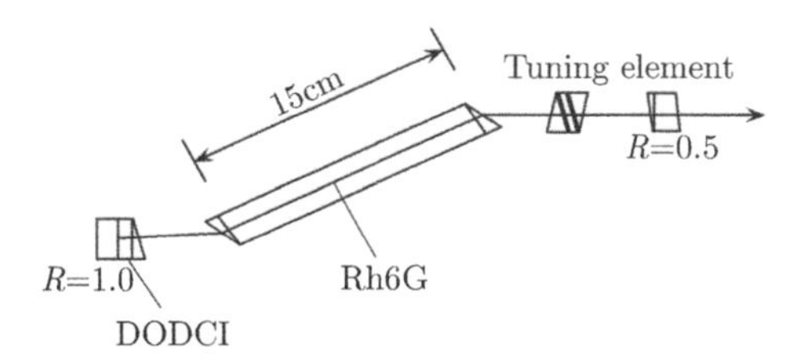

Fig. 3.3-4 The resonator structure dye laser pumped with flashlight

The process for the mode-locked dye laser to generate ultrashort pulses is similar to that of the solid passive mode-locked laser. First the strong rise and fall peak values are chosen from the rising and falling noise background through the non-linear absorption of dye absorber and the laser medium's amplifying action. Then, through the combined effect of the saturable absorption and the gain saturation state until the ultrashort pulses are finally formed, the difference being that the relaxation time of the upper energy level of the laser dye is short (of the nanosecond order of magnitude), so that the gain decay plays an important role in the generation of pulses. Usually the absorption cross-section of a dye absorber is greater than that of a gain medium. Making the energy that makes the absorber reach saturation smaller than the energy that makes the gain medium saturated and making the effective gain obtained by the pulse peak value greater than that obtained by the pulse leading edge will both be conducive to the formation of pulses.

Let's adopt the resonator structure shown in Fig. 3.3-5 to analyze the influence of various parameters in the resonator. By proceeding from the rate equation, have the continuous pump source pump the laser medium. The rate equation for the variation of the laser gain medium's inverted population is:

$$\frac{\partial n_3^a}{\partial t'} = \sigma_{14}^a I_p n_1^a - \sigma_{32}^a n_3^a I(z, t') \tag{3.3-1}$$

$$n_1^a + n_3^a = n^a \tag{3.3-2}$$

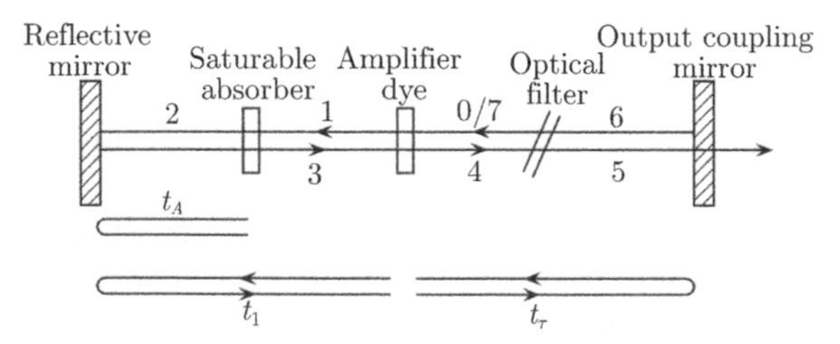

Fig. 3.3-5 The optical elements of a mode-locked dye laser (t_A, t_1, t_τ represent the time for the pulse to pass through the distances plotted; the numerals 0~7, the positions in the resonator)

The equation of variation of the pump light is

$$\frac{\partial I_p}{\partial z} = -\sigma_{14}^a n_1^a I_p \tag{3.3-3}$$

The equation of variation of population of the saturable absorber is

$$\frac{\partial n_1^b}{\partial t} = -\sigma_{13}^b n_1^b I \tag{3.3-4}$$

$$n_1^b + n_2^b = n^b \tag{3.3-5}$$

Using the above equations we can obtain the expression for the light intensities at different positions in the cavity. In the expression, $n_3^a, n_1^a, \sigma_{14}^a, \sigma_{32}^a$ and $n_1^b, n_2^b, \sigma_{13}^b$ are the populations and absorption cross-sections at different energy levels of the laser medium and saturable absorber, respectively. I is the light intensity in the cavity and I_p is the pump light intensity. When continuous light is used for pumping, I_p is a constant, $t' = t - \dfrac{z}{v}$. Using the self-consistent condition for laser oscillation in the resonator, we obtain the following equations:

$$I(7,t') = I(0,t'+h) \approx \left(1 + h\frac{\partial}{\partial t'} + \frac{h^2}{2}\frac{\partial^2}{\partial t'^2}\right) I(o,t') \tag{(3.3-6)}$$

$$\begin{aligned} I(7,t') = R\Bigg\{ & I(4,t') - \frac{4}{\Delta\omega}\frac{\mathrm{d}}{\mathrm{d}t'}I(4,t') \\ & + \frac{4}{\Delta\omega^2}\left[3\cdot\frac{\mathrm{d}^2 I(4,t')}{\mathrm{d}t'^2} - \frac{1}{2I(4,t')}\left(\frac{\mathrm{d}(4,t')}{\mathrm{d}t'}\right)^2\right]\Bigg\} \end{aligned} \tag{3.3-7}$$

$$I(4,t') = G(t')\frac{1}{R}I(0,t') \tag{(3.3-8)}$$

$$G(t') = \frac{V_\tau V_1 B_0 B_1 R_e E(t') k^{m-1}(t')\{1 - B_0 B_1[1 - k^m(t')]\}^{\frac{1-m}{m}}}{1 - V_1 + V_1\{1 - B_0 B_1[1 - k^m(t')]\}^{\frac{1}{m}}} \tag{(3.3-9)}$$

where $m = q^a\sigma_{13}^b/(q^b\sigma_{32}^a); B_0 = \exp(-n_1^b L^b \sigma_{31}^b)$ is the transmissivity of the dye absorber with respect to small signals; $B_1 = \exp\left\{-\sigma_{13}^b \int_0^z n_1^b(t' \to \infty, z')\mathrm{d}z'\right\}$ is the transmissivity of the pulse front edge passing through the dye absorber; V_τ is the gain of the pulse front edge; V_1 is the gain of the small signal; $V_1 = \exp(\sigma_{32}^a n_3^a L^a); k(t') = 1 + V_\tau\{\exp[E(0,t') - 1]\}; I(0,t'), I(1,.t'), I(4,t')$ and $I(7,t')$ are the light intensities at the positions 0, 1, 4, and 7 in the cavity, respectively; G (t') is the gain coefficient of the optical pulse; and h is the maximum time displacement likely to occur in the action of the active medium and saturable absorber. Suppose the active medium is located in the center of the resonator ($t_\tau = t_1$), and the saturable absorber is close to the total reflective mirror ($t_\mathrm{A} = 0$). Using the above-mentioned equations for calculation, we obtain the following conclusions:

(1) The range of the mode-locked stable region is proportional to the amount of the absorption loss and gain. A rather large loss will generate a wide stable region. It is now necessary to raise the pumping power.

(2) The range of the stable region very sensitively depends on T/T_{31}^a, the ratio of the period to the relaxation time in the resonator. In general it is necessary to satisfy the condition

$$0.1 < T/T_{31}^a < 10 \tag{3.3-10}$$

If T/T_{31}^a is too small, after the decay of the gain there will not be sufficient time for resetting up the inverted population while if too great, the rise and fall occurring at the front edge of the pulse will increase and lead to multiple pulses.

2. The structure of the passive mode-locked dye laser

Figure 3.3-6 shows the structure of a kind of continuous mode-locked dye laser, which mainly includes an optical resonator, a dye laser medium, a saturable absorber, a pumping source, etc. In general an argon ion laser or a flashlight is adopted as the pumping source to be input into the resonator after getting coupled via the quartz prism and focused on the freely spraying laser dye through a spherical reflective mirror. The saturable absorber dye cell is close to the total reflecting mirror; the box is 200~300 μm thick and slantingly placed. The laser radiation is focused on the absorber via the lens. If Rh6G is adopted as the laser medium and DODCI as the saturable absorber, the absorber box length is 0.5 mm, the output mirror transmissivity is 1%~6%, and the output pulse width can be as great as 1 ps.

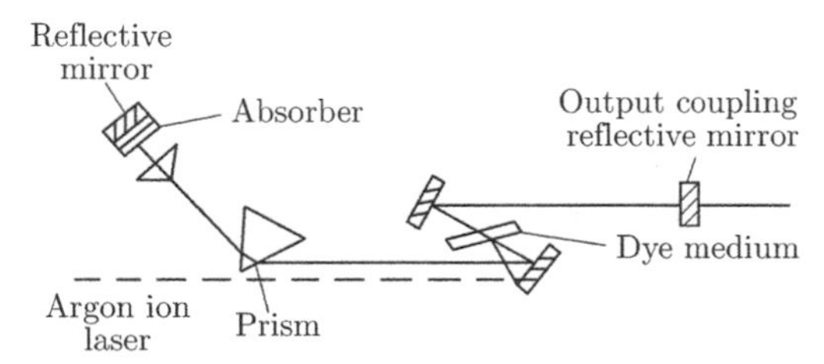

Fig. 3.3-6 The resonator structure of a passive mode-locked dye laser by CW pumping

A schematic diagram of the principle of the colliding mode-locked laser (CPM) that generates the femtosecond order of magnitude is as shown in Fig. 3.3-7. In a ring-shaped mode-locked laser there are two pulses propagating in opposite directions that accurately synchronously reach the saturable absorber generating the mutual superposition effect while making the light wave electric field (or light intensity) in the saturable absorber appear periodically distributed while generating spatial modulation of the light intensity to form the spatial "grating". In the course of the formation of the spatial grating, the energy leading edges of the two pulses are absorbed, its light intensity making the absorber saturated quicker than the single pulse. Furthermore, as the absorber's relaxation time is greater than the width of the light pulse, when the trailing edge of the pulse passes through, the grating is still modulated to a considerable degree, so that it will get compressed by the backward scattering. Therefore, in the time domain, every time the two pulses pass through the saturable absorption medium, their leading and trailing edges are cut. After many cycle times, the compression of the pulses will be quickened. In terms of the frequency domain, owing to the standing wave field formed by pulse coherence superposition, the intensity of the effective light field in the saturable absorber is appreciably increased. Because of the nonlinear self-phase modulating effect, the increase in the light field intensity is bound to lead to broadening of the frequency spectrum width, compensating in part for the restraint on the spectral width by gain dispersion, thereby forming narrower mode-locked pulses.

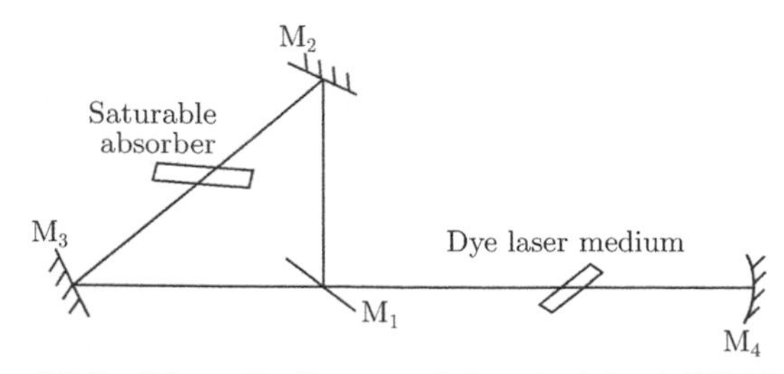

Fig. 3.3-7 Schematic diagram of the principle of CPM laser

In a CPM laser there exist four principal mechanisms, the saturable gain, saturable absorption, self-phase modulation, and dispersion, the balance among the four being the cruces for obtaining narrow pulse width stable mode-locking. The self-phase modulation and dispersion effect influence each other, making the pulse chirp compensated for. Therefore, by adopting the CPM laser, we can obtain stable femtosecond order of magnitude of pulse output. Figure 3.3-8 shows a CPM laser of 6-mirror ring-shaped structure, with the Rh6G gain jet-flow and the DODCI saturable absorber jet-flow placed at the focal points of M_1, M_2, M_3, and M_4, respectively. M_4 adopts the dichroic mirror to generate dispersion to compensate for the pulse chirp in the resonator and make the femtosecond pulse tunable. The laser's ring-shaped cavity length is 3 m and the distance between the Rh6G and DODCI jet-flows is 0.7 m. This laser can output stable optical pulses of dozens of femtoseconds.

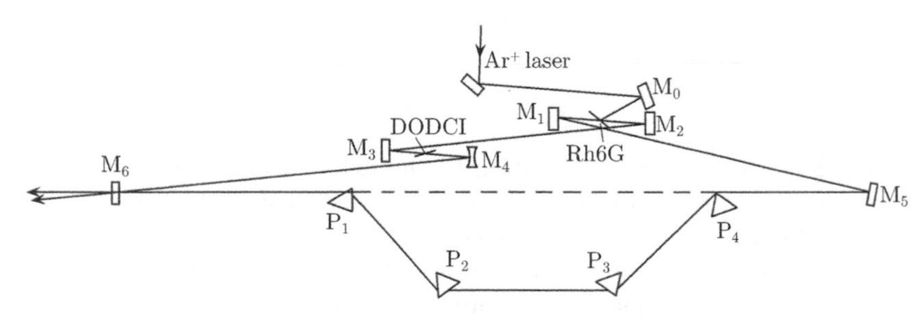

Fig. 3.3-8 The colliding mode-locked ring-shaped laser

The mechanism of the passive mode-locked dye laser pumped with a flashlight is different from that of generating the mode-locked pulses of the passive mode-locked solid lasers. As the lifetime of the laser dye upper energy level is rather short (of the nanosecond order of magnitude), the combined action of saturable absorption and gain attenuation will generate the fast pulse compressing process unrestrained by the absorber relaxation time.

As the dye's spectral line is broad, and the laser's upper energy level lifetime is short, the dye mode-locked laser can output narrower pulses than can the solid mode-locked laser. The CPM dye laser can output pulse sequences of dozens of femtoseconds. After the adoption of the optical pulse compressing technique, the narrowest 6-fs optical pulses ever have been obtained. This is a major breakthrough in the mode-locking technology in the 1980s.

3.4 The Synchronously pumped mode-locking

The synchronously pumped mode-locked laser consists in adopting the pulse sequence of a mode-locked laser to pump another laser, with mode-locking achieved by modulating the gain in the cavity. The key to realizing mode-locking by synchronous pumping is to make the resonator length of the pumped laser equal to that of the pumping laser or be its integer times. Under definite conditions, the gain is modulated, its modulating period equal to the cycling period of light in the resonator. Like loss modulation, in the maximum gain time domain, a short pulse will form, the pulse width being much narrower than the pumping pulse width. The synchronously pumped mode-locking is of practical significance for the dye laser since dyes have a very wide gain line width ($10^{13}\sim10^{14}$ Hz). The frequency of the ultrashort pulse generated by the synchronously pumping dye laser is continuously adjustable in the whole spectrum range.

3.4.1 The principle of synchronously pumped mode-locking

Synchronously pumped mode-locking is realized by modulating the gain in the cavity, as shown in Fig. 3.4-1.

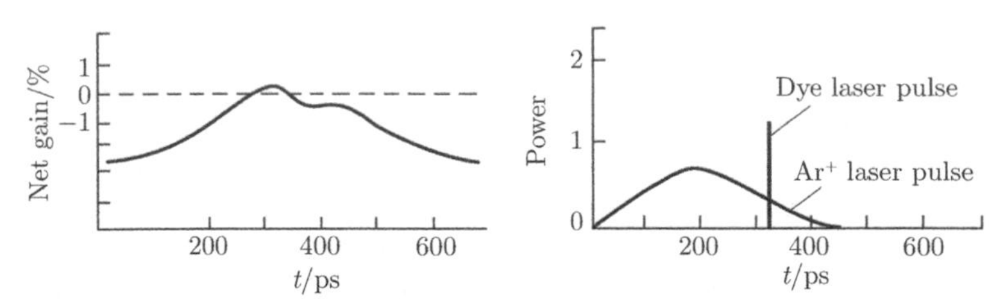

Fig. 3.4-1 Characteristics of a synchronously pumped dye laser

For instance, an active mode-locked argon ion laser is adopted to pump a dye laser. The width of the pumping pulse T_{p} is 100~200 ps while the relaxation time T_{31} of the laser upper energy level of the dye laser is of the nanosecond order of magnitude (eg for rhodamine 6G, $T_{31} = 5$ ns). T_{31} is greater than the pump pulse width T_{p} but smaller than the cycling period $T= 2L/c$ of light in the resonator, i.e.,

$$T_{\mathrm{p}} \ll T_{31} < T \tag{3.4-1}$$

Under this condition, the inverted population of the active medium is only dependent on the pump energy obtained in this instant, as shown in Fig. 3.4-1. The pumping makes its gain coefficient gradually increase until it exceeds the loss to reach the laser threshold and above. Beginning from this instant, stimulated radiation is generated, with the laser pulse energy rapidly rising. As the period of the pump pulse sequence is equal to the time for the photons to make a round trip in the dye laser cavity, the initial pulse in the resonator can be amplified only when it reach the dye cell at the same time as the pumping pulse. Thus, when the dye laser pulses generated by the preceding pumping reach the dye cell after its round trip transit in the cavity, the dye is exactly put in the state where the population is inverted by the pumping. Therefore, the energy of the laser pulses is amplified after passing through the gain medium. After many times of cycling, the pulses acquire rather great energy. This is the first stage of the formation of pulses, or the gain stage. The second stage is the pulse compression stage. When the pulses are rather strong, every time they pass through the active medium, because of the saturation effect, only the leading edge and the middle part get amplified while the trailing edge is restrained because it cannot be amplified. Hence, after many times of cycles, the pulse width is compressed until a stable pulse is formed in the end. Now, the pulse width, energy, and the position of the pumping pulse relative to the laser pulse remain unchanged.

Below we shall discuss the straight resonator, which includes the dye as the active medium and the elements for frequency modulation, as shown in Fig. 3.4-2. The active medium is a 4-energy level system. Suppose the bandwidth of the dye laser is greater than that of the frequency selecting element and the pumping laser. Compared with the optical filter, the active medium bandwidth related to the emission cross-section σ_{32} is negligible. According to the theory of interaction between the optical pulse and atomic system and Eq. (3.4-1), we obtain the following equations of the laser pulse photon flux density I (z, t') and the phase φ (z, t'):

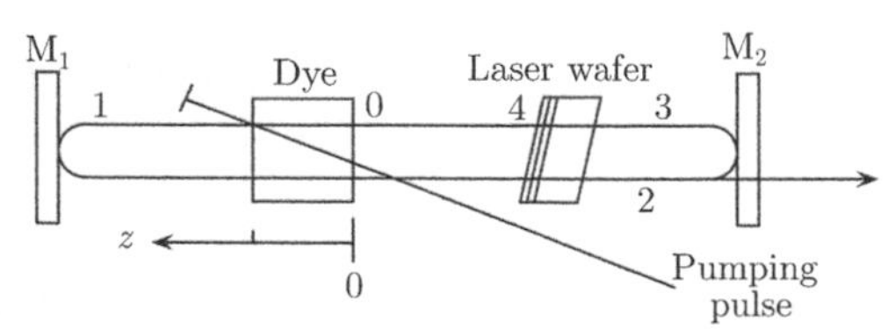

Fig. 3.4-2 Schematic diagram of the resonator of a synchronously pumped laser: M_1 is a 100% reflective mirror; M_2 an output mirror of reflectivity R; 0,1, $\cdots$, 4 represent the positions of the pulse in a cycling period

$$\frac{\partial I(z,t')}{\partial z} = \sigma_{32} n_3 I(z,t') \tag{3.4-2a}$$

$$\frac{\partial \varphi(z,t')}{\partial z} = -\frac{\Delta_{32}}{2}\sigma_{32} n_2 A(z,t') \tag{3.4-2b}$$

The equation of variation of the inverted population (spontaneous radiation neglected) is

$$\frac{\partial n_3}{\partial t'} = \sigma_{14} I_p n_1 - \sigma_{32} n_3 I(z,t') \tag{3.4-2c}$$

$$n_1 + n_3 = n$$

The equation of variation of the pumping pulse is

$$\frac{\partial I_{\mathrm{p}}}{\partial z} = -\sigma_{14} n_1 I_{\mathrm{p}} \tag{3.4-2d}$$

where n_1 and n_3 are the laser ground state and upper energy state population, respectively; n is the density of the total population; I_{p} is the photon flux density of the pumping light; $A(z, t')$ is the amplitude of the laser pulse; $\Delta_{32} = (\omega_0 - \omega_{32})\tau_{32}$; ω_0 is the central frequency of laser; and ω_{32} is the fluorescent light frequency. The same group velocity ($V_{\mathrm{p}} = V$) is adopted for both the laser pulse and pumping pulse; $t' = t - \frac{z}{v}$ is the auxiliary time coordinates, with the pulse center chosen as the origin of the coordinates. In the course of derivation, $\tau \gg \frac{l}{v}$ (l is the length of the laser medium) is taken as the precondition with the derivative term of the photon flux density with respect to time neglected. In Fig. 3.4-2, $I(z, t')$ and $\varphi\ (z, t')$ at positions 1, 2, 3, and 4 satisfy the following relation:

$$I(4,t') = I(0,t'+h) \tag{3.4-3}$$

That is, the pulse will automatically reappear every time it experiences a cycle and a stable state is obtained. The h is the maximum time displacement likely to be induced by the process of amplification and frequency choosing element and the difference between the effective period T of the pulse and the period T_0 of the un-pumped resonator is $h(T = T_0 - h)$. In order to attain synchronization, the time interval between the pump pulses T_{p} and T has to satisfy the condition $T = T_{\mathrm{p}} = T_0 - h$.

The laser resonator's equivalent length difference $\delta_L = \frac{hc}{z}$, called resonator mismatch. Suppose the bandwidth of the pulse is smaller than that of the frequency choosing element, and the time displacement h is smaller than the time duration of the pulse. Then from Eq. (3.4-3) and $I(4, t')$ and $\varphi\ (4, t')$, we have

$$I(0,t') + h\frac{\partial I(0,t')}{\partial t'} + \frac{h^2}{2}\frac{\partial^2 I(0,t')}{\partial t'^2} = I(0,t')G(t') - \frac{4}{\Delta\omega}\frac{\partial}{\partial t'}[I(0,t')G(t')]$$
$$+ \frac{12}{\Delta\omega^2}\frac{\partial^2}{\partial t'^2}[I(0,t')G(t')] - \frac{2\left\{\frac{\partial}{\partial t'}[I(0,t')G(t')]\right\}^2}{\Delta\omega^2 I(0,t')G(t')} \tag{3.4-4}$$

where $G(t')$ is the effective gain and $\Delta\omega$ is the bandwidth of the F-P etalon.

$$G(t') = \left\{\exp\left[\sigma_{32}\int_{-\infty}^{t'}(I(0,t'') + I_{\mathrm{p}}(0,t''))\mathrm{d}t''\right]\right\} \times [\gamma(L^a t')]^{-1} R \tag{3.4-5}$$

$$\Delta\omega = (c/d)[(1 - R_{\mathrm{FP}})R_{\mathrm{FP}}^{-1/2}] \tag{3.4-6}$$

where R is the reflectivity of the reflective mirror, R_{FP} is the combined reflectivity of the F-P etalon, d is the thickness of the F-P etalon, and $L^a = z$.

$$\gamma(L^a, t') = 1 + \sigma_{32} \int_{-\infty}^{t''} [I(0,t'') + I_{\mathrm{p}}(0,t'') \exp(-\sigma_{14} n L^a)]\mathrm{d}t'' \times \exp\left\{\sigma_{32} \int_{-\infty}^{t'''} [I(0,t''') + I_{\mathrm{p}}(0,t''')]\mathrm{d}t'''\right\} \tag{3.4-7}$$

The relationship between the parameters affecting the mode-locking effect can be obtained from Eq. (3.3-4) and will be briefly analyzed below.

1. The stable pulse state (mode-locked region)

When h reaches maximum or is in the finite region of the resonator mismatch amount δ_L, there exists a set of real number solutions for Eq. (3.4-4). Stable mode-locked pulses are obtained in this region. Within the range where the laser resonator length is variable, the cyclic periods of laser pulses and the pump pulses are identical because the frequency selecting element can make the pulse retarded, or the gain medium can make the pulse have a forward displacement. Because of the disappearance of gain, the trailing edge of a pulse is cut while the active medium can bring about pulse displacement. In the steady-state region, the condition for mode-locking of two laser systems is $T = T_{\mathrm{p}}$; that is, the systems can automatically regulate the interval between the laser pulse and pump pulse in each cycle.

The limit to the stable state region is related to the time interval between the dye laser pulse and the pumping pulse. When the laser resonator is rather short, the retardation time between the pump pulse and the laser pulse decreases. When the laser pulse reaches the active medium, the inverted population decreases, which makes the amplification of the pulse decrease. At a certain point in the laser resonator, as the laser pulse cannot obtain gain, the mode-locked pulse will vanish. This point is the left side limit to the steady state region. When the laser resonator is too long, all the pump energy is used for pulse amplification to make the laser pulse generate maximum time displacement through the gain medium. This is the right side limit to the steady state region. Based on the above analysis, the following conclusions can be drawn:

(1) The greater the pump energy or the reflectivity is, the greater the pulse energy while the narrower the pulse width will be. When the pump energy and reflectivity are rather small, the pulse width reciprocal and pulse energy exhibit the monotonic characteristic.

(2) The pump energy is different from reflectivity and so is the time delay of the narrowest pulse generated.

(3) The pulse shape is related to the pump energy and reflectivity. When the pump energy and reflectivity are rather small, the pulse is almost symmetric; on the contrary, with an increase of the pump energy and reflectivity, the pulses gradually appear to be asymmetric and the pump energy remains unchanged whereas the pulse width varies, which has no appreciable effect on the parameters of the laser pulse.

2. The phase modulated pulse

What is discussed above is the case of $\Delta_{32} = 0$. If $\Delta_{32} \neq 0$, then the resonator will be mismatched and the phase modulated pulse will be formed. According to the equation

$$\varphi(1, t') = -\frac{\Delta_{32}}{2} \ln\left[\frac{1}{R} G(t')\right] + \varphi(0, t') \tag{3.4-8}$$

it can be seen that the pulse possesses a phase related to time after passing through the active medium. This implies that the pulse frequency will vary and, in different portions of a pulse, the variation of the frequency is different. After phase modulation is formed, the spectrum of the dye laser pulse is broadened. So the frequency mismatch in the case of $\Delta_{32} \neq 0$, makes the steady-state single pulse oscillation range shrink. With an increase of the Δ_{32} value, the ratio of the maximum intensity of the major pulse to the intensity of the subpulse that might exist increases and the steady-state range diminishes. Meanwhile the pulse shape and the fluctuation of the pulse phase increase. Therefore, strictly speaking, there does not exist the steady state under definite conditions.

3.4.2 The structure of the synchronously pumping mode-locked laser

What is shown in Fig. 3.4-3 is a schematic diagram of the typical structure of the synchronously pumping dye laser, which includes a pump source, an optical resonator, the laser medium, and the modulating elements. In general, the active mode-locked argon ion laser or a solid mode-locked laser is adopted as the pump source, depending on the type of the laser medium. For a dye laser, usually a folded cavity structure made up of three reflective mirrors is adopted as the dye laser resonator. The reflective mirror M_1 feeds the pump optical pulse sequence into the dye laser cavity, with the pump light and laser passing through the dye at a small included angle at the same time. The reflective mirror M_2 makes laser reflected and then partly reflected and partly output at the output mirror M_4. The reflective mirrors are so arranged in the laser resonator that dye laser is made to oscillate between reflective mirrors M_2 and M_4 while the pump light injected at a small angle with the direction of the dye laser will leave the resonator after passing through the dye. But the beam waist of the pump light in the active medium must well overlap that of dye laser. For this reason, the included angle between the two light beams has to be as small as possible and a good astigmatic compensating device should be adopted.

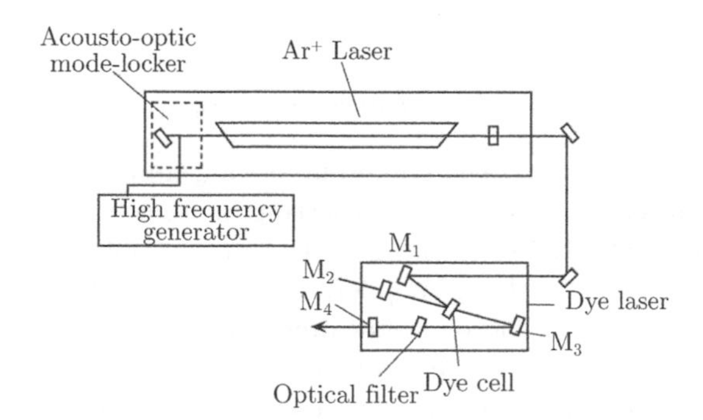

Fig. 3.4-3 Schematic diagram of the synchronously pumping dye laser

In the devices shown in Fig. 3.4-3, by means of the modulating elements such as the F-P etalon or the birefringence filter, the laser frequency can be continuously varied. If the prism is used as the modulating element, then the additional folding resonator structure can be adopted. By adopting different dyes, it will be possible to generate pulses of different pulse widths in the synchronously pumping laser. With the modulating element, the wavelength will be continuously tuned within a spectral range of 420~1000 nm.

If the stability of a synchronously pumping mode-locked laser is to be raised, an electronic feedback circuit can be adopted to control the frequency of the modulator, as shown in Fig. 3.4-4. This structure is capable of generating very stable ultrashort pulses with two kinds of control circuit. One is the fast control circuit used to measure the average power P_1 of laser pulses and the second harmonic average power P_2 generated by the KDP crystal.

The peak power of the second harmonic pulse is proportional to the square of the laser pulse peak power, that is, $P_2 \propto P_1^2\tau_p$, where τ_p is the width of the laser pulse, from which P_1^2/P_2 as the measure of the pulse width can be found. To facilitate control, the circuit adopts the autocorrelator to measure the pulse width. Experimental results show that the mismatch of the cavity has little influence on the average power P_1 of the fundamental wave, but P_2 and τ_p vary a great deal. By making use of the signal variation $[P_1^2/P_2]$ measured to change the oscillation frequency, pulses of minimum width can be generated. The other is the slow control circuit, which uses the pulse width calculated by microcomputer from the autocorrelator to find the direction of movement of the reflective mirror. Adoption of this kind of control system makes it possible to generate pulses of a width of 0.7ps that stay stable for a long time.

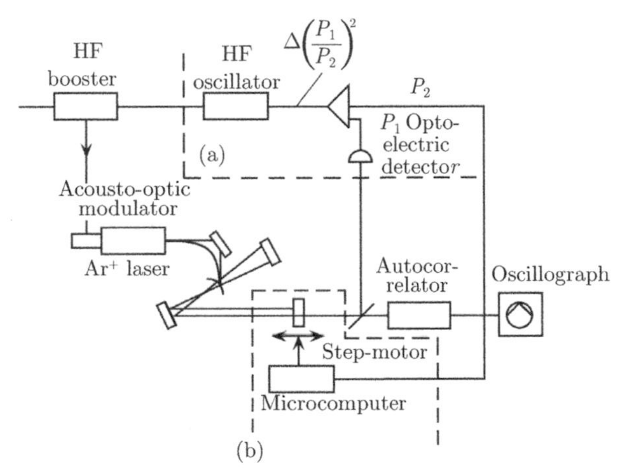

Fig. 3.4-4 Synchronous pumping dye laser adopting two control circuits

3.5 Self-mode-locking

By self-mode-locking is meant realization of mode-locking simply using the nonlinear effect of the laser medium itself without the need of inserting any modulating element in the laser cavity. As early as the 1960s-1970s of the 20th century, people observed the self-mode-locking phenomenon in the He-Ne laser, coppers team laser, Nd:YAG laser. But, as the self-mode-locked pulse sequence in these lasers cannot maintain itself, this phenomenon has never caught much attention. In 1991, self-mode-locking operation proved successful for the first time in the titanium sapphire (Ti:Sapphire) continuous laser, making the study of the solid laser self-mode-locking the hot topic in the field of the ultrashort pulse ever since. At present, the self-mode-locked pulse width can reach 8fs.

3.5.1 The mechanism of self-mode-locking

Regarding the principle of the self-mode-locking of the Ti:Sapphire laser (Ti:S), no unified theoretical explanation has been reported up to now. But it is believed by most scholars that its self-mode-locking phenomenon has something to do with the light beam self–focusing effect induced by the Kerr effect in the titanium-doped sapphire gain medium. The self-mode-locking of the Ti: Sapphire laser is passive mode-locking. From the point of view of the time domain, all mode-locked lasers that are passive in character have such elements existing in their cavity that choose pulses of rather great intensity from the noise as the seeds of the pulse sequence first and then, make gain to the leading and trailing edges of the pulse smaller than 1 by means of the nonlinear effect of the mode-locking device while making gain in the middle of the pulse greater than 1. In the back-and-forth process in the cavity, the pulse is steadily reshaped and amplified, and the pulse width is compressed, until stably mode-locked. In the titanium-doped sapphire self-mode-locked laser, the nonlinear effect of the Ti:Sapphire medium refractive index can be expressed as

$$n = n_0 + n_2 \cdot I(t) \tag{3.5-1}$$

where n_0 is the refractive index unrelated to the light intensity, n_2 is the nonlinear refractive index, and $I(t)$ is the light intensity of the pulse. Because of the Gaussian distribution of the light intensity, when it passes through the medium, the self-focusing effect will be generated. Taking a certain small segment ΔL of the medium, we can find the focal length of the self-focusing effect as

$$f_{\mathrm{m}} = \frac{\alpha w_{\mathrm{m}}^2}{4\Delta n_{\mathrm{m}} \Delta L} \tag{3.5-2}$$

where w_{m} is the size of the light spot injected into the said segment of medium; α is a constant, in general 5.6~6.7; and Δn_{m} is the variation of the refractive index of the incident light on the axial line.

$$\Delta n_{\mathrm{m}} = n_2 \cdot I_{\mathrm{m}}(t) \tag{3.5-3}$$

where $I_{\mathrm{m}}(t)$ is the light intensity near the axis of the light beam injected onto the medium ΔL. In terms of the time domain of the pulse envelope, the light intensity at the front and rear edges is smaller than that in the middle of the pulse. It is known from Eqs. (3.5-2) and (3.5-3) that corresponding quasi-lens focal length f_{m} of the pulse's middle part is smaller than the corresponding focal length of the pulse's front and rear edges. Thus, when an optical pulse has passed through the self-focusing medium, the Gauss light beam parameters of the pulse's front and rear edges will no longer be consistent with those in its middle. That is to say, after the pulse passes through the self-focusing medium, because of the self-focusing effect, the light intensity variation of the pulse in time can be represented in space. There are two beam waists in the Ti:Sapphire self-mode-locked laser, one in the Ti:Sapphire medium, and the other on the planar reflective mirror of the laser cavity or near it. If a diaphragm is incorporated near the waist, according to the above analysis, the loss on the front and rear edges of the optical pulse can be made to exceed that in the middle part; that is, owing to the self-focusing effect and the presence of the diaphragm in the cavity, the Ti:Sapphire self-mode-locked laser is modulated with respect to the loss related to light intensity, i.e.,

$$\alpha = \alpha_0 - \beta \cdot I(t) \tag{3.5-4}$$

Owing to the presence of the gain, when the pulses cycle in the cavity, those of small intensity will steadily be restrained and vanish while those of great intensity will steadily be strengthened and the leading and trailing edges keep suffering losses while the middle part of the pulses is amplified and the pulse width compressed. For an optical pulse, the combination of the self-focusing effect and the diaphragm in the cavity is equivalent to a fast saturated absorber, which can perform the function of compressing the front and rear edges of the optical pulse. The diaphragm can be an externally applied one such as the pinhole diaphragm applied near the position of the output mirror or it is possible to directly use the gain diaphragm (also called the 3-dimensional diaphragm or soft diaphragm) formed by the Gauss-distributed gain region in the Ti:Sapphire rod. The formation of the self-mode-locked pulses of the Ti:Sapphire laser is divided into the following stages:

1. The formation of the initial pulses

It has been proved by theoretical analysis and numerous experiments that, as the noise pulses in the continuously operating titanium-doped sapphire laser cannot reach the start threshold for mode-locking, this kind of laser cannot get started on its own. So it's necessary to introduce an instantaneous perturbation in the cavity first to bring about high losses. When the cavity mirrors are reset, the light intensity in the cavity will rise and fall vehemently. When passing through the gain medium, because of the self-focusing effect of the gain medium, its combination with the diaphragm in the cavity is equivalent to a saturable absorber. Following the self-amplitude modulation (SAM) and the linear amplification of the gain medium, the pulses are selected and amplified and preliminarily compressed to form the initial pulses.

2. The formation of stable mode-locked pulses

After the initial mode-locked pulses are formed in the cavity, as their peak power is rather large, because of the nonlinear Kerr effect in the gain medium, self-phase modulation (SPM) occurs with the pulses that seriously change the phase of the pulse. When passing through the titanium-doped sapphire rod, the optical pulse induces very great second-order positive group velocity dispersion (GVD) and third-order dispersion. At this stage, it's still the gain medium's self-amplitude modulation and gain amplification that play the dominant role. Unfortunately, owing to the increase in the pulse power, it's inevitable that self-phase modulation and very great positive group velocity dispersion will be generated, which is unfavorable to further compressing the pulse width but has to be compensated for by appropriate negative dispersion before the narrowest pulse width can be obtained.

It is shown by large numbers of experiments and analytic calculation that self-mode-locking should be started by taking additional measures (operating in the continuous state at first), the simplest method being lightly tapping the platform or a certain mirror to generate an intensity perturbation to start self-mode-locking. During its mode-locked stable operation, the started Ti:S laser is influenced by the surrounding environment. Once the lock is lost, it must be restarted. For this reason, many methods for actively and passively starting and maintaining Ti:S self-mode-locked operation have been invented.

(1) The scheme for starting by regeneration of the acousto-optic modulation. This method consists in introducing an acousto-optic modulator into the original self-mode-locked laser to make its frequency match the reciprocal of the resonator period. What the laser outputs are pulses of the order of magnitude of dozens to hundreds of ps, with the pulse repetition rate determined by the modulator's driving frequency. The laser is now in the active mode-locked state. Properly adjust the resonator to make the laser enter self-mode-locking to generate mode-locked pulses of a pulse width of the fs order of magnitude, with the average power almost unchanged. Now the pulse repetition rate is only determined by the cavity frequency, having nothing to do with the modulator's driving frequency. But when the two frequencies are precisely matched, the self-locked state is most stable. The frequency component of the laser cavity output signal can be taken (take the the second harmonic frequency and divide it into four frequency-division) to drive the modulator. This way the driving frequency of the modulator will be able to automatically match the cavity frequency.

(2) The self-mode-locked Ti:S laser can also be started using the saturable absorber. Insert a saturated absorber in the Ti:S laser cavity (such as HITCI) to change the dye concentration till the pulse width of the ultimately formed mode-locked pulse and average power are not related to the dye while only the time for the pulse to be set up is influenced by the concentration of the dye (the greater the concentration, the shorter the time for setting up) and, when the dye spray film is removed, the device is capable of generating self-locking. This shows that what the saturable absorber does is simply introduce the weakest

modulation to start self-mode-locking.

(3) Start self-mode-locking using the coupled cavity of the quantum well reflector. The degree of coupling of the active cavity and auxiliary cavity should be rather low ($R_L > 90\%$), with the output mirror replaced with a variable output coupler made up of a polarized beam splitter of a half wave plate.

(4) Start with a vibrating mirror. Use a linear outer cavity or directly vibrate a cavity mirror (in general a totally reflecting mirror), at a frequency of 25 Hz, and an amplitude smaller than 0.5 mm and the Ti:S laser can be made to enter the self-mode-locked state from the continuous state.

3.5.2 The ultrashort pulse compressing technique

When ultrashort optical pulses are transmitted in a medium, as they possess an extremely high peak power density after getting focused, the process of their transmission usually exhibits nonlinear effects, such as the self-phase modulation effect, self-focusing effect, the photoinduced birefringence effect, etc. Of these nonlinear effects, the nonlinear effect of the refractive index is fundamental.

$$\Delta n(t) = n - n_0 = n_2 I(t) \tag{3.5-5}$$

is the variation of the additional refractive index induced by the light field, which is a time function determined by the optical pulse envelope. Therefore, with the nonlinear effect of the refractive index taken into account, it's clear that the variation of the refractive index induced in different portions of the optical pulse is different.

(1) When the relaxation time T_r of n_2 is far faster than the pulse width τ_p, that is, $T_r \ll \tau_p$, this is a process of transient response and $\Delta n(t)$ can be written as

$$\Delta n(t) = n_2 I(t) \tag{3.5-6}$$

The variation of the additional phase induced in different portions of the pulse is

$$\Delta\varphi(t) = kL\Delta n(t) = kLn_2 I(t) \tag{3.5-7}$$

where k is the wave vector and L is the transmission length of the optical pulse in the medium. The instantaneous frequency of the pulse in different portions is

$$\omega(t) = \omega_0(t) + \Delta\omega(t) \tag{3.5-8}$$

$$\Delta\omega(t) = -\frac{\partial}{\partial t}\Delta\varphi(t) = -\frac{\partial}{\partial t}[kLn_2 I(t)] \tag{3.5-9}$$

where $\omega_0(t)$ is the instantaneous frequency without taking the self-phase modulation into consideration and $\Delta\omega(t)$ is the additional frequency induced by the self-phase modulation effect. The introduction of $\Delta\omega(t)$ makes the different portions of the pulse envelope possess different instantaneous frequencies. This phenomenon is called the chirp effect, which can be depicted by the following equation:

$$c(t) = \frac{\partial w(t)}{\partial(t)} = -\frac{\partial^2}{\partial t^2}[kLn_2 I(t)] \tag{3.5-10}$$

The above-mentioned effects are as shown in Fig. 3.5-1(a). $c(t)>0$ is the positive chirp and $c(t)<0$ is the negative chirp. The self-phase modulation effect makes the different portions of the pulse envelope possess different instantaneous frequencies. The leading and trailing edges of the pulse possess negative chirp while the middle part of the pulse possesses positive chirp. Because of the SPM effect, the spectral band is broadened, and is expanding toward the high side and low side of the original carrier frequency ω_0 at the same time.

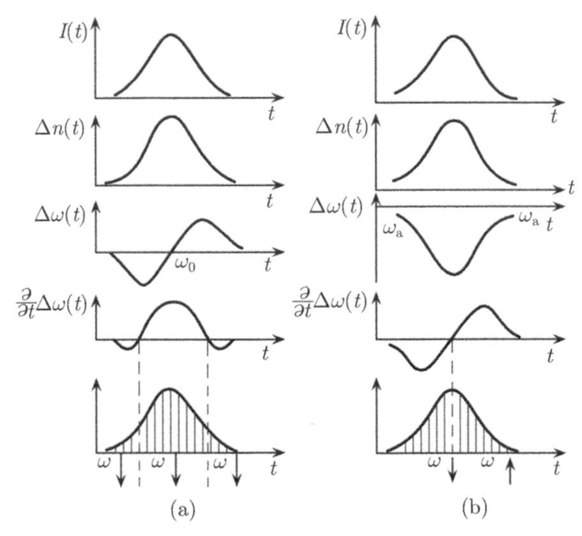

Fig. 3.5-1 The SPM effect of ultrashort optical pulse transmission in medium

(2) $T_r \gg \tau_p$, that is, T_r possesses a relaxation time much longer than τ_p. Now, to consider the refractive index in a portion of the pulse at time t, it is necessary to consider the remaining part of the refractive index induced at time t' prior to time t to relax to time t:

$$\Delta n(t', t) = \Delta n(t') \exp[(t - t')/T_r] \tag{3.5-11}$$

Hence the refractive index at time t should be written as

$$\Delta n(t) = \int_{-\infty}^{t} n_2 I(t') \exp[(t - t')/T_r] dt' \tag{3.5-12}$$

So, the frequency shift can be written as

$$\Delta\omega(t) = -\frac{\partial}{\partial t} \int_{-\infty}^{t} n_2 I(t') \exp[-(t - t')/T_r] dt' \tag{3.5-13}$$

In consideration of $T_r \gg \tau_p$, the above equation can be approximately written as

$$\Delta\omega(t) = -c' I(t) \tag{3.5-14}$$

Its chirp $c(t)$ can be written as

$$c(t) = -c' \frac{\partial}{\partial t} I(t) \tag{3.5-15}$$

The above-mentioned physical images can be expressed with Fig. 3.5-1(b). For the case of $T_r \gg \tau_p$, because of the SPM effect, different portions of a pulse possess different instantaneous frequencies and the leading and trailing halves of the pulse have chirps of opposite signs. The expansion of the pulse frequency spectrum is simply toward the $\omega < \omega_0$ end, i.e., the frequency expands toward the low frequency end.

In the above discussion, only the SPM effect in the process of propagation of the ultrashort optical pulses in a medium is considered. In reality, all mediums possess dispersion, the dispersion coefficient being defined as:

$$D = \lambda^2 \frac{\mathrm{d}^2 n}{\mathrm{d}\lambda^2} \tag{3.5-16}$$

Since the pulses already possess chirp, that is, the different portions of the pulse envelope possess different instantaneous frequencies, during the propagation of ultrashort optical pulses with chirp in the medium with dispersion, the propagation velocities in different portions are also different so that different portions of the pulse will have the broadening or narrowing effect. Broadening occurs when the chirp and dispersion are of the same sign while narrowing happens when the chirp and dispersion are of different signs. When the medium has positive dispersion, the leading and trailing edges of the pulse characterized by negative chirp get compressed while the middle part of the pulse characterized by positive chirp is broadened, the pulse waveform gradually becoming a square wave. When the medium possesses negative dispersion, the leading and trailing edges of the pulse with negative chirp get broadened while the middle part of the pulse is compressed, leading to the waveform of the whole pulse becoming narrower.

According to the above analysis, as long as a medium with negative dispersion can be chosen, the ultrashort pulse can be further compressed. There are at present the following two methods for compressing ultrashort pulses.

1. The optical fiber-grating pair method

Generally, further compression of ultrashort optical pulses using the optical fiber-grating pair is carried out outside the cavity; the process of compression is as shown in Fig. 3.5-2. After the laser pulses are injected into the optical fiber medium, the refractive index of the optical fiber is determined by Eq. (3.5-1). Then, the additional phase shift resulting from the nonlinear refractive index is

$$\Delta\varphi(t) = \frac{l}{\lambda} n_2 I(t) \tag{3.5-17}$$

where λ is the light wavelength in vacuum, l is the length of the optical fiber, and $I(t)$ is the light intensity. The above equation characterizes the SPM in the optical fiber, that is, the optical pulses that pass through generate the chirp effect. Chirped optical pulses have the characteristic of compressibility. Suppose the time duration of the pulses is τ_p, the central frequency is ω_0, and the pulses contain linear frequency shift (the chirp term). Therefore, the instantaneous frequency of the pulse is

$$\omega(t) = \omega_0 + \frac{\Delta\omega}{\Delta t}\tau_\mathrm{p} \quad (0 < \tau_\mathrm{p} < \Delta t)$$

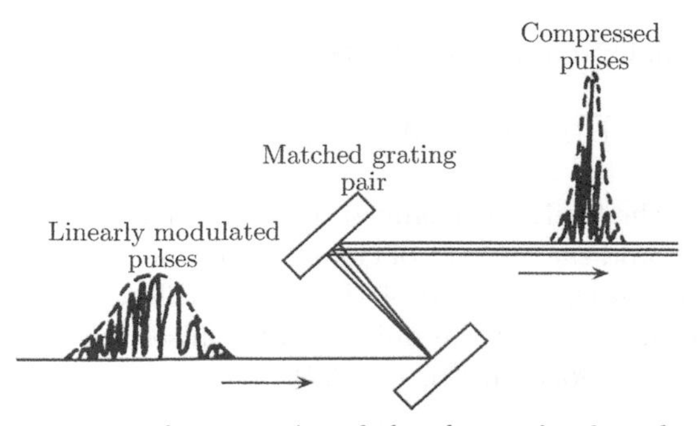

Fig. 3.5-2 Schematic diagram of compression of ultrashort pulses by reflection of grating pair

where $\Delta\omega$ is the maximum frequency shift. If $\Delta\omega > 0$, the leading end frequency of the pulse will be slower than the trailing end frequency. The optical pulse output from the single mode optical fiber is linearly modulated pulse, which will be acted on by the dispersive retardation line made up of the grating pair. As the optical path traversed by the rays between two gratings is related to the incident wavelength λ, the different frequency spectral components of the optical pulses will traverse different optical paths. The rays of a long wavelength will traverse a longer optical path than those of a short wavelength. Therefore, after passing through the grating pair, the trailing edge of the long wavelength will gradually catch up with the leading edge of the short wavelength, thereby making the waveform of the optical pulse compressed and the pulse's peak power, too, will be increased. Adoption of this technique has yielded ultrashort optical pulses of 6 fs.

2. Insertion of negative dispersion element in the laser cavity

It is known from the above analysis that, in the self-mode-locked laser, the optical pulse will generate very strong second-order positive group velocity dispersion (GVD) and third-order dispersion, which is unfavorable to further compressing the pulse width but needs the insertion of negative dispersion elements for making compensation before the narrowest pulse width can be obtained.

When ultrashort pulses are transmitted in a laser medium, the equation for depicting the transmission characteristics of the light wave field can be derived using the Maxwell equation.

$$\frac{\partial A}{\partial z}+\frac{\alpha}{2}A+\frac{\mathrm{j}}{2}\beta_2\frac{\partial^2 A}{\partial T^2}-\frac{1}{6}\beta_3\frac{\partial^3 A}{\partial T_3}=\mathrm{j}\gamma\left(|A|^2A+\frac{2\mathrm{j}}{\omega_0}\frac{\partial}{\partial T}(|A|^2A)+\mathrm{j}\frac{\alpha_2}{\gamma}A\frac{\partial|A|^2}{\partial T}\right) \tag{3.5-18}$$

where A is the amplitude of the light field, α is the absorption coefficient, and β is the mode transmission constant.

$$\beta(\omega)=n(\omega)\frac{\omega}{c}=\beta_0+\beta_1(\omega-\omega_0)+\frac{1}{2}\beta_2(\omega-\omega_0)^2+\cdots \tag{3.5-19}$$

$$T=t-\frac{z}{v_{\mathrm{g}}}=t-\beta_1 z \tag{3.5-20}$$

(1) The influence of second-order dispersion on the pulse. Suppose a laser medium absorbs very little for light wave ($\alpha \approx 0$) and the nonlinear effect can be neglected ($\gamma \approx 0$). For pulses of a pulse width greater than 100 fs, consideration is given only to the GVD effect of pulse transmission in the linear dispersive medium of the lowest order dispersion ($\beta_2 \neq 0$, $\beta_3 = 0$, $\beta_4 = 0$). Equation (3.5-18) becomes

$$\mathrm{j}\frac{\partial A}{\partial z}=\frac{1}{2}\beta_2\frac{\partial^2 A}{\partial T^2} \tag{3.5-21}$$

The solution for the normalized amplitude in its frequency domain is

$$\overline{U}(z,\omega)=\overline{U}(0,\omega)\exp\left(\frac{\mathrm{j}}{2}\beta_2\omega^2 z\right) \tag{3.5-22}$$

This equation shows that the GVD has changed the phase of each of the frequency spectrum components of the pulses, and the amount of change relies on the frequency and the distance of transmission. Although such a change of the phase will not affect the pulse frequency spectrum, it can change the shape of the pulse. GVD causes the pulses to broaden, as can be seen from the Fourier transform in Eq. (3.5-22):

$$U(z,t)=\frac{1}{2\pi}\int_{-\infty}^{\infty}\overline{U}(0,\omega)\exp\left(\frac{\mathrm{j}}{2}\beta_2\omega^2 z-i\omega T\right)\mathrm{d}\omega \tag{3.5-23}$$

For example, the initial waveform is the shape of a Gauss pulse:

$$U(0,T) = \exp\left(-\frac{T^2}{2T_0^2}\right) \tag{3.5-24}$$

where T_0 is the half width of its pulse. When its transmission distance is z, its waveform is

$$U(z,T) = \frac{T_0^2}{T_0^2 - \mathrm{j}\beta_2 z}\exp\left[-\frac{T^2}{2(T_0^2 - \mathrm{j}\beta_2 z)}\right] \tag{3.5-25}$$

Its waveform is still a Gauss form, but its width is

$$T_1 = T_0[1 + (z/L_D)^2]^{\frac{1}{2}} \tag{3.5-26}$$

where

$$L_D = \frac{T_0^2}{|\beta_2|} \tag{3.5-27}$$

is called the dispersion length. It is clear that the pulse broadening is related to the transmission distance, the initial width, and dispersion parameters.

(2) The influence of higher-order dispersion on the pulse. When the pulse is very narrow (smaller than 100 fs), the spectral component contained in it is already very broad, $\Delta\omega/\omega$ being already a quantity not to be neglected. Now, in the expanded form of $\beta(\omega)$, it is necessary to consider the higher-order dispersion term β_3, whose equation of transmission is

$$\mathrm{j}\frac{\partial U}{\partial z} = \frac{1}{2}\beta_2\frac{\partial^2 U}{\partial T^2} + \frac{\mathrm{j}}{6}\beta_3\frac{\partial^3 U}{\partial T^3} \tag{3.5-28}$$

The solution for this equation is

$$U(z,T) = \frac{1}{2\pi}\int_{-\infty}^{\infty}\overline{U}(0,\omega)\exp\left(\frac{\mathrm{j}}{2}\beta_2\omega^2 z + \frac{\mathrm{j}}{6}\beta_3\omega^3 z - \mathrm{j}\omega T\right)\mathrm{d}\omega \tag{3.5-29}$$

It is known from the above equation that the higher-order dispersion makes the pulse width broadened. Moreover, it causes oscillation to occur at the leading edge of the pulse ($\beta_3 < 0$) or its trailing edge ($\beta_3 > 0$), thus changing the shape of the pulse and making it symmetric.

(3) The compression of the self–mode-locked pulse. In a titanium-doped sapphire self-mode-locked laser, when the optical pulse passes through the operation medium, because of the SPM effect, the pulse becomes chirped pulse while generating very great second-order positive group velocity dispersion (GVD) and third-order dispersion. To further compress the width of the chirped pulse, it is necessary to introduce the negative group velocity dispersion for compensation before a narrow pulse width can be obtained.

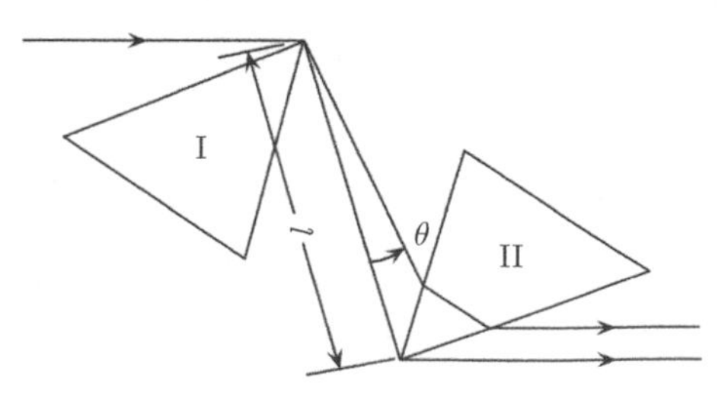

Fig. 3.5-3 The double quartz prism system

There are two methods of generating the negative group velocity dispersion, the diffraction grating method and the Gires-Tournois interferometer method. But their insertion loss is great and it's not easy to regulate the dispersion amount between positive and negative with the diffraction grating method. The currently most widely used method is to insert two prisms made of high dispersion SF14 glass, whose structure is as shown in Fig. 3.5-3.

The rays of the optical pulses with respect to both the incident angle and outgoing angle are Brewster angle and minimum deviation angle, thus reducing the insertion loss for the prisms as much as possible. When the optical pulse passes through the prism pair, the phase shifts of different frequency components are different. Expand the phase into Taylor series at the central frequency:

$$\varphi(\omega) = \varphi(\omega_0) + \left(\frac{\partial\varphi}{\partial\omega}\right)_{\omega_0}(\omega-\omega_0) + \frac{1}{2}\left(\frac{\partial^2\varphi}{\partial\omega^2}\right)_{\omega_0}(\omega-\omega_0)^2 + \frac{1}{6}\left(\frac{\partial^3\varphi}{\partial\omega^3}\right)_{\omega_0}(\omega-\omega_0)^3 + \cdots \tag{3.5-30}$$

where the first term represents phase translation and the second term only induces the time translation of the pulse envelope, also spoken of as group retardation, neither of which has any influence on the shape of the pulse envelope. It is the second-order term and even higher-order terms that influence the shape of the pulse, usually spoken of as dispersion phase shift.

The second-order dispersion of the prism pair is given by the following equations:

$$\frac{\mathrm{d}^2P}{\mathrm{d}\lambda^2} = 4l\left\{\left[\frac{\mathrm{d}^2n}{\mathrm{d}\lambda^2} + \left(2n - \frac{1}{n^3}\right)\left(\frac{\mathrm{d}n}{\mathrm{d}\lambda}\right)^2\right]\sin\theta - 2\left(\frac{\mathrm{d}n}{\mathrm{d}\lambda}\right)^2\cos\theta\right\} \tag{3.5-31}$$

$$\frac{\mathrm{d}^2\varphi}{\mathrm{d}\omega^2} = \frac{\lambda^3}{2\pi c^2}\frac{\mathrm{d}^2P}{\mathrm{d}\lambda^2} \tag{3.5-32}$$

The expressions for the third-order dispersion are

$$\frac{\mathrm{d}^3P}{\mathrm{d}\lambda^3} = 4l\left(\frac{\mathrm{d}^3n}{\mathrm{d}\lambda^3}\sin\theta - 6\frac{\mathrm{d}n}{\mathrm{d}\lambda}\frac{\mathrm{d}^2n}{\mathrm{d}\lambda^2}\cos\theta\right) \tag{3.5-33}$$

$$\frac{\mathrm{d}^3\varphi}{\mathrm{d}\omega^3} = -\frac{\lambda^4}{4\pi^2c^3}\left(3\frac{\mathrm{d}^2P}{\mathrm{d}\lambda^2} + \lambda\frac{\mathrm{d}^3P}{\mathrm{d}\lambda^3}\right) \tag{3.5-34}$$

The group velocity dispersion constant is

$$D = \frac{\mathrm{d}}{\mathrm{d}\lambda}\left(\frac{1}{v_g}\right) = -\frac{\omega}{\lambda l}\frac{\mathrm{d}^2\varphi}{\mathrm{d}\omega^2} = -\frac{\lambda}{cl}\frac{\mathrm{d}^2P}{\mathrm{d}\lambda^2} \tag{3.5-35}$$

where $P = l\cdot\cos\theta$, l is the distance between the apices of the two prisms, θ is the angle the ray deviates from the apex line after passing through the first prism, and n is the refractive index of the prism.

$\frac{\mathrm{d}^2P}{\mathrm{d}\lambda^2}$ determines the sign and size of the group velocity dispersion; $\frac{\mathrm{d}^2P}{\mathrm{d}\lambda^2} < 0$ is the negative group velocity dispersion, which will make the leading edge of the pulse red-shifted and the trailing edge blue-shifted so that the positively chirped pulse will be compressed. By choosing a suitable prism interval to balance the second-order dispersion and the SPM effect of the operation material in the cavity and adopting suitable prism material to reduce the third-order dispersion as much as possible, we shall be able to obtain self-mode-locked pulse width close to the Fourier transform limit.

3.6 The selection of single pulses and ultrashort pulse measuring technique

The ultrashort pulse technology consists of the generation of ultrashort pulses, the selection of single pulses, the compression of the pulse width, the amplification of pulses, and the measurement of ultrashort pulses. This section is devoted to a discussion on the techniques of selecting single pulses and measuring ultrashort pulses, respectively.

3.6.1 The selection of single pulses

For many applications, it is necessary to select single pulses from a mode-locked pulse sequence. The methods are mainly as follows:

1. The pulse transmittance method

The device is as shown in Fig. 3.6-1, in which M_1 is the totally reflecting mirror, M_2 the partially reflecting mirror ($R > 50\%$). The electro-optic switch can be made of a Pockels cell or Kerr cell. The polarizing prism and the electro-optic modulator make up the electro-optic switching system. At the beginning, the switch simply does not work. Regulate the polarizer to maximum transmittance and, after the pumping, the laser pulses propagate back and forth and a locked mode is formed. The radiation leaking out from M_2, after getting focused by the lens, breaks down the spark gap and forms a circuit through the pulses. Apply the high voltage pulse $V_{\lambda/4}$ on the electro-optic modulator. The mode-locked pulse twice passes through the the switch and the polarizing direction rotates 90°. Now the polarized light no longer gets through the polarizing mirror, while only capable of being reflected from its interface to reflect laterally output to obtain single pulses. The drawback of this method lies in the rather high operation threshold and the elements in the cavity are susceptible to damage. Moreover, the etalon effect is apt to occur owing to the presence of too many elements in the cavity, and light leak is serious and stability poor.

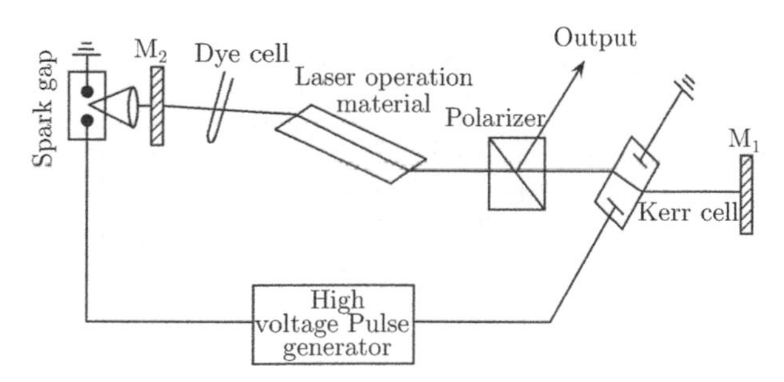

Fig. 3.6-1 The pulse transmittance method for selecting single pulses

2. method of single pulse selection outside the cavity

Such a device is as shown in Fig. 3.6-2. Place a Pockels cell (or Kerr cell) between two orthogonal polarizing prisms P_1 and P_2. Of a series of ultrashort pulses output from the mode-locked laser, when one pulse passes by, as the two polarizing prisms are orthogonal, the optical pulse is reflected by the polarizing prisms and, after being focused by the lens, break down the spark gap to apply the half-wave voltage ($V_{\lambda/2}$) rectangular pulse on the electro-optic switch. When the next pulse passes through the electro-optic switch, its polarized surface rotates 90°, just passing through the orthogonal prisms to output single pulses. As to which one in the sequence of pulses should be chosen, this can be done by adjusting the spark gap breakdown power and the time delay from the spark gap to the electro-optic switch. In general, the pulse with maximum energy in the sequence of pulses is chosen.

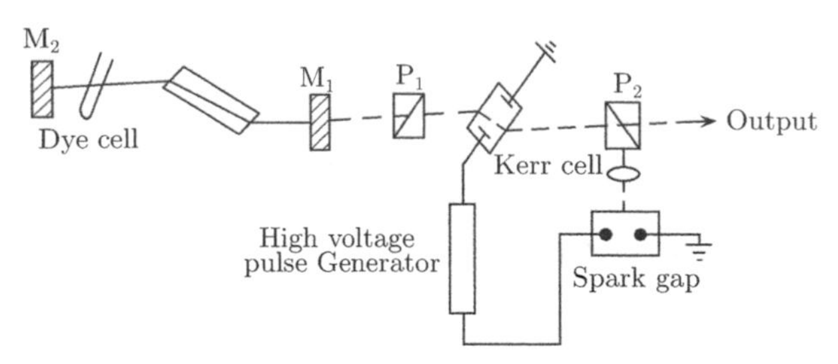

Fig. 3.6-2 Method of single pulse selection outside the cavity

The rising leading edge of the voltage pulse applied on the electro-optic switch should be as steep as possible while the width should be smaller than the mode-locked pulse interval to prevent selection of more than two ultrashort pulses. To reduce the loss, both the polarizing prisms and the electro-optic switch should be AR coated.

The shortcoming of this method lies in the need of applying the half-wave voltage on the Kerr cell and using two prisms, but it has the merits of using only a few elements in the cavity and convenient adjustment.

Both of the methods mentioned above require that rather high power be output, or the spark gap will not be able to activate.

3. The thyratron triggered circuit

The above-mentioned methods both consist in using optical signals to break down the spark gap followed by applying the specified voltage on the Kerr cell. This switch can be replaced with an opto-electric switch circuit, that is, a photodiode or avalanche photodiode to convert the optical pulses into electrical signals to push forward the thyratron for the realization of switch activation. The device is as shown in Fig. 3.6-3.

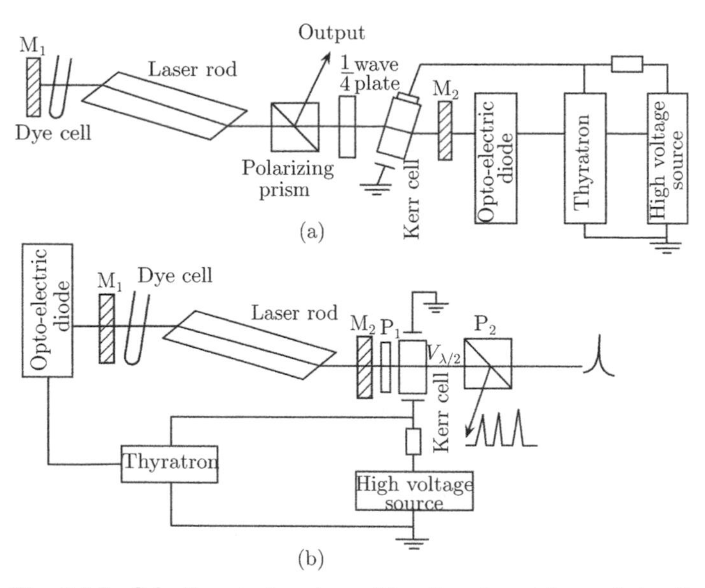

Fig. 3.6-3 Selecting single pulses with a thyratron triggered circuit: (a) intracavity method; (b) extracavity method

In Fig. 3.6-3(a), a neodymium glass laser is taken as an example, with the switching elements placed in the cavity. M_1 and M_2 are reflective mirrors with $R = 95\%$ and 99%, respectively. The time for the optical pulse to make a round-trip travel (about 16 ns) in the cavity is longer than the time needed for applying the quarter-wave voltage on the Kerr cell. The saturable absorption dye solution is the Eastman Kodak "9740". The surface of elements needed in the cavity is AR coated or cut into a Brewster angle. The principle of this device is similar to that of the method of pulse transmissivity. Prior to the output of the mode-locked pulse sequence, the quarter-wave voltage is applied on the Kerr cell, the resonator in the high Q state. When the first mode-locked pulse is output, the light signal is converted into electrical signal by the fast responsive photodiode to trigger the thyratron to short-circuit the quarter-wave voltage applied on the Kerr cell, laterally outputting single ultrashort pulses from the polarizing mirror. To ensure that single pulses are output, it is required that the switching time of the thyratron be shorter than the mode-locked pulse interval $2L/c$. By adjusting the retardation time from the photodiode to the thyratron, the peak value of the single pulse output can be changed.

Figure 3.6-3(b) is the improved form of (a), the difference being the Kerr cell is placed outside the cavity with a half-wave voltage ($V_{\lambda/2}$) applied on it.

3.6.2 The ultrashort pulse measuring technique

The accurate and reliable yet simple measurement of the pulse width of ultrashort pulses is very often the crux in the development of ultrashort lasers and their application. The chief measuring methods are the direct measuring method and the correlation measuring method, the former involving the use of the photoelectric detector in addition to the fast responsive oscilloscope for direct observation and streak photography while the latter often uses the two-photon fluorescence method and second harmonic method. Below is a description of each of these methods.

1. The direct observation method

Ultrashort optical pulses can be directly observed with a fast photoelectric detector and a high response time broad bandwidth oscilloscope. If the rising time of the oscilloscope is 1.3×10^{-10} s, the pulse width that can be measured is 10^{-10} s.

The waveform of a pulse can be directly observed with a streak camera, the streak diagram of whose structure is as shown in Fig. 3.6-4. It is composed of a fast streak image converter, a slow silicon target tube camera analyzer, and an electronic monitoring system. What is shown in Fig. 3.6-4(a) is the process of operation of the streak camera. The to-be-measured laser pulse passes through the narrow slit (S) and is focused on the cathode (K) of the streak tube by a lens to strike out optical electrons and, after getting accelerated by the accelerating grid (M) and focused by the clustered electrodes (F), is accelerated again by the net-shaped anode (A). The optical electrons are injected into the electrodes (A_b) in the deflection field. At this very instant, the pulse voltage generated by the deflecting voltage generator (G) is applied on the deflecting electrode, which can make the electrons scan from top to bottom at a high speed. After getting multiplied in the microchannel, the scanning electrons are shot onto the fluorescent screen (S_{ch}), with streak photographs appearing. As the deflection of the electrons is proportional to the voltage applied on the electrode the instant they pass through the deflecting electrode, the difference in the time for passing through the electrode can be recorded as the difference in position of streak imaging on the fluorescent screen. That is to say, this process has converted the temporal axis of the incident light into the spatial axis of the fluorescent screen. The luminous intensity on the screen reflects the intensity of the incident light. The length of the streaks correspond to the laser pulse width while the interval between streaks corresponds to the period of the optical pulses. This part of the device can exhibit ps optical pulses. Then the silicon target camera tube such as the CCD array can be used to take the pictures of the streaks. The optic multichannel analyzer (OMA) can also be used to analyze the image signals obtained to display the distribution of the incident light intensity. What is shown in Fig. 3.6-4(b) is a microdensitogram of the pulse pair taken with a streak camera, which can only distinguish ps pulses and is for the time being incapable of measuring fs pulses.

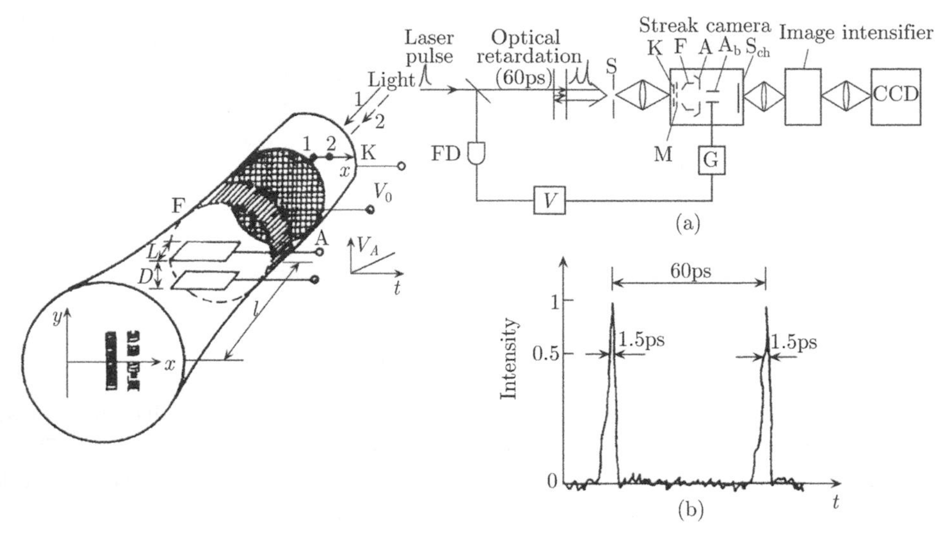

Fig. 3.6-4 The streak photography

2. The correlation measuring method

The correlation measuring method is currently fairly widely applied in measuring the width of ultrashort pulses. It is an indirect measuring method, that is, the correlation function curve (not the actual waveform of the pulse) is obtained by performing tests using the correlation function. Then, by means of conversion, the approximate value of the pulse width can be obtained. At present, it is customary for fs ultrashort pulses to adopt the correlation measuring method.

Suppose the light field measured is $E(t)$, the light intensity $I(t) \propto E(t)^2$. Then the normalized form of the second-order correlation function is defined as

$$G^2(\tau) = \frac{\langle I(t) \cdot I(t+\tau) \rangle}{\langle I^2(t) \rangle} = \frac{\langle E^2(t) \cdot E^2(t+\tau) \rangle}{\langle E^4(t) \rangle} \tag{3.6-1}$$

where $\langle \quad \rangle$ represents the average value in a sufficiently long time interval, τ is the time retardation. If the measurement of the optical pulse is related to $G^2(\tau)$, then the measurement is called nonlinear optical measurement. It can be seen from the above equation that no matter whether $I(t)$ is symmetric or not, $G^2(\tau)$ is always. Therefore, the asymmetry of the actual waveform cannot be found by measurement using the second-order correlation function. In order to accurately measure $I(t)$, it is necessary to adopt even higher-order correlation functions. The nth-order normalized correlation function is

$$G^n(\tau_1, \tau_2, \cdots, \tau_{n-1}) = \frac{\langle I(t)I(t+\tau_1) \cdots I(t+\tau_{n-1}) \rangle}{\langle I^n(t) \rangle} \tag{3.6-2}$$

It has already been shown mathematically that, if only $G^2$and G^3 are definitely known, then the knowledge is sufficient to depict all the higher-order correlation functions, thereby depicting the pulses themselves. In reality, if the accuracy of measuring the lower-order correlation function is inadequate, a higher-order measurement can be performed since it is more sensitive to the asymmetry of pulses.

$$G^n(0,0,\cdots,\tau) = \frac{\langle I^{(n-1)}(t)I(t+\tau)\rangle}{\langle I^{(n)}(t)\rangle} \tag{3.6-3}$$

When the order is increased, $I^{(n-1)}(t)$ becomes a sharper time function. Hence the value obtained by measurement is closer to the $I(t)$.

The two-photon fluorescence method and the second harmonic generating method are fairly mature correlation function measuring methods in current use and are discussed below.

(1) The two-photon fluorescence method

The basic principle of this method is based on the material fluorescence effect. After absorbing a photon ($h\nu_{\text{in}}$), a material atom (or molecule) jumps from the ground state to a certain excited state and emits fluorescence when jumping from the excited state to the low energy state. The wavelength of the fluorescence is longer than that of the incident light, that is, $h\nu_{f1} < h\nu_{\text{in}}$. When the incident light is one of a high power, the material can absorb two photons ($2h\nu_{\text{in}}$) simultaneously, jumping from the ground state to the excited state, called the two-photon absorption. Now the fluroescence emitted after transition to the low energy level is called the two-photon fluorescence, i.e., $h\nu_{f2} < 2h\nu_{\text{in}}$. This is a kind of nonlinear absorption effect. For instance, under the action of strong light the rhodamine 6G acetone solution generates two-photon fluorescence of 0.55 μm after absorbing two photons of 1.06 μm (See Fig. 3.6-5).

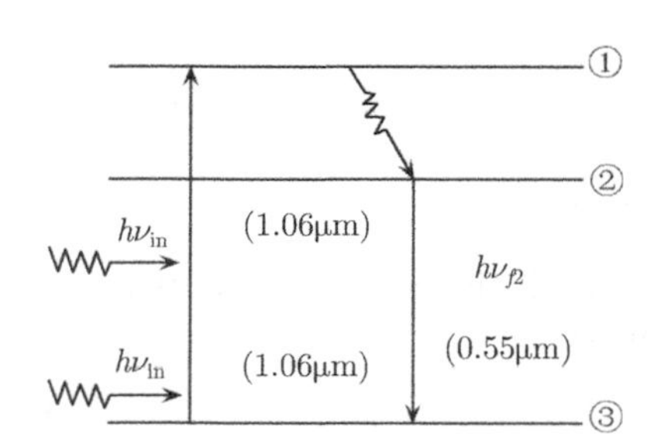

Fig. 3.6-5 The model for two-photon absorption energy levels of organic dyes

The measuring device is as shown in Fig. 3.6-6. The beam splitter splits the incident optical pulse into two beams of light of equal intensity, which are reflected by means of reflective mirrors M_1 and M_2 so that they are injected into the dye solution from two directions where, in the passage of the two pulses, they emit uniform weak fluorescent light. But at the position where the two pulses overlap, extremely bright strong fluorescent light will be emitted owing to the generation of two-photon absorption. If highly photosensitive film is used to take the picture of the spatial distribution of fluorescence brightness and place the exposed negative on the microscopic densitometer, the pulse width can be found according to the spatial distribution of the light density on the negative. The pulse width can also be directly measured using the CCD array.

two-photon absorption is a nonlinear phenomenon of being proportional to the square of the light intensity. This is a method of obtaining the data on the pulse width from the relation function of the optical pulse following an observation of the fluorescence generated by the two-photon absorption and then determining the mutual relation between two pulses split. Therefore, it is an intensity correlation measurement that is related not only to the intensity of the pulse but also to the phase between harmonic components making up the ultrashort pulse (every harmonic component is a mode). It's rather troublesome to determine the pulse width since similar graphs of fluorescence will be generated if the incident light is a regular pulse group or when there exists irregular noise.

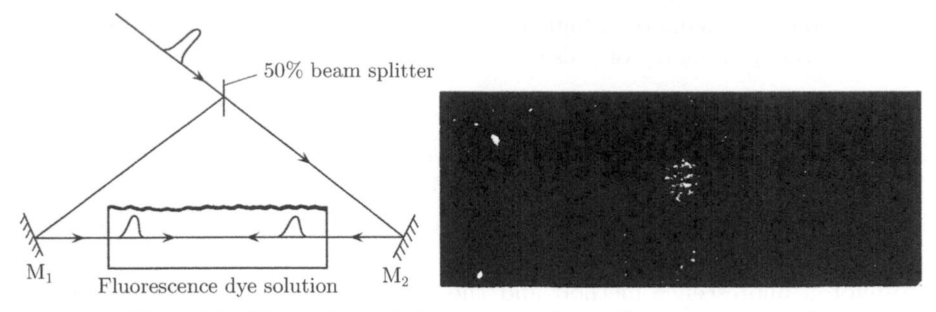

Fig. 3.6-6 The device and photo of two-photon fluorescence method

If it is under the ideal condition, that is, under the ideal mode-locked condition, of the N locked modes, their initial phases are identical, the amplitudes are equal, and the frequency difference between two adjacent modes is $\Delta\nu_q$, then the width of the fluorescent speckle $\Delta\tau = 1/N\Delta\nu_q$. Now the ratio of the fluorescent speckle brightness I_0 to the background brightness I_B is maximum. $I_0/I_B = 3$.

The contrast ratio is as follows: Suppose the electric field intensities of two counter-propagating optical beams are

$$\begin{aligned} E_1 &= E(t)\cos[\omega t + \phi(t)] \\ E_2 &= E(t-\tau)\cos[\omega(t-\tau) + \phi(t-\tau)] \end{aligned} \tag{3.6-4}$$

where τ is the relative retardation time the two pulses reach the center of the dye solution. Thus the dye-emitted instantaneous fluorescence intensity is

$$I_f \propto [E_1 + E_2]^4 \tag{3.6-5}$$

But in actual measurement, the response time of the receiving element is much longer than the pulse width. So what is actually measured is not the instantaneous intensity of the two photons, but the integral intensity $F(\tau)$. If the efficiency factor of the two photons fluorescence is not considered, then the fluorescence integral intensity received is

$$F(\tau) = A\left[\int_{-\infty}^{\infty} E^4(t)\mathrm{d}t + \int_{-\infty}^{\infty} E^4(t-\tau)\mathrm{d}t + \int_{-\infty}^{\infty} 4E^2(t)\cdot E^2(t-\tau)\mathrm{d}t\right] \tag{3.6-6}$$

where the former two terms are a sum of the fluorescence integral intensities of two beams, which is the background light during measurement. The third term is the overlapping part of the two pulses, which is the spatial graph of the two-photon fluorescence. Suppose

$$2F' = \int_{-\infty}^{\infty} E^4(t)\mathrm{d}t + \int_{-\infty}^{\infty} E^4(t-\tau)\mathrm{d}t \tag{3.6-7}$$

So

$$F(\tau) = A[2F' + 4\int_{-\infty}^{\infty} E^2(t)E^2(t-\tau)\mathrm{d}t] \tag{3.6-8}$$

where F' is the fluorescent integral intensity of a beam of light, and A is the coefficient. As the intensities of the two beams of light are equal, the fluorescent integral intensities of the two beams of light are equal. For Eq. (3.6-8), dividing the two sides with $2F'$, we have

$$\frac{F(\tau)}{2F'} = A[1 + 2G^2(\tau)] \tag{3.6-9}$$

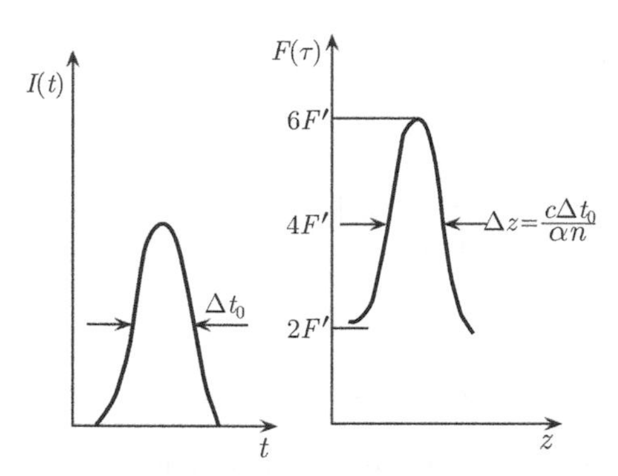

Fig. 3.6-7 The incident light waveform and two-photon fluorescence graph

where

$$G^2(\tau) = \frac{1}{F'}\int_{-\infty}^{\infty} E^2(t)\cdot E^2(t-\tau)\mathrm{d}t \tag{3.6-10}$$

When the two pulses completely overlap ($\tau = 0$), $G^2(\tau) = 1$ while if the two pulses are far apart ($\tau = \infty$), $G^2(\tau) = 0$. The fluorescence intensity now is the background light, so the ratio of the fluorescent light to the background light when the two pulses completely overlap is 3:1. The pulse width of the incident light can be found using the two-photon fluorescence graph $F(\tau)$. Shown in Fig. 3.6-7 are the waveform of the incident light $I(t)$ and the two-photon fluorescence graph $F(\tau)$.

In the two-photon fluorescence graph, the pulse width $\Delta\tau$ is the length of the spatial coordinates, which, converted into time, is $\Delta\tau_{\mathrm{p}} = \dfrac{\Delta z\cdot n}{c}$, where n is the refractive index of the dye solution, c is the velocity of the light in vacuum, while the ratio of the pulse width Δt_0 to the actually measured pulse width $\Delta\tau_{\mathrm{p}}\left(\dfrac{\Delta\tau_p}{\Delta t_0}\right) = \alpha$, α being the waveform coefficient of the pulse width. Thus the actual mode-locked laser pulse width can be found from Δz.

Measuring the pulse width using the two-photon fluorescence method has the following drawbacks: the measurement is trivial, the nonlinearity of the film photosensitivity will cause errors in measurement, and the resolving capability is low. For a pulse of 1 ps, its spatial fluorescence dimension is about 0.1 mm. So it's very difficult to obtain satisfactory resolution with the photographic method. The main problem lies in the presence of the background light and the fact that the waveform of the secondary related function $G^2(\tau)$ measured from the film does not completely correspond to the original pulse waveform.

(2) The second harmonic generation method

When two beams of light of frequency ω pass through a nonlinear crystal, if a definite phase matching condition is satisfied, frequency double light (second harmonic) of frequency 2ω can be generated. The intensity for generating the second harmonic is proportional to the square of intensity of the fundamental frequency light. Therefore, by means of a definite device, such as the Michelson interferometer shown in Fig. 3.6-8, we shall be able to split the fundamental frequency light into two equal beams of light to change the degree of overlapping of two beams in the nonlinear crystal, thus obtaining the second harmonic of a different intensity to measure the width of the optical pulse. This is also one of the methods of the second-order correlation measurement.

Suppose the electric field intensities of two beams of fundamental frequency light after going through the beam splitter are

$$E_1 = E(t)\cos(\omega t+\phi_1) \tag{3.6-11}$$

$$E_2 = E(t-\tau)\cos[\omega(t-\tau)+\phi_2] \tag{3.6-12}$$

The light intensity of the second harmonic generation is proportional to the intensity square of the fundamental frequency light, that is,

$$I_{2\omega}(t) = K[E^4(t)+E^4(t-\tau)+4E^2(t)E^2(t-\tau)] \tag{3.6-13}$$

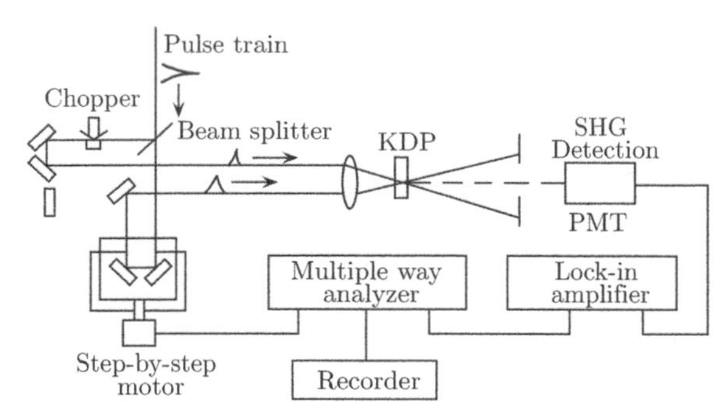

Fig. 3.6-8 The non-collinear correlation experimenting device of second harmonic

where K is the proportionality constant. As the response time of the photoelectric detector is much longer than the pulse width, what is received is not the instantaneous light intensity but the integral intensity.

$$S(\tau) = \int_{-\infty}^{\infty} I_{2\omega}(t)\mathrm{d}t = \int_{-\infty}^{\infty} K[E^4(t) + E^4(t-\tau) + 4E^2(t)E^2(t-\tau)]\mathrm{d}t$$
$$= 2W_{2\omega}K[1 + 2G^2(\tau)] \quad (3.6\text{-}14)$$

where

$$2W_{2\omega} = \int_{-\infty}^{\infty} E^4(t)\mathrm{d}t + \int_{-\infty}^{\infty} E^4(t-\tau)\mathrm{d}t \quad (3.6\text{-}15)$$

$W_{2\omega}$ is the second harmonic pulse light intensity acted on by single beams of light respectively; $G^2(\tau)$ is the second-order relation function, and

$$G^2(\tau) = \frac{1}{W_{2\omega}} \int_{-\infty}^{\infty} E^2(t) \cdot E^2(t-\tau)\mathrm{d}t \quad (3.6\text{-}16)$$

When two beams of light completely overlap ($\tau = 0$), $G^2(\tau) = 1$. When two pulses are completely separated from each other ($\tau = \infty$), $G^2(\tau) = 0$. So the ratio of the second harmonic to the background light when the two pulses completely overlap is 3:1. By making the step motor to shift the totally reflecting mirror, the movable arm of the Michelson interferometer can be changed. As the relative retardation of the two pulses has been changed, so has the intensity of the second harmonic. The second harmonic relation signal detected by the SHG, after undergoing mode-locked amplification, is sent into the function recorder or the multiple way analyzer to find out the magnitude of $S\ (\tau)/W_{2\omega}$. By means of multi-point measurement, the shape of $I(t)$ and the pulse width Δt_0 can be determined. The essence of the second harmonic method consists in transforming the measurement of time into that of length while transforming the measurement of $I(t)$ into that of $S\ (\tau)/W_{2\omega}$. The variation of 0.15 mm of the interferometer length l is equivalent to a variation of 1ps for τ, while to the interferometer, the regulation of an amount of 0.15 mm is easily controlled. In addition, the measurement of $S\ (\tau)/W_{2\omega}$ is measurement of the relative value of the secondary harmonic intensity when τ is a certain value, there being no need for a fast photoelectric detector or any fast electronic instrument. So the method under discussion is an ultrashort pulse measuring method that has a very high resolution and is simple and feasible.

In practice, the crucial device is the nonlinear crystal, such parameters as its type, phase matching angle, thickness, etc. being of paramount importance. In terms of the phase matching condition for the frequency doubling, the second harmonic method can be divided into the "background" measurement and "background-free" measurement. As stated previously, when $\tau \gg \Delta t_0$, the value of $S(\tau)/W_{2\omega}$ is equal to 1. This is the "background". When

$\tau = 0$, $S(\tau)/W_{2\omega} = 3$. If the measurement is accurate, the maximum ratio between the two is 1:3. So the presence of the background can be regarded as a basis of judging whether the measurement is accurate or not, and the pulse width can be determined by the curve. But the existence of the background will bring about errors in measurement. In order to realize "background-free" measurement, it is advisable to make the wave vector of E_1 and E_2 slightly deviate from the matching direction of the second harmonic. But the wave vectors synthesized should exactly satisfy the phase matching condition without generating the second harmonic effect. So $S(\tau)/W_{2\omega} = 0$, with the background term eliminated.

Another method of realizing "background-free" measurement consists in making use of the type II phase matching condition of frequency doubling.

$$\frac{1}{2}[n_o^{\omega} + n_e^{\omega}(\theta_m)] = n_e^{2\omega}(\theta_m) \tag{3.6-17}$$

Let's make both E_1 and E_2 injected into the frequency doubling crystal along one and the same direction (satisfying the second class of phase matching direction). If E_1 is o light, and E_2 is e light, it is obvious that, when $\tau > \Delta t_0$, there is only E_1 or E_2 that exists alone. In neither case will the second harmonic effect be generated. Hence $S(\tau)/W_{2\omega} = 0$. But when τ is within the range of $2\Delta t_0$, E_1 and E_2 coexist and overlap. As the type II phase matching condition is satisfied, the second harmonic effect is generated. Therefore, the $S(\tau)/W_{2\omega}$ value can be found, thereby determining the value of pulse width $\Delta t_0 = \dfrac{2\Delta l}{\beta c}$, where $2\Delta l = 2(l_1 - l_2)$ is the optical path difference, β is the waveform coefficient of the pulse, in general $\sqrt{2} \sim 2$ is taken, e.g., for the Lorentz-type pulse, $\beta = 2$, for the Gaussian pulse, $\beta = \sqrt{2}$.

Measuring the pulse width with the second harmonic method is fairly convenient for the continuous wave mode-locked measurement since what is measured is the regular pulse sequence. As the $I(t)$'s of the pulses are identical, there's no need for normalization of the measurement results. The difference between $S(\tau)$ and $W_{2\omega}$ is but a single constant, which does not affect the true value of the pulse width. Figure 3.6-9 shows the autocorrelation function curve of the femtosecond optical pulse using the second harmonic method. For the pulsed mode-locking, however, owing to the fact that the pulses in a column of pulses are different, i.e., $W_{2\omega}$ is no longer a constant, it is necessary to perform normalization of the measurement.

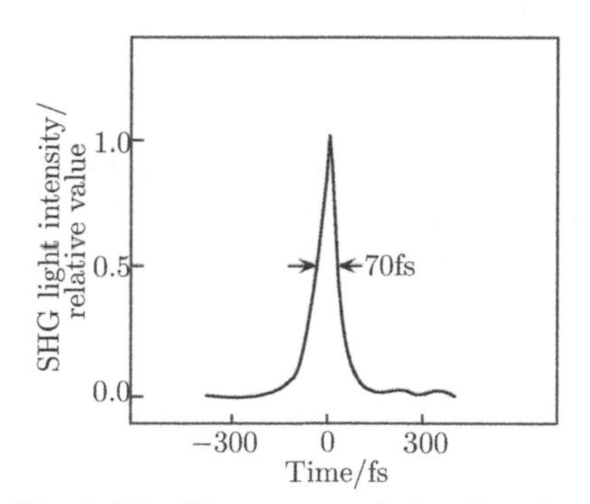

Fig. 3.6-9 The autocorrelation function curve of mode-locked pulse

The adoption of the intensity related second harmonic method will yield high precision in measurement and high resolution and data processing is convenient. But the structure is complicated, and the measurement must be carried out point by point. When the pulse width to be measured is rather narrow (of the fs order), it is advisable not to adopt the conventional function recorder whose sensitivity is inadequate but to adopt the equivalent width measuring method that can overcome the above-mentioned shortcomings.

3.7 Some typical mode-locked lasers

Since the implementation of the mode-locking technique in lasers in 1964, with the development of laser technology, the mode-locking technique, too, has made rapid progress.

Mode-locking has been realized one after another in solid-state lasers with a broad oscillation spectrum (such as ruby lasers, Nd:YAG lasers, neodymium glass lasers, titanium-doped sapphire lasers, fiber-optic lasers, semiconductor lasers, and dye lasers), with pulses of a duration of $10^{-12}\sim10^{-15}$ s. A number of typical mode-locked lasers will be discussed in this section.

3.7.1 The titanium-doped sapphire self-mode-locked laser

Since 1991, the most popular research topic concerning the titanium-doped sapphire laser is the ultrashort optical pulse technique. Because of the very broad fluorescence spectrum of the titanium-doped sapphire laser, if all the longitudinal modes of the laser are locked, a laser output of several fs can directly be generated in theory without having to adopt other pulse compressing techniques, which is beyond compare with any other mode-locked lasers. So far, quite a few mode-locking techniques have been adopted, such as active mode-locking, passive mode-locking, synchronously pumped mode-locking, additional pulsed mode-locking, seed-injection mode-locking, colliding mode-locking, as well as self-mode-locking, etc.

In the course of development of the titanium-doped sapphire laser technique, the availability of the titanium sapphire self-mode-locking technique marked another significant breakthrough in the titanium-doped sapphire laser technique. Research on self-mode-locking is not only of important academic significance and application value, but also, it is only by means of self-mode-locking that it will be possible to obtain the shortest mode-locked pulses. All the other means of mode-locking have in general to introduce various mode-locking elements into the resonator, which will no doubt limit the spectral width of the laser, thereby limiting the width of its output pulse. Moreover, once the self-mode-locked pulse sequence is maintained, its noise will be far lower than that with other mode-locked lasers while better stability is obtained. Below we will a look at the characteristics of the titanium-doped sapphire crystal and the structure of the self-mode-locked laser.

1. The characteristics of the titanium-doped sapphire laser

The Ti:S crystal is a titanium-doped Al_2O_3 single crystal of the hexagonal crystal system, whose spatial group is $R3\overline{C}-D_{3d}^6$ and materialized properties are similar to those of ruby. It has good stability, a thermal conductivity about three times that of Nd:YAG, high melting point (2050°C), and a hardness (grade 9) and refractive index of 1.76. In the structure of the crystal, the Ti^{3+} ions perform permutation at Al_2O_3 of the Al^{3+} ions in place C with triangular symmetry to place them in the center of a positive octahedron. The Ti^{3+} ions are acted on by the cubic field formed by six O^{2-} ions around them. The Ti^{3+} ions' electronic configuration is $1S^22S^22P^63S^23P^63d^1$ with only one unmatched $3d$ electron. Except for the $3d$ electron, we have a full shell layer (atom solid). Thus, the $3d$ orbit's unique valence electronic behavior determines the characteristics of the ions' absorption and emission spectra.

For the unmatched $3d$ electron in the outermost layer of the Ti^{3+} ions, its angular quantum number $l=2$, while the spin quantum number $S=\pm\dfrac{1}{2}$. $2l+1=5$ shows that the electron d has five orientations and a quintuplicate degeneracy of energy level. Under the action of the cubic field, ^2E is split into two energy levels, $E_{1/2}$and $E_{3/2}$. Owing to the action of the triangular field and the orbit-spin interaction, the triple degenerate ground state energy level is split into three energy levels, ${}_2E_{1/2}$, ${}_1E_{1/2}$, and $E_{3/2}$, as shown in Fig. 3.7-1.

The coupling of the Ti^{3+} ions, electron energy level with the vibration energy level of the surrounding sapphire lattice is the key to the generation of laser by the Ti:S crystal. The electron-phonon coupling action is so powerful it makes the ground state and excited state energy level distribution range very wide, as shown in Fig. 3.7-2. Therefore, both the Ti^{3+}'s absorption transition spectral band and the fluorescence transition spectral band are very wide. The rather widely distributed ground state energy level is the key to the tunable operation of the Ti:S laser. The particles at the vibration energy levels stimulated to the excited state reach the lowest energy level of the excited state via the fast relaxation process (the sub-ps order of magnitude). The lifetime of the lowest energy level is 3.3 μs (at room temperature). The storing function of the lowest active state energy level has brought about the population inverted distribution between it and the ground state vibration energy levels to form laser transition, the laser emission cross-section being (3~4) $\times 10^{-19}$ cm^2.

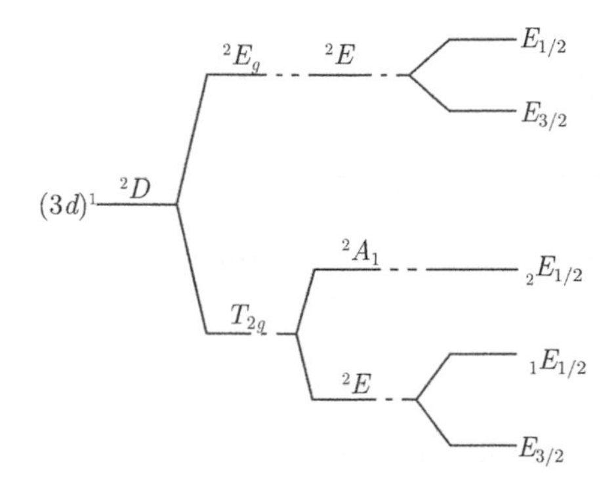

Fig. 3.7-1 The level graph of the Ti^{3+}ions in the titanium-doped sapphire crystal

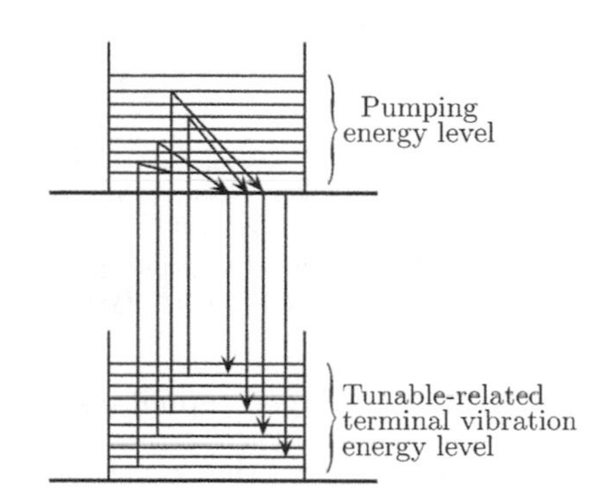

Fig. 3.7-2 The energy level transition graph of titanium-doped sapphire crystal

The Ti:S crystal's blue and green absorption bands possess different absorption cross-sections for different polarizations. π represents that the light's electrical vector is parallel to the optical axis (c-axis) of the crystal; σ represents that the light's electrical vector is perpendicular to the c-axis. The crystal will absorb much more of the π polarized light. For this reason, in order to enable the Ti:S crystal to absorb maximum pumping light, the pumping light's wave vector $\boldsymbol{k}$ should be made to be perpendicular to the c-axis, and the electrical vector E parallel to the c-axis (π polarized). Figure 3.7-3 shows the absorption spectrum of the Ti:S crystal, the absorption spectral range being 430~580 nm and the peak value 490 nm. So, it is appropriate to use laser of the blue-green wave band for pumping, such as Ar^+ laser, copper steam laser, frequency doubled YAG laser, and frequency doubled YLF laser. The important parameter for characterizing the Ti:S crystal is the FOM or quality factor. FOM $= \alpha_m/\alpha_r$, in which α_m is the absorption coefficient with respect to the blue-green light, and α_r the residual absorption coefficient, the absorption at 800 nm being in general used for the expression. So FOM $\approx\alpha_{490}/\alpha_{800}$. The FOM value of internationally high-level Ti:S can reach as high as 250 and 300, α_{490} can reach up to 0.7~2.5 cm^{-1}. Shown in Fig. 3.7-4 is the fluorescence spectrum of the titanium-doped sapphire crystal at room temperature, whose peak wavelength is about 745 nm, the fluorescence spectrum has very strong polarizing characteristics and the intensity of the π polarized light is greater than that of the σ polarized light. According to the relationship between fluorescence intensity and the gain coefficient, we can obtain the corresponding gain curve. The gain peak wavelength is in the vicinity of 795 nm, the spectral wavelength range of the gain is 650~1200 nm, and the bandwidth about 122 nm, the broadest ever obtained in all laser gain media so far. There is definite red shift for the rise of the fluorescence spectrum with the temperature and the semipolar large line width, too, will increase. The lifetime of the fluorescence increasing with temperature will rapidly shorten owing to the fluorescence quench induced by multiple photon non-radiation decay. For this reason, a water-cooling device should be adopted during the operation of the Ti:S crystal to prevent quenching and stabilize the output.

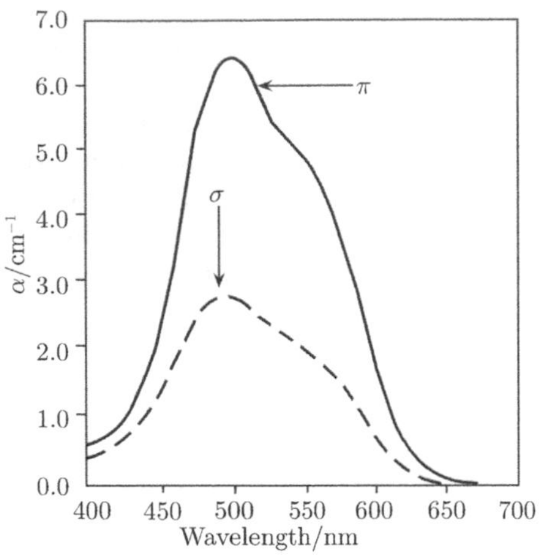

Fig. 3.7-3 The absorption spectrum of titanium sapphire crystal

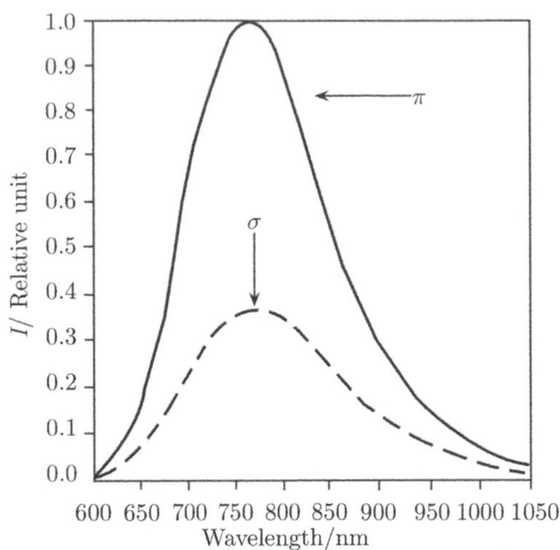

Fig. 3.7-4 The fluorescence spectrum of titanium sapphire crystal

2. The titanium-doped sapphire self-mode-locked laser

Figure 3.7-5 shows the structure of the self-mode-locked titanium-doped sapphire laser. The Ti:S crystal is 20 mm long with the end face cut into a Brewster angle and is placed in the center of a four-mirror folded resonator of length 1.5~2.0 m. The transmissivity of the planar output mirror M_0 is 3.5% within the spectral range 850~1000 nm. The curvature radius each of reflective mirrors $M_1$and M_2 is 10 cm; the all spectral line TEM_{00} mode argon ion (Ar^+) laser is in general used as the pumping source and the single wafer birefringence filter or variable narrow-seam diaphragm located between prisms P_1 and P_2 is employed for wavelength tuning. The self-mode-locked state is generally introduced by an external slight perturbation. Under the mode-locked condition, the wavelength can be within the range 845~950 nm. The narrowest pulse currently output by such lasers is 10.9 fs. By adopting the dispersion-controlled chirped medium mirror while removing the dispersion-compensating prism in the cavity, 8 fs ultrashort pulses can be obtained.

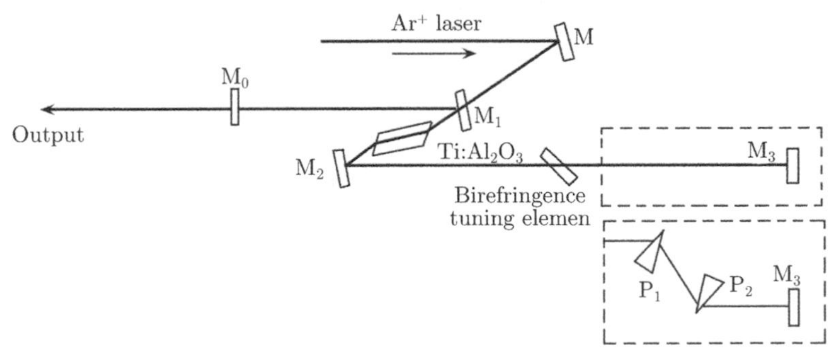

Fig. 3.7-5 The structure of the self-mode-locked titanium sapphire laser

3.7.2 The semiconductor mode-locked laser

The semiconductor laser ultrashort pulse is extensively applied in the field of information photoelectronics. For instance, it plays an important role in measuring the instantaneous characteristics of photoelectronic devices and of fiber-optic transmission. In particular, in the recently developed electro-optic sampling system, ultrahigh speed digital communication or time-division multiplex system, and optical soliton communication system, the semiconductor laser ultrashort pulse possesses incomparable advantages. There are two ways to generate semiconductor laser ultrashort pulses, that is, the gain switching method and the mode-locking method.

1. Gain switching

The gain switching method is also known as the method of directly modulating the dynamic single mode semiconductor laser (DFB, DBR), which is fairly simple and capable of generating single mode pulses of different repetition frequencies and is suitable for use in the fiber-optic communication system.

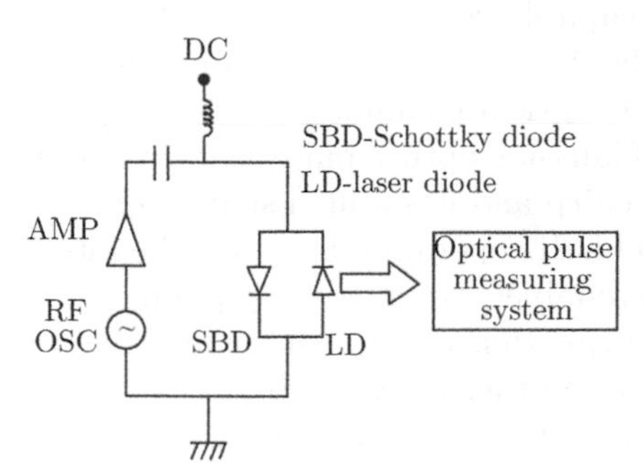

Fig. 3.7-6 Schematic of gain switch structure

A schematic diagram of the gain switch is shown in Fig. 3.7-6. When a semiconductor laser works in the single longitudinal mode, have the comb-shaped wave current injected into the laser via a 50-Ω microband matching line; that is, directly modulate the LD using the RF sinusoidal current. The ultrashort pulse is generated by using the gain switching effect. In order to protect the laser against breakdown by the voltage in the opposite direction, a Schottky diode is connected in parallel at the two ends of the laser. Pulses of 20~30 ps can be generated with this method.

The main shortcoming in using the gain switch to generate ultrashort pulses is the width of the pulse is rather large and the process is often accompanied by frequency chirp in different degrees. In order to compensate for the chirp of the pulse to attain the goal of compressing the pulse, usually the fiber-optic-grating pair compression or the G-T interferometer compression is adopted to obtain pulses of the ps order of magnitude.

2. The active mode-locking

For the active mode-locked semiconductor laser, the laser oscillation of itself may often have several longitudinal modes. So, very often the reflective mirror or mode-selecting device (e.g., grating, etalon, etc.) will be required to serve as the outer cavity. When the single sub-cavity mode (in reality a set of outer cavity longitudinal mode groups) is chosen, if the gain bandwidth can be effectively utilized, then optical pulses of the sub-ps order of magnitude can be generated.

The schematic diagram of the active mode-locking for a 1.3 μm F-P interferometer outer cavity semiconductor laser is as shown in Fig. 3.7-7. An F-P interferometer is adopted to

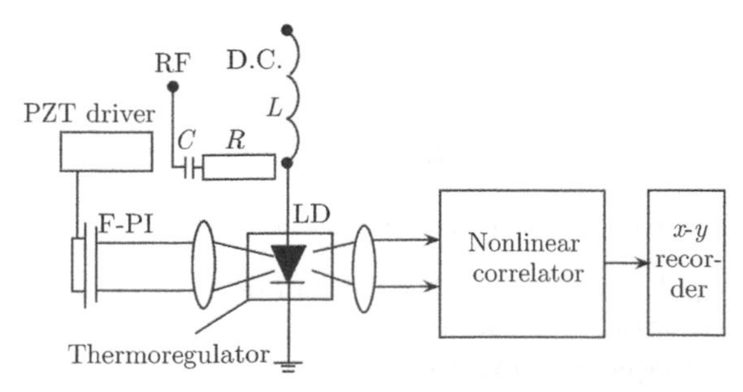

Fig. 3.7-7 Schematic diagram of the structure of an active mode-locked semiconductor lasers

serve as the outer cavity. The reflectivity of either mirror in the 1.3 μm waveband is 75% and the rear reflective mirror is fixed on the piezoelectric ceramic so that it will be able to control the source via the driving source to trim the F-P interferometer interval, with the laser output made from the other end via the microobjective in the form of coupled output. The end face of the semiconductor laser is not AR coated and its ambient temperature is controlled by the thermoregulator. The optical path of the outer cavity is regulated by the collimating lens. The radio frequency modulating signal is generated by the standard generator and injected into the laser through a 50-Ω microband matching line.

3.7.3 The erbium-doped fiber-optic mode-locked laser

The erbium-doped fiber-optic mode-locked laser is capable of generating ps and fs optical pulses. With its operating wavelength located at the 1.5 μm wavelength communication window, it is easily compatible and coupled with the optical communication system, and free from the insertion loss. The laser pulse it gives is Gaussian optical pulse that satisfies the Fourier transform relation. It is free from additional chirp and has a high signal-noise ratio plus a high laser conversion efficiency and low threshold. Its operation is stable, the integral machine exquisitely compact and flexible, stable, and reliable. And it is easily pumped using a semiconductor laser or made in the all-optic fibers form. Hence it has very good prospects for application in researches on ultrafast phenomena and other fields. In particular, if used as a soliton source in optical soliton communication, it will be highly competitive. It is known that mode-locked operation has been realized in the erbium-doped fiber-optic laser using active mode-locking, passive mode-locking, and synchronous pumping technology.

1. The active mode-locking

The basic structure of the erbium-doped fiber-optic active mode-locked ring-shaped laser is as shown in Fig. 3.7-8, the components being the erbium-doped optical fibers, the electro-optic or acousto-optic modulator, light isolator, the polarizers, the pumping wave division multiplexer, and the output coupler. Of the components, the pumping light is coupled into the erbium-doped optic fibers (EDF) by the wave division multiplexer (WDM), with laser output via the fiber optic coupler. The polarization characteristics of the light wave in the cavity are controlled by the polarizer (PC) and the structure of the unidirectional ring-shaped mode-locked laser is implemented by the light isolator ISO.

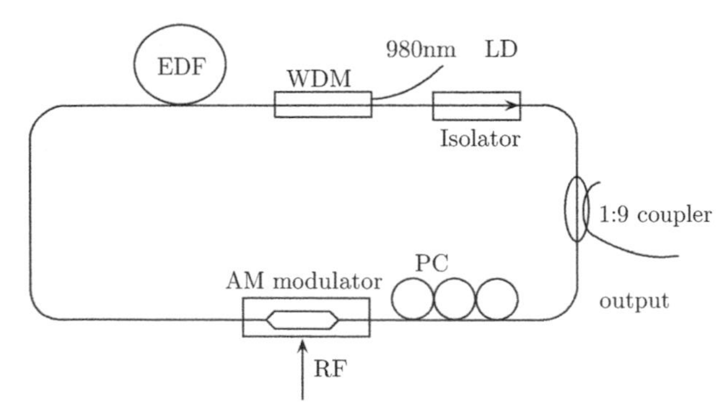

Fig. 3.7-8 Schematic diagram of erbium-doped fiber-optic active mode-locked laser

The crux of the principal technology of the erbium-doped fiber-optic active mode-locked laser is the erbium-doped optical fiber and the modulator. The frequency of the modulator and the laser oscillation frequency have to satisfy the mode-locking condition. To ensure that GHz repetition frequency will be generated in the long optical fiber ring for mode locking, the intracavity insertion loss of the optical elements should be small. Owing to the limited modulating capability of the active mode-locking, the width of the mode-locked pulse is usually of the ps order of magnitude.

2. The passive mode-locking–self-mode-locking

For the passive mode-locked erbium-doped fiber-optic laser, the operation of mode-locking is accomplished using the nonlinear optical effects of optical fibers or other elements. The structure of the laser is simple, with no need of inserting any elements. Under definite conditions, the laser can realize self-start up mode-locking. The erbum-doped fiber-optic self-mode-locked laser is composed of the pumping light wave division-multiplexer, the output coupler, the erbium-doped optical fibers, the polarizer, the light isolator, and the negative dispersion standard optical fibers, of which the erbium-doped optical fiber is the laser gain medium, with the second-order positive group velocity dispersion effect in the erbium-doped optical fiber being compensated for using the negative dispersion standard optical fiber. The self-mode-locking is started by properly adjusting the polarization controller. The mode-locked operation is implemented because of the optical fiber's nonlinear Kerr effect. Usually the erbium-doped fiber-optic self-mode-locked laser is capable of producing ultrashort optical pulses of the fs order of magnitude.

Exercises and problems for consideration

1. Consider a multi-longitudinal mode laser, in the frequency domain of which a distributed function can be used to replace the condition for the amplitudes of the modes to be equal. The output optical pulse obtained with the integral approximate summation method is Gaussian distributed in the time domain. Prove that in the frequency domain its frequency spectrum is also Gaussian distributed and find the relationship between the line width $\Delta\nu$ (total width of frequency spectrum semi-power point) and pulse width (total width of light intensity semi-power point).

2. We have a neodymium-doped yttrium-aluminum-garnet (YAG) laser of line width $\Delta\nu_{\mathrm{g}} = 120$ GHz, with a lithium niobate modulator inserted in the cavity. The dimensions of the modulator are 5 mm $\times$ 5 mm $\times$ 20 mm; the cavity length is 50 cm; the total cavity loss 0.05. When a sinusoidal modulating voltage of effective value 200 V is applied on the crystal, the x-axis is transmitting light, the voltage being applied along the z direction. When the electrical vector of light vibrates along the y-axis and z-axis, what is the mode-locked pulse width of each?

3. There is a multi-longitudinal mode laser with 1000 longitudinal modes; the cavity length is 1.5 m and the average power output is 1 W. It is considered that the amplitudes of the longitudinal modes are equal.

(1) Try to find the period, width, and peak power of the optical pulse under the mode-locked condition.

(2) When the acousto-optic loss modulating element is adopted for mode-locking, a voltage $V(t) = V_{\mathrm{m}}\cos(\omega_{\mathrm{m}}t)$ is applied on the modulator. How much is the frequency of the voltage?

4. There is a neodymium-doped YAG laser, of an oscillation line width (the range in the spectral line of fluorescence within which laser oscillation can be generated) $\Delta\nu_{\mathrm{osc}} = 12 \times 10^{10}$ Hz, the cavity length $L = 0.5$ m. Try to calculate the laser's parameters:

(1) the longitudinal frequency interval;

(2) the number of longitudinal modes that can be held in $\Delta\nu_{\mathrm{g}}$;

(3) suppose the amplitudes of the longitudinal modes are equal. Find the width and period of the pulse after mode-locking;

(4) the spatial distance occupied by the mode-locked pulse and the pulse interval.

5. For a neodymium-doped YAG laser, the KDP crystal loss modulation is adopted for mode locking. The voltage is applied along the x-axis with the z-axis transmitting light, the modulating voltage's effective value is 200 V, provided the cavity length $L = 60$ cm.

(1) Try to plot the method of placing the elements in the cavity, while marking the coordinates of the KDP crystal's main shaft.

(2) Try to find the period of the modulating voltage and the single-path phase retardation of the crystal.

6. Place a loss modulator at $L/2$ in the center of the resonator. To obtain mode-locked pulses, how long should the modulator's loss period T be? How is the energy of each pulse different from that when the modulator is placed immediately close to where the end mirror is?

7. The optical length of a neodymium-doped YAG laser is 40 cm, the single-path total loss coefficient $\delta = 0.1$. Perform mode-locking by loss modulation, the modulating depth $m = 0.2$. It is known that the gain line width is 190 GHz. Find the corresponding mode-locked pulse width.

8. For a mode-locked laser of cavity length $L = 1$ m, suppose all the longitudinal modes that exceed the threshold value are locked.

(1) If the operation material is ruby, rod length $l = 10$ cm, $n = 1.76$, $\Delta\lambda_{\text{g}} = 0.5$ nm, $\lambda = 694.3$ nm.

(2) If the operation material is neodymium glass, rod length $l = 10$ cm, $n = 1.83$, $\Delta\lambda_{\text{g}} = 28$ nm, $\lambda = 1.06$ μm.

(3) If it is an He-Ne laser, $l = 100$ cm, $\Delta\lambda_{\text{g}} = 2 \times 10^{-3}$ nm, $\lambda = 632.8$ nm. Find the output pulse width, the pulse interval, and number of modes. What conclusions can be drawn from the results of calculation?

(4) If KDP longitudinal application is adopted, what is the frequency of the externally applied modulating signal in both the amplitude and phase modulated modes? Draw a graph of principle of the structure.

9. For an Nd:YAG mode-locked laser of cavity length $L = 1.5$ m, is the structure shown in Graphic exercise 3-1 in the optimum and ideal operating state? For an ideal mode-locked laser, what are the characteristics of the laser output? The modulating signal: $V = V_{\text{m}} \cos(\omega_{\text{m}}t)$, $\omega_{\text{m}} = 2\pi \times 5 \times 10^7$, LN transversely applied, and the field applied along the x direction and transmitting light along the z direction, $\text{P}//x(y)$.

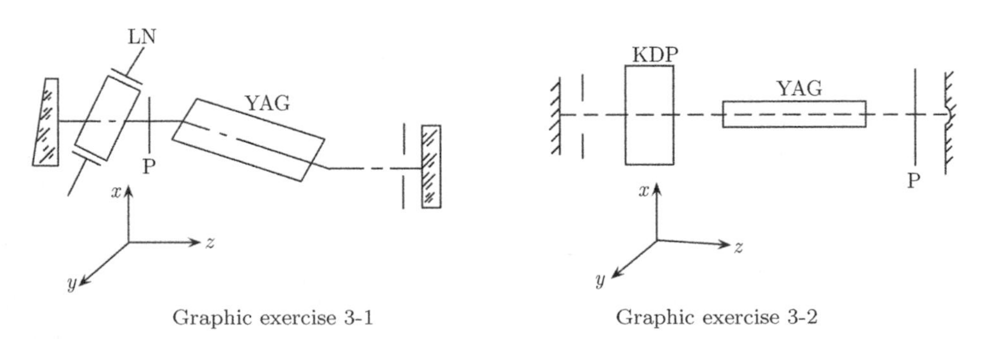

Graphic exercise 3-1 Graphic exercise 3-2

10. An Nd:YAG mode-locked laser has the KDP longitudinally applied. The vibrating direction of the polarizer is parallel to the inductive main shaft $x'(y')$ of the KDP crystal; the modulating signal applied on the crystal $V = V_m \sin(\omega_m t)$; the modulating frequency is 55×10^6Hz; and the cavity length $L = 1.5$ m, whose structure is s shown in Graphic exercise 3-2. Try to judge whether the above-mentioned structure is rational or not. Why?

11. Answer the following questions:

(1) Make a comparison between the dye laser and titanium-doped sapphire mode-locked laser by showing their similarities and dissimilarities.

(2) Why is the "FOM" value in the titanium sapphire crystal important?

(3) What is the mechanism of the self-mode-locked titanium sapphire laser?

References

[1] S. L. Shapioro. Ultrashort Light Pulses, PicosecondTechniques and Applications. New York: Springer-Verlag, 1977.

[2] P. W. Smith, M. A. Duguay, and E. P. Ippen. Mode-lockingof Lasers. Oxford: Pergamon Press LTD, 1974.

[3] D. J. Kuizenga and A. E. Siegman. FM and AM Mode-Lockingof the Homogeneous Laser-Part 1: Theory. *IEEE J. Q. E.*,1970, QE-6(11), p. 694.

[4] M. F. Becker, D. J. Kuizenga, and A. E. Siegman.Harmonic Mode Locking of the Nd:YAG Laser. *IEEE J. Q. E.*,1972, QE8(8), p. 687.

[5] J. Herrmann and U. Motschmann. Theory of the Synchronously Pumped Picosecond Dye Laser. *Appl. Phys.*, 1982, B27, p. 27.

[6] J. Hermann and B. William, Principles and Application of Ultrashort Optic Pulse Lasers, translated by Chen Xiue, Beijing, Science Press, 1991.

[7] S. L. Shapiro, The Picosecond Technique and Application of Ultrashort Light Pulses, translated by Zhu Shiqing, Beijing, Science Press, 1987.

[8] J. Mark, L. Y. Liu, K. L Hall. H. A. Haus, and E. P. Ippen. Femtosecond Pulse Generation in a Laser with a Nonlinear External Resonator. *Opt. Lett.*, 1989, 14(1), p. 48.

[9] J. Goodberlet, J. Wang, and J. G. Fujimoto. Femtosecond Passively Mode-Locked Ti:Al_2O_3 Laser with a Nonlinear External Cavity. *Opt. Lett.*, 1989, 14(20), p. 1125.

[10] P. M. French, J. A. R. Williams, and J. R. Tayior. Femtosecond Pulse Generation from a Titanium-Doped Sapphire Laser Using Nonlinear External Cavity Feedback. *Opt. Lett.*, 1989, 14(13), p. 686.

[11] D. E. Spence, P. N. Keen, and W. Sibbett. 60fs Pulse Generation from a Self-Mode-Locked Ti:Sapphire Laser. *Opt. Lett.*, 1991, 16(1), p. 42.

[12] K. F. Wall and A. Sanchez. Titanium Sapphire Lasers. *The Lincoln Laboratory Journal*, 1990, 3(3), p. 4470.

[13] N. Sarukura, Y. Ishida, H. Nakano, and Y. Yamamoto. CW Passive Mode Locking of a Ti: Sapphire Laser. *Appl. Phys. Lett.*, 1990, 56(9), p. 814.

[14] K. Naganuma and K. Mogi. 50 fs Pulse Generation Directly from a Colliding pulse Mode-Locked Ti:Sapphire Laser Using an Antiresonant Ring Mirror. *Opt. Lett.*, 1991, 16(10), p. 738.

[15] U. Keller, G. W. Hooft, W. H. Knox, and J. E. Cunning. Femtosecond Pulses from a Continuously Self-starting Passively Mode-Locked Ti:Sapphire Laser. *Opt. Lett.*, 1991, 16(13), p. 1022.

[16] J. Squier, F. Salin, and G. Mouyou. 100fs Pulse Generation and Amplification in Ti:Al_2O_3. *Opt. Lett.*, 1991, 16(5), p. 324.

CHAPTER 4

The Laser Amplifying Technology

4.1 An overview

In certain laser applications, the laser beam is often required to possess very high energy (or power). For instance, laser nuclear fusion needs at least an energy as high as ten thousand joules, lidar needs high power modulating laser, etc. But, if high energy laser is desired, it is in general very difficult if just a laser (oscillating) device is used. This is because the requirements on raising the output power (energy) of a laser and the indices (such as the light beam divergence angle, monochromaticity, pulse width, modulating performance, etc.) are contradictory. So, if the excellent characteristics of a laser beam are to be maintained, neither the caliber nor the length of its operation material should be too great. Furthermore, as the laser beam in the laser will pass through the operation back and forth many times, when the output power (energy) is very high, the operation material can be destroyed.

Using the Q-switching technology or mode-locking technology, extremely high peak power ($10^9 \sim 10^{12}$ W) can be obtained. The reason why the peak power is so astoundingly great is because energy is released in an extremely brief instant. But the energy actually output by such a high peak power laser is not necessarily very great. Therefore, in order to obtain high energy laser of excellent performance, the best choice is to apply the laser amplifying technology.

The laser amplifier and laser (oscillating) device are based on the same physical process (optical amplification of stimulated radiation), the main difference being the absence of a resonator for the laser amplifier (traveling wave). Under the action of optical pumping, the operation material is in the population inverted state. When the optical pulse signal generated from the laser (oscillating) device passes through it, as the frequency of the incident light overlaps the gain spectral line of the amplifying medium, violent stimulated radiation is generated by the particles in the excited state under the action of the incoming optical signal. Superposed on the incoming optical signal, such a radiation will get amplified. Therefore, the amplifier is now capable of outputting an outgoing light beam of a much greater brightness than that of the original laser. The laser amplifier requires that the operation material possess sufficient inverted population to ensure that the gain obtained when the optical pulse signal passes through it is greater than all the losses in the medium put together. In addition, in order to obtain resonant amplification, the amplifying medium is required to have an energy level structure to match the input signal.

Adoption of the traveling wave amplifying technology has the following advantages: first, as the laser beam passes through the amplifying medium in a single pass, the destruction threshold of the medium can be greatly increased. That is, under the condition of identical output power densities, the amplifier's operating medium is not easily destroyed; second, when high energy laser is needed, multi-stage traveling wave amplification can be adopted depending on necessity. The amplifier expands the aperture of the laser beam stage by stage while the length of the operation material at each stage can be shortened. This is helpful to prevention of destruction due to super-radiation and self-focusing; third, for the oscillator-amplifier system, the pulse width, spectral line width, light beam divergence angle, etc. can be determined by the oscillator while the amplifier will determine the energy and power of the pulse. So, by a combination of the two, not only can we obtain fairly good laser characteristics, we can also greatly increase the brightness of the output laser.

Figure 4.1-1 is a schematic diagram of the laser and amplifier operating in series. When the laser output from the first stage enters the amplifier, the active medium of the amplifier should exactly be stimulated and in the state of maximum population inversion, or get amplified upon generating resonant transition. In order to realize synchronous operation of two stages, a synchronous circuit is installed between the two stages' trigger circuits to perform control. Lasers with different delay times possess different values. Generally, the optimum delay time can be determined by experiment. For instance, the xenon lamp igniting time for the ruby amplifier is about several hundred ms ahead of the oscillation stage while the xenon lamp of the neodymium glass amplifier is almost ignited at the same time as that of the oscillation stage.

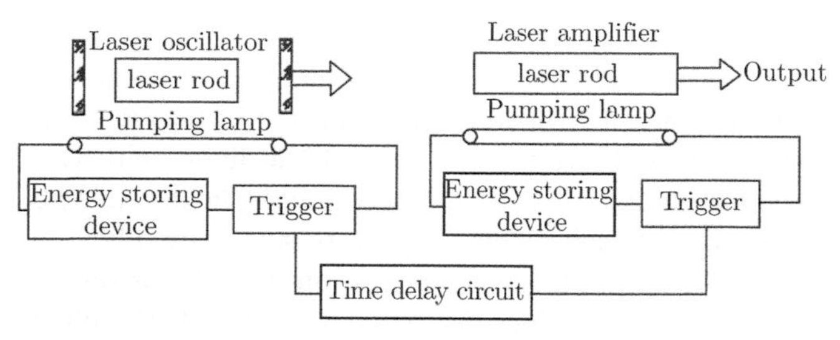

Fig. 4.1-1 Schematic diagram of laser amplifier operation

In terms of the difference in the amplifier pulse signal width, laser amplifiers can be divided into long pulse laser amplifiers (also called continuous laser amplifiers), pulse laser amplifiers, and ultrashort pulse laser amplifiers. For the laser amplifier, the particles (atoms, molecules, or ions) in the amplifying medium are in the excited state, owing to the fact that radiation transition has definite relaxation time, called the longitudinal relaxation time T_1, whose value is different for different amplifying media. For example, for solids like crystals and glass, T_1 is determined by the particle lifetime in the sub-steady state, which is 10^{-3} s; for gases and semiconductors, T_1 is determined by the lifetime of the allowable transitional energy level, which is $10^{-6}\sim10^{-9}$ s. In addition, the non-radiation transition induced by the process of exchange of energy between particles in the amplifying medium will make the induced dipole moment of particles in the excited state have definite relaxation time T_2, called the transverse relaxation time. For uniformly broadened operation material, T_2 possesses the order of magnitude of the reciprocal of the spectral line width. In a solid operation material, T_2 is of about the order of magnitude of 10^{-10} s. When the pulse width of the laser amplifier input signal is greater than the longitudinal relaxation time (or $\tau \gg T_1$; if the width of the pulse output by a common freely operating pulse laser can reach several ms, this condition can be satisfied), as the time of interaction between the optical signal pulse and the operation material is sufficiently long, and the inverted population consumed by the stimulated radiation can be quickly replenished by pumping excitation, the density of the inverted population can be maintained around a definite stable value. It can be approximately considered that the inverted population does not vary with time, or $\dfrac{\mathrm{d}\Delta n}{\mathrm{d}t} = 0$ while only related to the coordinates of the operation material, or $\Delta n = \Delta n(x)$. This makes it possible to study the process of amplification using the steady-state method. This class of laser amplifiers is called the long pulse amplifier.

When the pulse width of the input optical signal is rather narrow, when satisfying the condition $T_2 < \tau < T_1$, if the short pulse output by the Q-switched laser is only of dozens of ns, which is far shorter than the fluorescence lifetime of laser, as the inverted population consumed because of the stimulated radiation does not have sufficient time to be replenished by optical pumping and the inverted population such as the intracavity photon number density cannot reach the steady state in such an extremely brief time, the inverted population varies with time and space, or $\Delta n = \Delta n(x,t)$. Such laser amplifiers should be studied with the non-steady state method. They are pulsed laser amplifiers and are the focus of attention of this chapter.

Both the above two classes of amplifiers satisfy the condition $\tau > T_2$. The time needed for the induced dipole moment to be generated by the atom under the action of the light field can be neglected, so there is no hysteresis effect; that is, the macroelectric polarization of material can immediately follow the rapid variation of the optical field. So it is not necessary to consider the phase relation of the interaction between the atoms and the optical field. Hence the theory of the rate equation can be adopted to discuss the above-mentioned two classes of laser amplifiers.

In the case of laser amplification for ultrashort pulses (i.e., the pulse width $\tau < T_2$, e.g., the pulse output by the mode-locked laser is only of the order of magnitude of $10^{-11}\sim 10^{-12}$ s), during the interaction between light and material, as the macro-electric polarization of material cannot catch up with the rapid variation of the light field, the phase relation of the interaction between the atoms of material and the light field cannot be neglected. Such a coherence action makes the ultrashort pulse generate some new phenomena when passing through the amplifying medium, such as the generation of stable π pulse or 2π pulse in the medium, the generation of the "self-induced transparency" effect, the phenomenon of optical soliton wave therefrom, etc. So we just have to use the semiclassical theory to analyze the ultrashort pulse amplifier.

4.2 The theory of the pulse amplifier

4.2.1 The rate equation of the pulse amplifier

Suppose the length of the laser amplifier's operation material is L and the optical signal pulse is injected into the operation material along the x direction, as shown in Fig. 4.2-1. As the optical signal is steadily amplified in the course of its traversal while the inverted population is steadily consumed, both the photon number and inverted population in the unit volume are functions of time t and space x, denoted by $\phi(x,t)$ and $\Delta n(x,t)$, respectively. For simplicity, suppose the inverted population in the cross section of the amplifier's operation material is uniformly distributed, and the influence of the width of the spectral line and line type as well as that of optical pumping and spontaneous radiation on the inverted population are neglected. Then the inverted population density rate equations of the 3-level and 4-level systems are respectively

3-level:
$$\frac{\partial \Delta n(x,t)}{\partial t} = -2\sigma \Delta n(x,t) I(x,t) \tag{4.2-1}$$

4-level:
$$\frac{\partial \Delta n(x,t)}{\partial t} = -\sigma \Delta n(x,t) I(x,t) \tag{4.2-2}$$

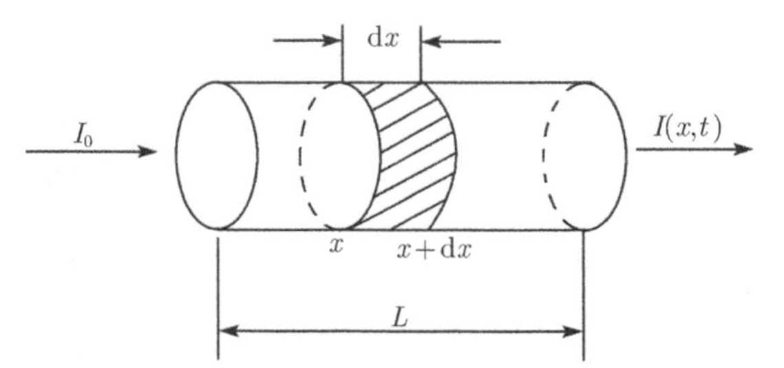

Fig. 4.2-1 Schematic diagram of the process of amplification

Below let us see how the photon number in the operation material varies in the volume element $x \to x+\mathrm{d}x$. There are two factors that cause changes in the photon number. One is because of the stimulated radiation, within time $\mathrm{d}t$, the photon number generated in $\mathrm{d}x$ $\sigma c\phi(x,t)\Delta n(x,t)\mathrm{d}x\mathrm{d}t$; second, within time $\mathrm{d}t$, the photon number entering the volume element at x is $\phi(x,t)c\mathrm{d}t$, while the photon number flowing out from $x+\mathrm{d}x$ is $\phi(x+\mathrm{d}x,t)c\mathrm{d}t$; therefore, the net photon number entering the volume element within time $\mathrm{d}t$ is $[\phi(x,t) - \phi(x+\mathrm{d}x)]c\mathrm{d}t$. Suppose all the other losses in the amplifier can be neglected. Then the rate of variation in the volume element within time $\mathrm{d}t$ should be the algebraic sum of the photon number generated by stimulated radiation and the net photon number entering the volume element, i.e.,

$$\frac{\partial\phi(x,t)}{\partial t}\mathrm{d}x\mathrm{d}t = [\phi(x,t) - \phi(x+\mathrm{d}x,t)]c\mathrm{d}t + \sigma c\phi(x,t)\Delta n(x,t)\mathrm{d}x\mathrm{d}t$$

So the variation rate of the photon number density can be expressed with a partial differential equation as

$$\frac{\partial\phi(x,t)}{\partial t} + c\frac{\partial\phi(x,t)}{\partial x} = \sigma c\phi(x,t)\Delta n(x,t) \tag{4.2-3}$$

The photon number flowing through the unit cross section within unit time is called the photon flow, denoted by $I(x,t)$, or $I(x,t) = c\phi(x,t)$. Hence the equation for depicting the variation rate of the intensity of the photon flow is

$$\frac{1}{c}\frac{\partial I(x,t)}{\partial t} + \frac{\partial I(x,t)}{\partial x} = \sigma\Delta n(x,t)I(x,t) \tag{4.2-4}$$

The variation rate equations of the level 3 and level 4 photon flow intensity are identical. Equations (4.2-1) to (4.2-4) are the basic equations for the pulse amplifier.

Suppose the initial photon flow intensity of the input signal to be amplified is $I_0(t)$, entering the operation material at $x = 0$. Also, suppose prior to the entry of the signal into the amplifier, the initial inverted population in the operation material is $\Delta n_0(x)$. Then the boundary conditions for the rate equation are

$$I(0,t) = I_0(t) \qquad (\text{at } x = 0)$$

$$\Delta n(x, t<0) = \Delta n_0(x) \qquad (\text{at } 0 < x < L)$$

According to the above-mentioned boundary conditions, solving the rate equations (4.2-1) and (4.2-4) simultaneously, we can find the variation of the intensity of photon flow and inverted population of the incident pulse signal entering the amplifier at any position x at any time t as well as the energy of the output pulse and the amplifier's gain.

4.2.2 Solving the rate equation

Equations (4.2-1) and (4.2-4) are a set of nonlinear partial differential equations. Here the variable separation method is adopted to solve the two equations without taking into account the loss of the amplifying medium.

Change Eq. (4.2-4) to

$$\Delta n = \frac{1}{c\sigma}\left(\frac{1}{I}\frac{\partial I}{\partial t} + \frac{c}{I}\frac{\partial I}{\partial x}\right) \tag{4.2-5}$$

Substitution of Eq. (4.2-5) into Eq. (4.2-1) yields

$$\frac{\partial}{\partial t}\left(\frac{1}{I}\frac{\partial I}{\partial t} + \frac{c}{I}\frac{\partial I}{\partial x}\right) = -2\sigma I\left(\frac{1}{I}\frac{\partial I}{\partial t} + \frac{c}{I}\frac{\partial I}{\partial x}\right) \tag{4.2-6}$$

Below we shall perform parametric transform to simplify the factor $\left(\frac{1}{I}\frac{\partial I}{\partial t}+\frac{c}{I}\frac{\partial I}{\partial x}\right)$. Let $\phi = x/c, p = t - x/c$. Then $I(x,t)$ becomes compound function $I[\varphi(x), p(x,t)]$. According to the differentiation of the compound function, we have

$$\frac{1}{I}\frac{\partial I}{\partial t}=\frac{1}{I}\frac{\partial I}{\partial p}$$

$$\frac{c}{I}\frac{\partial I}{\partial x}=\frac{1}{I}\left(\frac{\partial I}{\partial \varphi}-\frac{\partial I}{\partial p}\right)$$

By substituting the above two equations into Eq. (4.2-6) and performing reduction, we have

$$\frac{\partial}{\partial p}=\left(\frac{1}{I}\frac{\partial I}{\partial \varphi}\right)=-2\sigma\frac{\partial I}{\partial \varphi}$$

Exchanging the order of differentiation gives

$$\frac{\partial}{\partial \varphi}=\left(\frac{1}{I}\frac{\partial I}{\partial p}+2\sigma I\right)=0 \tag{4.2-7}$$

Perform integration over Eq. (4.2-7); the integration constant is but a function of p. Hence

$$\frac{1}{I}\frac{\partial I}{\partial p}+2\sigma I=c_1(p) \tag{4.2-8}$$

Further substitution will make this equation directly integrable. Let $\rho=\frac{1}{I}$, and Eq. (4.2-8) is transformed to

$$\frac{\partial \rho}{\partial p}+\rho c_1(p)=2\sigma \tag{4.2-9}$$

The general solution for this linear differential equation is

$$\rho=\exp\left[-\int c_1(p)\mathrm{d}p\right]\left\{\exp\left[-\int c_1(p)\mathrm{d}p\right]2\sigma\mathrm{d}p+c_2(\varphi)\right\}$$

where the integration constant $c_2(\varphi)$ is a function of φ. Again let

$$g'(p)=\frac{\mathrm{d}g(p)}{\mathrm{d}p}=\exp\left[\int c_1(p)\mathrm{d}p\right]$$

be substituted into the above equation and we have

$$\rho=\frac{2\sigma g(p)+c_2(\varphi)}{g'(p)}$$

The photon flow intensity

$$\begin{aligned}I(x,t)=&\frac{1}{\rho}=\frac{\frac{\mathrm{d}}{\mathrm{d}t}\left[g\left(t-\frac{x}{c}\right)\right]}{2\sigma g\left(t-\frac{x}{c}\right)+c_2\left(\frac{x}{c}\right)}\\ =&\frac{1}{2\sigma}\frac{\mathrm{d}}{\mathrm{d}t}\left\{\ln\left[2\sigma g\left(t-\frac{x}{c}\right)+c_2\left(\frac{x}{c}\right)\right]\right\}\end{aligned} \tag{4.2-10}$$

Using the boundary condition $I(0,t)=I_0(t)$, we find

$$I_0(t)=\frac{1}{2\sigma}\frac{\mathrm{d}}{\mathrm{d}t}\{\ln[2\sigma g(t)+c_2(0)]\} \tag{4.2-11}$$

As the initial photon flow intensity $I_0(t)$ is a known quantity, integrating the above equation, we have

$$g(t) = \left[-\frac{c_2(0)}{2\sigma}\right] + c_3 \exp\left[2\sigma \int_{-\infty}^{0} I_0(t')\mathrm{d}t'\right] \tag{4.2-12}$$

where c_3 is an arbitrary constant, t' the virtual variable of integration. Substituting Eq. (4.2-12) into Eq. (4.2-10), we have

$$I(x,t) = \frac{I_0(t)}{1 + \eta(x)\exp\left[-2\sigma \int_{-\infty}^{t-x/c} I_0(t')\mathrm{d}t'\right]} \tag{4.2-13}$$

where

$$\eta(x) = \frac{c_2\left(\frac{x}{c}\right) - c_2(0)}{2\sigma c_3}$$

Now substituting the photon flow intensity equation (4.2-13) into Eq. (4.2-5), we have

$$\Delta n(x,t) = -\frac{1}{\sigma}\frac{\partial\eta(x)/\partial x}{\eta(x)\exp\left[2\sigma \int_{-\infty}^{t-x/c} I_0(t')\mathrm{d}t'\right]} \tag{4.2-14}$$

By using the boundary condition $\Delta n(x, -\infty) = \Delta n_0(x)$ to determine $\eta(x)$ and taking into account $\int_{-\infty}^{t-x/c} I_0(t')\mathrm{d}t' = 0$, Eq. (4.2-14) can be reduced to

$$\Delta n_0(x) = -\frac{1}{\sigma}\frac{\partial\eta(x)/\partial x}{\eta(x)+1} = -\frac{1}{\sigma}\frac{\partial\eta(x)}{\partial x}\{\ln[\eta(x)+1]\}$$

Integration gives

$$\eta(x) = c_4 \exp\left[-\sigma \int_0^x \Delta n_0(x')\mathrm{d}x'\right] - 1 \quad (0 < x < L) \tag{4.2-15}$$

When $x = 0$, then $\eta(x) = 0$. Thus the integration constant $c_4 = 1$. Substituting $\eta(x)$ in Eq. (4.2-15) into Eqs. (4.2-13) and (4.2-14), we find the general solution for the rate equation (4.2-1) and that for (4.2-4) as

$$I(x,t) = \frac{I_0(t-x/c)}{1 - \left\{1 - \exp\left[-\sigma \int_0^x \Delta n_0(x')\mathrm{d}x'\right]\right\}\exp\left[-2\sigma \int_{-\infty}^{t-x/c} I_0(t')\mathrm{d}t'\right]} \tag{4.2-16}$$

$$\Delta n(x,t) = \frac{\Delta n_0(x)\exp\left[-\sigma \int_0^x \Delta n_0(x')\mathrm{d}x'\right]}{\exp\left[2\sigma \int_{-\infty}^{t-x/c} I_0(t')\mathrm{d}t'\right] + \exp\left[-\sigma \int_0^x \Delta n_0(x')\mathrm{d}x'\right] - 1} \tag{4.2-17}$$

4.2.3 An analysis of square pulse amplification

For the amplification of an incident pulse signal of an arbitrary shape and a traveling wave of an arbitrary initial inverted population density, not only should the relationship of variation of the amplifier's gain with the intensity of the incident signal, but also the changes experienced by the intensity and waveform of the incident signal in the process of amplification should be considered. To facilitate discussion, let's first look at an idealized amplification of the square pulse.

Suppose the amplitude of the incident square pulse signal is I_0 and the width is τ as shown in Fig. 4.2-2.

$$\begin{aligned} &\text{When } 0 < t < \tau, \quad I = I_0 \\ &\text{When } t < 0, t > \tau, \quad I = 0 \end{aligned} \tag{4.2-18}$$

In addition, suppose the density of impurities in the entire amplifying medium is homogeneous as is the stimulation by optical pumping. Then the initial inverted population Δn_0 in the amplifying medium can be regarded as a constant. Therefore,

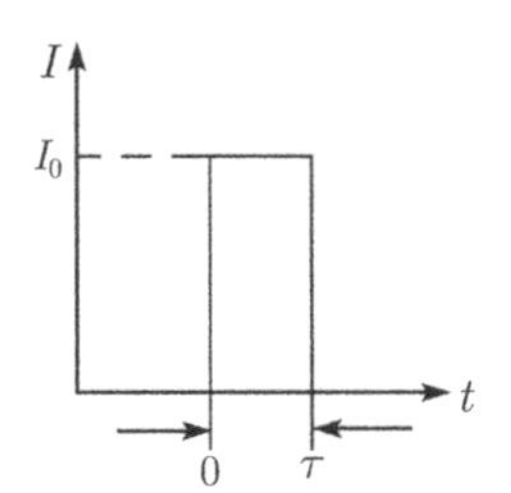

Fig. 4.2-2 The square pulse injected into the amplifier

$$\int_0^x \Delta n_0(x')\mathrm{d}x' = \Delta n_0 x \tag{4.2-19}$$

Substituting Eqs. (4.2-18) and (4.2-19) into Eq. (4.2-16), we can find the intensity of the photon flow in the interval $0 < t - \dfrac{x}{c} < \tau$:

$$I(x,t) = \frac{I_0}{1 - [1 - \exp(-\sigma\Delta n_0 x)]\exp[-2\sigma I_0(t - x/c)]} \tag{4.2-20}$$

Then the amplifier's power gain by one path can be obtained by calculating the intensity of $I(x,t)$ at $x = L$ and taking the ratio $\dfrac{I}{I_0}$, i.e.,

$$G_p = \frac{I(L,t)}{I_0} = \frac{1}{1 - [1 - \exp(-\sigma\Delta n_0 L)]\exp[-2\sigma I_0(t - L/c)]} \tag{4.2-21}$$

where G_p is called the power amplification coefficient that is related to the time and the intensity I_0 of the input signal. Below we shall make an analysis of the power amplification of the leading and trailing edges of the square pulse.

For the leading edge of the pulse, or when $t = x/c$, substitution into Eq. (4.2-21) yields

$$G_p = I(x, x/c)/I_0 = \exp(\sigma\Delta n_0 x) \tag{4.2-22}$$

That is, the leading edge of the pulse exponentially increases with an increase in the length, and the power amplification coefficient is unrelated to the intensity of the input signal pulse.

For the trailing edge of the pulse, or when $t = \dfrac{x}{c} + \tau$, substitution into Eq. (4.2-21) yields

$$G_p = \frac{I\left(x, \dfrac{x}{c} + \tau\right)}{I_0} = \frac{1}{1 - [1 - \exp(-\sigma\Delta n_0 L)]\exp(-2\sigma I_0 \tau)} \tag{4.2-23}$$

It is known from Eq. (4.2-23) that the necessary condition for obtaining an exponential increase is $2\sigma I_0 \tau < 1$, and $2\sigma I_0 \tau \ll \exp(-\sigma\Delta n_0 L)$. Now

$$G_p \approx \exp(\sigma\Delta n_0 L) \tag{4.2-24}$$

That is to say, only when we have a small signal (i.e., I_0 is very small) or a signal of an extremely narrow pulse width (i.e., τ is very small), will it be possible to obtain an exponential gain. Conversely, when the incident signal is very strong, or when the pulse width is rather great, the rear edge of the pulse cannot be amplified.

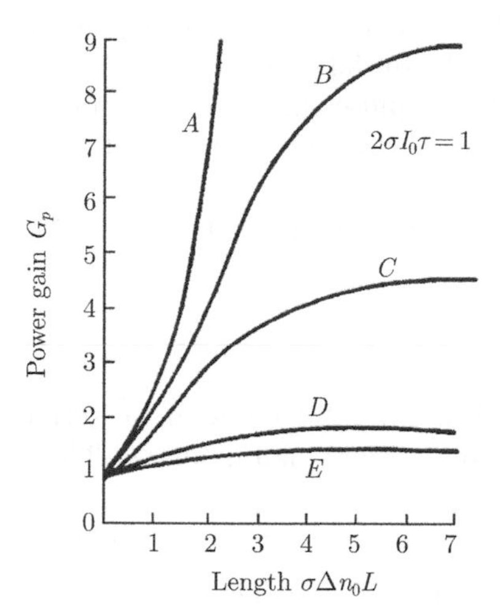

Fig. 4.2-3 The relationship between power gain in different portions of square pulse with L

In other words, when a square pulse passes through the amplifier, the gains obtained by different portions of the pulse are different. The front edge of the pulse has the greatest gain while the gain of some portions in the rear of the pulse will decrease with an increase in $t - \frac{L}{c}$, the gain at $t - \frac{L}{c} = \tau$ being minimum. Shown in Fig. 4.2-3 are the curves of relationship between the power gain of different portions of the square pulse and the length of the amplifier. A is the leading edge portion of the pulse, B is the 10% portion of the pulse, C is the 22% portion of the pulse, D is the 70% portion of the pulse, and E is the trailing edge portion of the pulse. It can be seen from the figure that power increases exponentially in the leading edge portion, while in the trailing edge portion, the gain tends to saturation. This is very apparent because, when the pulse leading edge enters the active medium, the inverted population density is maximum and very high gain can be obtained. But when the trailing portion enters the medium, the population at the upper level has almost been pumped empty, with only very little gain obtained. The result is the pulse becoming pointed in shape, its width becoming narrower. Figure 4.2-4 shows the variation of the square pulse shape in the course of amplification. Curve 1 represents the shape of the square pulse before entering the amplifying medium, curve 2 that after entry into the amplifying medium at $\sigma\Delta n_0 L = 1$, while curve 3 corresponds to the shape at $\sigma\Delta n_0 L = 2$.

Apart from the power gain, another important parameter is the gain to the pulse energy after the pulse passes through the amplifying medium. This can be obtained by performing integration over the intensity in time and taking the ratio of the output to input of the amplifier; that is,

$$G_g = \frac{\int_{-\infty}^{+\infty} I(L,t)\mathrm{d}t}{\int_{-\infty}^{+\infty} I(0,t)\mathrm{d}t} \tag{4.2-25}$$

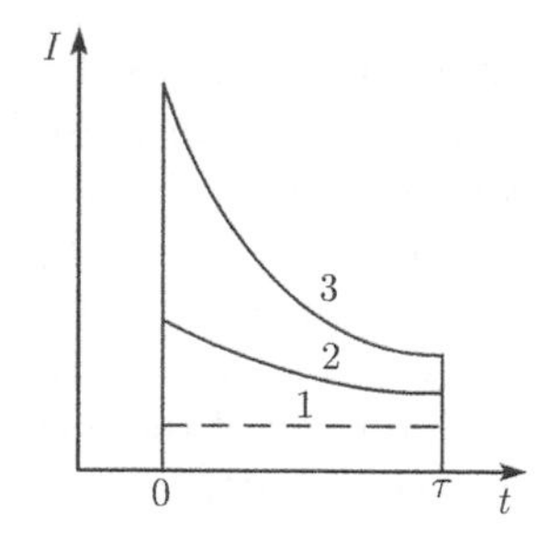

Fig. 4.2-4 Variation of square pulse after amplification

For the square pulse, as when $t < 0$ and $t > \tau$, $I(0,t) = 0$. Similarly, when $t < \frac{L}{c}$ and $t > \tau + \frac{L}{c}$, $I(L,t) = 0$. Thus the gain Eq. (4.2-25) can be written as

$$G_E = \frac{\int_{\frac{L}{c}}^{\tau+\frac{L}{c}} I(L,t)\mathrm{d}t}{\int_0^{\tau} I(0,t)\mathrm{d}t} \tag{4.2-26}$$

As $\int_0^{\tau} I(0,t)\mathrm{d}t = I_0\tau$, substitution from Eq. (4.2-20) into Eq. (4.2-26) and integration yields

$$G_E = \frac{1}{2\sigma I_0\tau}\ln\{1 + [\exp(2\sigma I_0\tau) - 1]\exp(\sigma\Delta n_0 L)\} \tag{4.2-27}$$

G_E is called the energy amplification coefficient. It can be seen from the above equation that the amplifier's energy gain is related to factors such as the initial inverted population, length of the amplifying medium, the amplitude of the incident pulse signal, and the pulse width. Below we shall discuss the relationship between the energy gain and the relevant parameters in three different cases.

(1) When the energy of the incident pulse signal is very small or the pulse is very short, with the relations

$$2\sigma I_0\tau \ll 1$$

$$2\sigma I_0\tau\exp(\sigma\Delta n_0 L) \ll 1$$

satisfied, perform serial expansion first of $\exp(2\sigma I_0\tau)$ in Eq. (4.2-27) followed by serial expansion of the logarithmic terms while neglecting the second-order small quantity. Then we have

$$G_E \approx \exp(\sigma\Delta n_0 L) \tag{4.2-28}$$

This is the expression for the energy gain of the small signal. It can be seen that its main characteristic is that the gain is unrelated to the intensity of the incident signal but will exponentially increase with an increase in the length of the amplifier and the inverted population density. In addition, when the small signal is being amplified, the entire pulse can be uniformly amplified. Hence no distortion of the pulse shape will be generated.

(2) When the incident pulse signal is very strong, with the condition

$$2\sigma I_0\tau \gg I$$

satisfied, through operation, Eq. (4.3-27) can be approximately written as

$$G_E \approx 1 + \frac{\Delta n_0 L}{2I_0\tau} \tag{4.2-29}$$

The above equation shows that when the incident signal is very strong (big signal), the gain will decrease with an increase in the incident signal, or there will appear the phenomenon of saturation. This is because when the incident signal is sufficiently large, the inverted population at the leading edge of the pulse will be pumped empty so that the gain at the trailing edge of the pulse will be far smaller than that at the leading edge, causing the pulse width to become narrower. Hence the distortion of the shape of the output pulse.

(3) The intensity of the incident signal is not too strong (medium), but the length of the amplifier is sufficiently long, with the condition

$$\sigma\Delta n_0 L \gg I$$

satisfied, there will still be the phenomenon of gain saturation. This is because when the optical pulse signal traverses in the amplifying medium, the gain will increase exponentially in the beginning portion. After propagation over a definite distance, as the energy of the optical pulse has already become sufficiently strong, the inverted population will abruptly decrease and enter the linearly increasing region till the stored energy is pumped empty. Suppose the exponential gain region is shorter than the linear gain region. Then its energy gain

$$G_E \approx \frac{\Delta n_0 L}{2I_0\tau} \tag{4.2-30}$$

It can be seen from the above analysis that both increasing the length L of the amplifier and enhancing the initial inverted population density Δn_0 can enhance the energy gain of the amplifier. But when it is considered that the amplifier does suffer definite loss and that energy will no longer increase after the length of the amplifying medium exceeds a definite limit, the best way is to increase the inverted population. Fig. 4.2-5 and Fig. 4.2-6 show the curves of relationship of the energy gain G_E with the incident photon density and with the length of the amplifier, respectively.

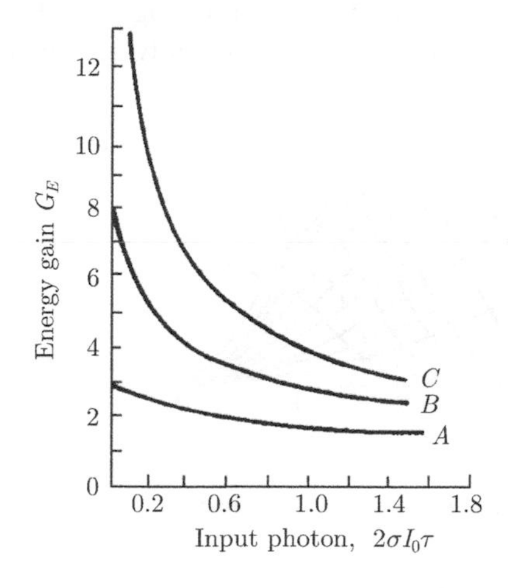

Fig. 4.2-5 Relation of G_E with incident photon (curve $A, \sigma\Delta n_0 L = 1$; curve $B, \sigma\Delta n_0 L = 2$; curve $C, \sigma\Delta n_0 L = 3$)

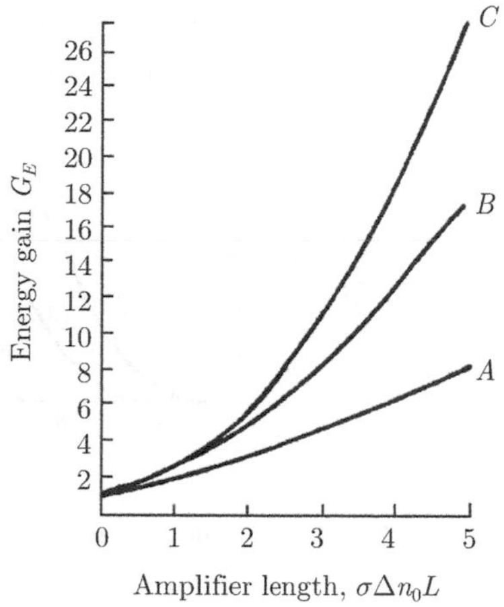

Fig. 4.2-6 Relation of G_E with length of amplifier (curve $A, 2\sigma I_0\tau = 0.6$; curve B, $2\sigma I_0\tau = 0.2$; curve $C, 2\sigma I_0\tau = 0.1$)

4.2.4 Amplification of other pulse waveforms

In fact, as none of the laser pulses output by the Q-switched laser is of square wave, a brief discussion will be made below of the amplification of the Gaussian type, Lorentzian type, and exponential type pulses. For the amplification of these pulse waveforms, Eqs. (4.2-1) and (4.2-2) can still be used to find the solution. But what should be pointed out is that both theoretical and experimental results show that after the laser pulse passes through the amplifier, the variation of its waveform is directly related to the law of variation of the leading edge of the incident signal pulse. The leading edge of the Gaussian type pulse varies as $\exp(-t^2/\tau^2)$, hence the rise of the leading edge is even faster than the exponential rise. So, after amplification the pulse width can be compressed. Fig. 4.2-7 shows the variation

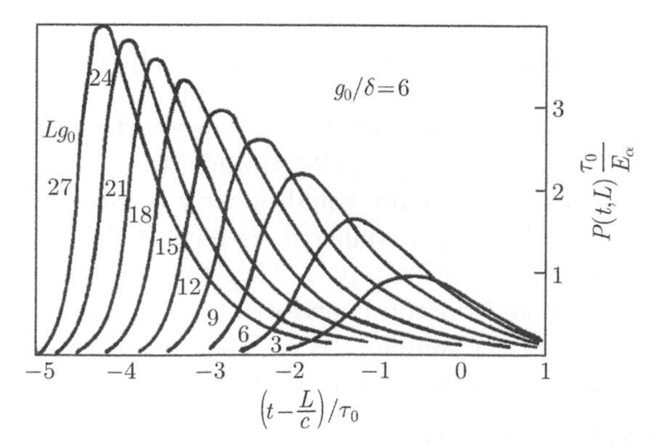

Fig. 4.2-7 The waveform of Gaussian type pulse after nonlinear amplification

of the waveform of the Gaussian type pulse due to nonlinear amplification. When passing through the amplifier, neither the shape nor the width of the exponential type pulse changes much. However, as the pulse leading edge has a greater gain than its trailing edge, its peak value moves forward because of an increase of the length L when penetrating the amplifying medium together with the pulse, the displacement amount being $\Delta\tau = \tau(g_0 - \delta)L$ (the meanings of symbols are the same as before). The variation is as shown in Fig. 4.2-8. If the rise of the input pulse front edge is slower than the exponential function, then the pulse will become broader after passing through the nonlinear amplifying medium. Fig. 4.2-9 shows the pulse shape $P_0[1+(t/\tau)^8]^{-1}$. After passing through the amplifier, not only has the pulse width not been compressed, but it has become broader.

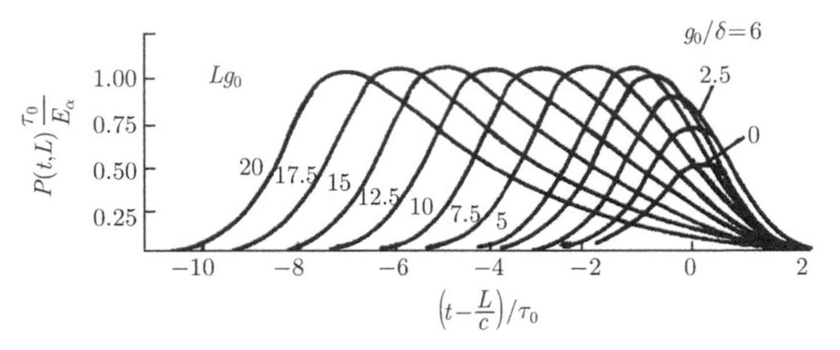

Fig. 4.2-8 The waveform of exponential type pulse after amplification

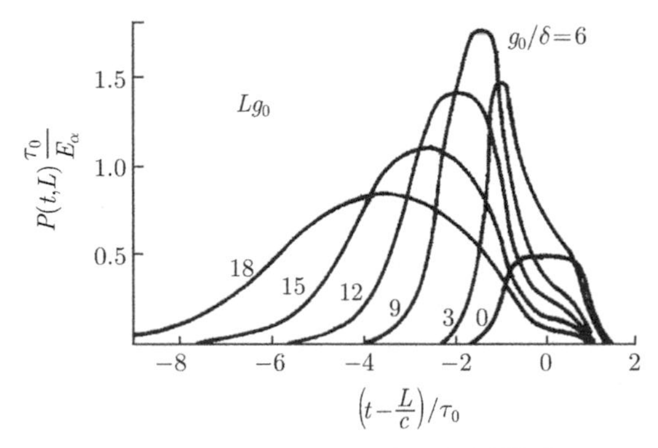

Fig. 4.2-9 Variation of pulse waveform $P_0[1+(t/\tau)^8]^{-1}$ after amplification

Generally, in order to obtain a high power and narrow width laser pulse, the clipping technique can be adopted to clip the slowly rising part of the pulse to make its front edge become steep before the signal enters the amplifier. In so doing, the pulse will get compressed.

4.2.5 Amplification of pulse signal in a medium with losses

All the amplification processes discussed above are those performed in lossless amplifying media that yield the gain calculating Eqs. (4.2-29) and (4.2-30). This result shows that, once gain saturation should occur, the gain would linearly increase with an increase in the length of the amplifier. In reality, an amplifier does have definite losses (e.g., the impurity absorption in the medium, scattering, etc.). Therefore, as the length of the amplifier increases, so will the total loss, and the total energy of the photon flow is made to decrease.

Suppose the loss coefficient of the amplifier is δ. Then a loss term δcI should be added to the rate Eq. (4.2-4) of the photon flow intensity and population inversion, i.e.,

$$\frac{\partial I(x,t)}{\partial t} + c\frac{\partial I(x,t)}{\partial x} = c\sigma\Delta n(x,t)I(x,t) - \delta cI(x,t) \tag{4.2-31}$$

$$\frac{\partial \Delta n(x,t)}{\partial t} = -2\sigma \Delta n(x,t) I(x,t) \tag{4.2-32}$$

To facilitate discussion, only the variation of the total energy after the pulse passes through amplifier will be involved without considering the variation in different portions of the pulse. We eliminate the time relation using the following integration condition:

$$I(x) = \int_0^{T} I(x,t) dt \tag{4.2-33}$$

The above equation shows that, at x of the amplifier, via the total photon number in unit cross section, substituting Eq. (4.2-32) into Eq. (4.2-31) and taking into account the condition for Eq. (4.2-33), we obtain

$$\frac{\mathrm{d}I(x)}{\mathrm{d}x} = \frac{1}{2}\int_{\Delta n_0}^{\Delta n(x,t)} \mathrm{d}\Delta n(x,t) - \delta I(x) = \frac{1}{2}[\Delta n_0 - \Delta n(x,t)] - \delta I(x) \tag{4.2-34}$$

where Δn_0 is the density of the initial inverted population; $\Delta n(x,t)$ can be found by integrating over Eq. (4.2-32) as

$$\Delta n(x,t) = \Delta n_0 \exp[-2\sigma I(x)] \tag{4.2-35}$$

Substitution of the above equation into Eq.(4.2-34) yields

$$\frac{\mathrm{d}I(x)}{\mathrm{d}x} = \frac{\Delta n_0}{2}\{1 - \exp[-2\sigma I(x)]\} - \delta I(x) \tag{4.2-36}$$

This is the expression for the process of pulse signal amplification with loss. After the initial inverted population density Δn_0 and the loss coefficient δ are determined, the transmission of the signal in the medium is completely determined. Performing numerical solution for the nonlinear differential Eq. (4.2-36), we can find the relation of variation of the output energy with the variation of the length of the amplifier.

For a small signal that is injected, satisfy the condition $\sigma I(x) \ll 1$. Thus $\exp[-2\sigma I(x)] \approx 1 - 2\sigma I(x)$. Expanding the exponential term in Eq. (4.2-36) and neglecting the second-order small quantity, we have

$$\frac{\mathrm{d}I(x)}{\mathrm{d}x} = \sigma \Delta n_0 I(x) - \delta I(x) \tag{4.2-37}$$

Thus, the expression for the variation of the small signal is found as

$$I(x) = I(0) \exp[\sigma \Delta n_0 - \delta] x \tag{4.2-38}$$

The above equation shows that when passing through the amplifier the small signal will get amplified in accordance with the exponential law.

For a strong signal, satisfy the condition $\sigma I(x) \gg 1$. Then $\exp[-2\sigma I(x)] \approx 0$. Thus, we obtain from Eq. (4.2-36)

$$\frac{\mathrm{d}I(x)}{\mathrm{d}x} = \frac{\Delta n_0}{2} - \delta I(x) \tag{4.2-39}$$

After integration the expression for the variation of the photon number density with the length of the amplifier is obtained:

$$I(x) = \frac{\Delta n_0}{2\delta}[1 - \exp(-\delta x)] + I(0)\exp(-\delta x) \tag{4.2-40}$$

The above equation shows that during saturation the loss has a very great influence on the output energy of the amplifier. The maximum output energy expected from the amplifier is determined by the term $\frac{\Delta n_0}{2\delta}$.

When $I(0), \Delta n_0$, and δ are given, find the numerical solution for Eq. (4.2-36), the result of which is as shown in Fig. 4.2-10. The curves A, B, and C represent the incident signal light energy surface densities $I(0)h\nu$ as 0.1J/cm^2, 1.0J/cm^2, and 2.5J/cm^2; $\delta = 0.05\text{cm}^{-1}$; and $\Delta n_0 h\nu = 2.0\text{J/cm}^3$.

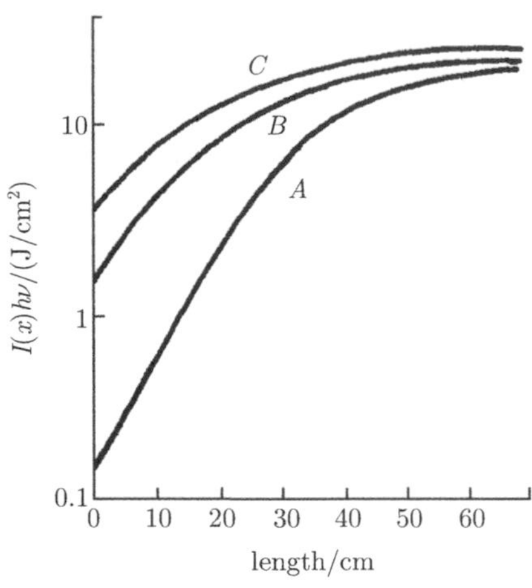

Fig. 4.2-10 Relation of output energy with the length of amplifier

It can be seen from Fig. 4.2-10 that when a small signal is injected, the amplification is fairly apparent (as curve A), and the rise of the small signal's gain near the input end of the amplifier is very rapid. With the strengthening of the incident signal, the gain rise becomes slower. If the amplifier is very long (i.e., x is very great), the output energy will be restrained by the loss, there quickly appearing the phenomenon of a tendency to saturation with the increase in the length (as shown by curve C).

4.3 The steady-state theory of long pulsed laser amplification

4.3.1 The rate equation of the steady state

When what is injected into the laser amplifier is a long pulse signal, or the duration of the optical pulse is greater than the longitudinal relaxation time (satisfying the condition $\tau \gg T_1$), it is necessary to adopt the steady-state theory to analyze the process of amplification. Because now as the population consumed because of the stimulated radiation can quickly be replenished by optical pumping to keep the inverted population maintained around a stable numerical value, we can approximately consider that $\frac{\mathrm{d}\Delta n}{\mathrm{d}t} = 0$. So, the impact of optical pumping and spontaneous radiation on population inversion should be counted in the rate equation (assuming the incident signal possesses sufficiently wide spectral line so that the inverted population within the range of the entire gain line width all contribute to the output without the hole burning effect ever occurring). The rate Eqs. (4.2-31) and (4.2-32) can be expressed as (to facilitate writing, $I(x,t)$ is written as $I, \Delta n(x,t)$ as Δn)

$$\frac{1}{c}\frac{\partial I}{\partial t} + \frac{\partial I}{\partial x} = \sigma \Delta n I - \delta I \tag{4.3-1}$$

$$\frac{\partial \Delta n}{\partial t} = -2\sigma \Delta n I - \frac{1}{T_1}(\Delta n - \Delta n_0) \tag{4.3-2}$$

where δ is the loss coefficient of the amplifier and T_1 is the longitudinal relaxation time representing the lifetime of the atoms in the excited state. Under the steady-state condition, the left side of the equal sign of Eq. (4.3-2) is zero $\left(\frac{\partial \Delta n}{\partial t} = 0\right)$. Thus we have

$$\Delta n = \frac{\Delta n_0}{1 + 2\sigma I T_1} \tag{4.3-3}$$

Adopting total differential operators $\frac{\mathrm{d}}{\mathrm{d}x}=\frac{1}{c}\frac{\partial}{\partial t}+\frac{\partial}{\partial x}$ for Eq. (4.3-1), we obtain

$$\frac{1}{I}\frac{\mathrm{d}I}{\mathrm{d}x}=\sigma\Delta n-\delta$$

If Eq. (4.3-3) is substituted into the above equation, then

$$\frac{1}{I}\frac{\mathrm{d}I}{\mathrm{d}x}=\frac{\sigma\Delta n_0}{1+2\sigma IT_1}-\delta \tag{4.3-4}$$

where $\frac{1}{I}\frac{\mathrm{d}I}{\mathrm{d}x}$ represents the gain coefficient with loss on unit length which, denoted by k, can be written as

$$k=\frac{\sigma\Delta n_0}{1+2\sigma IT_1}-\delta$$

Then, by introducing $k_0=\sigma\Delta n_0$ and $I_0=\frac{1}{2\sigma T_1}$, Eq. (4.3-1) becomes

$$k=\frac{k_0}{1+\dfrac{I}{I_0}}-\delta \tag{4.3-5}$$

When the signal intensity I does not increase any longer after having made $\frac{k_0}{1+I/I_0}-\delta=0$, it is called the saturated light intensity I_s.

$$I_\mathrm{s}=I_0\left(\frac{k_0}{\delta}-1\right) \tag{4.3-6}$$

Now, the gain provided by the inverted population in the amplifying medium will be completely consumed in the losses of the cavity and operation material.

4.3.2 The influence of spectral line profile on the gain coefficient

In the afore-going discussion, no consideration has been made on the impact of the spectral line profile and line width of the active ion on the gain coefficient. In fact, different spectral line broadening mechanisms will lead to different gain results. Below is a discussion of this impact.

The impact of the spectral line profile is mainly manifested in the emission cross section in Eqs. (4.3-1) and (4.3-2). For the uniformly broadened Lorentz spectral line shape,

$$\begin{aligned}\sigma(\nu,\nu_0)&=\frac{h\nu}{c}B_{21}g(\nu,\nu_0)\\&=\frac{h\nu}{c}B_{21}\frac{1}{\pi}\frac{\dfrac{\Delta\nu}{2}}{\left(\dfrac{\Delta\nu}{2}\right)^2+(\nu-\nu_0)^2}\\&=\sigma_0\frac{T_2^{-2}}{T_2^{-2}+(\nu-\nu_0)^2}\end{aligned} \tag{4.3-7}$$

where $\sigma_0=\frac{h\nu}{c}B_{21}\frac{T_2}{\pi}$; $T_2^{-1}=\frac{\Delta\nu}{2}$, T_2 is the coherence time of the spectral line, or the transverse relaxation time; $\Delta\nu=\frac{2}{T_2}$ is the uniformly broadened width; ν_0 is the central frequency; ν is the frequency of the amplified signal; and σ_0 the maximum emission cross section at $\nu = \nu_0$. For the non-uniformly broadened spectral line, all ions each have their different central spectral lines.

Therefore, the total population n can be classified according to their central frequencies. Suppose the population with a central frequency ν' is $\Delta n(\nu')$, whose contribution to the gain is $\Delta n(\nu')\sigma(\nu, \nu')$. Thus the total gain of all kinds of ions is

$$\sum_{\nu'} \sigma(\nu, \nu')\Delta n(\nu') = \sum_{\nu'} \frac{\sigma_0 T_2^{-1} \Delta n(\nu')}{T_2^{-1} + (\nu - \nu')^2} \tag{4.3-8}$$

The distribution of the inverted population according to ν' in the absence of the action of the optical signal is of the Gaussian type, i.e.,

$$\Delta n(\nu') = \frac{2}{\Delta\nu_D \sqrt{\pi}} (\ln 2)^{\frac{1}{2}} n \exp\left\{-\left[\frac{2(\nu - \nu_{12})}{\Delta\nu_D}\right]^2 \ln 2\right\} \Delta\nu' \tag{4.3-9}$$

where ν_{12} is the central frequency of atomic transition and $\Delta\nu_D$ is the spectral line's non-uniformly broadened width.

In general, the impact of spectral line broadening on the total gain obtained from Eqs. (4.3-1) and (4.3-2) is

$$\left(\frac{\partial}{\partial t} + c\frac{\partial}{\partial x}\right) I(\nu) = \sum_{\nu'} \sigma(\nu, \nu')\Delta n(\nu') I(\nu) \tag{4.3-10}$$

$$\frac{\partial}{\partial t}\Delta n(\nu') + \frac{1}{T_1}[\Delta n(\nu') - \Delta n_0(\nu')] = -2\sigma(\nu, \nu')\Delta n(\nu') I(\nu) \tag{4.3-11}$$

For the steady-state case, $\dfrac{\partial \Delta n(\nu')}{\partial t} = 0$. Thus, from the above equation we have

$$\frac{1}{I}\frac{\mathrm{d}I}{\mathrm{d}x} = \sum_{\nu'} \sigma(\nu, \nu')\Delta n(\nu') \tag{4.3-12}$$

$$\Delta n(\nu') = \frac{\Delta n_0(\nu')}{1 + 2T_1 \sigma(\nu, \nu') I(\nu)} \tag{4.3-13}$$

Substituting Eqs. (4.3-13), (4.3-8), and (4.3-9) into Eq. (4.3-12), we obtain the total gain

$$\begin{aligned} k &= \sum_{\nu'} \sigma(\nu, \nu')\Delta n(\nu') = \sum_{\nu'} \frac{\Delta n_0(\nu')}{\dfrac{1}{\sigma(\nu, \nu')} + 2T_1 I(\nu)} \\ &= \sum_{\nu'} \frac{\dfrac{2(\ln 2)^{\frac{1}{2}}}{\Delta\nu_D\sqrt{\pi}} n \exp\left\{-\left[\dfrac{2(\nu - \nu_{12})}{\Delta\nu_D}\right]^2 \ln 2\right\}}{\dfrac{1}{\sigma_0}[(\nu - \nu')^2 T_2^2] + 2T_1 I(\nu)} \Delta\nu' \\ &= k_0 \frac{1}{\pi} \int_{-\infty}^{\infty} \frac{T_2 \exp\left\{-\left[\dfrac{2(\nu' - \nu_{12})}{\Delta\nu_D}\right]^2 \ln 2\right\}}{1 + [(\nu - \nu')T_2]^2 + 2\sigma_0 T_1 I(\nu)} \mathrm{d}\nu' \end{aligned} \tag{4.3-14}$$

where $k_0 = \sigma_0 \dfrac{2\sqrt{\pi}}{\Delta\nu_D T_2} (\ln 2)^{\frac{1}{2}} n$. For simplicity, introduce variable

$$x = \nu T_2, \quad x_0 = \nu_{12} T_2, \quad x' = \nu' T_2, \quad \varepsilon = (\ln 2)^{\frac{1}{2}} \frac{2}{\Delta\nu_D} \frac{1}{T_2}, \quad 2\sigma_0 T_1 = \frac{1}{I_0}$$

Then

$$k_0 = \sqrt{\pi}\varepsilon n \sigma_0 \tag{4.3-15}$$

which, substituted into Eq. (4.3-14), gives the expression for the total gain as

$$k = \sum_{\nu'} \sigma(\nu, \nu')\Delta n(\nu') = \frac{k_0}{\pi} \int_{-\infty}^{\infty} \frac{\exp[-\varepsilon^2(x' - x_0)^2]}{1 + (x - x')^2 + \dfrac{I}{I_0}} \mathrm{d}x' \tag{4.3-16}$$

The above equation represents the total gain coefficient with the spectral line broadening (including uniform and non-uniform broadening) taken into account. The physical meanings of the parameters ε, k_0, and I_0 in the equation are: ① ε is proportional to the ratio of the uniformly broadened spectral line width $\Delta\nu_N = \dfrac{1}{T_2}$ to the non-uniformly broadened spectral line width $\Delta\nu_D$, i.e., $\varepsilon \propto \dfrac{\Delta\nu_N}{\Delta\nu_D}$; ② σ_0 is the absorption cross section at the central frequency (ν_{12}). So $k_0 \propto \sigma_0 \Delta n_0$ represents the gain coefficient at the central frequency.

When $\varepsilon \to 0$, that is $\dfrac{\Delta\nu_N}{\Delta\nu_D} \to 0$, this means that the spectral line's non-uniform broadening is dominant and the expression for its gain coefficient is reduced to

$$k = \frac{k_0}{\left(1 + \dfrac{I}{I_0}\right)^{1/2}} \tag{4.3-17}$$

In the steady-state case, by taking the loss into account, there is

$$k = \frac{1}{I}\frac{\mathrm{d}I}{\mathrm{d}x} = \frac{k_0}{\left(1 + \dfrac{I}{I_0}\right)^{1/2}} - \delta \tag{4.3-18}$$

When $\varepsilon \to \infty$, that is, $\dfrac{\Delta\nu_N}{\Delta\nu_D} \to \infty$, the spectral line's uniform broadening is dominant. Then the gain coefficient obtained from integration of Eq. (4.3-16) is

$$k = \frac{k_0}{\varepsilon\sqrt{\pi}\left(1 + \dfrac{I}{I_0}\right)} = \frac{\Delta n_0 \sigma_0}{1 + \dfrac{I}{I_0}} \tag{4.3-19}$$

Similarly, by taking the loss into account, there is

$$k = \frac{1}{I}\frac{\mathrm{d}I}{\mathrm{d}x} = \frac{\Delta n_0 \sigma_0}{1 + \dfrac{I}{I_0}} - \delta \tag{4.3-20}$$

This is the relation of variation of the gain coefficient with loss on unit length with the intensity of the incident signal shown in Eq. (4.3-5). When $I = I_0$, the gain coefficient will decrease to half the maximum gain coefficient of the small signal.

4.4 Problems for consideration in designing a laser amplifier

Although the laser amplifier and laser (oscillating) device are both based on one and the same physical process, the laser amplifier has its own specific problems. It is noted that, in designing a laser amplifier, it is necessary to consider the following points:

4.4.1 The selection of the amplifier operation material

The operation material of the amplifier should match the oscillation stage (geometric dimension, the active ion concentration, etc.). We already know that active ion concentrations are different, as is the energy stored in unit area. In general, the higher the concentration, the greater the energy stored. But if it's too high, the phenomenon of "concentration quench" is apt to appear, which can only reduce the efficiency. When the concentration of the active ions is too low, the efficiency will not be high. For example, for neodymium glass, 2%~4% is in general adopted as the Nd^{+3} ion weight percentage. In neodymium glass of a concentration like this, the density of the stored energy is 2~4 J/cm^3. In addition, taking into consideration the laser's output average power $W = \dfrac{EA}{\tau}$ (E is the energy on unit area, A is the laser rod's cross section, τ is the pulse width), to obtain a rather high power, apart from raising the energy on the unit area, it is advisable to appropriately enlarge the caliber of the rod as the energy that can be borne by the unit area of the operation material is restrained by the destructive threshold. When the density of the stored energy in unit volume and the caliber of the rod are decided on, it is necessary to choose an adequately long rod since only by so doing will it be possible to attain the required energy output. For instance, in certain applications the multistage (traveling wave) amplifier is often needed to obtain great energy output. Then, when making an overall consideration, usually a rod of a rather high doped ion concentration is selected for the front stage amplifier and the caliber of the rod is rather small, the optical pumping easily gets uniform; for the relay amplifier, the rod caliber should properly be enlarged; the light beam output by the front stage amplifier is effectively coupled to the relay amplifier through the beam extending telescope and is further amplified. As the caliber of the rod of the final amplifier is rather large, the concentration of the rod's doped ions should not be too high; in general a weight percentage of 2% or so will do and multi-lamp pumping should be adopted to achieve uniform illumination. For instance, for the multistage Nd^{+3} glass amplifier, the calibers of the different levels of amplifier and the energy that can be attained are shown in Tab. 4.4-1. The length of the Nd^{+3} glass rod of amplifiers of different calibers is taken as 500 mm and the effective gain length is calculated as 480 mm. So that all the amplifiers will possess the gain in their own right, it is necessary to reach sufficient pump energy density and the optical pumping should be uniform. Thus, for the $\phi 20 \times 500$ mm amplifier, two $\phi 20 \times 500$ mm xenon lamps are used; for the $\phi 35 \times 500$ mm amplifier, four $\phi 20 \times 500$ mm xenon lamps are used; for the $\phi 45 \times 500$ mm amplifier, four $\phi 30 \times 500$ mm xenon lamps are used for pumping, 1/3~1/2 of the rated load energy being taken as the pump energy of all the xenon lamps.

Tab. 4.4-1 The parameters of different levels of amplifiers

	Amplifier caliber	Energy density taken	Amplifier cross section	Reachable energy	Remarks
Preamplifier	ϕ20 nm	$\leqslant$ 3 J/cm^2	3.14 cm^2	$\leqslant$ 10 J	Stage 3
Intermediate amp.	ϕ35 nm	$\leqslant$ 3 J/cm^2	9.60 cm^2	$\leqslant$ 30 J	stage 3
Final amplifier	ϕ45 nm	$\leqslant$ 4 J/cm^2	15.9 cm^2	$\leqslant$ 64 J	stage 2

4.4.2 Elimination of the end face feedback of the amplifier operation material

To enable the traveling wave amplifier to operate stably, the generation of self-excited oscillation must be prevented; that is, care should be taken to make all the incident pulse signals leave the amplifier after getting amplified. But, owing to the action of reflection of the end face of the amplification medium, a small part of the light will be fed back to the amplifier. When the gain is rather great, self-excited oscillation may be generated. This will not only destroy the effect of amplification, but also, instability will result and damage to the amplification medium may often be done. So measures should be taken to eliminate the feedback. Usually, one may either plate the end face of the amplification medium with a layer of transmission increasing film to reduce the reflection from the end face. But this method is suitable only for devices of a not-too-high power density. Or, one may so grind the end face of the amplification medium that it has a rather small oblique angle (in general $2°\sim3°$) or a Brewster angle. But the latter will increase the difficulty in use and adjustment, and the astigmatism will affect the directionality of the light beam. In addition, as the angle is too big, the deviation of the light beam from the rod's axis after getting refracted will be great. Thus, the room taken by the device will be large and adjustment inconvenient. The common practice is to grind the two end faces of the rod in such a way that it will have a rather small oblique angle so that the rays reflected from the rod end faces will not return to the rod. Suppose the rod caliber is ϕ, the length L, and the oblique angle α. Then $\alpha = \arctan\ (\phi/L)$. For example, if the size of the rod is $\phi 20 \times 500$ mm, then angle α will be $2°\sim3°$. Thus, not only is the influence of the end face feedback eliminated, but also no phenomenon of astigmatism will appear with the light beam.

4.4.3 The problem of interstage decoupling

The interstage coupling of an amplifier will bring about self-excited oscillation or, after getting amplified by the traveling wave, induce the phenomenon of strong superradiation, not only reducing the gain of the amplifier, but also affecting the stable operation of the laser oscillator. For this reason, in a high gain multistage amplifying system, isolating elements must be inserted to prevent the feedback between the amplifiers, the purpose being to permit the optical signal to go only from the oscillator to the amplifier, or from a front stage amplifier to a rear stage amplifier while not allowing reverse traversal of the optical signal so as to attain the goal of making the amplified optical signal do its one-way passage.

Commonly used optical isolating elements include the Faraday optical isolator, the electro-optic isolator, the saturable absorption isolator, etc.

1. The Faraday optical isolator

The Faraday optical isolator is made by employing the Faraday magneto-optic rotational effect (see Chapter 1), whose structural principle is as shown in Fig. 4.4-1. P_I and P_{II} are polarizers, the polarization axes of the two forming an angle of 45° with each other. The Faraday rotator is composed of laser medium and the surrounding coils. The direction of the magnetic field is along the transmitting light direction. After the incident light verdet constant V of the rotatory material with respect to the given wavelength λ and the length L of the rotatory medium are determined, an appropriate magnetic field intensity H can be selected to make the polarized surface rotate an angle of 45° when the incident polarized light passes through the rotatory device. Therefore, when passing through the polarizer P_I, the incident light becomes linearly polarized light, whose polarized plane is as shown by the arrows in the figure. When passing through the rotator, this beam of light rotates an angle of 45° clockwise along the given direction of the magnetic field. As its polarized plane is in agreement with the polarized axis of polarizer P_I, the optical signal can pass through the P_{II} unobstructed. Conversely, if the optical signal should return from an opposite direction, as the Faraday effect is not related to the direction of light propagation, after passing through the polarizer P_{II} and the magneto-optic medium, its polarizing direction will rotate 45° clockwise again while normal to the polarizing axis of P_I. So it cannot get through P_I but is reflected out of the optical path, thus achieving the goal of isolating the light from the opposite direction.

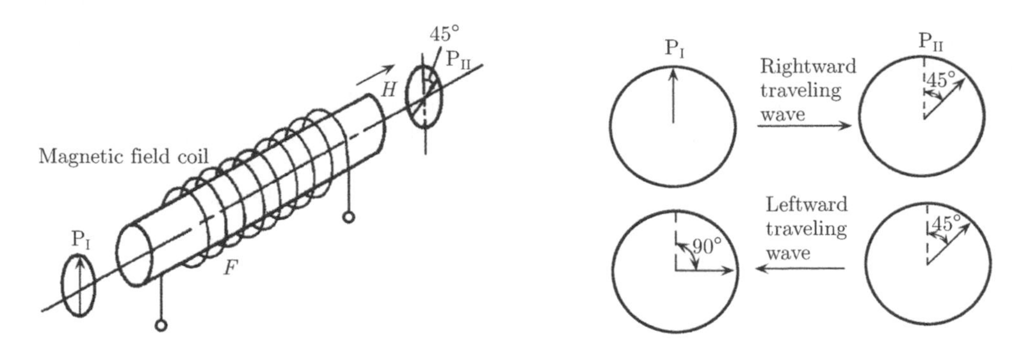

Fig. 4.4-1 The magneto-optic isolator

2. The electro-optic isolator

The electro-optic isolator mainly makes use of the short duration characteristic of the high power pulse, that is, an electro-optic switch isolating element is inserted between amplifiers. Within the time before the arrival of the optical pulse signal, the electro-optic switch is in the off state and it is only at the transient instant the optical pulse arrives that the switch is quickly turned on to make the pulse pass through and enter the rear stage amplifier. Its structure (as shown in Fig. 4.4-2) is composed of two polarizing prisms P_I and P_{II} (orthogonally placed), the KD*P Pockels cell K, the discharge spark gap G, and the collecting lens L. Prior to the arrival of the optical pulse, the optical path is in the off state; when the optical pulse arrives, the small part of the light beam reflected by the spectroscope $M_1$is focused by the lens L on the discharge spark gap to punch through the spherical gap air. Now the circuit is made to immediately apply the half-wave voltage on the KD*P crystal to make the incident light beam's polarized plane rotate 90° to become parallel to the polarizing axis of P_{II}. Now the optical pulse will be able to enter the rear stage amplifier via the switch. As the switching time is very short (the rise time about several nanoseconds), the switch immediately turns off after the passage of the optical pulse, playing the role of allowing only the light beam to pass through.

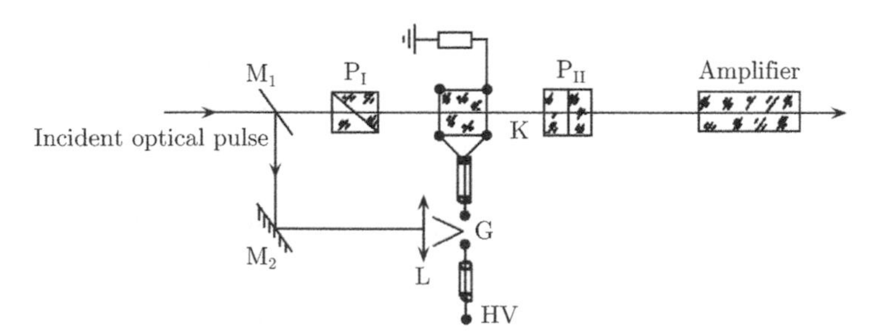

Fig. 4.4-2 Schematic diagram of electro-optic isolation: P_I and P_{II} are polarizing prisms, P_I and P_{II} are orthogonalized; K is Pockels cell or Kerr cell; G the discharge spark gap; M_1 partially reflecting splitting plate; M_2 totally reflecting plate; L collecting lens

The advantage of the electro-optic switch isolator is its capability of both isolating and performing clipping the front edge of the incident pulse. In certain high-power multistage amplifying systems, very often the electro-optic clipping switch and electro-optic isolator are adopted in combination, sharing a high-voltage pulsed driving source. During operation, the high-voltage electric pulse first drives the electro-optic clipping switch, then following transmission by a length of cable, drives the isolator. As the loss of the electro-optic isolator is rather large and it's not easy to get large-caliber component, it can only be used for isolation of the front stage amplifier.

4.4.4 The problem of interstage aperture matching

In order to raise the total energy output, the medium cross section of all the stages of amplifiers is increased stage by stage (but should not be too large lest the pumping would be nonuniform) so as to make all stages of amplifiers work below the laser destructive threshold. For this reason, the apertures of all stages have to match each other to attain the goal of making full use of the amplifying medium. To realize interstage aperture matching, it is customary to insert a beam-expanding telescope between two stages to make the light beam output by the front stage well coupled to the rearward stage amplifier. In the ten thousand MW level tunable Nd^{3+} glass laser system developed by Chinese University of Science and Technology, a device called "image transmitting spatial filter" is inserted between the front stage amplifier and the intermediate amplifier, whose structure is as shown in Fig. 4.4-3. It is composed of two flat convex lenses L_1 and L_2 of a suitable caliber, and a pinhole diaphragm (optional apertures). The spatial filter is used mainly to improve the transverse uniformity of a light beam, restrain self-focusing from occurring, and carry out image transmission diffraction-free. Moreover, it is capable of playing the function of light beam expanding and coupling between amplifiers of different calibers. The caliber of the front stage amplifier in the amplifying system is $\phi 20$, that of the mediate stage amplifier $\phi 35$, so the relevant parameters to be selected for the spatial filter are as shown in Tab. 4.4-2.

Tab. 4.4-2 Parameters of the spatial filter

L_1's focal length f_1/mm	L_2's focal length f_2/mm	Hole diameter /mm	Beam expansion ratio	Caliber of incident light beam/mm	Energy transmissivity/%
688	1050	0.5, 1.5 1.0, 2.0	1.53	$\phi 20$	78

It has thus been found by calculation that the caliber of the output light beam of this spatial filter is $1.53 \times \phi 20 = \phi 30.6$, which is advisably to be input into an amplifier of $\phi 35$.

4.4.5 The matching between the pumping times of different stages

The adoption of the traveling wave amplification system is for the purpose of making the amplifier operate highly efficiently in the hope that both the oscillating stage and all the amplifying stages will be enabled to attain maximum population inversion. Hence the need to match the ignition times of the pumping lamps of different stages.

As the caliber of the amplifier is rather large, so is the pump input electric energy; the time for discharge to attenuate to $1/e$ the peak value $t = RC$ (R is the impedance of the lamp), the time of the discharge pulse is rather long. For instance, a $\phi 23 \times 500$ mm straight tubular xenon lamp of atmospheric pressure 20 kPa is adopted for the oscillation stage, with 12 kJ energy input, the discharge time is about 1ms while a $\phi 32 \times 500$ mm straight tubular xenon lamp of atmospheric pressure 20 kPa for the amplifying stage, with 24 kJ energy input, the discharge time is about 2 ms. The discharge times of the two are different, as shown in Fig. 4.4-4, in which curve A represents the xenon lamp discharge waveform of the oscillating stage while curve B represents that of the amplifying stage. It can be seen from the figure that discharge is triggered at both the oscillating and amplifying stages at the same time, but the times of discharge do not match each other. The gain of the amplifier is not high, so delayed triggering should be adopted instead.

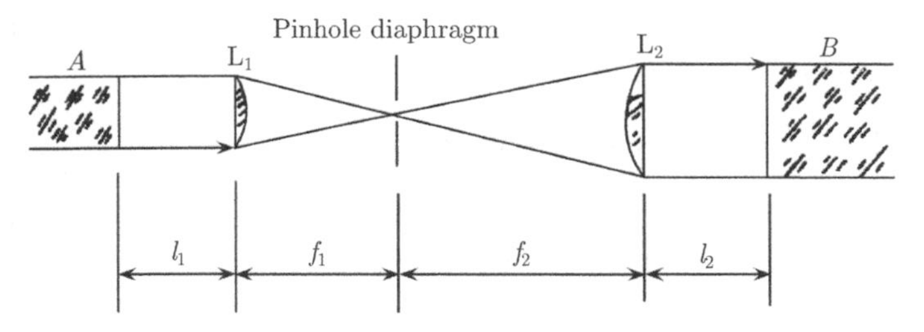

Fig. 4.4-3 Schematic diagram of image transmitting spatial filter

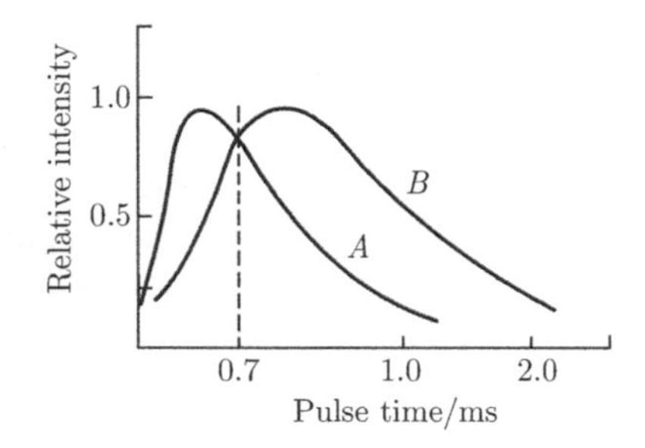

Fig. 4.4-4 Xenon lamp discharge waveforms of oscillating and amplifying stages

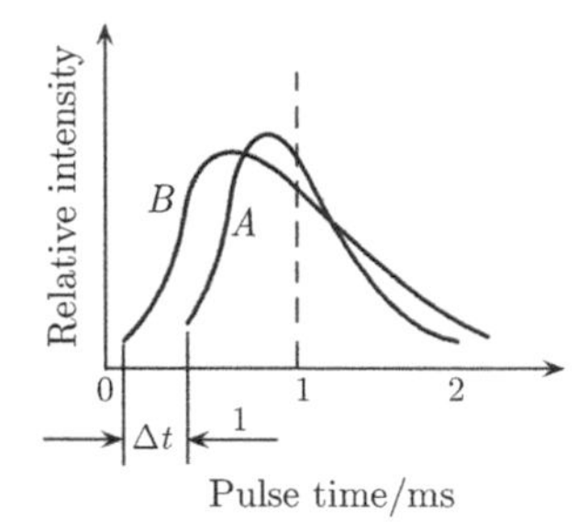

Fig. 4.4-5 Discharge waveforms after retardation

To attain matching in time of different stages of pumping, a trigger sync circuit should be incorporated between the stages to make the xenon lamp ignition time at the oscillating stage slightly retarded, as it is possible to find the optimum retardation time by experiment. Change the retardation time Δt during experiment while measuring the output energy at the amplifying stage. Fig. 4.4-5 shows the discharge at the two stages after experiencing the relative retardation time Δt. It can be seen that the pumping times of the two stages are fairly well matched, so the gain of the amplifier is rather high. When the number of amplifying stages is increased, the multichannel retardation trigger discharge should be adopted.

4.4.6 Elimination of the nonuniformity influence

To obtain high output energy, a traveling wave amplifier with a solid as its operation material often uses an operation material of a large caliber. But this will bring about problems. To begin with, it is very difficult achieving uniform pumping; the difference between the outer edge of the rod and the central portion can be very great, which will induce thermal distortion of the operation material, as shown in Fig. 4.4-6. Obviously, when the optical signal passes through such an amplifier, the directionality of the light beam will deteriorate; second, in the course of pumping, the distribution of the rod's inverted population is not uniform along the cross section; that is, the density of the inverted population in the central portion is smaller than that on the outer edge. Therefore, the gain of the optical pulse signal is also nonuniform; the gain in the central portion is lower than that on the outer edge.

It is thus clear that the presence of thermal distortion not only restrains the increase of the output energy and power of the amplifier, but also causes the deterioration of the directionality of the light beam. Judging by the operation of many solid laser amplifiers, in the course of optical pumping, both the optical distortion and stress destruction brought about by the rod's geometric shape and the heat conducting characteristics of material are rather serious, as shown in Tab. 4.4-3.

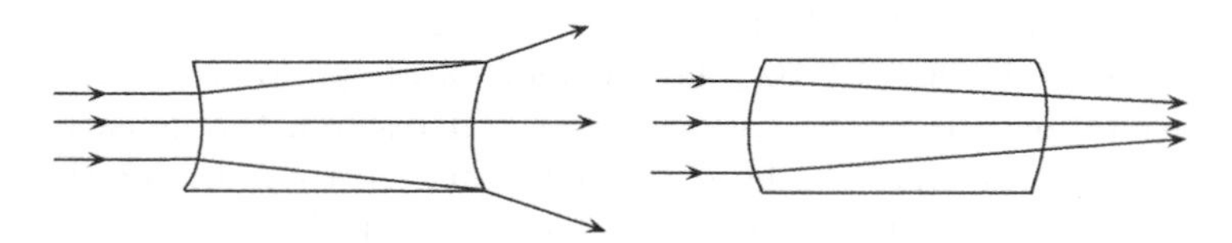

Fig. 4.4-6 The phenomenon of thermal distortion

Tab. 4.4-3 Thermo-optic characteristics and stress distortions of several solid operation materials

Operation material	Lens effect /[W/cm·(x)]	Stress /(W/cm)	Heat conductivity /(10^3J/cm·C)	$\frac{dn}{dt}$/C^{-1}
Nd^{3+}glass (normal state)	32.2	15.7	0.008	(−)2.2×10^{-6}
Nd^{3+} glass (prestressed state)	17.7	>57.2	0.012	(+)1.2×10^{-6}
ruby	785	1045	0.352	1.3×10^{-4}
Nd:YAG	193	130	0.12	7.4×10^{-4}

In order to eliminate the above-mentioned phenomenon, in certain actual applications, the method of adopting a multipath light-splitting system for further amplification is used, following which the various paths of amplifying laser pulses are gathered through an optical system. Thus, the total energy of the optical pulses will be equal to the sum of the energy of each of the paths. Not only can high output energy be obtained, but also the influence due to inhomogeneity of the single path amplifying system can be avoided.

In addition, according to literature, there is a kind of "slab laser amplifier" that can eliminate the above-mentioned phenomenon. The structure of such an amplifier consists in recombining the operation material made into thin layers together. As the thin layers look like disks, it is also called the disk amplifier. As the amplification medium is a series composed of a number of thin layers, each of the layers can be surface cooled; that is, the radial refractive index gradient with respect to the light beam can be removed. Thus, the directionality of the light beam will not deteriorate after passing through the amplifier. But the efficiency of the laminated amplifier is lower than that of the rod shaped amplifier. For example, for an amplifier made up of 20 disks, the thickness of each layer is 2.54 cm, the usable cross section for light beam transmission is 29 cm^2, the energy of the input optical pulse is 7.8 J, the total gain is sixfold, and the efficiency reaching a mere 0.05%, over twice lower than an Nd glass rod-shaped amplifier of equivalent dimensions. Therefore, how to raise the efficiency of the laminated combined amplifier is a problem of importance.

An example Consider an Nd:YAG laser amplifier.

To facilitate calculation, transform the previous Eq. (4.2-27) into an expression of which the parameters can be directly measured. The incident energy on unit area is $E_{\text{in}} = I_0 h\nu\tau$; the saturated energy density is defined as

$$E_{\text{s}} = h\nu/\sigma$$

Multiply the right end of the above expression with Δn and we can see the physical meaning of the saturated energy E_{s}, which is just the ratio of the energy E_{ex} extracted from the amplifying medium to the small signal's gain coefficient g_0. Substituting the above equation into Eq. (4.2-27), we obtain the formula for calculating the energy amplification coefficient:

$$G_E = \frac{E}{E_{\rm in}} \ln\left\{1 + \left[\exp\left(\frac{E_{\rm in}}{E_{\rm s}}\right) - 1\right] \exp(\beta E_{\rm st} l)\right\}$$

where $E_{\rm st}$ represents the stored energy in unit volume and β the parameter representing the relationship between gain and stored energy, or $g_0 = \beta E_{\rm st}$.

Because of the high gain of Nd:YAG, the phenomena of spontaneous radiation amplification and self-excitation are apt to appear. This will restrain the density of the stored energy and also the energy that can be extracted from the rod. Under the small signal gain condition, we can find from the relevant table that, for Nd:YAG, $\beta = 4.73\ \text{cm}^2/\text{J}$. If it is desired to extract 500 mJ of energy from a YAG rod of $\phi 6.3 \times 75$ mm, the density of the stored energy should at least reach $E_{\rm st} = 0.2\ \text{J/cm}^3$. Now the gain of the medium $G_E = \exp(\beta E_{\rm st} L) = 1720$.

As Nd:YAG has a high gain, once the gain reaches a definite value, the amplification of spontaneous radiation will greatly consume the population of the laser upper energy level. Meanwhile, scant reflection from the rod end face or the other elements in the optical path will also induce oscillation. These losses will make the curves of relation between the output energy of the YAG amplifier and the input pumping energy become flat. Figure 4.4-7 shows the curves of relation between the output energy of the YAG amplifier and the pumping input energy. It can be seen from these curves that the maximum energy that can be extracted from different amplifying media has a saturated value. Experimental data show that when the rod length is greater than 50 mm, the saturated value of its output energy density is unrelated to the rod length while the density of the stored energy is directly related to the diameter of the rod. Measurement shows that for rods of diameters 5 mm, 6.3 mm, and 9 mm, the maximum energy that can be taken out will be 0.3 J, 0.5 J, and 0.9 J, respectively.

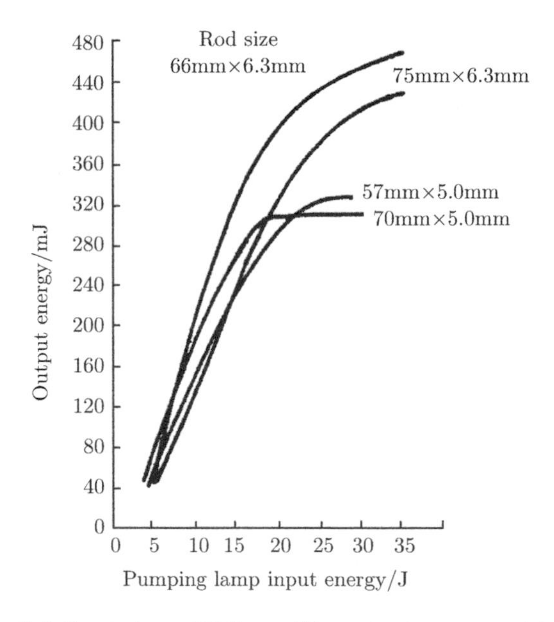

Fig. 4.4-7 Relation between YAG amplifier output energy and input energy

4.5 The regenerative amplifying technology

What has been discussed is laser pulse traveling wave amplifying technology that deals with the single-pass passage of the optical signal to be amplified through the amplifier in the traveling wave mode, the times of amplification being limited. For example, the

Nd:YAG laser amplifier can only amplify the pulse energy 3~6 times. Therefore, if it is required that more amplification times should be made, multistage amplification has to be adopted; Furthermore, in certain applications, a so-called multi-channel traveling wave amplifying system should be adopted; that is, the high-quality light beam output by laser (oscillating) device should be split into several channels which, after getting amplified by the corresponding amplifiers, are gathered together via an optical system. By so doing, high energy laser of fairly good performance can be obtained. For instance, the experimenting device for laser thermonuclear fusion often adopts such an amplifying system.

In the 1980s, a novel regenerative amplification technology was developed, which consists in injecting a faint signal of a light beam of good quality into a laser (oscillating) device. The injected optical signal serves as a "seed" for controlling the generation of laser oscillation; that is, laser oscillation is made to develop on the basis of this "seed" rather than from noise as before and be output from the cavity after getting amplified so that laser of excellent light beam performance and high power can be obtained. Regenerative amplification can be divided into two classes, the externally injected regenerative amplification and self-injected regenerative amplification.

4.5.1 Externally injected regenerative amplification

Optical amplification using this kind of technology is achieved by making a laser (called the driving oscillator) generate faint optical signals of excellent performance and injected into another laser (called the driven oscillator). In terms of its operating characteristics, there are two cases. One is when the gain of the driven oscillator is rather low while the injected optical signal is rather strong. Now, if the frequency ν_1 of the injected signal is very close to the free oscillation frequency ν of the driven laser, then in the course of laser oscillation, the intensity of the injected signal is far greater than the noise of the spontaneous radiation. It appears to be superior in the competition with the laser free oscillation mode, making the frequency of the oscillation mode jump to ν_1 while the free oscillation mode of frequency ν is restrained. The frequency of the output light beam is determined by the externally injected signal. Now, the light intensity output by the driven laser exceeds that during its free oscillation, showing that the injected optical signal has been regeneratively amplified in the driven laser. The other case is one in which the driven laser (e.g., the Q modulated laser) has a rather high gain while the injected optical signal is, by comparison, rather weak. Thus the injected signal and the spontaneous radiation noise in the cavity will increase at the same time, only, in the course of amplification, this injected signal experiences a fast phase shift and moves to the closest longitudinal mode, making this longitudinal mode dominant in the competition with other noises, with the medium gain quickly reaching saturation while restraining the growth of the other modes. Hence, the frequency characteristics of the finally output laser are determined by the driven laser. Such externally injected amplification is usually called injected locking technology.

4.5.2 The injected locking technology

For a low-power laser, a spectrum selecting element can be inserted in its cavity to compress the line width so as to obtain laser of a narrow line width operating in single mode, and at a stable frequency. By contrast, a high-power laser often has a rather broad line width, and operates in multiple modes and at an unstable frequency. By utilizing the injected locking technology, it will be possible to have a low-power laser control a high-power laser so as to obtain high-power laser output of a narrow line width and in single longitudinal mode.

Usually a laser that provides the injected "seed" signal is called the driving oscillator while the "seed" signal accepting laser is called the driven oscillator, as shown in Fig. 4.5-1.

Suppose the frequency of the injected signal is ω_i, and the longitudinal mode frequency closest to the injected signal in the driven oscillator is ω_c. When the injected "seed" signal enters the driven oscillator, the Q switch is on. The eigenmodes of both the injected signal and the driven oscillator itself will form oscillation. If the line width of the injected signal is sufficiently narrow, or much smaller than the interval between longitudinal modes of the driven cavity (Fig. 4.5-2), then the longitudinal mode closest to the injected signal, getting stimulated, will occur resonance with it, which makes it reach saturation earlier than the other longitudinal modes and amplified by extracting energy from the gain medium while the other longitudinal modes not influenced by the injected signal field will still begin oscillation from the spontaneous radiation noise. As the intensity of the injected signal field is much grater than that of the noise field, the ω_c mode forms oscillation first, thereby leading to a decrease in the gain coefficient. Now, because of the competition mechanism of the uniform broadening medium mode, the other longitudinal modes are restrained to be output in single longitudinal mode in the end.

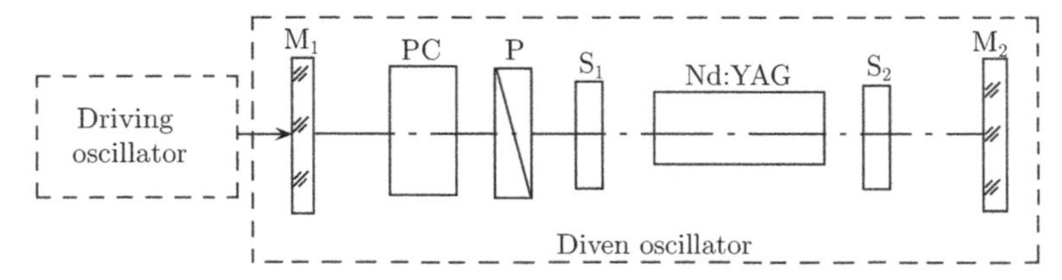

Fig. 4.5-1 Schematic diagram of structure of an injected locking device: M_1, M_2 are cavity mirrors; PC is Q switch crystal; P polarizing prism; S_1, S_2 are 1/4 wave plate

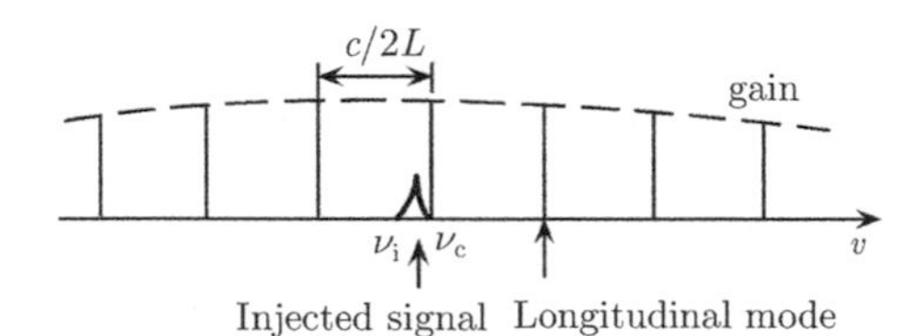

Fig. 4.5-2 Schematic diagram process of operation after injection of "seed"

When the frequency of the injected signal is not in complete agreement with the excited longitudinal mode frequency of the driven oscillator (within the range of mismatch $\Delta\omega$), the transient oscillation at the beginning is at the frequency of the injected signal ω_i. Then, in the course of oscillation, it moves onto the closest resonant longitudinal mode $\omega_c = \omega_i + \Delta\omega$ after experiencing a fast phase shift. So the output single longitudinal mode laser frequency is the longitudinal mode frequency closest to the injected signal rather than the frequency of the injected signal. It has been proved by experiment that the mismatch range of the single longitudinal mode operation of the high gain Q switch is wider than continuous steady-state injected locking range. Therefore, it is possible to satisfy injected locking within a fairly wide range of frequency. At the same time, the range of longitudinal mode selection will increase with an increase in the gain and the intensity of the injected signal.

It is known from the above analysis that the factors affecting the effect of injected locking are mainly:

(1) To realize injected locking, definite requirements are made on the power density of the injected signal, the mismatch amount, and the time at which the Q switch is to be turned on. If the power density of the injected signal is too low, lock losing will occur and multiple longitudinal mode oscillation will be generated. This is not hard to understand in terms of the physical meaning because when the injected signal field is too weak, its effect on the cavity mode field close by will decrease. When the Q switch is turned on, it cannot reach

the degree high enough to restrain the other longitudinal modes. So, in the competition between modes, it cannot win superiority while leading to multimode-oscillation output.

(2) For a definite injected power density, the influence of the injected field on the cavity mode field closest to it will decrease with an increase in the mismatch amount $\Delta\omega$. When the mismatch amount is increased to a definite degree, there will occur lock losing till the appearance of multiple longitudinal mode oscillation.

(3) For a definite intensity of the injected signal and amount of mismatch, turning on the Q switch early or late will affect the effect of the injected locking. Only when the peak value of the injected pulse signal is so controlled that it can realize optimum match with the time at which the Q switch is turned on, will it be possible to achieve optimum efficiency. If the seed signal is injected before the Q switch is turned on, then the net gain in the optical field of the driven cavity is equal to zero and the injected signal will get attenuated, the subsequent behavior being identical with that when there is no injected signal. Hence the occurrence of multiple mode oscillation. On the contrary, if the seed signal is injected too late, that is to say, injected after the Q switch is turned on, then the optical fields at the longitudinal modes in the driven cavity will be stronger than the injected signal, or comparable with it. Now, the injected signal is already incapable of restraining the other longitudinal modes through mode competition, so locking cannot be realized.

Thus two conclusions can be obtained. First, under the condition the frequency width of the injected signal is smaller than the longitudinal mode frequency interval of the driven oscillator, to effectively realize injected locking, it should be noted that the cavity mode match between the driving oscillator and driven oscillator is the necessary condition for generating injected locking since only under such a condition will it be possible to make injected field play the role of resonance with longitudinal mode field close to it. Now the injected field plays the role of coherent pumping in forcing the mode to oscillate. Second, to realize injected locking, it's important to inject sufficient seed power density and ensure small mismatch amount and appropriate switching time[13].

4.5.3 The self-injected amplifying technology

While externally injected amplifying technology is designed to help obtain high-power laser output of good quality light beam by making a laser generate the "seed" pulse signal to be injected into another laser, the self-injected amplifying technology works by using a laser itself to generate the "seed" signal to be self-injected into the cavity to realize regenerative amplification, thus greatly reducing the volume of the laser device. Figure 4.5-3 shows the schematic diagram of an intracavity self-injected amplifying device and the operation principle. Insert a Pockels cell PC_2 in a depressurized Q modulated laser and, with PC_2 as the demarcation, the resonator is divided into two sections L_1 and L_2; $L = L_1 + L_2$ is the length of the resonator, Pockels cell PC_1 is the Q-switching device, PC_2 is used to generate the "seed" pulse for injected amplification, and P is the polarizing prism, located in section L_2. M_1 and M_2 are cavity mirrors. At the beginning, during the time the xenon lamp is pumping the operation material to store energy, apply $V_{\lambda/4}$ on PC_1, to put the resonator in the "off" state so as not to form oscillation. When the stored energy reaches maximum, or at time t_0, remove $V_{\lambda/4}$ on PC_1, turn on the Q switch, and begin to set up laser oscillation. After an appropriate delay time t_d (before laser oscillation reaches peak value) apply half-wave voltage ($V_{\lambda/2}$) on Pockels cell PC_2. Now, the linearly polarized light originally located in section L_1 passes through PC_2 once, the polarizing direction changes by 90°, and, upon reaching the polarizing prism P, gets refracted and escapes from the cavity while the light originally located in section L_2 twice passes through crystal PC_2 with a half-wave voltage applied on it in its round trip in the cavity and, with the polarizing direction not changed, it remains in the cavity and, as the seed pulse, passes through the laser operation material many times and gets amplified. Thus a sequence of pulses is output at the end of the output mirror M_2, its envelope being in the shape of Q-switched waveform. If the voltage $V_{\lambda/4}$ on PC_1 is restored at time t_1, then cavity dumped single pulse, output can be obtained.

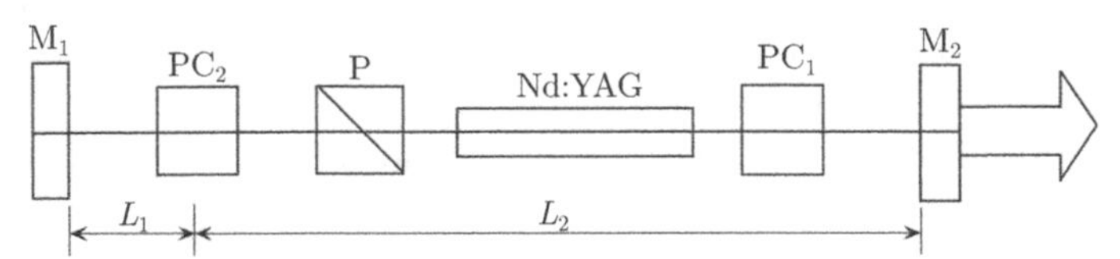

Fig. 4.5-3 Diagram of the self-injected amplifying device and principle: M_1, total reflection mirror; M_2, output mirror; P, polarizing prism; PC_1, Q modulating electro-optic crystal; PC_2, self-injection electro-optic crystal

Suppose the half-wave voltage applied on Pockels cell PC_2 is an ideal square wave whose front and rear edges are both very steep without considering the variation of the pulse width in the course of amplification. It has been found from an analysis that, when selecting the different cavity structural parameter values L_1, L_2, L, and T_g (width of seed pulse), different output pulse widths Δt can be obtained, while the condition for the seed pulse width to be minimum is

$$\frac{2L}{c} \leqslant T_g < \frac{2(L+L_1)}{c}$$

Then the corresponding output pulse width $\Delta t = \dfrac{2(L+L_1)}{c} - T_g$. So, by changing the cavity structural parameters of the self-injected regenerative amplifier, it will be possible to obtain narrow pulse laser output. In addition, the delay time t_d from turning on the Q switch at the beginning to the application of the half-wave voltage on PC_2 should appropriately be controlled. When t_d is great, the superiority won by stimulated radiation will be apparent and the signal-noise ratio high. But t_d should not be greater than the delay time needed by critical avalanche, otherwise what is obtained will be Q-switched laser pulse. Therefore, it is necessary to apply the self-injected pulse voltage on the Pockels cell PC_2 before the Q-switched laser reaches peak value, which is the important condition for obtaining sequential pulse output.

Example Choose the resonator structural parameters $L = 180$ cm, $L_1 = 10$ cm, $T_g = 16$ ns, and $t_d = 17$ ns. Under the condition of a pumping energy of 74.4 J, the values of energy and power in three different cases, viz. the Q-switched pulse, self-injected sequential pulse, and self-injected cavity dumped single pulse, have been obtained, respectively, as shown in Tab. 4.5-1.

Tab. 4.5-1 Comparison between energy and power

	Q-switched	Self-injected sequential pulse	PTM single pulse
Energy/mJ	74.6	70.8	47.8
Peak power/MW	1.78	5.90	16.0
Pulse width/ns	42	3	3

It can be seen from the table that the energy of the self-injected sequential pulse and that of PTM single pulse are 94% and 64% that of the Q-switched pulse, respectively, but the peak power has been increased multiple times. It is clear that the self-injected regenerative amplifier is a laser capable of obtaining narrow pulse width and high peak power.

4.6 The semiconductor laser amplifier

With the development of fiber-optic communication, there's an urgent need for increasing the intermediate distance and capacity of communication. For conventional long-distance fiber-optic transmission of information, it is required that a regenerative relay be mounted a definite distance apart and the transmission can be continued only after the "opto-electro-optic" conversion of an optical signal has been performed for regeneration. Obviously, such a mode is tedious. It has all along been imagined that the signal can be directly amplified on the optical path to realize "all optical" communication. Hence the emergence of quite a few optical amplifying technologies in recent years. So far, there have appeared mainly the semiconductor laser amplifier, fiber Raman amplifier, and rare earth-doped (mainly erbium-doped) fiber amplifiers. This section will first of all discuss the semiconductor laser amplifier.

The semiconductor laser amplifier for optical fiber communication is mainly divided into two forms, the Fabry-Perot semiconductor laser amplifier (FP-SLA) and the traveling wave semiconductor laser amplifier (TW-SLA); the former is in effect a semiconductor laser biased below the threshold and what is amplified is an incoming optical signal. The photons go back and forth in the resonator of the laser many times and can obtain a rather high gain, but the bandwidth of the gain is rather narrow (basically the line width of an F-P cavity longitudinal mode). To ensure that the incident optical signal will obtain a rather great gain, it is required that the device have a sufficiently high temperature stability, while the latter has in effect performed ideal antireflection of the cleavage surface of the semiconductor laser, with the incident signal getting amplified only by one path in such an amplifier. So it is required that the gain be high. This is realized by relying on the driving current of the amplifier to increase to 2~3 times the pre-antireflection threshold current. It possesses a very wide gain bandwidth (3 orders of magnitude higher than that of FP-SLA). Therefore, the requirement on temperature stability of the amplifier is lower than on that of the former. The role of optical signal amplification by the semiconductor laser amplifier is essentially obtained from the stimulated emission mechanism generated by the interaction between the photons and electrons in the gain medium. The cavity gain obtained by the optical signal in the traveling wave semiconductor laser amplifier can be represented by

$$G(\nu) = \frac{(1-R_1)(1-R_2)G_\mathrm{s}}{\left(1-\sqrt{R_1R_2}G_\mathrm{s}\right)^2 + 4\sqrt{R_1R_2}G_\mathrm{s}\cdot\sin^2(\pi\Delta\nu/\Delta\nu_\mathrm{s})} \tag{4.6-1}$$

where R_1 and R_2 are reflectivities of the laser amplifier's incident plane and outgoing plane, respectively; $\Delta\nu$ is the cavity's longitudinal mode interval; G_s is the gain by one path undergone on the optical signal.

$$G_\mathrm{s} = \exp[(\Gamma g - \alpha)L] \tag{4.6-2}$$

where Γ is the light field limiting factor in the active region of the amplifier; g and α are the active medium's gain coefficient and loss coefficient; and L is the length of the gain medium. In an ideal case, $\nu = \nu_0$ and $R_1 = R_2 = R$. Thus Eq. (4.6-1) can be reduced to

$$G(\nu) = \frac{(1-R)^2G_\mathrm{s}}{(1-RG_\mathrm{s})^2} \tag{4.6-3}$$

It can thus be seen that when two cleavage surfaces are completely antireflected (that is, $R_1 = R_2 = 0$), $G(\nu) = G_\mathrm{s}$, or in the traveling wave amplifier, the incident light signal can only obtain gain by one path in the amplifier while the traveling wave amplifier can permit operation under the condition of threshold value current 2~3 times higher than that before antireflection, so very high gain by one path can be obtained.

Although the semiconductor laser amplifier has very high gain, it suffers very great loss getting coupled with optical fiber (reaching up to 5 dB or so). Furthermore, the gain is very sensitive to the polarization of optical fiber and the ambient temperature. So its stability is poor and it is suitable for use in combination with the opto-electrically integrated circuit.

Below we shall describe the saturation characteristics and output power of the semiconductor laser amplifier using the theory of the rate equation. With the single mode coupled rate equations we shall be able to describe the relation of variation of the carrier concentration N with the injected current I and signal power P, that is,

$$\frac{dN}{dt} = \frac{I}{q} - \frac{N}{\tau_c} - R_{st}P \tag{4.6-4}$$

$$\frac{dP}{dt} = R_{st}P + R_{sp} - \frac{P}{\tau_p} \tag{4.6-5}$$

$$R_{st} = \Gamma v_g g = R_N(N - N_0) \tag{4.6-6}$$

where R_{st} is the rate of the stimulated radiation, v_g is the group velocity, R_{sp} is the rate at which the spontaneous radiation enters the signal mode, g is the peak gain, q is the quantity of electron charge, τ_c is the lifetime of carrier, and τ_p is the lifetime of photon, which can be expressed as

$$\tau_p = \left\{c\left[(1/L)\ln\left(1/\sqrt{R_1R_2}\right) + \alpha_{fc}\right]\right\}^{-1} \tag{4.6-7}$$

where c is the speed of light, L the length of the amplifier, and α_{fc} the absorption loss of the free carrier.

First let's find the steady-state solution for the set of equations. For continuous wave amplification or when the pulse width (τ) of the light wave is far greater than τ_c, the steady-state solution can be used for description. Now, letting $dN/dt = 0$, we can obtain the steady-state solution for N as

$$N = \tau_c\left(\frac{I}{q} - R_{st}P\right) \tag{4.6-8}$$

We can suppose from experience that the peak gain is in the following linear relation with the carrier concentration N:

$$g = (\Gamma\sigma_g/V)(N - N_0) \tag{4.6-9}$$

where Γ is the limiting factor, σ_g is the differential gain coefficient or the gain cross section, V is the volume of the active layer, and N_0 is the concentration of the transparent carrier, representing the concentration of the carrier when the stimulated radiation rate is equal to the stimulated absorption rate. Substituting Eq. (4.6-8) into Eq. (4.6-9), we find that when

$$g = \frac{g_0}{1 + (P/P_s)} \tag{4.6-10}$$

the optical gain will get saturated. This shows that the gain coefficient of the semiconductor optical amplifier has a saturation mechanism similar to that of the gain coefficient of the homogeneously broadened two-level laser system. In Eq. (4.6-10), g_0 is the small signal gain, whose expression is

$$g_0 = (\Gamma\sigma_g/V)(I\tau_c/g - N_0) \tag{4.6-11}$$

where P_s is the saturation power, whose expression is

$$P_s = h\nu\sigma_m/\sigma_g\tau_s \tag{4.6-12}$$

where σ_m is the cross section of the waveguide mode.

Proceeding from Eq. (4.6-10), we can calculate the output power and saturated gain of the amplifier. Suppose the variation rate of the optical power of the semiconductor laser amplifier with time is $dP/dt = gP$. Suppose the length of the gain medium is L, the initial input power $P(0) = P_{in}$, and the output power $P_{out} = P(L) = GP_{in}$. After simple integration, the expression for the large signal gain G is found to be

$$G = G_0 \exp\left(-\frac{G-1}{G} - \frac{P_{out}}{P_s}\right) \tag{4.6-13}$$

where $G_0 = \exp(g_0 L)$ is the small signal gain when $P_{out} \ll P_s$. It can be seen from the above equation that G decreases with the output power, as shown in Fig. 4.6-1.

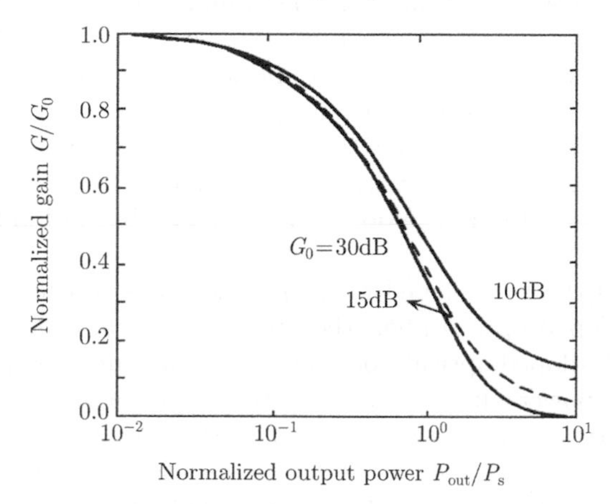

Fig. 4.6-1 Variation of normalized gain of semiconductor laser amplifier with normalized output power

4.7 The rare earth element-doped fiber amplifier

The rare earth element-doped fiber amplifier realizes light amplification by using the gain mechanism induced by doping rare earth elements (such as erbium, praseodymium, neodymium, etc.) in optic fibers. The emergence of the erbium-doped fiber amplifier in the mid-1980s is a significant breakthrough in the development of the fiber amplifier. Its operation wavelength exactly falls within the region of optimum length (1.31~1.55 μm) of fiber optic communication, the gain is rather high, and the pumping power needed is fairly low (<100 mW). As the erbium-doped optical fiber itself is a gain medium, the loss in coupling with a line is very small and the noise low.

Because of its unique merits, the fiber amplifier is pretty suitable for relay amplification in the lines of the fiber optic communication system, optical power amplification of the transmitter, and the preamplification of the receiver. By relay amplification is meant insertion of the optical fiber to amplify the signal to compensate for the transmission loss of the optical fiber so as to lengthen the transmission distance between two relay stations, thus replacing the conventional "light-electricity-light" conversion. Thus, not only can cost be reduced and trouble mitigated, but also, in the event the system's transmission rate and mode of modulation should change, there is no need to change the line. The end optical power amplifier of the transmitter is directly arranged in the rear of the laser, whose main function is amplify the signal optical power to maintain the intensity of the modulating signal. As further development of the optical fiber communication is aimed at realizing the comprehensive transaction net, large numbers of power dividers are needed to dispatch information. In order to compensate for the insertion losses of the couplers, optical power amplifiers are also needed. The receiver's preamplifier is used to place the fiber amplifier in front of the receiver, its function being to raise the sensitivity of the receiver and improve the minimum detectable power. As in ordinary detecting systems, the sensitivity of a receiver is limited by the thermal noise of the device itself and the electronic circuit. The addition of a preamplifier will make things better.

4.7.1 The erbium-doped fiber-optic amplifier

The operation mechanism of the erbium-doped fiber amplifier lies in the stimulated emission of the Er^{+3} ions (the substrate material of the fiber amplifier is silicon-based glass or fluoride glass plus other oxides doped). Erbium is a 3-level system, of which level 1 is the ground state level, level 2 is the metastable state, and level 3 the high-energy state. When a high energy pump laser is used to stimulate erbium-doped optical fiber, the bond electron absorption energy of the erbium ions can be made to jump from the ground state to the high energy state and quickly relax to the metastable state with a long lifetime. As population inversion distribution is formed between level 2 and level 1, stimulated radiation will occur, emitting a light band with a very wide range of spectrum. If the wavelength of the incident signal light falls exactly in the above-mentioned light-emitting band, the signal light will be amplified after obtaining energy via stimulated radiation. As the signal light and the pump light are both transmitted in the optical fiber in the form of guide wave, energy can be effectively exchanged.

The basic structure of the Er^{3+}-doped fiber amplifier is as shown in Fig. 4.7-1, which is composed of the Er^{3+}-doped optical fiber, the pump source, and the wave splitting multiplexer, of which the Er^{3+}-doped optical fiber is the gain medium; the pump source provides sufficiently strong pump power while the wave splitting multiplexer is a device for increasing the capacity of information transmission, which performs the mixed function of the signal light and pump light. There are three modes of configuration for the fiber amplifier in terms of the transmission direction of the signal light relative to the pump light. When the signal light and pump light are injected into the optical fiber along the same direction, the pumping is called same-directional pumping. When the signal light and pump light enter the optical fiber from two different directions, the pumping is called reverse pumping. When two streams of identical pump light are made to enter the optical fiber from two directions, it is called bidirectional pumping. This mode possesses very high output signal power and the performance of the fiber amplifier is unrelated to the transmission direction of the signal at that.

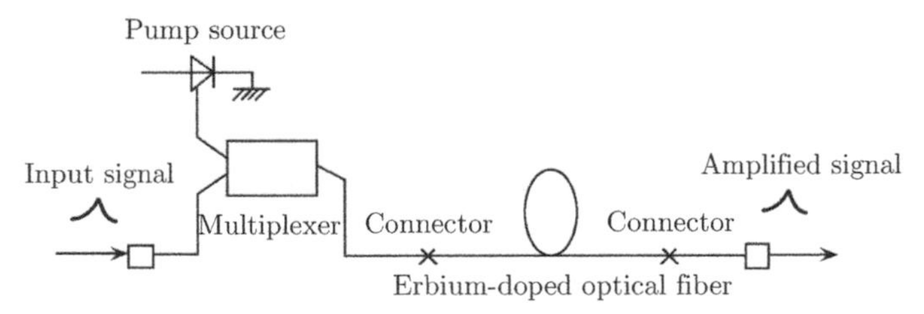

Fig. 4.7-1 The structure of erbium-doped fiber-optic amplifier

The principal technical linchpins of the Er^{3+}-doped fiber-optic amplifier are the Er^{3+}-doped optical fiber and the pumping source. For the fiber amplifier, apart from the requirement for a high gain coefficient and pumping efficiency and high saturated output power, a rather wide and flat gain spectrum is also desirable to adapt to the wave splitting multiplexing system and the high temperature stability. Therefore, in addition to the Er active ions doped in the optical fiber, other oxides such as Al_2O_3 and GeO_2 should also be doped. Meanwhile, under the given pumping power, the length of the optical fiber should be so chosen that it is within the optimum range to enable the signal light to effectively extract energy from the pumping light. An optical fiber exceeding this length, however, will form re-absorption with respect to the signal, thereby limiting the gain of the optical fiber. In general, the length of the optical fiber from several meters to dozens of meters is chosen. The pumping source is required to be of high power and a long lifetime. By taking into consideration the pumping efficiency and the currently available semiconductor laser pumping source, 980-nm and 1480-nm wavelengths are preferable, depending on the core diameter of the active optical fiber, the way doping is made, the refractive index between the fiber core and envelope, etc.

4.7.2 The praseodymium-doped fiber amplifier

The Er^{3+}-doped fiber amplifier has provided a fiber low loss window of 1550 nm wavelength with good quality devices of high gain, high power, high broad band, low noise, and found wide application in the modern optical fiber communication system. It has been found from research that Nd-doped optical fiber exhibits the capability of stimulated amplification in the 1300-nm wavelength band. But as it has very strong spontaneous radiation at 900 nm and 1050 nm and very strong absorption around 1300 nm wavelength, the development of the fiber amplifier as one of a wavelength of 1300 nm is restrained. It has been found by investigation that the Pr-doped optical fiber also has the capability of stimulated amplification of the optical signal with a wavelength of 1300 nm as well as the characteristic of broad band and high gain amplification. Meanwhile, 1300 nm wavelength is the communication window in the optical fiber communication system; the R&D of fiber amplifiers of 1300 nm wavelength is of great significance.

Figure 4.7-2 shows the schematic diagram of the structure of an engineering Pr-doped fiber amplifier developed by the BT Laboratory in the U.K. in which the pumping source adopts a GaAs broad strip-shaped waveguide structured laser of wavelength 800 nm and output power 4 W to pump an Nd:YLF laser of wavelength 1047 nm, output power 900 mW. The amplifying medium is Pr-doped optical fiber of concentration 1000 ppm, and length 25 m. Similar to the Er-doped fiber amplifier, the Pr-doped fiber amplifier also possesses excellent characteristics such as high gain, broad frequency band, and low noise. But its conversion efficiency is rather low. When making Pr-doped optical fiber design, adopting high numerical value aperture and low loss, the small signal gain can reach up to 30 dB and the optical bandwidth up to 30 mm, and the highest small signal conversion efficiency up to 0.22 dB/mW.

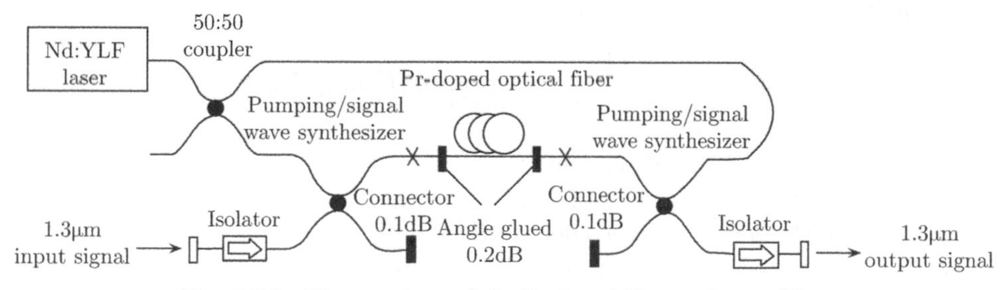

Fig. 4.7-2 The structure of the Pr-doped fiber-optic amplifier

4.8 The distributed fiber amplifier

The rare earth elements-doped fiber amplifiers discussed in the preceding section possess such characteristics as high doping concentration (10~1000 ppm), and short length (1~10 m). Compared with the fiber-optic communication system, the length is very short. Hence the name of the lumped fiber-optic amplifier. When accessing such an amplifier into a fiber amplifying system, the result will be very great fluctuation of the signal, which is apt to the nonlinear effect, reduce the signal-to-noise ratio causing crosstalk. In order to reduce such harmful effects, another kind of fiber amplifier known as the distributed fiber-optic amplifier has been proposed, the amplifying medium of which can be the erbium-doped optic fiber with a very low doping concentration (0.05~0.1 ppm) and a very long length (1000 m~100 km), with the stimulated amplifying effect used to amplify the optical signal. Or, it can also be ordinary optic fiber based on the nonlinear optical effect, such as Raman effect and Brillouin effect to amplify the optical signal. Such a distributed fiber amplifier is capable of effectively overcoming the shortcomings of the centralized fiber-optic amplifier to enable the signal in fiber-optic communication to be "transparently" transmitted loss-free. Below we shall focus on the distributed fiber-optic amplifier based on the nonlinear effect.

4.8.1 The fiber-optic Raman amplifier

As a nonlinear medium, optic fiber is capable of confining the interaction between the strong laser field and the medium within a very small cross section, thus greatly increasing the optic power density of the incident light field. The length of action of light with medium can be maintained at a very long distance to enable the energy to be fairly sufficiently coupled. The stimulated Raman scattering in optic fiber possesses the characteristic of low threshold and high gain. If an incident light signal is transmitted in the optic fiber together with the pump light, and the wavelength of the signal light is obtained from the gain mechanism generated when falling exactly in the Raman scattering effect region, the case is called fiber-optic Raman amplification. The structure of the fiber-optic Raman amplifier is just the same as that of a fiber laser without reflective mirrors. In a structure with pumping in the same direction, the pump light and signal light propagate along the same direction whereas in a structure of pumping in opposite directions, the propagating directions of the two beams of light are opposite.

If the incident signal light intensity I_s is much smaller than the pumping light intensity I_0, and pumping exhaustion can be neglected, then the signal output at L is:

$$I_s(L) = I_s(0)\exp(g_R I_0 I_{eff} - \alpha_s L) \tag{4.8-1}$$

In the absence of the pump light, then

$$I_s(L) = I_s(0)\exp(-\alpha_s L) \tag{4.8-2}$$

So, the gain of the fiber Raman amplifier is

$$G_s = \frac{I_s(L)}{I_s(0)\exp(-\alpha_s L)} = \exp(g_R P_0 L_{eff}/A_{eff}) \tag{4.8-3}$$

where I_s is the intensity of the Stokes' light; g_R is the Raman gain coefficient; α_s is the loss of light; L is the actual length of optical fiber; L_{eff} is the effective length of optical fiber when the loss of the pump light is not zero; I_0 is pump light intensity at the input end of the optic fiber; $I_s(0)$ is the Stokes' light intensity at the optical fiber input end; and $P_0 = I_0 A_{eff}$ the amplifier's input pump power. For instance, if $\lambda_p = 1.017$ μm is used as the pump source to amplify λ_s to an optical signal of 1.064 μm, with the optical fiber $L = 1.3$ km, the curve of relationship between the Raman amplifier's gain and pumping power is as shown in Fig. 4.8-1. This figure shows that, at the beginning, G_s increases exponentially with an increase in P_0, but later, it deviates from the exponential curve when $P_0 > 1$ W, which is gain saturation induced by exhaustion due to pumping. And it can be seen that, when $P_0 \approx 1$ W, the gain of the fiber optic Raman amplifier is very high.

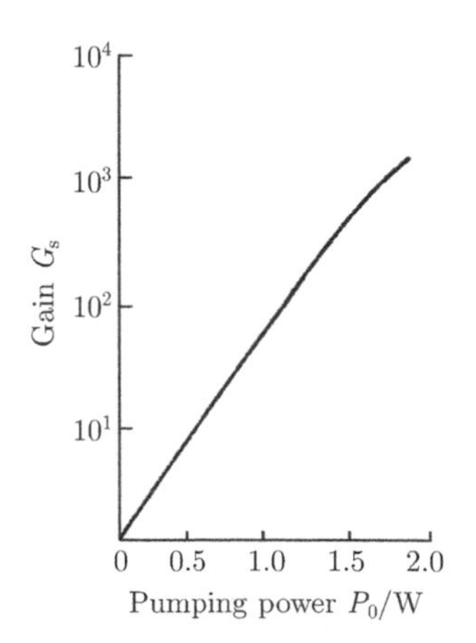

Fig. 4.8-1 The relationship between G_s of fiber-optic Raman amplifier and P_0

The advantage of the fiber-optic Raman amplifier lies in that the transmission line and amplification proceed in the optical fiber alike. Hence the coupling loss is very small, the noise rather low, and the gain fairly stable. But rather high pump power (several hundred mW and above) and very long optical fibers (several km) are needed.

4.8.2 The fiber-optic Brillouin amplifier

The stimulated Brillouin scattering is another important nonlinear process similar to stimulated Raman scattering. The incident pump power required by such a nonlinear process is far lower than that required by stimulated Raman scattering. Once the stimulated Brillouin threshold is reached, stimulated Brillouin scattering will transform most of the input power into backward Stokes' light that will be detrimental to the fiber-optic communication system. But such a scattering effect can also be used to construct a fiber-optic Brillouin amplifier to be used in the fiber-optic communication system as a distributed fiber-optic amplifier.

In the process of stimulated Brillouin scattering, the signal light derives energy from the pumping light via the scattering gain to get amplified while getting attenuated when absorbed by the optical fiber. The pump light gets attenuated when transferring energy to the signal light through the scattering process. The coupled equations for the interaction between the pump light and Stokes' light are

$$\frac{\mathrm{d}P_{\mathrm{s}}}{\mathrm{d}z} = -\left(\frac{g_{\mathrm{B}}}{\alpha}\right) P_{\mathrm{p}} P_{\mathrm{s}} + \alpha P_{\mathrm{s}} \tag{4.8-4}$$

$$\frac{\mathrm{d}P_{\mathrm{p}}}{\mathrm{d}z} = -\left(\frac{g_{\mathrm{B}}}{\alpha}\right) P_{\mathrm{p}} P_{\mathrm{s}} - \alpha P_{\mathrm{p}} \tag{4.8-5}$$

where P_{p}, P_{s}, g_{B}, and α stand for the pumping light power, Stokes' light power, Brillouin scattering gain, and the fiber-optic absorption coefficient, respectively. Under the small signal condition $P_{\mathrm{s}} \ll P_{\mathrm{p}}$, neglecting the first term on the right-hand side of Eq. (4.8-5) and solving Eqs. (4.8-4) and (4.8-5) simultaneously, we obtain

$$P_{\mathrm{s}}(0) = P_{\mathrm{s}}(L)\exp\{[g_{\mathrm{B}} P_{\mathrm{p}}(0) L_{\mathrm{eff}}/\alpha] - \alpha L\} \tag{4.8-6}$$

where $L_{\mathrm{eff}} = 1/\alpha$. It can be seen that the backward Stokes' light power rises exponentially with the propagation distance.

With the amplification of the backward Stokes' light by the pump light, a greater part of the power of the pump light is transferred to the Stokes' light. Now the small signal approximation can no longer be satisfied. For this reason, it is necessary to accurately solve the coupled Eqs. (4.8-4) and (4.8-5), whose general solutions are

$$P_{\mathrm{s}}(z) = \frac{b_0(1-b_0)}{G(z)-b_0} P_{\mathrm{p}}(0)\exp(-\alpha z) \tag{4.8-7}$$

$$P_{\mathrm{p}}(z) = \frac{(1-b_0)G(z)}{G(z)-b_0} P_{\mathrm{p}}(0)\exp(-\alpha z) \tag{4.8-8}$$

where

$$G(z) = \exp\{(1-b_0)(g_0/\alpha)[1-\exp(-\alpha z)]\} \tag{4.8-9}$$

$$b_0 = P_{\mathrm{s}}(0)/P_{\mathrm{p}}(0), \quad g_0 = g_{\mathrm{B}} P_{\mathrm{p}}(0)/\alpha \tag{4.8-10}$$

The parameter b_0 stands for Brillouin scattering efficiency, showing how much pump power has been transformed into Stokes' light power. Fig. 4.8-2 shows the law of variation of pump light and Stokes light power with the variation of the transmission distance.

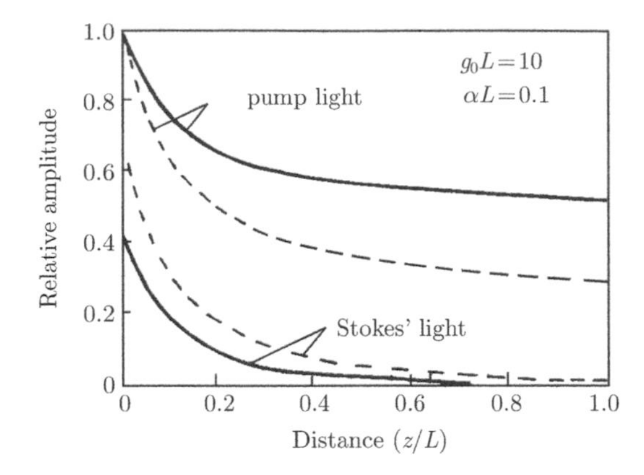

Fig. 4.8-2 Pump light and Stokes' light vary with distance in fiber-optic Brillouin amplifier: $b_{\text{in}} = P_{\text{s}}(L)/P_{\text{p}}(0) = 0.001$ and 0.01 corresponds to the solid and dashed lines

The fiber-optic Brillouin amplifier is a kind of amplifier for tunable light of high gain, low power, and narrow band. Low power is a very great merit, which makes it possible to adopt the low-power semiconductor laser diode as the source of pump light to obtain a gain of dozens of db. The adjustable narrow bandwidth enables it to be used in the optical frequency split multiplexing system as a narrow band filter, frequency selecting light amplifier, and photolysis multiplexer, and in the coherent and multi-channel light wave communication system. Although the low threshold value and high gain characteristics of the fiber-optic Brillouin amplifier are favorable factors, they are detrimental to the multi-channel system as it will restrain the channel interval from becoming too narrow and the number of channels from getting too big. Also, it will restrain the signal power from getting too high, thus restraining the increase in the communication distance. It is customary to adopt the method of increasing the width of the spectral line of the pump light source and raising the threshold power of the stimulated Brillouin scattering to reduce this influence.

Exercises and problems for consideration

1. Compare the similarities and dissimilarities between the laser oscillator and amplifier.

2. Make an analysis of the characteristics of small and large signals after going through an amplifier (analyze the physical meanings of the energy gain formulae in three different cases).

3. Why can an amplifier narrow the pulse width? How is it different from narrowing the pulse width by mode-locking?

4. A square optical pulse signal of 694.3 nm is input in a ruby laser amplifier. It is known that the signal satisfies $2\sigma I_0 \tau = 1$, and the amplifier medium length $L = 2/\sigma\Delta n_0$. What is the power gain at the leading edge and what is that at the trailing edge of the pulse after passing through the amplifier?

5. A YAG laser amplifier with $\Delta n_0 = 6 \times 10^{-17}\text{cm}^{-3}$, $\sigma_{12} = 5 \times 10^{-19}\text{cm}^{-2}$ is about to amplify a square optical pulse. It is known that the cross section of the light beam is 0.5 cm^2, the photon energy $h\nu = 1.86 \times 10^{-19}$ J, and the pulse width is 10 ns, energy is 50 mJ. If it is required to be amplified to 200 mJ, how long should the amplification medium be?

6. There is a neodymium glass long pulse amplifier, whose rod is 20cm long ,the medium's loss coefficient $\delta = 0.005\text{cm}^{-1}$, the stimulated radiation cross section $\sigma_{21} = 3.03 \times 10^{-20}$ cm^2, spontaneous radiation probability $A = 3.3 \times 10^4$ s^{-1}, the total population density $n = 2.83 \times 10^{20}$ cm^{-3}, the population density of the upper energy level $n_2 = 2.83 \times 10^{18}$ cm^{-3}, and the ratio of the pumping probability W_{p} to A is approximately equal to n_2/n. Try to find the small signal gain.

References

[1] Laser Physics, compiled by Compiling Group of Laser Physics, Shanghai, Shanghai People's Press, 1976.

[2] W. Koechner, Solidf-state Laser Engineering, New York, Springer Verlag, 1976 (translated by Huaguang, Beijing, Science Press, 1983).

[3] F. T. Arechi, E. O. Schulz-Dubois, Laser Handbook, North Holland Publishing Company, 1972.

[4] Laser Technology, compiled and translated by Tianjin University, Beijing, Science Press, 1972.

[5] An Introduction to Solid-state Laser, compiled by Group of Solid-state Laser, Shanghai, Shanghai People's Press, 1974.

[6] L. S. Earl and C. D. Walter, Laser Amplifiers, *J. Appl. Phys.*, 1965, 36(2), pp. 348–351.

[7] A. Y. Cabezas, G. L. Mallister, Gain Saturation in Neodymium: Glass Laser Amplifiers, *J. Appl. Phys.*, 1967, 38(9), pp. 3487–3491.

[8] Huang Dexiu et al., Theory and Experimental Study of Semiconductor Laser Amplifiers, Laser in China, 1987, 15(7).

[9] A. E. Siegman, Lasers, Printed in the United Stat es of America, 1986.

[10] Wang He and Fan Dianyuan, Generation of Pulse-width Adjustable Nanosecond Optic Pulses with the Intracavity Self-injection Technique and Ferrite Line Technique, Laser in China, 1985, 3(12).

[11] Wu Hongxing et al., Research on Self-injected Regeneratively Amplified Nd:YAG Lasers, J. of Chinese University of Science and Technology, 1993.

[12] W. H. Lowdermilk et al., The Regenerative Amplifier, Laser and IR, 1980.

[13] Lan Xinju et al., Laser Devices and Techniques (II), Wuhan, Huazhong University of Science and Technology Press, 1991.

[14] Y. Yamamoto, Characteristics of AlGaAs Fabry-Perot Cavity Type Laser Amplifiers, *IEEE J. Quantum Electronics*, 1980, 16(10), pp. 1047–1058.

[15] Yang Xianglin et al., Light Amplifiers and Their Applications, Beijing, Electronic Industry Press, 2000.

[16] T. J. Whitley, A Review of Recent System Demonstration Incorporating 1.3 μm Praseodymium-doped Fluoride Fiber Amplifiers, *J. Lightwave Technology*, 1995, 13(5), pp. 744–760.

CHAPTER 5

The Mode Selecting Technology

5.1 Overview

Many application fields of laser require that the laser beam possess very high beam quality (i.e., very good directionality and monochromaticity). But ordinary lasers can hardly meet such a requirement. The method of further enhancing the light beam quality is to select the mode of the laser resonator. The mode selecting technology can be divided into two major classes. One is the transverse mode selecting technology, which can choose the fundamental transverse mode TEM_{00} from among the oscillation modes while restraining other higher-order mode oscillations. The fundamental mode diffraction loss is minimum and energy is concentrated around the cavity axis to make the light beam's angle of divergence compressed, thereby improving the directionality. The other class is longitudinal mode selecting technology, which can limit the number of oscillating frequencies in multiple longitudinal modes and choose the single longitudinal mode oscillation, thus improving the monochromaticity of laser.

5.1.1 The concept of mode, the general association of laser resonator with mode

Only a specific light beam with definite oscillation frequency and definite spatial distribution is capable of forming the "self-reconstruction" oscillation in the cavity.

In the terminology of laser technology, such a specific light beam that may likely exist in an optical resonator is generally referred to as the mode of the cavity.

Different resonators possess different oscillation modes, hence selecting different resonators enables one to obtain different forms of the output light beam.

The stable field distribution that exists within the transverse X-Y plane perpendicular to the propagation direction Z of the intracavity electro-magnetic field is called the transverse mode. Different transverse modes correspond to different transversely stable light field distributions and frequencies.

The mode of laser is in general marked by TEM_{mn}, where TEM represents the transverse electro-magnetic field, while m and n are the ordinal numbers of the transverse mode, denoted by positive integers, which depict the number of nodal lines on the mirror surface.

We call $m = n = 0$, or TEM_{00} the fundamental mode, which is the simplest structure of the speckle with the field of the mode concentrated in the center of the reflective mirror. The other transverse modes are spoken of as higher-order transverse modes, or the field nodal lines (i.e., the position of amplitude 0), which will appear as a field on the surface of the mirror and the "center of gravity" of field distribution will be close to the edge of the mirror.

What is shown in Fig. 5.1-1 is a number of photographs of low-order transverse mode light field intensity distribution. It is not hard to see that the higher the order of the transverse mode, the more complicated the light intensity distribution and the greater the range of distribution will be, hence the greater the light beam divergence angle. Conversely, if the light intensity distribution pattern of the fundamental mode (TEM_{00}) appears to be circular and the range of distribution is very small, its light beam divergence angle is minimum and the power density maximum. Hence the highest brightness. In addition, the radial intensity distribution of such a mode is uniform. There are many applications that not only require a high laser power, but also a small divergence angle. So it is necessary to endeavor to choose single mode laser. For instance, in refined laser machining applications (e.g., welding, drilling, trimming, etc), it is required that the laser beam after getting focused should possess very small light spot. It is known from applied optics that its spot diameter $d = f\theta$ (f is the focal distance of the lens, θ is the divergence angle of the light beam). As the lens focal distance is limited, to reduce the d value, the light beam divergence angle θ should be reduced as much as possible. Also, in such application fields as laser communication, radar, and ranging, it is hoped that the distance of action should be as large as possible. Theoretical analysis shows that the action distance d is inversely proportional to the square root of the light beam's divergence angle θ (i.e., $S \propto \theta^{-1/2}$). So the divergence angle θ should preferably be small. After mode selection, it is possible that the output power may decrease. But because of the improvement in the degree of divergence, the brightness can be increased by several orders. In addition, after getting focused, a light spot of diffraction limit can be generated.

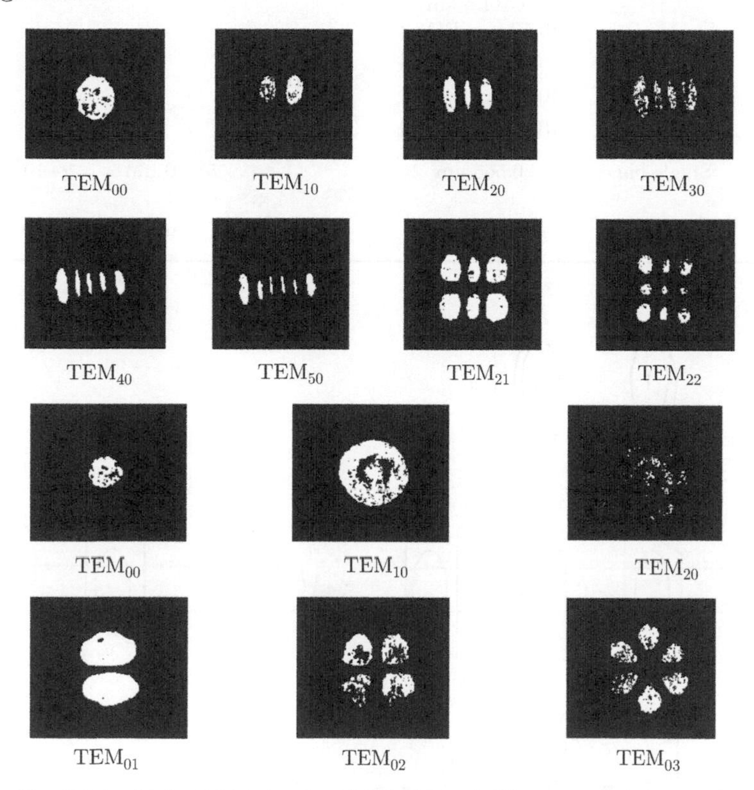

Fig. 5.1-1 Light field intensity distribution of different transverse modes

By the longitudinal mode is meant the laser light field distribution along the resonator axial line direction. For a laser of ordinary cavity length, often several and even several hundred longitudinal mode oscillations may be generated at the same time, the number of the longitudinal modes depending on the width of the gain curve of laser and the frequency interval between two adjacent longitudinal modes. For example, if the cavity length $L = 1$ m, $\Delta\nu = c/2nL = 150$ MHz, then the number of longitudinal modes for various classes of lasers that are likely to generate oscillation are as shown in Tab. 5.1-1. Many applications (such as precision interferometric length measurement, holography, high resolution spectroscopy, etc.) all require that laser of extremely good monochromaticity and coherence be used as the light source; that is, the single frequency laser is required, while the longitudinal mode selecting technique is the essential means of single frequency laser operation.

Tab. 5.1-1 **Spectral Characteristic of Various Classes of Lasers**

Operation material	Number of operating fluorescence spectral lines	Central wavelength of main operating fluorescence spectral line	Fluorescence spectral line width	Frequency range of laser oscillation	Number of longitudinal modes ($L = 1$ m)	Remarks
Ruby (crystal)	Single (R_1 line)	0.6943 μm	0.3~0.5 nm	0.02~0.05 nm	$\sim 10^2$	300 K
YAG:Nd^{3+} (crystal)	Single	1.06 μm	0.7~1 nm	0.03~0.06 nm	$\sim 10^2$	300
Neodymium glass	Single	1.06 μm	20~30 nm	5~13 nm	$\sim 10^4$	300
He-Ne (atomic gas)	Single or multiple	0.6328 μm 1.153 μm	1~2 GHz	1 GHz	~ 6	
Ar^+ (ionic gas)	Multiple	0.4880 μm 0.514 5 μm	5~6 GHz	4 GHz	~ 25	
CO_2 (molecular gas)	Multiple	10.57 μm [P(18)] 10.59 μm [P(20)] 10.61 μm [P(22)] 10.63 μm [P(24)]	52 MHz			
Rhd6G (organic dyeliquid)	Single band	0.565 μm		5~10 nm	$\sim 10^4$	
GaAs (PN junction diode)	Single band	840 nm	17.5 nm	3 nm	~10	77 K L~0.1 cm

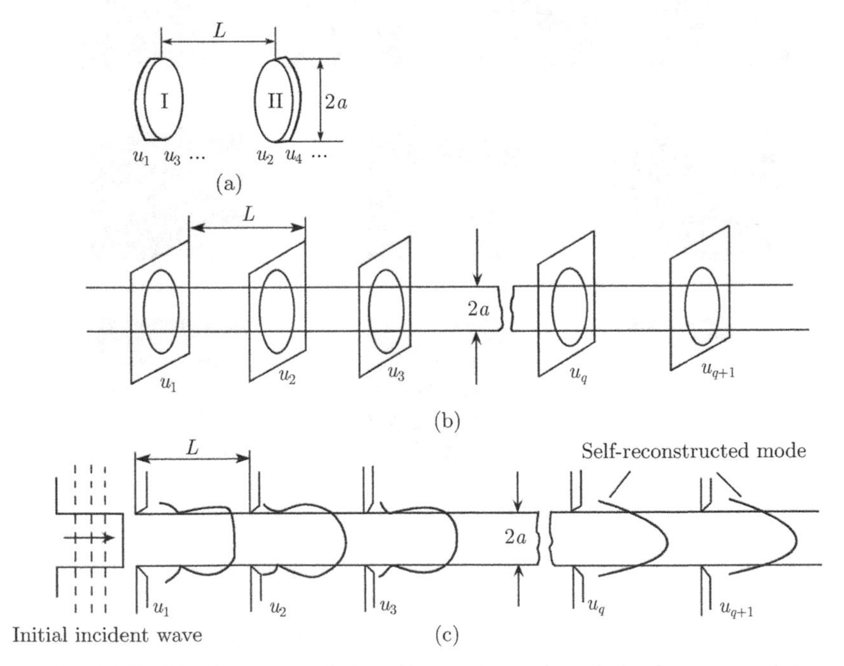

Fig. 5.1-2 The formation of the self-reconstructed mode in the open cavity

5.1.2 The formation of the transverse mode

Suppose the parallel plane resonator is as shown in the figure, in which the diameter of both reflective mirrors is $2a$ and the distance is L (i.e., the cavity length). As the geometric dimensions of the reflective mirror are limited, only the part of the light beam that falls on the surface of the mirror can be reflected back. That is, when the light beam propagates back and forth between the two mirrors, there will surely be loss due to the diffraction effect at the edge of the mirrors. Suppose at the initial moment there is a certain field distribution u_1 on mirror 1. Then, when the light wave is transmitted from mirror 1 to mirror 2 in the cavity, there will be generated on mirror 1 a new field distribution u_2, and field u_2 after propagation for the second time will generate a new field distribution u_3 on mirror 1. Every time a propagation is made, the light wave loses part of its energy because of the diffraction, which will also induce the variation of the energy distribution. Therefore, the field u_3 generated after propagation back and forth once will not only have an amplitude smaller than that of u_1, but its distribution may be different from that of u_1. Later, u_3, in turn, will generate $u_4, \ldots$; this back-and-forth process will continue without stop. Theoretical analysis shows that after sufficiently many times of back and forth propagations, there will be formed in the cavity such a steady-state field, whose relative distribution will no longer be affected by diffraction and which, after a round trip in the cavity, will "self-reconstruction" the field distribution at the departure time. The only possible variation of such a field is simply the attenuation of the field amplitudes in identical proportion at the various points on the mirrors and the retardation of identical magnitude of the phases at the points. The stable field distribution that can be "self-reconstructed" after propagating back and forth on the face of the mirror in the cavity is spoken of as the self-reconstructed mode or transverse mode. If the two mirror surfaces are completely identical symmetric cavities, then such a stable field distribution will realize reconstruction only after propagation by single path.

The shape of the initial incident wave is of no importance in a definite sense. In principle, the other initial incident waves can also form self-reconstructed modes. Of course, the ultimate steady-state field distribution obtained from different initial incident waves may be different from one another, which indicates the diversity of the cavity mode.

The actual physical process is that all oscillations in the cavity begin with a certain accidental spontaneous radiation while spontaneous radiation obeys the law of statistics and is therefore capable of providing different kinds of initial distribution. Diffraction here plays the role of a certain "screen" that screens out the self-reconstructed mode that can exist therein.

5.1.3 The transmission law of an actual laser beam

The mode of a laser's output light beam is first of all dependent on the specific form of the resonator while also on many other factors such as the dimensions and distribution of the active medium, the adjustment of the optical cavity, the actual quality of the optical components, etc. For an actual laser, it is often very difficult to give the light beam's light field distribution in theory and also very difficult to find the light beam's light field distribution by actual measurement.

Then how does one judge a laser's actual output light beam after all in actual work? This is a fundamental problem for all technical specialists engaged in laser application and also a very-hard-to-exactly-answer question. Any one technical specialist engaged in laser application will often hear or use the term "light beam quality". But, over the years, there has been no unified exact physical meaning of this term. And the meaning is often rather vague. For instance, for a specialist engaged in laser holography application, a light beam of good coherence is a light beam of good quality. For an expert engaged in lidar, laser remote sensing and laser weaponry that require long-distance transmission of laser energy, a light beam of good directionality is a light beam of good quality while for an expert engaged in laser cutting and laser drilling applications, the closer to the fundamental mode Gaussian light beam the better the light beam will be.

For quite some time, specialists engaged in theoretical and actual application have put forward many parameters trying to give a clear definition of "light beam quality", the parameters that have been adopted including the laser mode composition, laser coherence, focusing light spot dimensions, laser angle of divergence, the Strehl ratio, the β value, the K parameter, the diffraction limit multiple and M^2 parameter, and so on and so forth. Of all these parameters, the parameter M^2, owing to its greater strictness and integrity in theory and its greater capability of truthfully representing the quality of a light beam in actual application, has won greater approval. The parameter M^2 has now become one of the standard parameters of the laser.

1. The definition of parameter M^2

The definition of the parameter M^2 can be stated as

$$M^2 = \frac{\text{(real light beam waist spot diameter} \times \text{real light beam far field divergence angle)}}{\text{(fundamental mode Gauss light beam waist spot diameter} \times \text{fundamental mode Gauss light beam far field divergence angle)}}$$

The physical essence of parameter M^2 lies in comparing the actual light beam with the fundamental mode Gaussian light beam. The value of parameter M^2 represents the degree in which the actual light beam deviates from the fundamental mode Gaussian light beam.

Defining the actual light beam's spot size and divergence angle by adopting a unified method is also a problem in actual work that should be solved. The basis for defining parameter M^2 is to define the light beam's near field spot size and far field divergence angle with the second-order moment method. In order to better understand this defining method, it will be helpful to first review the characteristics of the fundamental mode Gaussian light beam. The light intensity distribution in the cross section of the one-dimensional fundamental mode Gaussian light beam transmitted along the z-axis can be expressed as

$$I(x,z) = E(x,z)E^*(x,z) = \sqrt{\frac{1}{2\pi}\frac{1}{w(z)}}\exp[-2x^2/w^2(z)] \tag{5.1-1}$$

where $w(z)$ is the spot radius of the fundamental mode Gaussian light beam. Suppose the waist spot radius of the fundamental mode Gaussian light beam is w_0, and the position of the waist spot is z_0. Then there is

$$w^2(z) = w_0^2 + \frac{\lambda^2}{\pi w_0^2}(z - z_0)^2 \tag{5.1-2}$$

The important characteristic of the ideal fundamental mode Gaussian light beam is that its Fourier transform angular spectrum distribution still possesses Gaussian distribution:

$$I(\theta_x) = E(\theta_x)E^*(\theta_x) = 2\sqrt{2\pi}w_0\exp\left(-\frac{2\pi^2}{\lambda^2}w_0^2\theta_x^2\right) \tag{5.1-3}$$

where θ_x is the included angle between the plane wave vector and x-axis. According to Fourier optics, there is

$$E(\theta_x) = \int_{-\infty}^{+\infty} E(x,z)\exp\left(-\mathrm{i}2\pi\frac{\theta_x}{\lambda}x\right)\mathrm{d}x \tag{5.1-4}$$

The spatial frequency angular spectrum distribution $I(\theta_x)$ represents the light beam's far field divergence angle distribution. The mean square deviations of Gaussian distribution $I(x,z)$ and $I(\theta_x)$ are, respectively,

$$\begin{aligned}\sigma_x(z) &= w(x)/2\\ \sigma_{x0} &= w_0/2\\ \sigma_{\theta_x} &= \lambda/2\pi w_0\end{aligned} \tag{5.1-5}$$

Or the light beam diameters can probably be rewritten as

$$\begin{aligned} D_x(z) &= 2w_x(z) = 4\sigma_x(z) \\ D_{x0} &= 2w_{x0} = 4\sigma_{x0} \end{aligned} \tag{5.1-6}$$

Correspondingly, the total angle of divergence of the far field fundamental mode Gaussian light beam can be written as

$$\theta_x = 4\sigma_{\theta_x} = \frac{2\lambda}{\pi w_{x0}} \tag{5.1-7}$$

Multiply the waist spot of the fundamental mode Gaussian light beam by the far field divergence angle and we have

$$D_{x0} \cdot \theta_x = 4\lambda/\pi \tag{5.1-8}$$

It can be seen from the above simple analysis that the mean-square deviation value of the light field distribution is in one-to-one correspondence with the light beam's spot size and the far field divergence angle. Of course this result is no surprise because the light beam's light field distribution mean-square deviation value represents the degree of deviation of the light beam from the center. The degree of deviation of the light beam from the center in the cross section represents the light beam's spot size. And the degree of deviation of the light beam far field propagation direction from the central propagation direction represents the far field's divergence angle of the light beam. The above-mentioned characteristics of the light beam reminds people of utilizing the mean square deviation value to represent common light beam near field spot size and far field divergence angle, thus giving the unified definition of the common light beam's near field spot size and far field divergence angle.

Consider an arbitrary light beam with light field distribution $I(x, y, z)$ in the cross section and the corresponding light field distribution $I(\theta_x, \theta_y)$ in the spatial frequency angular spectrum domain. Their mean-square deviation values can be found respectively as:

$$\begin{aligned} \sigma_x^2(z) &= \frac{\iint (x-\overline{x})^2 I(x,y,z)\mathrm{d}x\mathrm{d}y}{\iint I(x,y,z)\mathrm{d}x\mathrm{d}y} \\ \sigma_y^2(z) &= \frac{\iint (y-\overline{y})^2 I(x,y,z)\mathrm{d}x\mathrm{d}y}{\iint I(x,y,z)\mathrm{d}x\mathrm{d}y} \\ \sigma_{\theta_x}^2 &= \frac{\iint (\theta_x-\overline{\theta}_z)^2 I(\theta_x,\theta_y)\mathrm{d}\theta_x\mathrm{d}\theta_y}{\iint I(\theta_x,\theta_y)\mathrm{d}\theta_x\mathrm{d}\theta_y} \\ \sigma_{\theta_y}^2 &= \frac{\iint (\theta_y-\overline{\theta}_y)^2 I(\theta_x,\theta_y)\mathrm{d}\theta_x\mathrm{d}\theta_y}{\iint I(\theta_x,\theta_y)\mathrm{d}\theta_x\mathrm{d}\theta_y} \end{aligned} \tag{5.1-9}$$

where

$$\overline{x}(z) = \frac{\iint xI(x,y,z)\mathrm{d}x\mathrm{d}y}{\iint I(x,y,z)\mathrm{d}x\mathrm{d}y}$$

$$\overline{y}(z) = \frac{\iint yI(x,y,z)\mathrm{d}x\mathrm{d}y}{\iint I(x,y,z)\mathrm{d}x\mathrm{d}y}$$

$$\overline{\theta}_x = \frac{\iint \theta_x I(\theta_x,\theta_y)\mathrm{d}\theta_x\mathrm{d}\theta_y}{\iint I(\theta_x,\theta_y)\mathrm{d}\theta_x\mathrm{d}\theta_y} \tag{5.1-10}$$

$$\overline{\theta}_y = \frac{\iint \theta_y I(\theta_x,\theta_y)\mathrm{d}\theta_x\mathrm{d}\theta_y}{\iint I(\theta_x,\theta_y)\mathrm{d}\theta_x\mathrm{d}\theta_y}$$

represent the light beam central coordinates and the light beam central transmission direction angle, respectively. It can be proved that

$$\begin{aligned}\overline{x}(z) &= \overline{x}_0 + \overline{\theta}_x \cdot z \\ \overline{y}(z) &= \overline{y}_0 + \overline{\theta}_y \cdot z\end{aligned} \tag{5.1-11}$$

That is, the center of the light beam is linearly transmitted.

In view of the results of the fundamental mode Gaussian light beam, the International Standardization Organization (ISO) had the spot diameter and far field divergence angle of an arbitrary light beam defined as

$$\begin{aligned}D_x(z) &= 4\sigma_x(z) \\ D_y(z) &= 4\sigma_y(z) \\ \theta_x &= 4\sigma_x \\ \theta_y &= 4\sigma_y\end{aligned} \tag{5.1-12}$$

It can be proved using the Collins formula that, in free space, for an arbitrary light beam, the following equations hold:

$$\begin{aligned}D_x^2(z) &= D_{x0}^2 + \theta_x^2(z-z_0)^2 \\ D_y^2(z) &= D_{y0}^2 + \theta_y^2(z-z_0)^2\end{aligned} \tag{5.1-13}$$

where D_{x0} and D_{y0} are waist spots of the light beam, or the minimum values of $D_x(z)$ and $D_y(z)$. This formula is completely consistent with the transmission formula for the fundamental mode Gaussian light beam in free space. It can be seen from this that in the way the above-mentioned second-order moment is defined, as long as D_0, θ, and z_0 are known, the transmission of the light beam in free space is completely determined.

The waist spot and divergence angle of an arbitrary light beam can be defined by adopting the above-mentioned second-order moment method. On this basis the parameter M^2 of an arbitrary actual light beam can be written as

$$\begin{aligned}M_x^2 &= \frac{D_{x0}\cdot\theta_x}{D_{0\mathrm{TEM}_{00}}\cdot\theta_{\mathrm{TEM}_{00}}} = \frac{D_{x0}\cdot\theta_x}{4\lambda/\pi} \\ M_y^2 &= \frac{D_{y0}\cdot\theta_y}{D_{0\mathrm{TEM}_{00}}\cdot\theta_{\mathrm{TEM}_{00}}} = \frac{D_{y0}\cdot\theta_y}{4\lambda/\pi}\end{aligned} \tag{5.1-14}$$

It can be proved that the parameter $M^2 \geqslant 1$, and parameter $M^2 = 1$ represents an ideal fundamental mode Gaussian light beam. As an actual laser beam cannot be an ideal fundamental mode Gaussian light beam, the parameter M^2 of an actual light beam is always greater than 1. The minimum parameter M^2 of currently reported actual laser products is 1.1. It is shown that the greater the parameter M^2 is, the more the actual light beam will deviate from the fundamental mode Gaussian light beam, as can be easily inferred. It can be seen from Eq. (5.1-14) that, under the same waist spot diameter condition, the diffusion of light beam diffraction of the said actual light beam would be quicker. Therefore, from this perspective, the parameter M^2 characterizes the characteristics of transmission of an actual light beam. Hence the parameter M^2 is also called the light beam transmission factor.

For a centrally circularly symmetric light beam,

$$\begin{aligned} \sigma_r^2(z) &= \sigma_x^2(z) = \sigma_y^2(z) \\ \sigma_\theta^2 &= \sigma_{\theta_x}^2 = \sigma_{\theta_y}^2 \end{aligned} \tag{5.1-15}$$

Correspondingly, there will be

$$\begin{aligned} D(z) &= 4\sigma_r \\ \theta &= 4\sigma_\theta \\ D^2(z) &= D_0^2 + \theta^2(z - z_0)^2 \\ M^2 &= \frac{D_0 \cdot \theta}{4\lambda/\pi} \end{aligned} \tag{5.1-16}$$

2. Generalized ABCD of light beam parametric transmission under second-order moment definition

Suppose an arbitrary light beam passes through a thin lens of focal distance f, as shown in Fig. 5.1-3. D_0 and D_0' are the waist spot diameters of two light beams in the front and rear of the lens, respectively, θ and θ' the far field divergence angles of the two light beams in the front and rear of the lens, and L and L' the distances from the waist spot positions of the two light beams in the front and rear of the lens to the thin lens, respectively. For convenience, suppose the light beam is a circularly symmetric light beam.

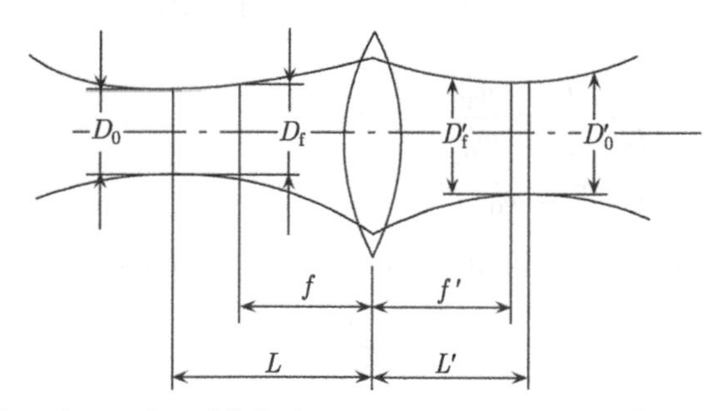

Fig. 5.1-3 Transformation of light beam parameters passing through the thin lens

According to the Fourier transform law of the thin lens, the light field of point (x_f, y_f) on the front focal plane of the lens corresponds to the plane wave spectral component with direction angle ($\theta_x' = x_f/f$, $\theta_y' = y_f/f$) in the light beam in the rear of the lens. It is stated in general physics that "the light field of point (x_f, y_f) on the front focal plane of the lens becomes a plane wave with direction angle ($\theta_x' = x_f/f$, $\theta_y' = y_f/f$) after passing through the lens". Thus we can obtain

$$D_f = f \cdot \theta' \tag{5.1-17}$$

where D_f is the spot diameter at the front focal plane of the lens.

$$D_f^2 = D_0^2 + \theta^2(L - f)^2 \tag{5.1-18}$$

Combining the above two equations gives

$$\theta'^2 = \left(\frac{D_0}{f}\right)^2 + \theta^2\left(\frac{L-f}{f}\right)^2 \tag{5.1-19}$$

Similarly, according to Fourier transform, there is

$$D'_f = f \cdot \theta \tag{5.1-20}$$

where D'_f is the spot diameter at the rear focal plane of the lens.

$$D'^2_f = D'^2_0 + \theta'^2(L'-f)^2 = (f \cdot \theta)^2 \tag{5.1-21}$$

According to the definition of the thin lens, the light beam amplitude distributions on the lens front and rear incoming and outgoing surfaces are identical. Therefore,

$$D_{\text{in front of lens}} = D_{\text{in rear of lens}}$$

that is,

$$D_0^2 + \theta^2 \cdot L^2 = D'^2_0 + \theta'^2 \cdot L'^2 \tag{5.1-22}$$

By collation of the above equations, we obtain

$$\begin{aligned} &\frac{1}{D'^2_0} = \frac{1}{D_0^2}\left(1-\frac{L}{f}\right)^2 + \frac{1}{(f\theta)^2} \\ &\frac{L-f}{L'-f} = \frac{D_0^2}{D'^2_0} \\ &D_0 \cdot \theta = D'_0 \cdot \theta' \end{aligned} \tag{5.1-23}$$

The transform of the fundamental mode Gaussian light beam when passing through a thin lens can be analyzed via the ABCD rule. Via the ABCD rule we can obtain

$$\begin{aligned} &\frac{1}{w'^2_0} = \frac{1}{w_0^2}\left(1-\frac{L}{f}\right)^2 + \frac{1}{f^2}\left(\frac{\pi w_0^2}{\lambda}\right)^2 \\ &\frac{L-f}{L'-f} = \frac{w_0^2}{w'^2_0} \\ &w_0 \cdot \theta = w'_0 \cdot \theta' \end{aligned} \tag{5.1-24}$$

taking into consideration that, for the fundamental mode Gaussian light beam, there is

$$\theta = \frac{\lambda}{\pi w_0}$$

After the actual light beam passes through the thin lens, the variation of the light beam parameters D_0, θ, and L is completely consistent with that of the light beam parameters w_0, θ, and M, L after the fundamental mode Gaussian light beam passes through the thin lens. This reminds us that, for an actual light beam, there should also exist a generalized ABCD rule just like that for the fundamental mode Gaussian light beam.

After the definition of the fundamental mode Gaussian light beam, we define an arbitrary light beam

$$\begin{aligned} &R(z) = z + \frac{1}{z}\left[\frac{\pi}{4\lambda}\left(\frac{D_0}{M}\right)^2\right]^2 \\ &\frac{1}{q(z)} = \frac{1}{R(z)} - \mathrm{i}\frac{4\lambda}{\pi}\left(\frac{M}{D(z)}\right)^2 \end{aligned} \tag{5.1-25}$$

Then the parameter $q(z)$ of the light-beam, after passing through an optical system that can be described by an optical matrix, satisfies the ABCD rule.

When actually making an analysis of the transmission parameters of a light beam, the following method can be adopted:

First construct a virtual fundamental mode Gaussian light beam, whose waist spot diameter is $2w_0 = D_0/M$; the position of the waist spot is consistent with that of the actual light beam waist spot. Then find the waist spot radius w_0' and position after the fundamental mode Gaussian light beam passes through the optical system before finally obtaining the waist spot diameter directly following the passage of the actual light beam through the optical system:

$$D_0' = 2Mw_0' \tag{5.1-26}$$

The position of the waist spot is consistent with that of the virtual fundamental mode Gaussian light beam's waist spot. Thus the far field divergence angle after the actual light beam passes through the optical system is

$$\theta' = 2M\lambda/\pi w_0' \tag{5.1-27}$$

From the above analysis we can obtain the following conclusions:

The waist spot diameter of the oscillating light beam in an actual laser resonator is M times that of the oscillating ideal fundamental mode Gaussian light beam waist spot in the said laser resonator. The far field divergence angle of the oscillating light beam in the actual laser resonator is M times that of the oscillating ideal fundamental mode Gaussian light beam in the said laser resonator. The waist spot position of the oscillating light beam in the actual laser resonator is consistent with that of the oscillating ideal fundamental mode Gaussian light beam in the said laser resonator. The parameter M^2 of the oscillating light beam in the actual laser resonator can be obtained by actually measuring the light beam parameters (the waist spot diameter and far field divergence angle) of the said laser's output light beam.

The measurement of the light beam parameters (the waist spot diameter and far field divergence angle) can be made by using Eq. (5.1-13). During measurement, measure the diameter of the spot in more than three different positions along the light beam propagating direction. Substitute the results into Eq. (5.1-13) to form an equation set concerning D_0, θ, and Z_0, to fit out a quadratic curve closest to the results of measurement and identical with Eq. (5.1-13), thereby determining the three parameters D_0, θ, and w_0. The above-mentioned method has bypassed direct calculation of θ and is a standard measuring scheme approved by the International Standardization Organization (ISO).

From the above-mentioned inference, another important conclusion can also be obtained:

$$D_0 \cdot \theta = D_0' \cdot \theta'$$

That is, after an actual light beam passes through a linear optical system, the M^2 parameter remains unchanged. It can be seen from this that it is impossible for the characteristics of a light beam to be improved by means of a simple linear optical system.

5.2 The transverse mode selecting technology

5.2.1 The principle of transverse mode selection

It is known from the principle of laser that it's possible for a resonator of a laser to have a number of stable oscillating modes. As long as the single pass gain of a certain mode is greater than the single pass loss, the laser oscillation condition is met. And it is possible for this mode to be excited to oscillate. Suppose the reflectivity of the reflectors at the two ends of the resonator are r_1 and r_2, respectively; the single pass loss is δ, the single pass gain coefficient is G; and the laser operation material length is L. Then, for the light of a certain transverse mode (TEM_{mn}) with an initial light intensity I_0, after a round-trip in the resonator, affected by both the loss and gain factors, its light intensity becomes

$$I = I_0 r_1 r_2 (1 - \delta)^2 \exp(2GL) \tag{5.2-1}$$

The threshold value condition is

$$I \geqslant I_0$$

Thus we obtain

$$r_1 r_2 (1 - \delta)^2 \exp(2GL) \leqslant 1 \tag{5.2-2}$$

Now examine the two lowest-order transverse modes TEM_{00} and TEM_{10}, whose single pass losses are expressed by δ_{00} and δ_{10}, respectively, and it is considered that the gain coefficients of the active medium are identical for all transverse modes. When the following two inequalities are satisfied at the same time,

$$\sqrt{r_1 r_2}(1 - \delta_{00}) \exp(GL) > 1 \tag{5.2-3}$$

$$\sqrt{r_1 r_2}(1 - \delta_{10}) \exp(GL) < 1 \tag{5.2-4}$$

the laser will be able to realize single transverse mode (TEM_{00}) operation.

Then, how does one satisfy the above-mentioned conditions? There exist two kinds of loss of different properties in the resonator. One kind of loss is unrelated to the order of the transverse mode, such as the transmission loss of the cavity mirrors, the absorption and scattering losses of the intracavity components, etc., while the other is the diffraction loss closely related to the order of the transverse mode. In a stable resonator, the diffraction loss of the fundamental mode is minimum. With an increase in the order of the transverse mode, its diffraction loss will gradually increase. In order to find the $\delta_{\min}$ value of transverse mode under general conditions, numerical solution can be found using the computer. What is shown in Fig. 5.2-1 is a set of curves of single pass loss of two lowest-order modes in symmetric round mirrors' stable spherical surface cavity obtained using the numerical solving method. It can be seen from the figure that, under the condition of identical Fresnel number N values, the diffraction loss of the symmetric stable cavity falls with a decrease of $|g|$. The performance of the resonator, that it has different diffraction losses for transverse modes of different orders, is the physical basis of realizing transverse mode selection while appropriately choosing the Fresnel number N value to make it satisfy Eqs. (5.2-3) and (5.2-4) will make it possible to attain the goal of realizing single transverse mode selection. In consideration of the competition between the modes, the condition for selecting the single transverse mode can be more liberal. That is, in the beginning the laser has quite a few transverse modes that satisfy the threshold condition. If the gains of the modes are identical, as the diffraction loss of the fundamental mode is minimum, it will be dominant in mode competition. Once the fundamental mode sets up oscillation the first, it will extract energy from the active medium. Furthermore, because of the gain saturation effect, the gain of the operation material will decrease as a result. When the following condition is satisfied:

$$\sqrt{r_1 r_2}(1 - \delta_{00}) \exp(GL) = 1 \tag{5.2-5}$$

oscillation tends to be stable. Now the other transverse modes will be restrained for their failure to continue to satisfy the threshold condition, so the laser can still operate in the single transverse mode.

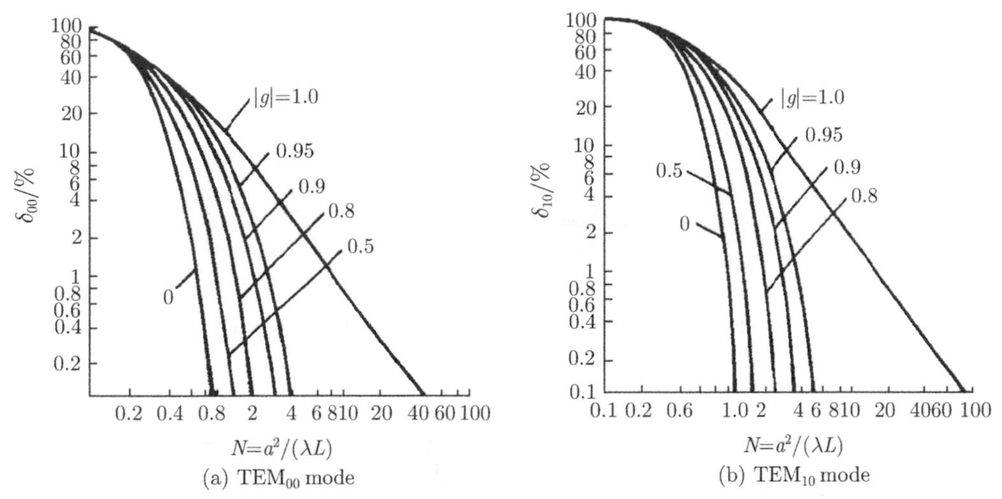

Fig. 5.2-1 Variation of diffraction loss with N for symmetric resonators of different configurations

In order to effectively select the transverse mode, it is also necessary to consider two problems. One is the necessity to consider, apart from the magnitude of the absolute values of the diffraction loss of the transverse modes, the discriminating ability of a transverse mode. That is to say, only when the difference in the diffraction loss between the fundamental mode and the higher-order transverse mode is sufficiently great (i.e., the ratio δ_{10}/δ_{00} is great), will it be possible to distinguish one mode from the other so as to facilitate mode selection. Otherwise, it will be rather difficult selecting the mode. The difference in transverse mode diffraction loss is related to not only the structure of different types of resonator but also the Fresnel number N of the resonator. Figure 5.2-2 shows the relationship between the δ_{10}/δ_{00} value and Fresnel number N of various g factor symmetric cavity. Figure 5.2-3 shows the relationship between the δ_{10}/δ_{00} value and N of the plane-concave cavity, the dashed lines in the figure representing the equal-loss lines of various loss values of the TEM_{00} mode. Different N and g resonators with equal loss values all correspond to one and the same dashed line. It can be seen from the above figures that the transverse mode discriminating ability becomes greater with an increase in N, but the diffraction loss decreases with an increase in N. So the N value should be appropriately selected to effectively select the transverse mode. In addition, although δ_{10}/δ_{00} of the confocal cavity is great while that of the parallel plane cavity is small, yet when the N value is not too small, the diffraction losses of the transverse modes of the confocal cavity are in general very low and are rather small compared with the other non-selective losses in the cavity. Therefore it is inadvisable to carry out selection of modes. Moreover, the volume of the confocal cavity fundamental mode is very small, so the single mode output power is rather low. In contrast, although the δ_{10}/δ_{00} value of the plane cavity is rather low, the diffraction loss absolute values of the different modes are rather great. As long as rather large N values are chosen, the fundamental mode can be selected. And the volume of the fundamental mode is fairly large at that. Once single mode oscillation is formed, the output power will be rather high. In other words, to effectively select the mode, it is necessary to consider selecting appropriate cavity type structure and proper Fresnel number N value (to be analyzed in the mode selecting method below). A second consideration is that the diffraction loss must occupy a position in the total loss of the mode so important that it can be compared with the other non-selective losses. For this reason, it is advisable to reduce as much as possible the absorption loss, scatter loss, and other losses of different components in the cavity, thus relatively increasing the proportion of the diffraction loss in total loss. In addition, this goal can be attained also by reducing the Fresnel number N in the cavity.

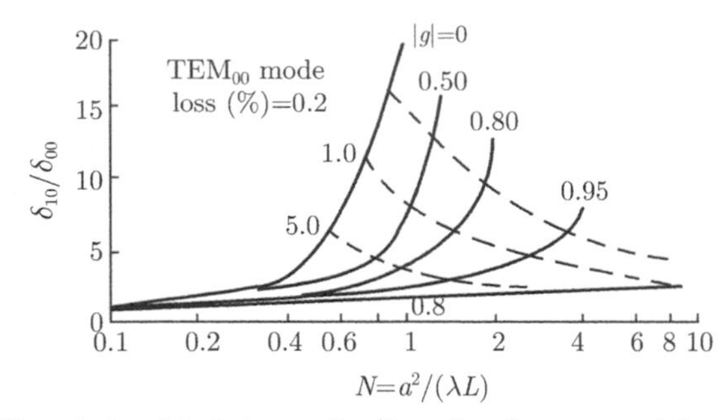

Fig. 5.2-2 The relationship between δ_{10}/δ_{00} of various symmetric cavities and N

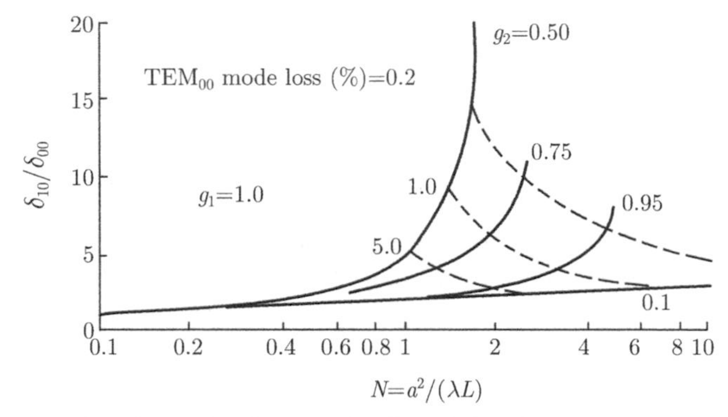

Fig. 5.2-3 The relationship between δ_{10}/δ_{00} of the plane concave cavity and N

5.2.2 The method of selecting the transverse mode

The method of selecting the transverse mode can be divided into two classes. One consists in changing the resonator structure and parameters to obtain a rather big difference in the diffraction loss of all modes and enhance the mode selecting performance of the resonator, and the other in inserting an additional mode selecting component in a definite resonator to enhance the mode selecting performance. Most gaseous lasers adopt the first method, it being often that, during resonator design, cavity type and resonator parameters g, N values are appropriately chosen to realize fundamental mode output while for solid lasers, it is necessary to adopt the second method as solid operation material has a large caliber. To reduce the Fresnel number N, it is necessary to insert a mode selecting component in the cavity.

1. The method of selecting the resonator parameters g and N

Figure 5.2-2 shows the relationship between the confocal cavity's ratio δ_{10}/δ_{00} and Fresnel number N. It can be seen from the figure that, when N is definite, the parameter $|g|$ is small, δ_{10}/δ_{00} is great, but the δ_{00} and δ_{10} values are also small. If it is desired that the fundamental mode should be selected and the higher-order mode be restrained, the only way to do so is to reduce the Fresnel number N to raise the mode loss value. Therefore, from the point of view of fundamental mode selection, it is hoped that small $|g|$ and N values will be chosen. But if the N value is too small, the mode volume will be very small and the output power will be very low as a result. So, in order to obtain both fundamental mode oscillation and a rather high output power, that is, under the precondition of ensuring fundamental mode operation, the N value should be properly increased until it satisfies Eqs. (5.2-3) and (5.2-4) simultaneously. For the commonly used large curvature radius dual-concave spherical surface stable cavity, it is suitable to choose the Fresnel number N between 0.5~2.0. Take the smaller value for a low gain device and the higher value for a high gain device.

Then how should we choose g and N to help with transverse mode selection? It is known from the principle of laser that, in the graph of the resonator stability region, the demarcation line between the stable region and the unstable region is to be determined by $g_1g_2 = 1$ or $g_1 = 0$, $g_2 = 0$. Appropriately choose the resonator parameters R_1, R_2, and L to make them operate at the border of the stable region, that is, operating in the critical operation state, which will be beneficial to mode selection. As of all orders of transverse modes the diffraction loss of the lowest-order mode (the TEM_{00} mode) is minimum. When changing the parameters of the resonator to make its operation points change from the stable region to the unstable region, the diffraction loss of all orders of mode will rapidly increase, but the diffraction loss of the fundamental mode increases the most slowly. Therefore, when the operation point of the resonator shifts to a certain position, all higher-order modes may get suppressed for the high diffraction losses suffered, with only the fundamental mode left to operate in the end.

Take the TEM_{00} mode and TEM_{01} mode as examples. Figure 5.2-4 shows the relationship of the difference in the single path diffraction loss between the two modes with the variation of $|g|$ when the Fresnel number N values are different. Here, there are two cases of the variation of the critical region $|g|(|g| \leqslant 1)$. One is the case in which the g stable region value approximates to -1, which corresponds to the case of the concentric cavity, and the other is the case where g approximates to 1, which corresponds to the case of the parallel plane cavity. It can be seen from the figure that with the $|g|$ value tending to 1, the velocity with which the TEM_{01} mode single pass diffraction loss increases is far greater than that of the TEM_{00} mode.

In actual application, it is customary to adopt a stable cavity with one cavity mirror that is a plane mirror while the other a spherical mirror, with the mode selected by making the reflective mirror distance L gradually approximate to R. The relationship between the mode diffraction loss difference of such resonators and L can also be found from Fig. 5.2-4. Sometimes, the curvature radii of the two reflective mirrors should be as great as possible, that is, the method of the critical region with $g \to 1 (g \leqslant 1)$ is also adopted. It can be seen from Fig. 5.2-5 that, in the nearly semi-spherical cavity, the curvature center 0 of the reflective mirror M_1 is in the vicinity of the plane mirror M_2; the slight mismatch of the reflective mirror makes the light transmission aperture (shown by the dashed line in figure) somewhat decrease, which will make the intracavity light beam loss greatly increase, causing the output power to fall or even leading to the phenomenon of no oscillation occurring.

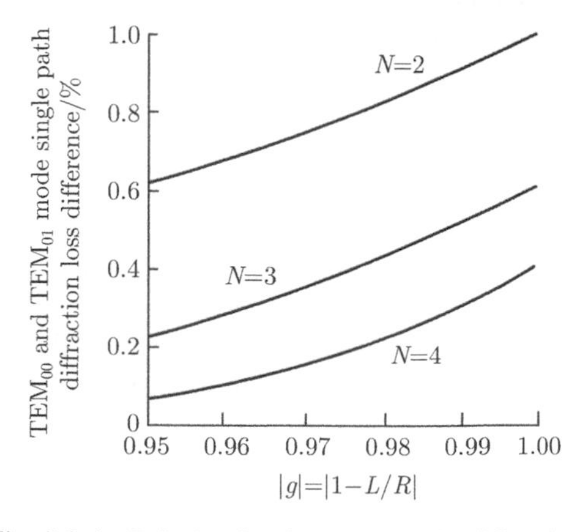

Fig. 5.2-4 Relationship between mode diffraction loss and $|g|$ when N values are different

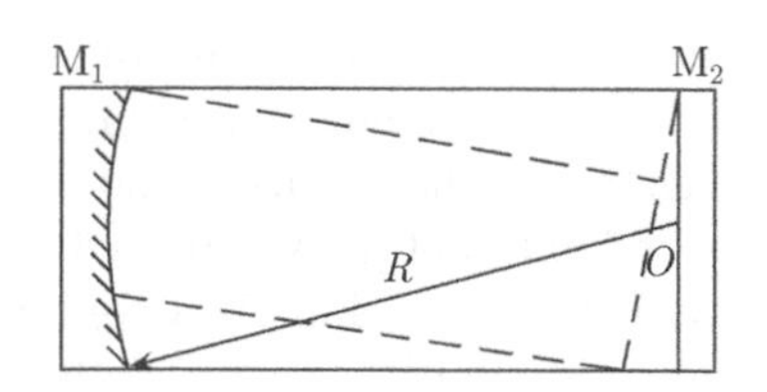

Fig. 5.2-5 Influence of mismatch on semi-spherical cavity

In addition, by adopting different cavity types and parameters g, N, the fundamental mode can also be chosen. However, the fundamental mode's output power (or energy) will differ with the difference in the cavity type and the parameters g, N, since the fundamental mode mode volume varies with the variation of the cavity type and the parameters g, N. So, in order to obtain high power output, when designing a resonator, it is advisable to consider the mode volume of the fundamental mode. It is known from theoretical analysis of the resonator that, for the stable spherical cavity TEM_{00} mode, the effective light beam radius $w(z)$ is in general transmitted along the cavity axis z direction in obedience to the law of double curves, as shown in Fig. 5.2-6, in which w_0 is the intracavity minimum effective light beam radius (called the beam waist), and $\varDelta$ is the distance from the beam waist position to the origin. If only the case of the symmetric cavity is considered, its expression is

$$w_0 = (\lambda/2\pi)^{\frac{1}{2}}[L(2R - L)]^{\frac{1}{4}} \tag{5.2-6}$$

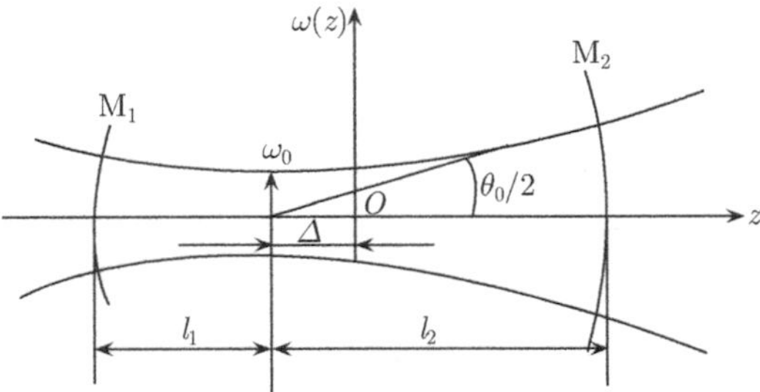

Fig. 5.2-6 The effective light beam radius of stable spherical cavity TEM_{00} mode

The following properties can be obtained from the above equation:

(1) When increasing the radius of curvature R of the cavity mirror, the light beam radius w_0 of the lowest order mode, too, will increase with it, and so will the volume of the fundamental mode.

(2) When the curvature radius R is definite, there exists a maximum value of w_0 that varies with the cavity length L. Differentiating Eq. (5.2-6) with respect to L and letting it be zero, we can find the condition for the maximum to be $L = R$, or the semi-concentric cavity.

It can be seen from (1) that, other things being equal, to increase the volume of the fundamental mode, it is advisable to choose a rather large curvature radius R as best one can. Such a cavity will become a parallel plane cavity under extreme conditions. It can be seen from (2) that, under the condition where the cavity mirror R is determined, to obtain a fundamental mode volume as big as possible, the cavity length L should be appropriately elongated. But, while suitable for the plane cavity and the big curvature radius spherical cavity, this is unsuitable for the spherical cavity of a small curvature radius.

For instance, there is a He-Ne laser of cavity length $L = 1$ m, whose output mirror transmissivity $T = 1.5\%$. In order to obtain fundamental mode output, different g, N parameters can be chosen. The relevant parameters are listed in Tab. 5.2-1, of which the gain G is found by calculation using the empirical formula $G = 1.25 \times 10^{-4} \times L/a$ (L is the cavity length, a is the radius of the discharge capillary tube). There are in all five groups of resonators of different g, N parameters.

For cavity 1, $g = 0$. This is a confocal cavity. To obtain fundamental mode output, it is necessary to satisfy $G \leqslant \delta_{10} + T$ (other losses neglected). For this purpose, it is advisable to choose the capillary radius $a = 0.6$ mm and find $G = 21\%$. According to parameters g, N, we can find that δ_{10} is 25%, thus satisfying condition $G \leqslant \delta_{10} + T$. So the fundamental mode output can be obtained. It can be found using Eq. (5.2-6) that $w_0 = 0.33$ mm. Clearly, the volume of this fundamental mode is rather small.

Tab. 5.2-1 Tab. of relevant parameters for fundamental mode selection

Cavity number	1	2	3	4	5
N	0.6	0.9	0.9	1.6	5.1
a/mm	0.6	0.74	0.74	1.0	1.8
G/%	21	16	16	12	6.6
δ_{10}/%	25	3	14.5	10.5	5.1
g	0	0	0.5	0.9	0.99
R/mm	1	1	2	10	100
w_0/mm	0.33	–	0.43	0.70	1.20
δ_{00}/%	3	0.2	2	3	2
δ_{10}/δ_{00}	8	15	7.3	3.5	2.6
mode	TEM_{00}	multiple mode	TEM_{00}	TEM_{00}	TEM_{00}
cavity type	Confocal	confocal	general	general	parallel plane①

① To be exact, the said cavity is still a general stable cavity, but very close to the parallel plane cavity.

If in order to increase the mode volume of the fundamental mode, increase the radius of the capillary to $a = 0.74$ mm (as cavity 2). Because of the growth of the Fresnel number N, δ_{10} falls (to only 3%), which cannot satisfy the above-mentioned condition. Hence the emergence of the multi-mode oscillation output.

To restrain higher-order modes, while increasing the N value for cavity 3 through cavity 5 on the basis of cavity 1, proper increase of the value of parameter g can also ensure fundamental oscillation with the mode volume of the fundamental mode, too, somewhat increased. For the He-Ne laser, increasing the capillary diameter to increase the Fresnel number will make the small signal's gain coefficient decrease. Furthermore, the increase of parameter g results in a decrease of δ_{10}/δ_{00}, in which the stability would deteriorate and adjustment will also be rather difficult.

It is known from the above discussion that, by changing the resonator parameters g, N, the laser oscillation mode and mode volume can be controlled. In all the above cases, the cavity length L is a definite value. In reality, the cavity length L is sometimes increased to reduce the Fresnel number N, that is, to enhance the mode selecting performance by augmenting the mode loss difference. This is mainly used in solid lasers and extracavity gas lasers.

2. Mode selecting with the pinhole diaphragm method

Adopting a pinhole diaphragm as the mode selecting component to be inserted in the cavity is a frequently used mode selecting method for solid lasers, as shown in Fig. 5.2-7. For a concentric cavity $R_1 + R_2 = L$, this method is especially effective. As the light waist of a higher-order transverse mode is greater than that of the fundamental mode, if the aperture of the diaphragm is appropriately selected, it will be possible to shield part of the light beam of the higher-order transverse mode while the fundamental mode can pass through without a hitch. Then, from the theory of diffraction, it is known that the insertion of a pinhole diaphragm is equivalent to reducing the cross-section area of the cavity mirror, or a reduction in the cavity's Fresnel number N, thus increasing the diffraction loss of all orders of mode. As long as the pinhole diaphragm's aperture is properly chosen and the TEM_{00} mode and TEM_{10} mode satisfy Eqs. (5.2-3) and (5.2-4), the fundamental mode can be selected. Figure 5.2-8 shows the effects of inserting diaphragms of different apertures in the concentric cavity on the TEM_{00} and TEM_{10} mode diffraction losses. The Ns marked on the curves are the corresponding Fresnel numbers of the reflective mirrors radius. It can be seen from the figure that when the pinhole diaphragm aperture r is very small, the losses of both modes are very great and the difference between the two is very little. With an increase in r, the δ_{10}/δ_{00} value of the two modes increases. At $ra/(\lambda L) = 0.3$, it reaches maximum (a being the radius of the circular reflective mirror). Now the TEM_{10} mode loss is about 20% while the fundamental mode has only lost 1%. Now the diaphragm aperture is the optimum value. If the diaphragm aperture is further increased, the loss of both modes will decrease as does the ratio. When $ra/(\lambda L) > 0.5$, the mode loss is basically the same as when no diaphragm was applied.

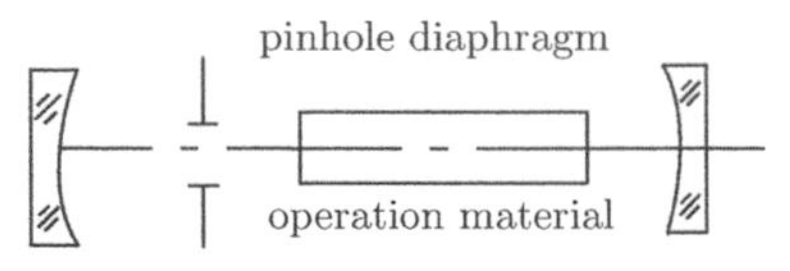

Fig. 5.2-7 pinhole diaphragm mode selection

Figure 5.2-9 shows the relationship of the diffraction loss ratio (δ_{10}/δ_{00}) between two lowest-order modes in the same resonator with the Fresnel number N. It can be seen from the figure that, for a fixed N value, the δ_{10}/δ_{00} *value* has a maximum value for a certain diaphragm aperture. It is most beneficial selecting the mode with this aperture. For a concentric cavity of $N = 2.5\sim20$, $ra/(\lambda L) = 0.28\sim0.36$ is preferable.

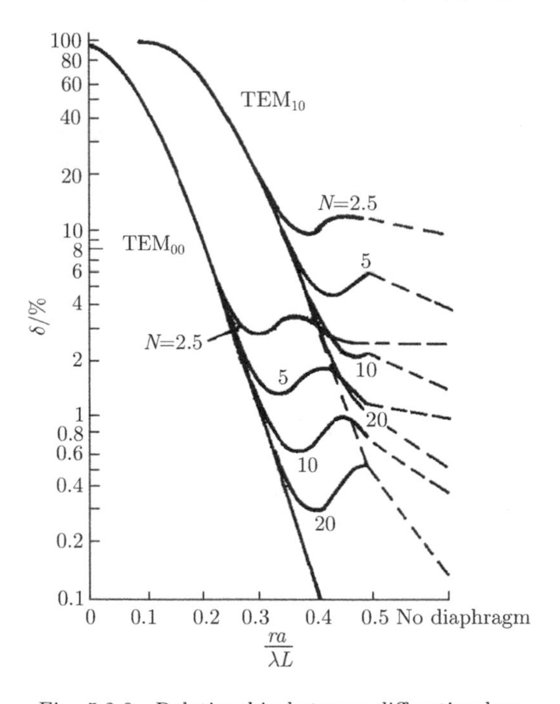

Fig. 5.2-8 Relationship between diffraction loss of two concentric cavity lower-order modes and diaphragm aperture

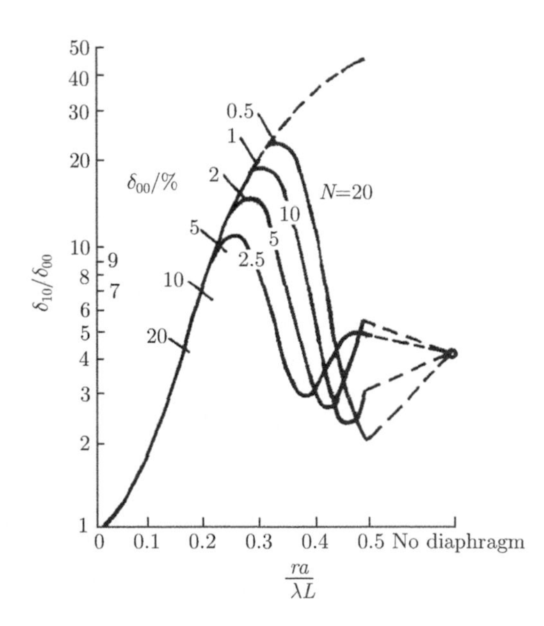

Fig. 5.2-9 Relationship between concentric cavity δ_{10}/δ_{00} and diaphragm aperture

In practice, it is customary to first choose a pinhole radius according to the above-mentioned theory and then determine the dimensions of the diaphragm by experiment. Or, a variable diaphragm can be used to choose a suitable pinhole depending on specific requirements. Despite the simple structure and convenient adjustment using the pinhole diaphragm for mode selection, limited by the pinhole, the small volume of the fundamental mode in the cavity, the volume of the operation material cannot be made full use of, the laser power output is rather low. When the density of the power in the cavity is rather high, the pinhole is damage prone.

3. Insertion of a lens in the cavity for selecting the transverse mode

This method consists in inserting a lens or lens combination for mode selection in collaboration with the pinhole diaphragm. The diaphragm is to be placed on the focus of the lens. Thus, when propagating in the cavity, the light beam can possess a rather large mode volume. What is shown in Fig. 5.2-10 is a schematic diagram of mode selecting cavity type with two lenses incorporated in the cavity. Take the parallel plane cavity as an example. Because of the focusing action of the lens, the light beam is a plane wave when passing through the operation material. Therefore, the mode volume occupies the volume of the entire active medium. When the light beam passes through the pinhole diaphragm, the higher-order mode in the edge part of the light beam, blocked by the diaphragm, suffers loss and is suppressed. So, such a focusing diaphragm device not only retains the mode selecting characteristics of the pinhole diaphragm, but has also increased the volume of the fundamental mode, making it possible to enhance the output power of laser. The size of the diaphragm aperture is related to the focal distance f of the lens. If the focal distance is short, the pinhole diameter should preferably be small while if the focal distance is long, so should the pinhole diameter.

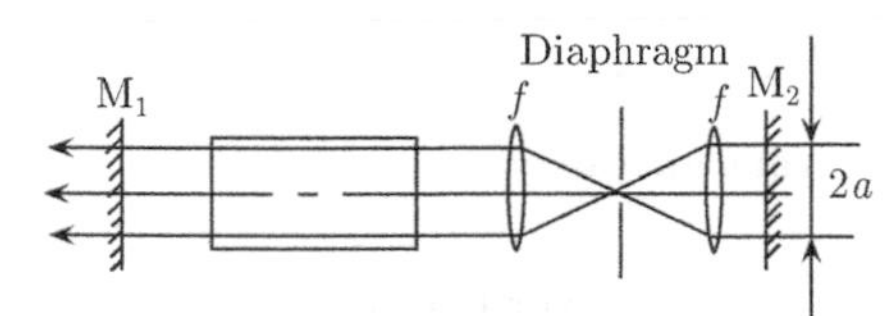

Fig. 5.2-10 Mode selection by focused diaphragm method

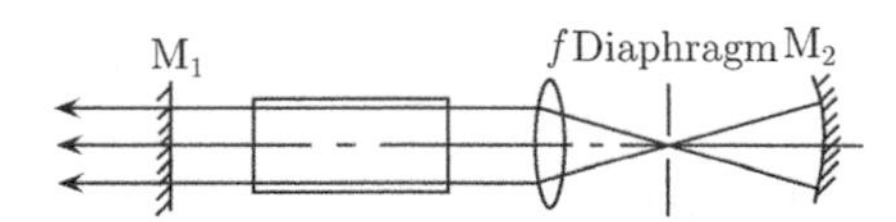

Fig. 5.2-11 Single lens focused mode selection

Although the fundamental mode mode volume can be expanded using this method, the incorporation of two lenses has increased the insertion loss to the cavity in addition to the difficulty in adjustment. In order to simplify the system and reduce the loss, a concave reflective mirror can be adopted to replace the lens on the right side and a plane reflective mirror, as shown in Fig. 5.2-11. But it is required that the curvature center of the concave reflective mirror overlap the focus of the lens.

On the basis of inserting a lens and disphragm in the cavity for mode selection, a method of incorporating a telescope in the cavity has been developed, the structure of which is as shown in Fig. 5.2-12. A telescope composed of a concave and convex lens is inserted in the cavity and the diaphragm is placed on the left side of the concave lens. It can be seen that, as the real focus has been bypassed, the requirement on the diaphragm material is somewhat lowered. Because of the beam expanding action of the telescope, when passing through the laser operation material, the light beam can have its mode volume increased M^2 times (M is the amplification of the telescope). As the focal distance of the telescope's eyepiece is very short, it has a very powerful focusing (or divergent) action on the light beam so that the divergence angle of all orders of mode is amplified M times. As the divergence angle of the fundamental mode is minimum, if the aperture for selecting the diaphragm can be compared with the spot of the fundamental mode, then the higher-order mode will be blocked by the diaphragm and consumed while only the fundamental mode can be retained. Moreover, the position of the concave lens can be regulated relative to the convex lens. When choosing an appropriate misfocus amount, the thermal lens effect of the laser rod can be compensated to obtain the thermally stable fundamental mode laser output.

In addition, there is a so-called "cat's eye resonator" method of mode selection, in effect also a form of pinhole diaphragm mode selection. Its structure is as shown in Fig. 5.2-13, in which M_1 and M_2 are both plane reflective mirrors. In the resonator is placed a lens of focal distance f and cavity length $2f$. Such a cavity structure in the geometric optical path is equivalent to a confocal cavity. Experiment shows that, when the diaphragm at M_2 closes up, the mode volume can fill the entire laser operation material. In this situation, the selectivity of the mode is basically close to a confocal cavity. Thus, not only is the divergence angle reduced, but also the output power is increased.

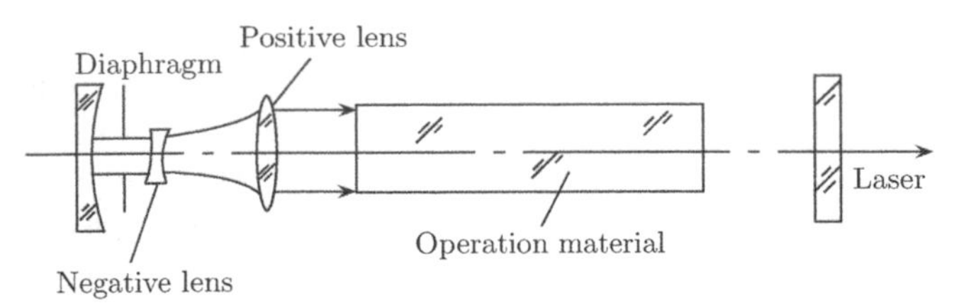

Fig. 5.2-12 Schematic diagram of mode selection by telescope

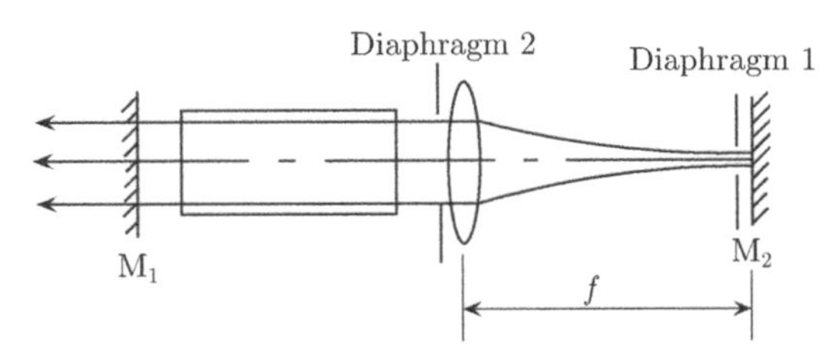

Fig. 5.2-13 Mode selection with the cat's eye cavity

Several typical mode selecting methods have been briefly described above. As other methods of stable resonator mode selection such as the mode selection with the concave–convex cavity, incorporation of critical angle reflector in the cavity for mode selection, mode selection using Q modulation etc., are already treated in publications on relevant laser physics, their discussion is here omitted.

4. Unstable resonator mode selection

It is known from the principle of laser that, as long as a spherical resonator satisfies the non-stable condition $g_1g_2 > 1$ or $g_1g_2 < 0$, it is called an unstable resonator. It is characterized by the presence of inherent ray divergent loss. When experiencing continuous reflection on the face of the cavity mirror, the paraxial rays are always deflected outward, getting farther and farther away from the cavity axial line till running outside the cavity. So such a cavity is also called the high loss cavity.

There is a great variety of forms of constructing unstable resonators, but those in application are mainly as follows:

(1) The double convex cavity, which is made up of two convex mirrors (Fig. 5.2-14(a)).

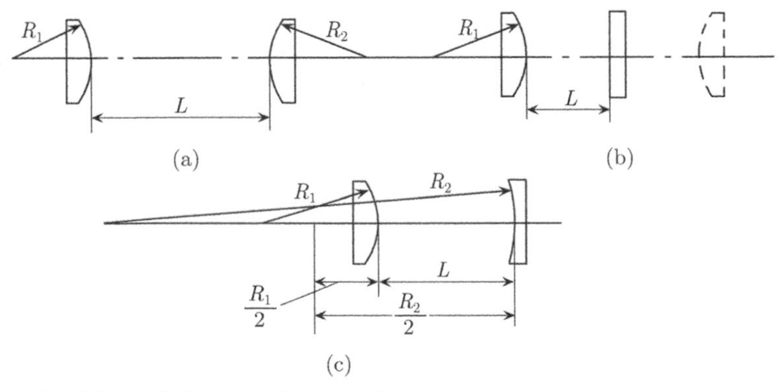

Fig. 5.2-14 Schematic diagram of three unstable resonators structures

(2) The plane-convex cavity, made up of a plane mirror and a convex mirror (Fig. 5.2-14(b)).

(3) The virtual confocal telescopic cavity, made up of a concave mirror and a convex mirror and the two mirrors have a common (virtual) focus outside the cavity (Fig. 5.2-14(c)). Such a cavity can output a light beam of very small divergence angle and is a frequently adopted unstable resonator.

Compared with the stable resonator the unstable resonator has some remarkable merits:

(1) The unstable resonator has a rather large mode volume. The light beam in the stable cavity is completely concentrated around the cavity axis and confined in the limited volume in the cavity, restraining the effective utilization of the laser operation material volume while the light beam in the unstable resonator cavity is divergent; the oscillating light beam can fill the entire operation material in the cavity, raising the availability of the operation material volume. Even if it's single mode output, there is still a very large mode volume, thus making it possible to obtain a high power output.

(2) It is easy to realize fundamental mode oscillation with the unstable resonator. For the stable cavity, the diffraction loss of different oscillation modes does not differ much and single mode oscillation cannot be realized unless mode selection measures are taken. But, for the non-stable cavity, the geometric deflection loss is great and the difference between different oscillating modes is very great, that is, the mode discriminating capability is high. Therefore, the higher-order modes are easily suppressed while allowing only for fundamental mode oscillation of the lowest order.

(3) As the divergence angle of the confocal resonator is small, the diameter of its output light spot is over one order greater than that of the beam waist light spot of the stable cavity and so the divergence angle is one order smaller. Moreover, the unstable resonator has such advantages as uniform light beam in the cavity and the ease with which the output coupling is regulated. Therefore, such a cavity is most suitable for high gain and large operation volume gas and solid lasers.

Something about the amplification and energy loss of unstable resonators. The propagation of light in a non-stable cavity can be regarded as geometric imaging of a pair of conjugate image points P_1 and P_2 in a reflective mirror system M_1 and M_2, as shown in Fig. 5.2-15. Thus the two image points on the axis satisfy the self-consistent condition of back and forth imaging. The two conjugate image points can be regarded as two geometric light emitting points. After the two spherical waves emitted by them travel back and forth in the cavity once, the wave surface of the spherical wave remains self-consistent or realizes self-reproduction. The feature of such a spherical wave oscillation mode is that, after a number of travels back and forth in the cavity, the central position of the spherical wave remains unchanged. When such a self-reproducted spherical wave makes a back and forth travel in the cavity, the transverse dimensions of its wave face are amplified M times. Suppose $m_1 = \dfrac{a'_1}{a_1} = \dfrac{l_1 + L}{l_1}$ is the single path amplification of mirror M_1, and $m_2 = \dfrac{a'_2}{a_2} = \dfrac{l_2 + L}{l_2}$

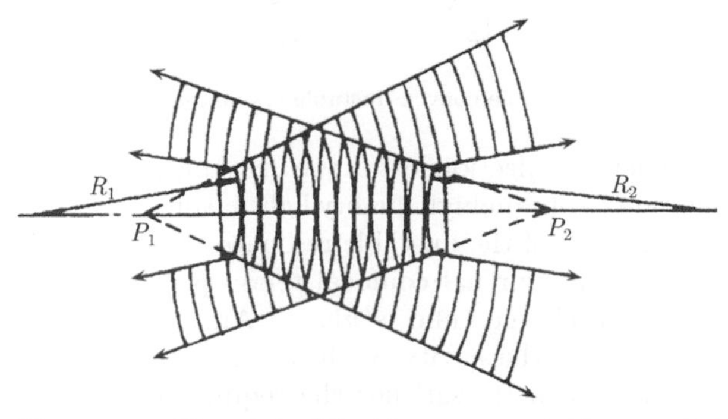

Fig. 5.2-15 The mode of double convex unstable resonator

that of mirror M_2. Then the back and forth amplification

$$M = m_1 m_2 = (l_1 + L)(l_2 + L)/l_1 l_2 \tag{5.2-7}$$

where l_1 is the distance from image P_1 to mirror surface M_1 and l_2 that from image P_2 to mirror surface M_2. Similarly, the amplification of the telescope-type unstable resonators can be found.

$$M = m_1 m_2 = \left|\frac{R_1}{R_2}\right| \tag{5.2-8}$$

The amplification M is an important parameter of the unstable resonators. It is known from the above analysis that it is related to only the cavity length L and the cavity mirror curvature radii R_1, R_2 while unrelated to the transverse dimensions a_1, a_2 of the cavity mirror. As the transverse dimensions of the light beam will increase M times, every time it travels back and forth once in the cavity, after many times of travel, the mode volume of the light beam in the cavity will fill the entire active medium, making full use of its operation material.

The energy loss of the unstable resonators is closely related to the geometric amplification, as shown in Fig. 5.2-16. When the spherical wave equivalent to being emitted from the image point P_1 and completely covering M_1 reaches M_2, its wave face dimensions are already beyond the range of M_2, of which only the part intercepted by M_2 is reflected back while the wave face exceeding the M_2 range will run away from the cavity resulting in energy loss. As the self-reproduced waveform is a uniform spherical wave, the proportion occupied by the energy loss will be determined by the area of the wave face exceeding M_2 to the area of the whole wave face and this value is related to the geometric amplification. It is found by inference that the energy loss rate for light to travel back and forth once in the cavity (virtual confocal cavity).

$$\delta = 1 - \frac{1}{M^2} = 1 - \left(\frac{R_2}{R_1}\right)^2 \tag{5.2-9}$$

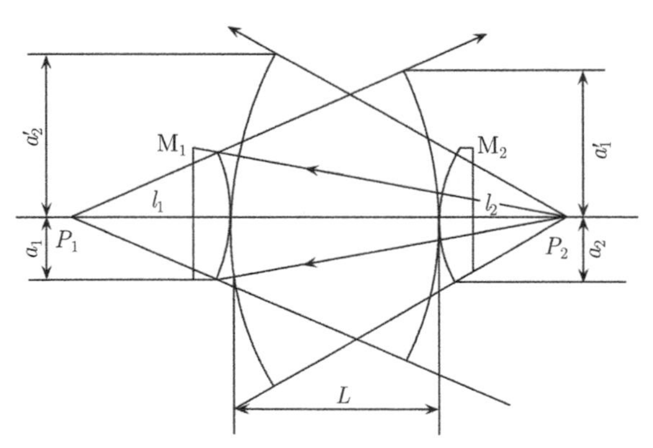

Fig. 5.2-16 The double convex unstable resonators amplification

Because of the great geometric deflection loss of the unstable resonators, it is easy to realize fundamental mode oscillation to obtain high quality light beam of single mode output close to the diffraction limit. An example of the unstable resonator laser is as shown in Fig. 5.2-17, which displays the structure of a virtual confocal telescope-type unstable resonator CO_2 laser. The material of the discharge tube is glass. An inclined (in general 45°) plane reflective mirror M_3 is placed in the cavity, with a light transmitting hole in the center of the mirror, whose diameter exactly satisfies the requirement that the oscillating laser beam agree with the light transmitting aperture (Fig. 5.2-18). Now satisfy $M = a_0/a_2$ (a_0

is the radius of the light spot on the concave mirror and a_2 the radius of the central light transmitting hole of the 45° plane reflective mirror M_3), with the laser beam laterally output. The inner diameter of the laser discharge tube is 2.54 cm, with a 5-cm ring-shaped aluminum electrode each mounted on either end. The distance between the electrodes is 120 cm. The operation current of the DC source is 40 mA, the voltage 10 kV, and the resonator length 160 cm while the curvature radii of the convex mirror and concave mirror are −27 m and +30 m, respectively, where three coupled reflective mirrors are placed, the light transmitting apertures (diameters) being 1.43 cm, 1.85 cm, and 2.00 cm, respectively. The corresponding Fresnel numbers are 3.6, 6.1, and 7.1, respectively. By adopting the mode of lateral output, when choosing a 2.00-cm aperture coupling reflector, the fundamental mode output power is 22 W.

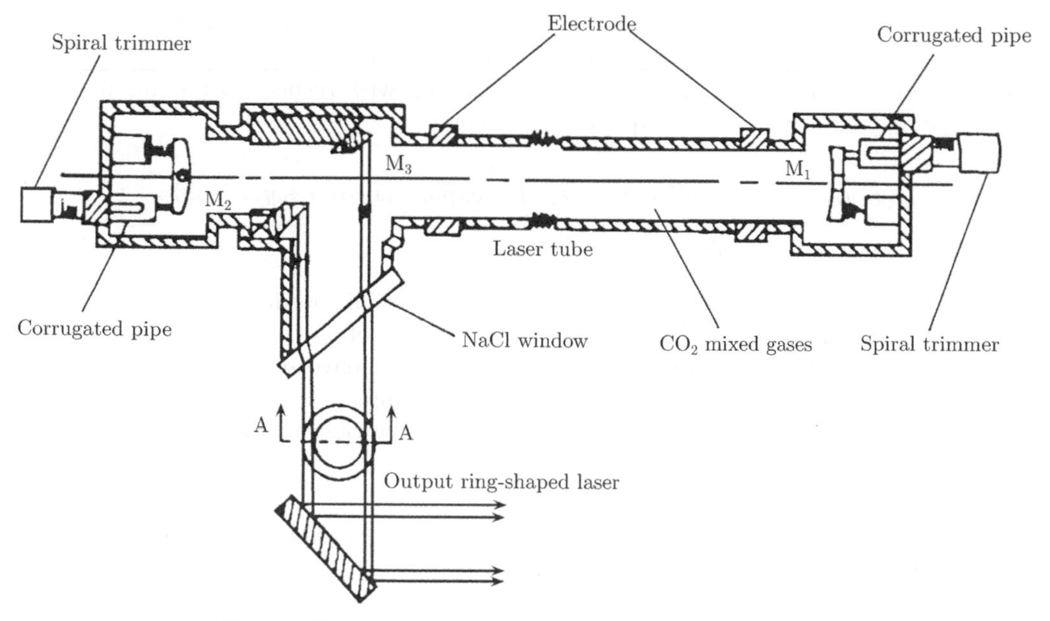

Fig. 5.2-17 The structure of a unstable resonator CO_2 laser

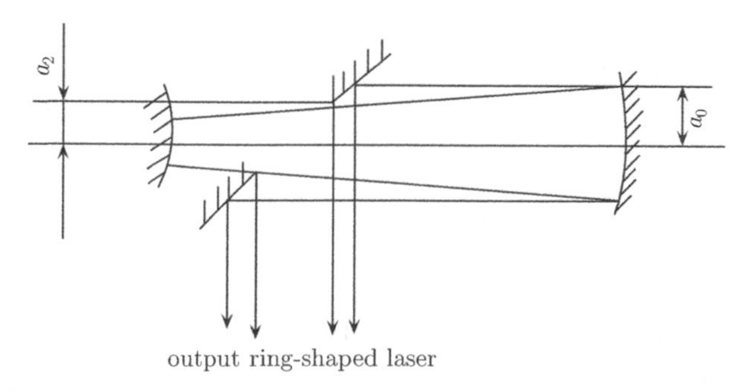

Fig. 5.2-18 The light transmitting aperture matches light beam

5.2.3 New development in research on the form of laser beam—the Bessel light beam

J. Durnin suggested that the scalar wave equation of the electromagnetic field

$$\left(\nabla^2 - \frac{1}{c^2}\frac{\partial^2}{\partial x^2}\right) E(r,t) = 0 \tag{5.2-10}$$

set of rigorous solutions

$$E(x, y, z, t) = \exp[\mathrm{i}(k_z z - \omega t)] \int_0^{2\pi} A(\phi) \exp[\mathrm{i}k_\perp (x \cos\phi + y \sin\phi)]\mathrm{d}\phi \tag{5.2-11}$$

represents a light beam transmitted along the z axis, where $k_\perp$-k_z represent the transverse and z oriented wave vector components, respectively.

$$k_\perp^2 + k_z^2 = (2\pi/\lambda)^2 \tag{5.2-12}$$

As the z coordinate and x, y coordinates of the light field represented by Eq. (5.2-11) are separated from each other, the cross-section distribution of the light beam does not vary with z. Therefore, when k_z is a real number, Eq. (5.2-11) represents a light beam transmitted along the z direction and whose light field cross-section distribution does not vary with the transmission distance. J. Durnin calls it the nondiffraction beam.

The $A(\phi)$ in Eq. (5.2-11) can be an arbitrary function with respect to the angular coordinate ϕ. By letting $A(\phi) = 1/2\pi$, Eq. (5.2-11) becomes

$$\begin{aligned} E(x, y, z, t) &= \frac{1}{2\pi} \exp[\mathrm{i}(k_z z - \omega t)] \int_0^{2\pi} \exp[\mathrm{i}k_\perp (x \cos\phi + y \sin\phi)]\mathrm{d}\phi \\ &= \exp[\mathrm{i}(k_z z - \omega t)] J_0(k_\perp \rho) \end{aligned} \tag{5.2-13}$$

where $J_0(x)$ is the first class of Bessel function; ρ is the radial coordinate, $\rho^2 = x^2 + y^2$.

Therefore, the nondiffraction light beam is also spoken of as the "Bessel light beam". When $k_z = 0$, the above is reduced to the plane wave. Conversely, the above equation represents a radially attenuating "nondiffraction light beam".

The light intensity distribution in the cross section of the Bessel light beam is

$$I(x, y) = J_0^2(k_\perp \rho) \tag{5.2-14}$$

The Bessel light beam in the above equation is in reality composed of many rings, the radius of whose central spot is

$$\rho_0 = \frac{2.4}{k_\perp} \sim \lambda \tag{5.2-15}$$

Thus it can be seen that the Bessel light beam possesses a central spot only the wavelength order size. And, according to its "nondiffraction" characteristic, the central spot that is so small does not vary with the transmission distance. This characteristic of the Bessel light beam is very attractive and, precisely because of this, as soon as the Bessel light beam was put forward, it received immediate recognition.

But, very unfortunately, the Bessel light beam is non-square integrable. Hence an ideal Bessel light beam possesses infinitely large power. This is clearly impossible. Therefore, in reality it is impossible for an ideal Bessel light beam to exist. An actual Bessel light beam can only be a finite caliber approximate Bessel light beam.

According to an analysis, for an approximate Bessel light beam with an aperture radius R_0, within the distance

$$Z_{\max} = R_0 \left[\left(\frac{2\pi}{k_\perp \lambda} \right)^2 - 1 \right]^{\frac{1}{2}} \approx R_0 \frac{2\pi}{k_\perp \lambda} \tag{5.2-16}$$

the spot in the center of the light beam can approximately remain undiffused. The Rayleigh distance is one of the most well-known parameters of wave optics:

$$Z_R = \frac{\pi r_0^2}{\lambda} \tag{5.2-17}$$

which represents the transmission distance over which a light beam with a spot radius r_0 and wavelength λ can stay undiverged. This distance is the greatest distance the light beam is capable of transmission beginning from the waist spot under the condition the light beam spot area is doubled. The Rayleigh distance of a light beam with a spot radius of only $\rho_0 = \dfrac{2.4}{k_\perp} \approx \lambda$ is

$$Z_{R\rho_0} = 18.2\frac{1}{\lambda k_\perp^2} \tag{5.2-18}$$

As $\dfrac{Z_{\max}}{Z_{R\rho_0}} = 0.35R_0 \cdot k_\perp \gg 1$, judging merely by the transmission features of the Bessel light beam's central spot, the distance it can be transmitted is far greater than the light beam transmission distance represented by the Rayleigh distance.

Of course, the above conclusion is not in contradiction with the light beam transmission features characterized by $M^2 \geqslant 1$ since only the central spot of the Bessel light beam is considered in the above analysis while in reality, for the Bessel light beam, there exists a large amount of energy apart from the central spot, which must be considered in the second-order moment integration.

For quite some time, people have proposed different forms of the light beam, attempting to break through the constraint brought about by the Rayleigh distance and even the restraint by $M^2 \geqslant 1$. As the definition of the Rayleigh distance itself is rather qualitative with respect to the so-called "indistinct divergence" of the light beam, the breakthrough itself of the Rayleigh distance restraint, too, cannot be explicitly defined. But $M^2 \geqslant 1$ can be rigorously proved.

5.3 The longitudinal mode selecting technology

5.3.1 The principle of longitudinal mode selection

The oscillation frequency range of a laser is dependent on the width of the gain curve of the operation material, while the generation of the number of multi-longitudinal mode oscillations is dependent on the gain line width and the frequency interval between the resonator's two adjacent longitudinal modes. That is, within the gain line width, if only several longitudinal modes reach the oscillation threshold value at the same time, oscillation can in general form. If $\Delta\nu_0$ is used to represent the width of the part of the gain curve that is higher than the threshold, the frequency interval of the adjacent longitudinal modes is $\Delta\nu_q$. Then the number of longitudinal modes that are likely to oscillate at the same time is

$$n = \frac{\Delta\nu_0}{\Delta\nu_q}$$

For general stable cavities, it is known from the theory of diffraction that different transverse modes (TEM_{mn}) have different resonance frequencies. So, the greater the number of transverse modes that take part in oscillation is, the more complicated the total oscillating frequency spectral structure will be. It is when there exists only single transverse mode (TEM_{00}) oscillation in the cavity that its oscillating frequency spectral structure will be rather simple, being a series of separate oscillating frequencies with an interval of $\Delta\nu/2nL$.

If the laser operation material possesses multiple laser spectral lines, in order to realize single longitudinal mode selection, it is first necessary to reduce the fluorescent spectral lines of the operation material that may generate laser to make it retain only one fluorescent spectral line. So the frequency rough choice method must be used to restrain the unwanted spectral lines. Then the transverse mode selecting method can be used to choose the TEM_{00} modes, on the basis of which the longitudinal modes can be chosen.

The basic philosophy of longitudinal mode selection: Whether a certain longitudinal mode in the laser can begin to oscillate or not and maintain the oscillation mainly depends on the relative magnitude of this longitudinal mode's gain and loss. Therefore, by controlling one of the two parameters so that only one of the longitudinal modes that may exist in the resonator will satisfy the oscillation condition, the laser will be able to realize single longitudinal mode operation. For different longitudinal modes of one and the same transverse mode, their losses are the same, but there exists some difference in the gain between different longitudinal modes. By making use of the difference in the gain between different longitudinal modes, we can introduce a definite selective loss into the resonator (e.g., insert an etalon) to make the to-be-selected longitudinal mode loss minimum while the additional loss of the remaining ones will be rather great, that is, increasing the difference in the net gain between the longitudinal modes, with oscillation set up for only the few high gain longitudinal modes around the central frequency. Thus, in the course of formation of laser, via the mode competition mechanism between multiple longitudinal modes, what is finally formed and amplified is the corresponding single longitudinal mode of the central frequency of maximum gain.

5.3.2 The method of longitudinal mode selection

1. Dispersion cavity frequency rough choice

If the laser operation material possesses the capability of emitting multiple spectral lines of different wavelengths, for example, the He-Ne laser can emit three spectral lines, of 632.8 nm, 1.15 μm, 3.39 μm, then, prior to selection of longitudinal modes, it is necessary to perform rough choice of the frequency to suppress the unwanted spectral lines. This is usually done by making use of the spectral characteristics of the cavity mirror's reflective film or by inserting such dispersion elements as a prism or a grating to separate the light beams of different wavelengths emitted by the operation material in space. Then effort should be made to allow only light beams in the rather narrow wavelength region to form oscillation in the cavity, while the light beams with other wavelengths, for lack of feedback capability, will be suppressed.

Figure 5.3-1 shows the rough choice equipment for inserting dispersion prisms in the cavity. In this case, the minimum wavelength range the resonator can select for oscillation is determined by the angular dispersion of the prism and the divergence angle of the oscillating light beam in the cavity. Suppose the incident angle α_1 at which the ray enters the prism is equal to the outgoing angle α_2 at which the ray leaves the prism, or $\alpha_1 = \alpha_2 = \alpha$.

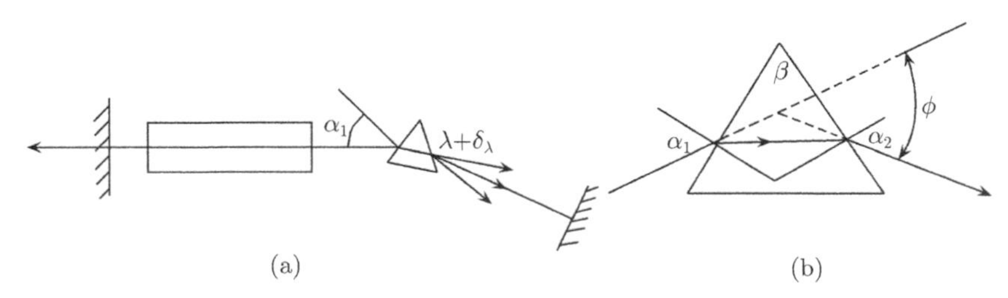

Fig. 5.3-1 The prism dispersion rough choice devices

According to a physical optics analysis, there is

$$n = \sin\alpha / \sin\frac{\beta}{2} = \sin\left(\frac{\phi+\beta}{2}\right) \Big/ \sin\frac{\beta}{2} \tag{5.3-1}$$

where α_1 is the incident angle, n is the refractive index, β is the apex angle of the prism, and

ϕ is the deviation angle. The angular dispersion power of the prism is defined as $D_\lambda = \dfrac{\mathrm{d}\phi}{\mathrm{d}\lambda}$, that is, the amount of variation of the deviation angle every time the wavelength varies by 0.1 nm. Substitution of Eq. (5.3-1) after differentiation yields

$$D_\lambda = \frac{\mathrm{d}\phi}{\mathrm{d}n}\frac{\mathrm{d}n}{\mathrm{d}\lambda} = \frac{2\sin\dfrac{\beta}{2}}{\sqrt{1-n^2\sin^2\dfrac{\beta}{2}}}\frac{\mathrm{d}n}{\mathrm{d}\lambda} \tag{5.3-2}$$

where $\mathrm{d}n/\mathrm{d}\lambda$ represents the derivative of the refractive indices of different materials with respect to wavelength variation. Suppose the divergence angle allowed by the light beam in the cavity is θ. Then because of the spectroscopic action of the dispersion prism, the minimum wavelength splitting range that can be allowed by the laser wavelength in the cavity is

$$\Delta\lambda = \frac{\theta}{D_\lambda} = \frac{\sqrt{1-n^2\sin^2\dfrac{\beta}{2}}}{2\left(\sin\dfrac{\beta}{2}\right)\dfrac{\mathrm{d}n}{\mathrm{d}\lambda}}\cdot\theta \tag{5.3-3}$$

For a prism made of glass material and the visible light waveband, when $\theta \approx 1$ mrad, the reachable $\Delta\lambda \approx 1$ nm. This prism dispersion method is highly effective for some lasers to select oscillation. For instance, the two powerful operating spectral lines 488 nm and 514.5 nm of the argon ion laser can be selected by adopting such a dispersion.

Another dispersion cavity uses a reflective grating instead of a reflective mirror in the resonator, as shown in Fig. 5.3-2. Suppose d is the grating distance (grating constant), α_1 the incident angle at which the ray is injected, and α_2 the reflection angle from which the ray is reflected. Then the condition for forming the grating diffraction main maximum value is

$$d(\sin\alpha_1 + \sin\alpha_2) = m\lambda \tag{5.3-4}$$

where $m = 0, 1, 2, \ldots$ is the level of diffraction. It can be seen from Eq. (5.3-4) that, when the incident angles are identical, the level-0 spectral lines ($m = 0$) of different wavelengths overlap one another without the dispersion light splitting action. For all the other levels of spectral lines, the grating's angular dispersive power can be found by the following equation:

$$D = \frac{\mathrm{d}\alpha_2}{\mathrm{d}\lambda} = \frac{m}{d\cos\alpha_2} = \frac{\sin\alpha_1 + \sin\alpha_2}{\lambda\cos\alpha_2} \tag{5.3-5}$$

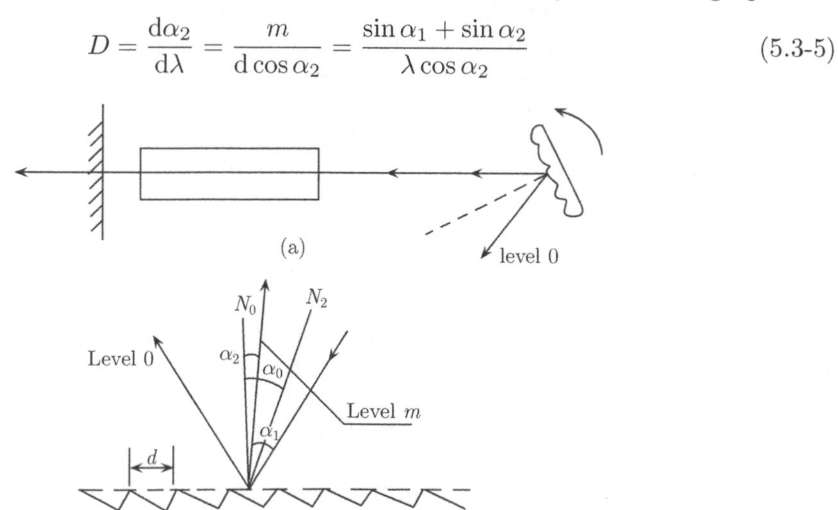

Fig. 5.3-2 The grating dispersion cavity

Usually the grating operates in the autocollimating state, that is, $\alpha_1 = \alpha_2 = \alpha_0$ (α_0 is the blazed angle of the grating, i.e., the included angle between the normal N_0 of the grating plane and the normal N_2 of each seam's plane). Thus the grating's angular dispersive power is

$$D_0 = \frac{2\tan\alpha_0}{\lambda} \tag{5.3-6}$$

Suppose the allowable light beam divergence angle in the cavity is θ. Then, because of the grating dispersion, the allowable minimum split wavelength range is

$$\Delta\lambda = \frac{\theta}{D_0} = \frac{\lambda}{2\tan\alpha_0}\theta \tag{5.3-7}$$

For the visible spectral region, suppose $\alpha_0 = 30°$ and $\theta = 1$ mrad. Thus, $\Delta\lambda$ is not up to the 1-nm order. It can thus be seen that its dispersion selection capability is even higher than that of the prism. As there does not exist the light beam's transmission loss with the grating method, it is applicable to rather wide spectral region lasers. Moreover, when properly turning the position of the grating's angle, the grating dispersion cavity can change the oscillation spectral region needed.

Although the dispersion cavity method is capable of selecting rather narrow oscillating spectral lines from a wide range of spectral lines to realize single fluorescent spectral line oscillation, this is as yet a comparatively rough selection. Within the range of the fluorescent line width of the said spectral line, there still exists a series of separate oscillating frequencies with a frequency interval of $\Delta\nu = c/2nL$, that is, multiple longitudinal modes. How to further select single longitudinal modes from a single spectral line calls for adoption of the following methods.

2. The short cavity method

The possible number of laser oscillation longitudinal modes is mainly determined by the gain line width $\Delta\nu_0$ of the operation material and the resonator's longitudinal modes interval $\Delta\nu_q$, while the longitudinal modes interval $\Delta\nu_q = c/2nL$ is inversely proportional to the cavity length. Therefore, one of the methods of selecting the single longitudinal mode is to shorten the length L of the resonator to increase $\Delta\nu_q$ such that there exists only one longitudinal mode within the range of $\Delta\nu_0$ while all the remaining longitudinal modes are located outside $\Delta\nu_0$, as shown in Fig. 5.3-3. This is what is called the short cavity method for selecting the longitudinal mode. Simple and practical, this method is especially suitable for small power gas lasers. Take the He-Ne laser. When its cavity length $L = 1$m, its longitudinal modes interval $\Delta\nu_q = c/2nL = 150$ MHz (suppose $n = 1$). Within the range of gain line width $\Delta\nu_0 = 1500$ MHz, there may be ten longitudinal modes oscillating. When the cavity length L is shortened to 10 cm, $\Delta\nu_q = 1500$ MHz. Now there can be only one longitudinal mode oscillation. In fact, some lasers just cannot adopt this method, for example, the Ar^+ laser, whose gain line width $\Delta\nu_0 = 5500$ MHz. If single longitudinal mode oscillation is desired, then the cavity length is required to be below 3 cm. Also, for solid lasers such as the YAG, the gain line width $\Delta\nu_0 = 200$ GHz, and if single longitudinal mode oscillation is required, the cavity length L is required to be 0.4 mm. When a watt order continuous laser diode is adopted for pumping, a 0.5~1 mm YAG thin chip can output single longitudinal mode laser greater than 100 mW. So the short cavity method is suitable only for lasers of a rather narrow gain line width. At the same time, as the cavity length is shortened, the laser output power is bound to be restrained. Hence on occasions for high power single longitudinal mode output, this method is unsuitable.

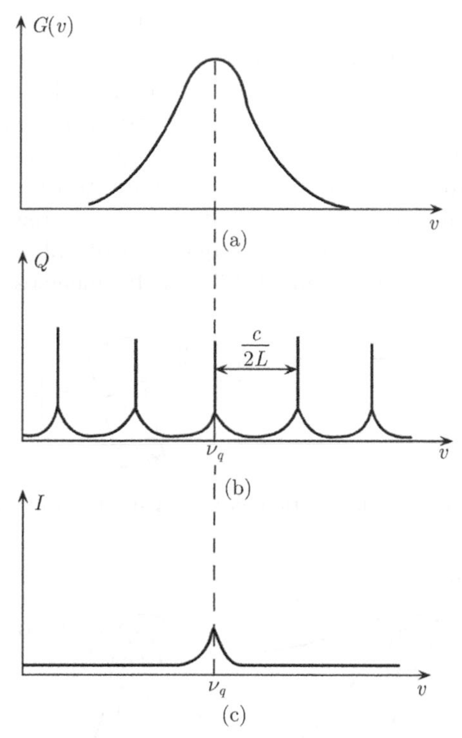

Fig. 5.3-3 The principle of mode selection with short cavity method

3. The F-P etalon method

What is shown in Fig. 5.3-4 is a device schematic diagram for longitudinal mode selection with the etalon. The Fabry-Perot (F-P) etalon possesses different transmissivities for light beams of different wavelengths, which can be expressed with the following equation:

$$T(\lambda) = \frac{1}{1 + F\sin^2\left(\frac{\varphi}{2}\right)} = \frac{1}{1 + F\sin^2\left(\frac{2\pi d}{\lambda}\right)} \tag{5.3-8}$$

where $F = \dfrac{\pi\sqrt{R}}{1-R}$ is the etalon's fineness, R is the reflectivity of etalon to light, d is the thickness of the etalon (or the interval between the two parallel planes), and φ the phase difference of the two and adjacent rays in the etalon that take part in the multi-light beam interfering effect, i.e., $\varphi = \dfrac{2\pi}{\lambda}2nd\cos\alpha'$ (where n is the refractive index of the etalon medium, α' is the refractive angle after the light beam enters the etalon, which is in general very small, $\cos\alpha' \approx 1$), $T(\lambda)$ is a function of the wavelength or φ and R). Figure 5.3-5 shows the curves of variation of $T(\lambda)$ and φ when R takes different values. It can be seen from the figure that, the greater the reflectivity R of the etalon, the narrower the transmission curve, and the better the selectivity. The interval between the maximum values of two adjacent transmissivities is

$$\Delta\nu_{\mathrm{m}} = \frac{c}{2nd\cos\alpha'} \approx \frac{c}{2nd}$$

The above equation is usually spoken of as the free spectral region of the etalon. It can be seen that the thickness d of the etalon is much smaller than the resonator length L. Therefore, its free spectral region is much larger than the longitudinal modes interval of the resonator. Thus, insert an etalon in the resonator of the laser and select appropriate thickness and reflectivity to make $\Delta\nu_{\rm m}$ equivalent to the laser operation material's gain line width, as shown in Fig. 5.3-6. It can be seen from the figure that the longitudinal mode located at the central frequency is consistent with the $\nu_{\rm m}$ at the location of the etalon's maximum transmittance. So this mode has the least loss, or the Q value is maximum and can begin to oscillate while the remaining longitudinal modes, owing to the excessive additional losses and too low a Q value, cannot form laser oscillation. Regulating the inclined angle of the etalon to change α can make $\nu_{\rm m}$ overlap the frequencies of different longitudinal modes so that single longitudinal mode laser output of different frequencies can be obtained.

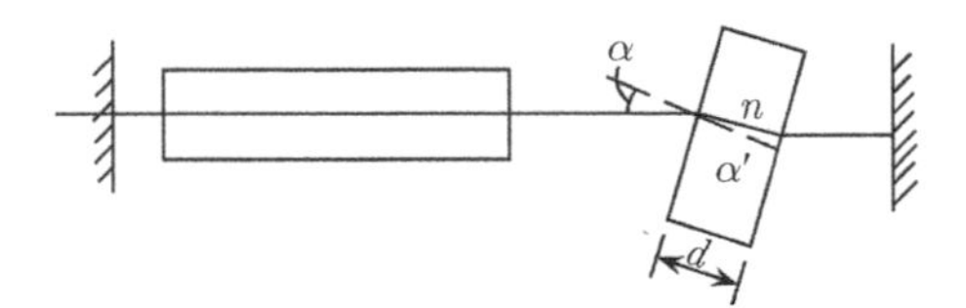

Fig. 5.3-4 Mode selection with the F-P etalon method

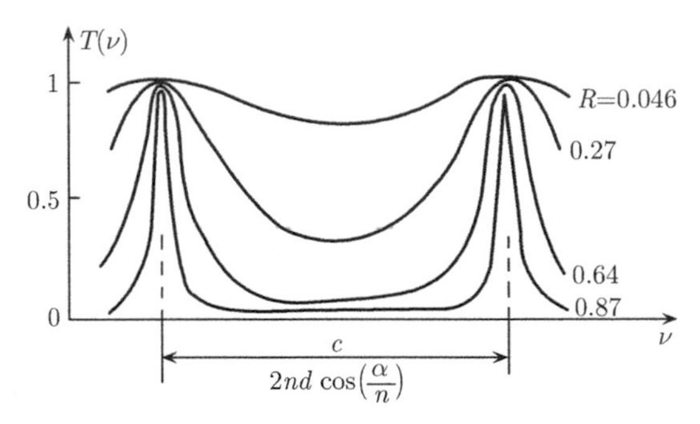

Fig. 5.3-5 The transmissivity of F-P etalon

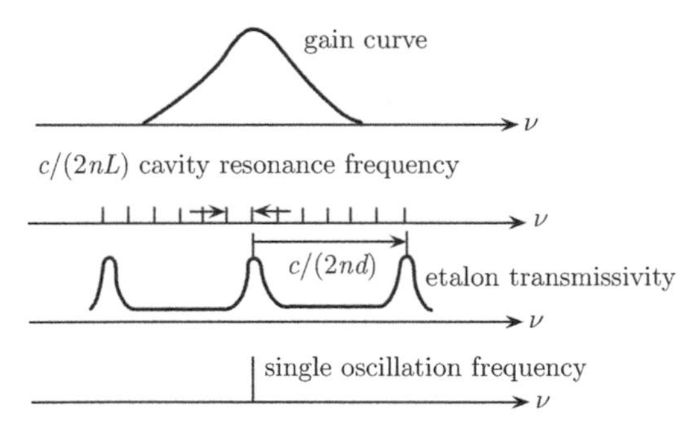

Fig. 5.3-6 Single longitudinal mode selection with F-P etalon

The advantage of longitudinal mode selection using the F-P etalon is that the thickness between the etalon's parallel plane plates can be made very thin. Hence for laser operation materials with a very wide gain line width, such as Ar^{+}, Nd:YAG, ruby lasers, etc., single longitudinal mode oscillation can always be obtained. In addition, as the cavity length has not been shortened, the output power can still be very high.

The fluorescence line width of gas lasers is in general rather narrow. When selecting the longitudinal mode with the etalon, only one etalon will do. For solid lasers, however, as the fluorescence line width is very wide, just one etalon is often insufficient for realizing single longitudinal mode oscillation because, restricted by technological factors, it's impossible for

F to have very great numerical values. When a laser has a rather long cavity length, the interval between longitudinal modes is rather small. If the etalon's free spectral region is very large, so will be its bandwidth. Therefore, it's difficult to ensure single longitudinal mode oscillation. So only by inserting a second etalon with a relatively small free spectral region will it be possible to obtain the single longitudinal mode (Fig. 5.3-7). We have an example below.

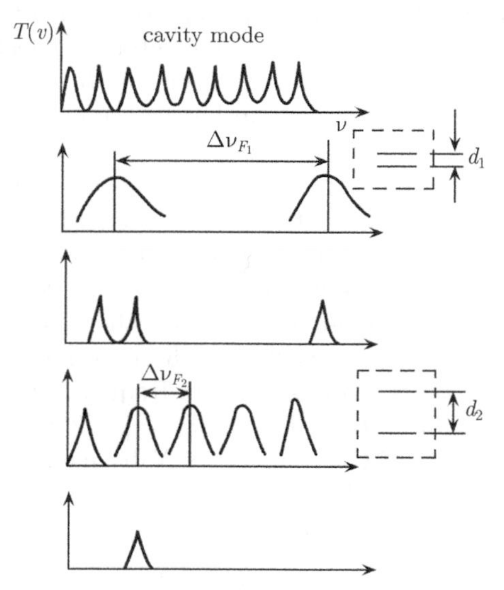

Fig. 5.3-7 Selection of single longitudinal mode with dual etalon

Suppose there is a solid laser with an Nd:YAG as its operation material, whose fluorescent line width $\Delta\nu_0$ is 2×10^{11} Hz, resonator length $L = 850$ mm, from which it is found by calculation that the longitudinal mode interval $\Delta\nu_{\mathrm{q}} = c/2nL = 1.7 \times 10^8$ Hz (supposing $n = 1$). In order to select the single longitudinal mode, the etalon is required to have a sufficiently narrow bandwidth. Let $\Delta\nu_{t_1} = 2\Delta\nu_{\mathrm{q}}$, that is, $\Delta\nu_{t_1} = 3.41 \times 10^8$ Hz. Suppose the surface planeness of the etalon is $\lambda/100$ and reflectivity $R = 94\%$. Then $F_{\mathrm{p}} = m/2 = 100/2 = 50$, nad $F_{\mathrm{r}} = \pi R^{1/2}/(1 - R) \approx 50$. From the formula $\dfrac{1}{F^2} = \dfrac{1}{F_{\mathrm{r}}^2} + \dfrac{1}{F_{\mathrm{p}}^2}$ (F_{r} is the fineness determined by the etalon reflectivity, F_{p} is that determined by the etalon's planeness m), it can be found that $F = 35$. The free spectral region of this etalon $\Delta\nu_{m_1} = F \cdot \Delta\nu_{t_1} = 35 \times 3.41 \times 10^8 \text{ Hz} \approx 1.2 \times 10^{10}$ Hz, from which we can find that the thickness of the etalon (supposing $n = 1.5$) is

$$d = c/(2n\Delta\nu_{m_1}) = 3 \times 10^{11}/(2 \times 1.5 \times 1.2 \times 10^{10}) \text{ mm} = 0.83 \text{ mm}$$

When choosing the second etalon, there should be $\Delta\nu_{t_2} \leqslant 2\Delta\nu_{m_1}$. Take $\Delta\nu_{t_2} = 2\Delta\nu_{m_1} = 2.4 \times 10^{10}$ Hz. Let $\Delta\nu_{m_2} = \Delta\nu_0$. Then $F_2 = \Delta\nu_{m_2}/\Delta\nu_{t_2} = \Delta\nu_0/\Delta\nu_{t_2} \approx 10$, which can satisfy the condition for selection of the single longitudinal mode. By finding the second etalon's surface planeness to be $\lambda/30$ and $R_2 = 75\%$, then the etalon's thickness $d_2 = c/(2n\Delta\nu_{m_2}) = 0.41$ mm.

From the analysis of the above example, we have the following results:

(1) The parameters of the first etalon: thickness $d = 0.83$ mm (can choose 1 mm), planeness is $\lambda/100$, $R = 94\%$.

(2) The parameters of the second etalon: thickness $d_2 = 0.41$ mm (can choose 0.5 mm), planeness is $\lambda/30$, $R_2 = 75\%$.

The above calculations are all made under definite assumptions. In reality, it is necessary to modify some of the parameters by experiment before the actual requirements can be met.

4. The compound resonator method

If a reflective interferometer system is used to replace a reflective mirror in the resonator, then the combined reflectivity is a function of the optical wavelength (frequency). What is shown in Fig. 5.3-8 is a diagram of the principle of two kinds of combined interferometer compound resonators. Its characteristic lies in the combined reflectivity R varying periodically with the frequency. At certain particular frequencies, R possesses maximum values. It is shown by analysis that the frequency interval between maximum values can be changed by adjusting the length of the compound resonator.

Figure 5.3-8(a) shows a Michelson interferometer-type compound resonator, which is constructed by replacing a reflective mirror of the resonator with a Michelson interferometer. This compound resonator can be regarded as made up of two subresonators. The totally reflecting mirrors M and M_1 form a subresonator, of cavity length $L + l_1$, and resonant frequency $\nu_{1i} = \{c/[2(L+l_1)]\}q_i$ (suppose $n = 1$). The other subresonator is made up of the totally reflecting mirrors M and M_2, the cavity length being $L + l_2$, and resonant frequency $\nu_{2j} = \{c/[2(L+l_2)]\}q_j$ Therefore, the laser's resonant frequency has to satisfy the above two conditions at the same time, that is, $\{c/[2(L+l_1)]\}q_i = \{c/[2(L+l_2)]\}q_j$. Furthermore, the frequency of the first subresonator light beam after N frequency intervals is exactly equal to that of the second subresonator light beam after $N + 1$ frequency intervals once again. From this the frequency interval of the compound resonator can be obtained:

$$\Delta\nu = c/[2(l_1 - l_2)] \tag{5.3-9}$$

It can be seen from the above equation that appropriate selection of l_1 and l_2 can make the compound resonator's frequency interval sufficiently great; that is, the interval between two adjacent longitudinal modes is sufficiently great. When compared with the gain line width, single longitudinal mode operation can be realized.

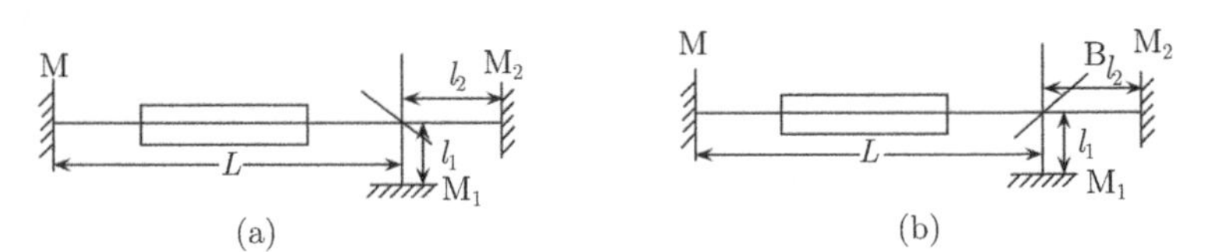

Fig. 5.3-8 Mode selection with the compound resonator

What is shown in Fig. 5.3-8(b) is a Fox-Smith interferometer type compound resonator. The resonator is also composed of two subresonators, one of which is made up of reflective mirrors M and M_2, the cavity length is $L + l_2$, and the other is made up of reflective mirrors M and M_1. But the process of light beam propagation of this subresonator consists in the light getting through mirror B to reach mirror M_2 and then reflected by M_2 and mirror B to mirror M_1. The process of return propagation is also one in which the light is reflected by mirror B to reach mirror M_2 first and then reflected by M_2 through mirror B to mirror M. Hence the length of this subresonator is $L + 2l_2 + l_1$. The resonant frequencies of the two subresonator are, respectively,

$$\nu_{1i} = \{c/[2n(L + l_2)]\}q_i$$

and

$$\nu_{2j} = \{c/[2n(L + l_2 + l_1)]\}q_j$$

Thus the resonant frequency of the compound resonator must satisfy the above two equations at the same time, that is,

$$\nu = \nu_{1i} = \nu_{2j} \tag{5.3-10}$$

In this case, the intensity of light output from mirror B is zero, and the interferometer has the greatest reflectivity for the light beam in the resonator. Supposing in the compound resonator another longitudinal mode frequency adjacent to ν is ν', then it can be proved that in the compound resonator the interval between the two adjacent frequencies is

$$\Delta\nu = \frac{c}{2(l_1 + l_2)} \tag{5.3-11}$$

It is obvious that when suitable l_1 and l_2 are so chosen that $\Delta\nu$ and the gain line width are made comparable, we can obtain the single longitudinal mode output.

For instance, there is a kind of composite cavity similar to the Fox-Smith type, as shown in Fig. 5.3-9. This is a compound resonator made up of a prism, one of whose sides is equivalent to a spectroscope and the other two plated with totally reflecting film and equivalent to two totally reflecting mirrors. Such a structure has two advantages. First, what plays the role of beam splitting is a single plane, thus avoiding the multiple beam interference due to two planes of the commonly used beam splitter; second, the prism can be placed in a thermostat groove, thus ensuring the stability of l_1 and l_2. Used in the Ar^+ laser, such a structure can maintain the resonant frequency within the range of ±3 MHz to select the single longitudinal mode output.

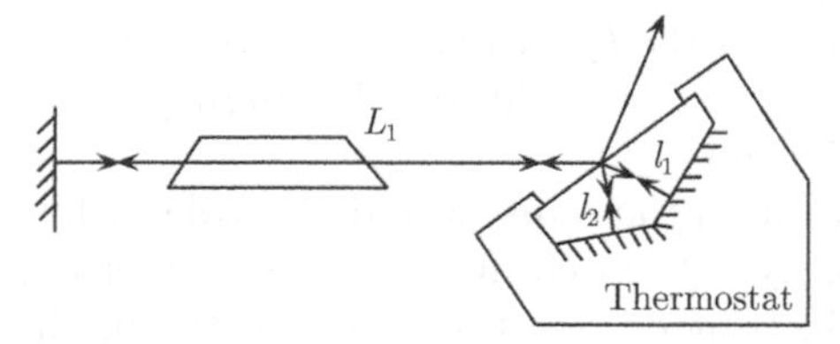

Fig. 5.3-9 Mode selection with a prism compound resonator

5. Other longitudinal mode selecting methods

(1) Longitudinal mode selection with the ring-shaped traveling wave resonator in a uniformly broadened laser. The ring-shaped traveling wave resonator can be adopted to obtain single longitudinal mode oscillation, the structure of whose device is as shown in Fig. 5.3-10. As in common vertical resonators the oscillating light field is a standing wave field, in which the light is most intensive at the antinode while weakest at the node, forming the so-called standing wave effect that causes spatially longitudinal inhomogeneity of light intensity distribution in the cavity, which, in turn, leads to the spatial inhomogeneity of population inversion or the spatial hole-burning effect. In order to enable the homogeneously broadened active medium to realize single longitudinal mode oscillation through mode competition, it is necessary to eliminate the standing wave field that causes the spatial inhomogeneity effect in

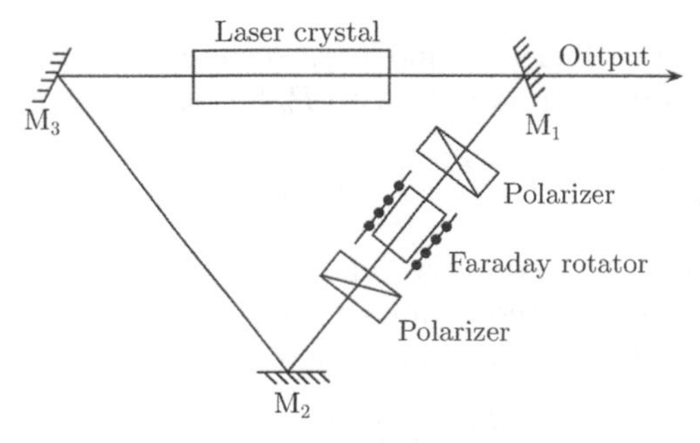

Fig. 5.3-10 Schematic diagram of the ring-shaped traveling wave resonator laser

the cavity while adopting the ring-shaped structure and placing in the cavity an optical isolator made up of a polarizer, a Faraday rotator, and the quartz crystal wafer to make the laser beam only propagate in the form of a traveling wave in one direction. In the traveling wave resonator, instead of being fixed at a certain location in space, the maximum value of the light intensity varies with the propagation of the light wave. Hence the excited radiation will be able to uniformly consume the medium inverted population, thereby eliminating the standing wave effect. Owing to the gain saturation, in the mode competition between the longitudinal modes, the single longitudinal mode of the central frequency is dominant and obtains longitudinal laser output in the end.

(2) Single longitudinal mode selection using Q-switching. This is a method of single longitudinal mode selection based on the difference in gain between different longitudinal modes (one and the same transverse mode). At the beginning, the Q switch is in a not completely turned-off state (called the pre-laser state). Under a definite pump power, a few of the longitudinal modes with a large gain near the central frequency set up oscillation while none of the remaining ones with a small gain and unable to reach the threshold value can initiate oscillation. Thus, on the one hand, very few longitudinal modes initiated oscillation at the start and, on the other, it was under the critical oscillating condition that the few longitudinal modes oscillated. The laser forming process is rather long and the competition between the longitudinal modes is sufficient. So what can finally form laser are only the longitudinal modes with the maximum gain near the central frequency. After the formation of the single longitudinal mode laser, turn on the Q switch in time to enable the already-formed single longitudinal mode laser to get fully amplified till a high power single longitudinal mode optical pulse is output in the end.

In addition, Soffer, Sooy, and others discovered the role of longitudinal mode selection of the saturable absorption dye Q switch in the 1960s. According to them, when a pulse is set up in noise, longitudinal mode selection is generated in the laser. During the time of setting up, the amplitude of the modes with a high gain and low loss increases more quickly. Furthermore, apart from the difference in gain and loss between the modes, there is an important parameter that determines the spectral characteristics of the laser, that is, the number of back-and-forth travels of the light pulse in the process of setting up in noise. If the number of back-and-forth travels is rather great, then the difference in amplitude between two modes will become large. Therefore, after the difference in loss between two different modes is determined, what is important is to increase the times of back-and-forth travels as much as possible to obtain good mode selection results. Sooy obtained two results after an analysis. One is that the increase of power P_q of the qth longitudinal mode with time t is

$$P_q(t) = P_{0q} \exp[k_q(t - t_q)^2] \tag{5.3-12}$$

where P_{0q} is the noise power when this mode is just set up; t_q is the time for the net gain of mode q to reach 1; $k_q = \frac{1}{2T}\frac{\mathrm{d}g_n}{\mathrm{d}t}$ (where T is the time for the light to travel back-and-forth in the resonator, g_n is the gain coefficient of mode q); second, after n back-and-forth travels in the setting up process, the ratio of power P_{q+1} to P_q has the following approximate relation:

$$\frac{P_q}{P_{q+1}} = \left(\frac{1-\delta_q}{1-\delta_{q+1}}\right)^n (1-\delta_q)^{n[(g_{n+1}/g_n)-1]} \tag{5.3-13}$$

where δ_{q+1} and δ_q represent the loss every time $q+1$ modes and q modes make a round trip, respectively; g_{n+1} and g_n are the gain coefficients of the two modes. The first part of Eq. (5.3-13) is equivalent to loss discrimination and the second part to gain discrimination.

$$\frac{P_q}{P_{q+1}} = \left(\frac{R_q}{R_{q+1}}\right)^n \tag{5.3-14}$$

where $R_q = 1 - \delta_q R_{q+1} = 1 - \delta_{q+1}$. In Fig. 5.3-11 is plotted the function of the output power of two modes and the times of back-and-forth travels.

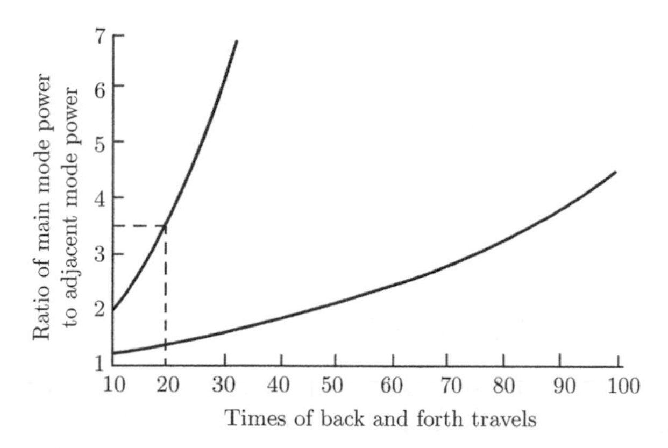

Fig. 5.3-11 The mode selecting role in the pulse setting up process

5.4 Methods of mode measurement

Whether a laser has realized single mode (longitudinal or transverse mode) operation and whether the operation is stable or not has to be judged with a suitable observing and measuring method. A description of several frequently used methods is made in this section.

5.4.1 The direct observing method

It is well known that the light intensity of different transverse modes possesses different distributions in a cross section. For medium-sized and small power lasers of continuous visible light waveband, this method is suitable as it is only necessary to place a screen in the laser output passage to directly observe the pattern (light spot) of the transverse mode with the naked eye. But the discriminating capability of such a method is rather poor and is inapplicable to intensive and invisible light at that. For infrared laser of medium power, an ablation method can be adopted, that is, observing the patterns formed by laser ablation with wood blocks, organic glass, refractory bricks, etc. to distinguish the pattern of the transverse mode. For 1.06-μm near-infrared laser, thin wafers made of upconversion materials (the wavelength changed from long to short) can be adopted to convert the near-infrared light into visible light, which makes it very convenient to observe the pattern of the transverse mode spot. For infrared lasers of medium and small power, an image converter tube or a CCD camera can be used to observe the transverse mode, the structure of the image converter being as shown in Fig. 5.4-1. It is composed of an opto-electric cathode, a control grid, an anode, and a fluorescent screen. After beam expansion and attenuation, the laser beam is injected onto the receiving surface of the image converter. The optical cathode emits photoelectrons which, under the action of the intensive electric field between the cathode, control electrode, and anode, move toward the anode till finally getting shot onto the fluorescent screen to give forth fluorescent light, displaying the pattern of the transverse mode intensity distribution of the laser beam. By choosing different photoelectric cathodes, the laser transverse modes in the near-ultraviolet to near-infrared waveband can be observed. By virtue of the high sensitivity of the image converter, its pattern recognition capability is better than that of the preceding few. In a word, the direct observation method is simple and intuitive, but of poor discriminating ability, and is a method only for rough observation.

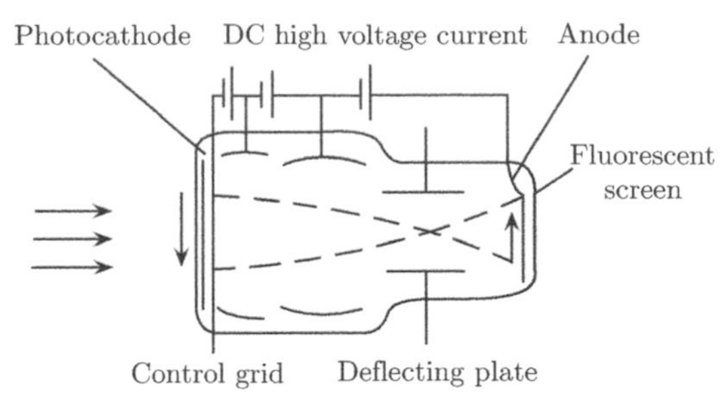

Fig. 5.4-1 Schematic diagram of the structure of an image converter

5.4.2 The light spot scanning method

This method is mainly used for observation of a continuous laser using light spot scanning to record the light intensity distribution curve, from which to find the corresponding transverse mode, whose testing device is as shown in Fig. 5.4-2. After beam expansion, laser is projected onto the motor-driven rotating mirror and, after getting reflected, projected onto the photo-detector. Then, after getting amplified by the electronic circuit, it is sent to the oscillograph for waveform display. This method is capable of transforming the two-dimensional image of the laser transverse mode light intensity distribution onto the oscillograph to display a one-dimensional image of the light intensity distribution. The fundamental mode will appear as a smooth Gauss curve while the higher-order transverse mode will exhibit more than two wave crests. What is shown in Fig. 5.4-3 is a number of light intensity distribution curves (waveforms) of low-order modes of the symmetric stable

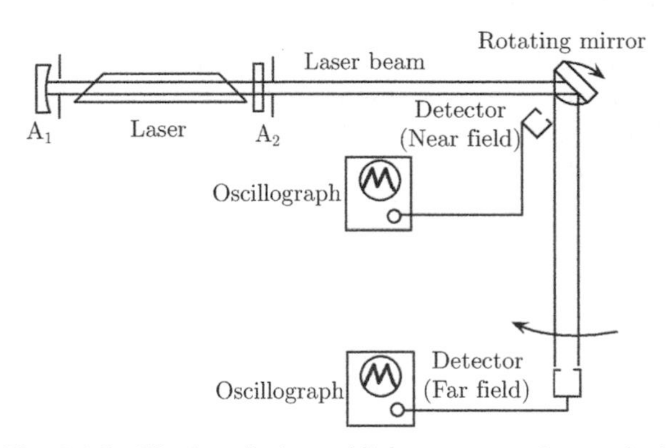

Fig. 5.4-2 The installations of light spot scanning method

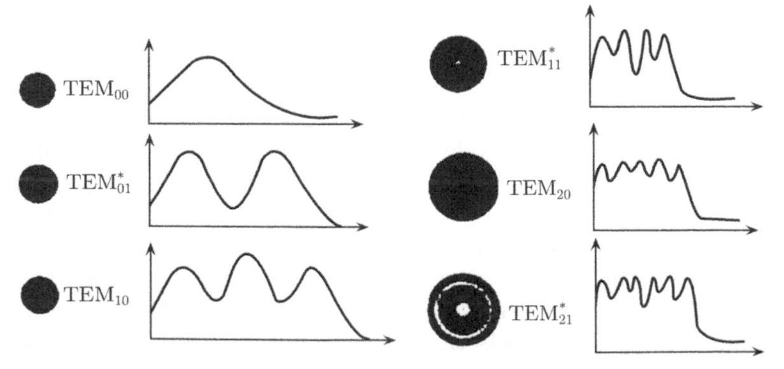

Fig. 5.4-3 Corresponding mode patterns of certain modes' light intensity distribution

cavity and their corresponding patterns of transverse mode spot. Of the transverse modes given here, some are circularly symmetric modes, such as TEM_{00}, TEM_{10}, TEM_{20}, etc., the mode spot central region of which is the peak value of the light intensity. The others are mixed modes, such as $TEM_{01}{}^*$, which is a mixture of the TEM_{10} and TEM_{01} modes. The central region of these modes is the valley value region (dark region) of the light intensity. During measurement, if the distance from the rotating mirror to the detector and the rotating speed of the mirror are given, the dimensions of the light spot can be found. In addition, during observation, make sure that the scanning line does get through the center of the light spot as only than will it be possible to obtain fairly accurate results.

5.4.3 The method of scanning interferometer

It is known from the theory of the resonator that different modes (transverse or longitudinal modes) each have their own different frequency spectra. For this reason, the frequency tunable F-P scanning interferometer can be adopted to find the various frequency distributions and distinguish the laser mode. Owing to its high resolution, convenient adjustment, and easy coupling, the confocal scanning interferometer is often used to measure the laser transverse mode. What is shown in Fig. 5.4-4 is a schematic diagram of the principle of transverse mode measurement using a scanning interferometer. Two concave mirrors plated with highly reflective film and of identical curvature radii make up a passive cavity. The testing device is divided into two parts. One is an optical system consisting of a convergent lens, a scanning interferometer, and a photodiode, the other an electronic mode measuring system composed of a sawtooth wave generator, an amplifier, and an oscilloscope.

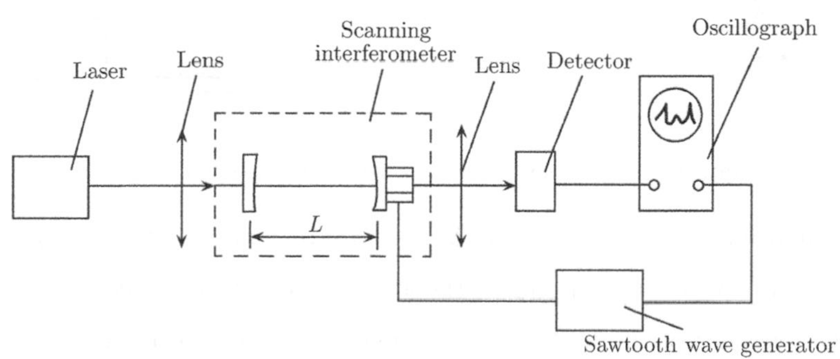

Fig. 5.4-4 Schematic diagram of principle of transverse mode measurement with a scanning interferometer

The resonant frequency (eigenmode) of the scanning interferometer passive cavity

$$\nu_{mnq} = \frac{c}{2L}\left[q + \frac{1}{\pi}(m+n+1)\arccos\sqrt{g_1 g_2}\right]$$

where L is the length of the passive cavity, $g_1 = 1 - L/R_1$, and $g_2 = 1 - L/R_2$, where R_1 and R_2 are the curvature radius of the two reflective mirrors, respectively; m and n the ordinal numbers of the transverse modes; and q is the ordinal number of the longitudinal modes. It is known from the principle of the interferometer that only the part of the laser light field that is consistent with the eigenmode of the interferometer passive cavity can be output resonance coupled. This refers to the modes in Eq. (5.4-1). If a pinhole diaphragm is incorporated in the passive cavity to increase the diffraction loss of the higher-order transverse modes, what is output resonance coupled via the interferometer passive cavity is only the fundamental mode that satisfies

$$\nu_{00q} = \frac{c}{2L}\left(q + \frac{1}{\pi}\arccos\sqrt{g_1 g_2}\right)$$

In order to determine what specific light fields are contained in the laser to be measured, it is necessary to make man-made changes in the frequency of the interferometer, that is, to make it perform frequency scanning to obtain the spectrum graph of the laser light field, thus determining the corresponding transverse mode. The implementation of frequency scanning can change the index of refraction in the interferometer, the to-be-measured laser's incident angle, and the passive cavity length. Transverse mode observation is realized by changing the length of the cavity, the method being to adhere a piezoelectric ceramic ring to a cavity mirror in the interferometer passive cavity. When a sawtooth wave voltage is applied on the piezoelectric ceramic, the cavity length will undergo periodic linear variation, making the eigenfrequency of the interferometer undergo periodic linear variation, that is, perform periodic frequency scanning of the laser that passes through it. The modes that fall within the range of the scanning period frequency, when received via the photoelectric receiver, can be displayed on the oscilloscope. What is shown in Fig. 5.4-5 is a graph of the laser spectrum measured with the scanning interferometer. Fig. 5.4-5(a) shows the graph of the spectrum before insertion of the diaphragm while Fig. 5.4-5(b) shows the graph of the spectrum after insertion of the diaphragm.

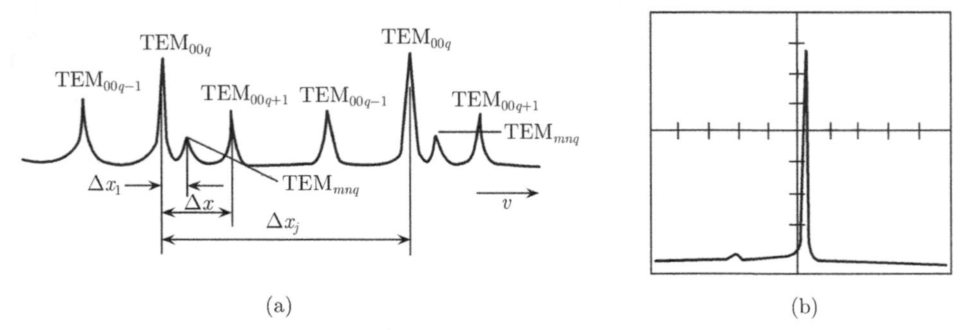

Fig. 5.4-5 Graph of the laser spectrum measured with the scanning interferometer

It can be seen from Fig. 5.4-5(a) that Δx_j is proportional to the free spectral region $\Delta\nu_j(\Delta\nu_j = \frac{c}{4L})$, and Δx is proportional to the adjacent longitudinal mode frequency interval $\Delta\nu_q$. In the presence of the higher-order transverse mode, the higher-order transverse mode TEM$_{mnq}$ can be seen by the side (Δx_i) of the fundamental mode (TEM$_{00q}$). In the figure, Δx_1 is proportional to $\Delta\nu_{nm,00}$. It has been found by experiment that

$$\frac{\Delta\nu_{nm,oo}}{\Delta\nu_{q,q\pm1}} = \frac{\Delta x_1}{\Delta x}$$

By comparing the value $\frac{\Delta x_1}{\Delta x}$ measured with the theoretically calculated value, the higher-order mode can be determined.

For instance, for a plane concave cavity He-Ne laser, the cavity length is 210 mm and the curvature radius is 1 m; in the graph of spectrum obtained with the scanning interferometer, $\Delta x_1 = 1.6$ mm and $\Delta x = 10.7$ mm. Then the experimental value is

$$\frac{\Delta x_1}{\Delta x} = \frac{1.6}{10.7} = 0.149$$

while the theoretical value

$$\Delta\nu_{q,q\pm1} = \frac{c}{2nL} = \frac{3 \times 10^8}{2 \times 0.21}\text{MHz} = 7.14 \text{ MHz}$$

According to the formula

$$\Delta\nu_{00,01} = \frac{c}{2nL}\left[\frac{1}{\pi}(\Delta m + \Delta n)\arccos\sqrt{g_1 g_2}\right]$$

by calculation, we find

$$\Delta\nu_{00,01} = \frac{3\times 10^8}{2\times 0.21}\left[\frac{1}{\pi}\arccos(0.89)\right]\text{MHz} = 1.07\ \text{MHz}$$

Thus

$$\frac{\Delta\nu_{00,01}}{\Delta\nu_{q,q\pm 1}} = 0.15$$

So it is found by judgment that the position Δx in the spectrum is TEM_{01} mode. Hence what this He-Ne laser outputs are two transverse modes TEM_{00} and TEM_{01}.

During transverse mode observation, the free spectral region of the interferometer should be made to be larger than the gain line width of the operation material. To enable the laser to be measured effectively coupled into the passive cavity of the interferometer, generally a positive lens can be used to achieve matching between the laser beam and the interferometer.

The scanning interferometer is highly accurate in mode observation and is capable of discriminating all the effects of certain factors on the spectrum. So it's an important measuring instrument in laser technology.

5.4.4 The method of F-P photography

Despite the excellent performance of the above-mentioned scanning interferometer, this instrument can only observe the pattern of the continuous laser. If what should be observed is pulsed laser, it can do nothing because no sooner had the interferometer begun to scan than the pulse of laser came to an end. For this reason, the F-P etalon photography can be adopted to observe the pulsed laser. Fig. 5.4-6 shows the schematic diagram of the principle of the F-P photography method. A beam of laser of diameter D, after getting converged by lens L_2, is projected onto the etalon F-P, which transforms the light beams injected at different angles into parallel streaks of light of different directions. In other words, the rays injected at different angles after many times of reflection by two planes of the etalon have become a set of parallel light beams at angles different from that of the optical axis. The transmitted light via L_1 forms equal inclination interference fringes on the focal plane of lens L_1. The transmissivity of the F-P etalon is

$$T(\lambda) = \frac{1}{1 + F\sin^2\dfrac{\varphi}{2}}$$

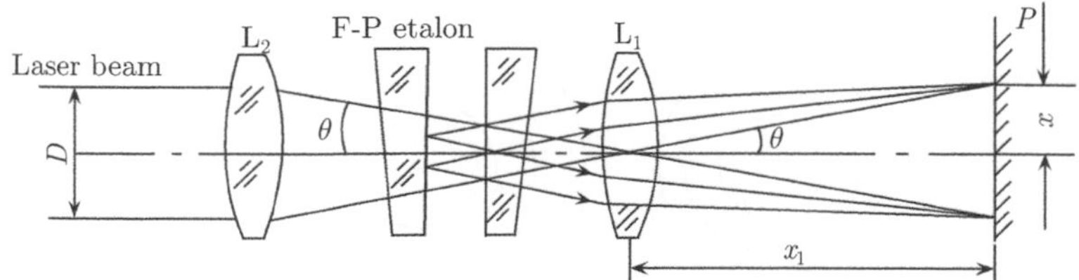

Fig. 5.4-6 The schematic diagram of the principle of F-P photography method

The condition for appearance of bright fringes (T value is maximum) should be

$$\sin^2\frac{\varphi}{2} = \sin^2\left(\frac{2\pi\Delta\delta}{2\lambda}\right) = 0$$

that is,

$$\pi\Delta\delta/\lambda = m\pi \qquad (m = 0, 1, 2, \cdots) \tag{5.4-1}$$

while

$$\Delta\delta = 2\pi d\cos\theta$$

Then there is

$$2\pi nd\cos\theta = m\pi\lambda$$

that is

$$2nd\cos\theta = m\lambda \tag{5.4-2}$$

It can be seen that the bright fringes are a series of θ valued concentric rings and, when the to-be-measured laser wave length has a definite line width $\Delta\lambda$, the angle θ of the concentric ring has a range of variation $\Delta\theta$, after which getting focused, there is also a range of variation Δr in the position of the interference fringe on the focal plane P. Under the condition of near-axis rays approximation, there is

$$r/f_1 \approx \tan\theta \approx \theta \tag{5.4-3}$$

where f_1 is the focal distance of lens L_1. Combining Eq. (5.4-2) with (5.4-3) and performing derivation, we have

$$\Delta\nu = \nu_r\Delta r/f_1^2 \tag{5.4-4}$$

where r is the radius of the bright interference fringe at a certain level, and Δr the width of the interference fringe at the said level. The width Δr of the interference fringe on the screen can be directly measured by photography. Then via Eq. (5.4-4) the line width of the said laser can be found. As the above equation is derived under the condition of the near-axis rays approximation, the interference fringes near the center should be chosen for calculation.

The F-P photography method can not only measure the width of the spectral lines of laser, but also distinguish the laser mode. When the laser is operating in the single mode state, there is only one wavelength in the output light beam. It is known from Eq. (5.4-2) that there is now a series of concentric interference fringes of different θ values on the screen, as shown in Fig. 5.4-7(a) while if laser is operating in the two-mode state, two sets of different interference fringes will be generated, as shown in Fig. 5.4-7(b). Therefore, just by referring to the number of sets of interference fringes, we can judge the state the laser is in. If there are too many modes that, worse still, are very close to one another, then the interference fringes will become indistinct and very thick concentric rings.

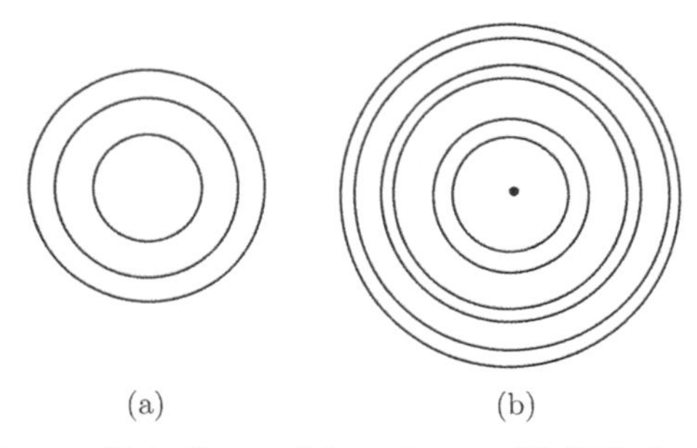

Fig. 5.4-7 Pictures of interference fringes taken with F-P photography method

Exercises and problems for consideration

1. Say something about the significance and role of selecting the single (transverse, longitudinal) mode.

2. Analyze the principles of using different diffraction losses for selecting the fundamental mode (TEM_{00}).

3. Why do we say that for one and the same transverse mode, the smaller the Fresnel number N is, the greater will the loss be? Why for one and the same N value, the higher its mode order is, the greater will its loss be?

4. For a neodymium laser operation material, its fluorescent light line width $\Delta\lambda_D = 24.0$ nm, refractive index $n = 1.50$. If a short cavity is to be used to select the single longitudinal mode, how long should the cavity be?

5. There's a ruby laser, with a cavity length $L = 500$ mm, oscillation line width $\Delta\nu_D = 2.4 \times 10^{10}$ Hz. Insert an F-P etalon to select the single longitudinal mode ($n = 1$). Try to find its interval d and the reflectivity R of the parallel plates.

6. Suppose we have a circular mirror confocal cavity of length $L = 1$ m. Find the longitudinal interval $\Delta\nu_q$ and transverse mode interval $\Delta\nu_m \cdot \Delta\nu_n$. If the gain line width above the oscillation threshold value is 60 MHz, is it possible that there are more than two longitudinal modes oscillating at the same time? Is it possible that more than two different transverse modes are oscillating at the same time? Why?

7. In order to suppress higher-order transverse modes, a pinhole diaphragm is placed at the reflective mirror of a confocal cavity. The cavity length L is 1 m, laser wavelength λ is 632.8 nm. In order that only the TEM_{00} mode is allowed to oscillate, how big should the pinhole be? (In general, the diameter of an orifice is 3~4 times that of the fundamental mode spot size on the surface of the mirror.)

8. There is a square aperture confocal cavity He-Ne laser, the cavity length $L = 30$ cm, diameter $d = 2a = 12$ cm, laser wavelength $\lambda = 0.6328$ μm, the reflectivity $r_1 = 1$, $r_2 = 0.96$, the other losses being negligible. Can the laser perform single mode operation? If it is desired that a square pinhole diaphgram should be placed near the surface of the confocal mirror to choose the TEM_{00} mode, how long should the side of the pinhole be? (The gain of He-Ne should be calculated using the formula $\exp(Gl) = 1 + 3 \times 10^{-4} L/d$.)

References

[1] Lan Xinju and Zhang Yunan, Laser Technology, Changsha, Science and Technology Press, 1988.

[2] Laser Physics, compiled by the Laser Physics Group, Shanghai, Shanghai People's Press, 1975.

[3] W. Koechner, Solid-state Laser Engineering, New York, Springer Verlag, 1976.

[4] Zhou Bingkun et al., Principles of Laser, Beijing, National Defense Industry Press, 1984.

[5] Gas Laser (first half). Gas Laser Compiling Group, Shanghai, Shanghai People's Press 1975.

[6] A. E. Siegman, Unstable Optical Resonators, *Appl. Opt.*, 1974, 13, pp. 353–367.

[7] Ma Zuguang et al., Laser Experimenting Methods, Shanghai, Shanghai Science and Technology Press, 1987.

[8] D. C. Hanna et al., Large Volume TEM_{00} Mode Operation of Nd:YAG Laser, *Opt. Commun.*, 1981, 37(5), pp.359–362.

[9] J. Dembowski, H. Weber, Optical Pinhole Radius for Fundamentgal Mode Operation, *Opt. Commun.*, 1982, 42(2), pp. 133–135.

[10] Lan Xinju et al., Compressing the Linewidth with the Laser Injection Locking Technique, J. Huazhong University of Science and Technology, May1991.

[11] H. P. Kortz and H. Weber, Diffraction Losses and Mode Structure of Equivalent TEM_{00} Optical R4esonators, *Appl. Optics*, 1981, 20(11), pp. 1936–1940.

[12] F. W. Smith, Mode Selection in Lasers, *Proc. IEEE*, 1972, 40, pp. 422–440.

CHAPTER 6

The Frequency Stabilizing Technology

6.1 An overview

By virtue of its good monochromaticity and coherence, laser is widely used in such fields as precision metrology, optical communication, optical frequency standards, high resolution spectroscopy, etc. For instance, in precision interference measurement, the laser wavelength is used as a "ruler" that uses the principle of optical interference to measure various kinds of physical quantities (e.g., length, displacement, velocity, etc.). So, the accuracy of the laser wavelength (or frequency) will directly influence the accuracy in measurement. In laser communication, in order to enhance its receiving sensitivity, generally the coherent heterodyne receiving method is adopted; whether the laser frequency is stable or not will directly influence the receiving quality. For this reason, if it is desired to use laser in the above-mentioned fields, not only is the laser required to implement single frequency output, but also the laser frequency itself is required to be stable. However, for ordinary freely operating lasers, affected by the operation environment and relevant factors, the laser output, as an irregularly fluctuating quantity varying with time, is often unstable. To make laser frequency stable, we have to resort to the frequency stabilizing technology. Hence frequency stabilization has become an indispensable means of modern precision measuring technology. This chapter will mainly describe the methods and principles of frequency stabilization of a number of commonly applied He-Ne lasers as well as several newly developed frequency stabilizing methods.

6.1.1 The stability and reproducibility of frequency

Perturbed by its surroundings and various other factors, a freely operating laser often has its frequency altered with time. If definite frequency stabilizing measures are taken to automatically compensate for the frequency fluctuation induced by outside disturbance, the variation of the output frequency can be greatly diminished. In reality, however, different frequency stabilizing techniques adopted will yield different effects. In order to measure the frequency stability, our description can be made in terms of the time domain and frequency domain; that is, the discussion can be made both in terms of its variation with time and its frequency spectral distribution. But, for the noise spectral density of unstable frequency, it's always rather difficult no matter whether in the development of the conception or in the testing technique. Therefore, this chapter will represent the stability of laser frequency by making a description of the time domain with the two physical quantities, the stability and reproducibility.

The frequency stability usually refers to the ratio of the frequency's average value $\overline{\nu}$ within a definite time of observation τ during the continuous operation of a laser to the amount of frequency variation $\Delta\nu$ within the said time, i.e.,

$$S_{\nu}(\tau) = \frac{\overline{\nu}}{\Delta\nu(\tau)} \tag{6.1-1}$$

Obviously, the smaller the variation amount $\Delta\nu(\tau)$, the greater will S be, showing that the frequency stability is even better. Customarily, the reciprocal of S is sometimes used as a measure of the stability, i.e.,

$$S_{\nu(\tau)}^{-1} = \frac{\Delta\nu(\tau)}{\overline{\nu}} \tag{6.1-2}$$

For instance, some people often say that the stability is 10^{-8}, 10^{-9}, etc., which is a case in point.

The variation of frequency or wavelength with time can be manifested as short-term tremble or long-term drift. Therefore, if the time of observation and measurement of the frequency is different, so, too, will the results of measurement be. So the frequency stability can be divided into short-term stability and long-term stability. The criterion for the division is determined by the relationship between the detecting system's time of response (distinguishing) τ_0 and the sampling time τ by the measurer. When $\tau \leqslant \tau_0$, the stability measured is spoken of as the short-term stability while when $\tau > \tau_0$, it is called the long-term stability. An appropriate method of representation is marking the sampling time value τ after the value of the stability, for example, $S_\nu(\tau) = 10^{-10}$ ($\tau = 10$ s).

For a laser as the criterion for frequency or wavelength, not only is it required that the stability be high, but also the frequency reproduction, too, should be high. For instance, as is the wont, when measuring a length with a ruler, not only is it required that the length of the ruler be stable, but also the length of the ruler itself should be up to standard. This is true of precision measurement using laser. For instance, the frequency of laser A may be different from that of laser B that is supposed to stabilize the frequency using the same method (despite the fact that both the structure and operation condition of the two lasers are identical); or using the same frequency stabilizing laser, the stability is 10^{-6} when used in position A and the frequency is maintained at ν_1 when used in position B, with the stability remaining unchanged, but the frequency is maintained at ν_2; or, when measuring in the same place, the frequency is maintained at ν' on a certain day, which remains unchanged a few days later, but the frequency is maintained at ν'', and so on and so forth. As the values of the frequencies stabilized for every time are slightly different from each other, the values measured are inaccurate. The ratio of the amount of deviation from frequency stabilization in different locations and at different times to their average frequency is spoken of as the frequency reproducibility, expressed by

$$R_\nu = \frac{\delta\nu(\tau)}{\overline{\nu}} \tag{6.1-3}$$

where $\overline{\nu}$ is the average frequency of the laser measured or the standard frequency of one and the same laser (or the original operation frequency); $\delta\nu$ is the deviation amount of the frequency. It can thus be seen that the stability and reproducibility of the frequency are two different conceptions. For a frequency stabilized laser, not only its stability but also its reproducibility should be considered.

6.1.2 Factors affecting the frequency stability

It is known from the principle of laser that the laser oscillation frequency is influenced by both atomic transition spectral frequency ν_{m} and the optical resonator resonance frequency ν_{c}. If the width of the atomic transition spectral line is $\Delta\nu_{\text{m}}$, and the width of the resonator spectral line is $\Delta\nu_{\text{c}}$, then the laser resonance frequency can be expressed as

$$\nu = \frac{\nu_{\text{m}}\nu_{\text{c}}(\Delta\nu_{\text{m}} + \Delta\nu_{\text{c}})}{\nu_{\text{m}}\Delta\nu_{\text{m}} + \nu_{\text{c}}\Delta\nu_{\text{c}}} \tag{6.1-4}$$

This is the primary approximation at a small amplitude with the saturation effect neglected. In the near-infrared and visible light waveband, its $\Delta\nu_{\text{m}}$ (Doppler line width) is in general no smaller than $10^8 \sim 10^9$ Hz while the oscillation line width $\Delta\nu_c$ of the resonator is of the order of $10^6 \sim 10^7$ Hz. So $\Delta\nu_m \gg \Delta\nu_c$. Then Eq. (6.1-4) can be reduced to

$$\nu = \nu_c + (\nu_m - \nu_c)\frac{\Delta\nu_c}{\Delta\nu_m} \tag{6.1-5}$$

The above equation shows that the oscillation frequency of a laser is jointly determined by the frequency of the atomic transition spectral line and the resonance frequency of the resonator. The variations of the two can both induce the instability of the laser frequency. The effect of the spectral line on the oscillation frequency is expressed by the second term in Eq. (6.1-5) as the frequency pulling effect, the scale factor of which is $\frac{\Delta\nu_c}{\Delta\nu_m}$. In general, the frequency pulling effect is very small, while the resonance frequency of the resonator is very sensitive to the environmental influence. So the stability of the laser frequency is mainly dependent on that of the resonator's resonance frequency.

When the slight variation of the spectral line frequency of the atomic transition is not taken into consideration, the laser oscillation frequency is mainly determined by the resonator's resonance frequency, that is,

$$\nu = q\frac{c}{2nL} \tag{6.1-6}$$

where L is the cavity length, c is the light velocity, n is the refractive index of the medium in the cavity, and q is the ordinal number of the longitudinal modes. It can be seen from the equation that if the cavity length or the refractive index n in the cavity should vary, so will the laser oscillation frequency.

$$\Delta\nu = -qc\left(\frac{\Delta L}{2nL^2} + \frac{\Delta n}{2n^2L}\right) = -\nu\left(\frac{\Delta L}{L} + \frac{\Delta n}{n}\right)$$

or

$$\left|\frac{\Delta\nu}{\nu}\right| = \left|\frac{\Delta L}{L}\right| + \left|\frac{\Delta n}{n}\right| \tag{6.1-7}$$

So the problem of laser frequency stability can be boiled down to the problem of how to best keep the cavity length and the refractive index stable. The factors from the outside that affect frequency stability are mainly as given below:

1. The influence of change in temperature

The fluctuation of ambient temperature or the heating of the laser tube during operation can make the cavity material stretch or contract with the change in temperature and can lead to the drift of the frequency, that is,

$$\alpha\Delta T = \frac{\Delta L}{L} = \frac{\Delta\nu}{\nu} \tag{6.1-8}$$

where ΔT is the amount of variation of temperature, α is the linear expansion coefficient of the resonator's partitioning material, and the magnitude of the coefficient is related to the kind of material. For example, for ordinary hard quality glass, $\alpha=10^{-5}/°C$; for quartz glass, $\alpha = 6 \times 10^{-7}/°C$; for invar, $\alpha = 9 \times 10^{-7}/°C$. So, frequency stabilizing lasers all adopt quartz glass with small thermal expansion coefficient for their tubes and invar material for making the support and have the entire laser system thermostatically controlled so as to reduce the effect of temperature variation as much as possible. But even so, it's very hard to obtain a frequency stability better than 10^{-8}.

2. The influence of atmospheric variation

For an extracavity gas laser, suppose the resonator length is L and the length of the discharge tube is L_0. Then the relative length of the part exposed to the atmosphere is $(L - L_0)/L$. The variation of atmospheric temperature, pressure, and humidity will all induce a variation of the atmospheric refractive index, leading to an alteration of the laser oscillation frequency. Suppose the ambient temperature $T = 20°\text{C}$, atmospheric pressure $P = 1.01 \times 10^5$ Pa, and humidity $H = 1.133$ kPa. Then the refractive index variation coefficients of atmosphere with respect to a 633-nm wavelength light are, respectively,

$$\beta_T = \frac{1}{n}\left(\frac{\mathrm{d}n}{\mathrm{d}T}\right) = -9.3 \times 10^{-7}/°\text{C}$$

$$\beta_P = \frac{1}{n}\left(\frac{\mathrm{d}n}{\mathrm{d}P}\right) = 5 \times 10^{-5}/\text{Pa}$$

$$\beta_H = \frac{1}{n}\left(\frac{\mathrm{d}n}{\mathrm{d}H}\right) = -8 \times 10^{-6}/\text{Pa}$$

Then suppose the time variation rates of temperature, atmospheric pressure, and humidity in measurement are $\frac{\mathrm{d}T}{dt} = \pm 0.01°\text{C}/\min$, $\frac{\mathrm{d}P}{\mathrm{d}t} = \pm 133.3\text{Pa/h}$, and $\frac{\mathrm{d}H}{dt} = \pm 656.6\text{Pa/h}$. Then the changes in the laser wavelength induced are, respectively,

$$\begin{aligned} \left|\frac{\Delta\lambda(\tau)}{\lambda}\right|_T &= \beta_T \tau \frac{\mathrm{d}T}{\mathrm{d}t} = \pm 9.3 \times 10^{-9}\tau \\ \left|\frac{\Delta\lambda(\tau)}{\lambda}\right|_P &= \beta_P \tau \frac{\mathrm{d}P}{\mathrm{d}t} = \pm 6 \times 10^{-9}\tau \\ \left|\frac{\Delta\lambda(\tau)}{\lambda}\right|_H &= \beta_H \tau \frac{\mathrm{d}H}{\mathrm{d}t} = \times 4.8 \times 10^{-9}\tau \end{aligned} \tag{6.1-9}$$

where τ is the time of measurement. For the oscillograph, $\tau = 3\sim5$ s, for the XY recorder, $\tau \leqslant$ 1min. It has been proved by experiment that the ventilation-induced air flutter in the extracavity gas laser can generate high speed pulsation of several MHz in seconds. So it is required that the part of an extracavity gas laser exposed to the atmosphere be as small as possible and direct ventilation should be avoided.

3. The influence of mechanical vibration

Mechanical vibration is also an important factor that leads to a variation of the optical cavity resonance frequency. It can be transmitted from ground surface or air onto the cavity support. For instance, the vibration of a building, the passage and sound of vehicles, all can cause the support of a cavity to vibrate, thus causing the cavity's optical length to change and the oscillation frequency to drift. For an optical cavity of $L = 100$ cm, when a 10^{-6} cm change in cavity length is caused by mechanical vibration, there will be a variation of the frequency of 1×10^{-8}. Therefore, to overcome the influence of mechanical vibration, frequency stabilized lasers should adopt good vibration-proof measures.

4. The influence of the magnetic field

In order to reduce the influence of temperature, the isolators for laser resonators are often made of invar, but the magnetostrictive effect of this kind of material may cause a variation in the cavity length. For example, a He-Ne laser of a wavelength of 1.15 μm may generate a frequency drift of 140 kHz simply because of the geomagnetic field effect. Therefore, the geomagnetic field effect and those from the speromagnetic field of the surrounding electronic instruments on the highly stabilized lasers have to be considered.

The above are factors from the outside that cause the frequency to be unstable. Apart from these, such internal factors as the variation of the atmospheric pressure and discharge current in the laser tube, the random noise due to spontaneous radiation, etc. can also affect the stability of the frequency. The former can be controlled with pressure and current stabilizing devices while the latter just cannot be completely controlled. Hence it is the inherent factor that limits the stability of laser frequency.

To sum up, the variation of the ambient temperature and the interference from the outside including mechanical vibration do exert a very great influence on the stability of laser frequency. So it's only natural to have the association that the very direct method is the adoption of thermostat, shockproof, sealing, and sound-proof devices and stabilizing of the power source. Shown in Fig. 6.1-1 is a set of shockproof and thermostat devices for a CO_2 laser. The thermostat measure taken enables the temperature to be fixed at $35 \pm 0.03°C$ while the temperature fluctuation of the cooling water is no greater than 0.5°C. To prevent shock, foam-rubber cushions are placed between all components and the entire device is placed on a solid and stable shockproof platform (preferably a compressed air shockproof platform, which has very good shockproof capability for mechanical vibration of a very low frequency) and a voltage and current stabilizing power source is adopted. It has been proved by experiment that, with the adoption of the thermostat and shockproof devices, the long-term frequency stability of the CO_2 laser can reach as high as the 10^{-7} order of magnitude. But if it is desired to raise to the 10^{-8} order of magnitude and above, it just won't do to simply rely on the passive frequency stabilizing method. Rather, it is necessary to adopt a servo-control system to perform automatically controlled frequency stabilization of the laser, or the method of active frequency stabilization.

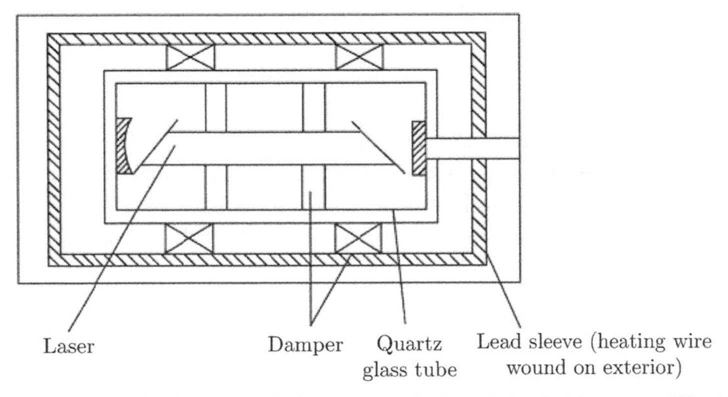

Fig. 6.1-1 The shockproof and thermostat device of single frequency CO_2 laser

6.1.3 The method of active frequency stabilization for a laser

The essence of frequency stabilization is maintaining the stability of the length of a resonator's optical path. Active frequency stabilizing technology lies in selecting a stable reference standard frequency. When an external influence makes the laser frequency deviate from this specific standard frequency, there is a way of its detection followed by man-made automatic regulation of the cavity length via the control system to restore the laser frequency to the specific standard frequency to finally attain the goal of frequency stabilization.

The method of active frequency stabilization is approximately divided into two kinds. One uses the central frequency of the atomic spectral line as a discriminator for frequency stabilization, such as the Lamb dip frequency stabilizing method, and the other makes use of an external reference frequency as a discriminator, such as the saturated absorption method. The Lam dip frequency stabilizing device is rather simple and is a gas laser frequency stabilizing method applied fairly early, whose degree of frequency stabilization is of the order of magnitude of 10^{-9}. Its main shortcoming is its poor reproducibility, the reason being the presence of about 150 kHz/Pa pressurized frequency shift. Owing to the difference between the gas charging pressures of the laser tubes and the variation of the gas in the course of use, the reproducibility can only be ensured to be in the order of 10^{-9}, limiting its application in measurement requiring higher accuracy. After 1966, the method of frequency stabilization by saturated absorption was developed, that is, a method of adopting the absorption lines of some molecular gases as the reference frequency. Thus, the absorption lines in the ground state can avoid discharge disturbance and the low pressure gases get less pressure widening. Its long energy level lifetime means less natural widening, all contributing to the narrowing of the width of the absorption peak, which is conducive to improving the frequency stability. The commonly used saturated absorption molecules are mainly iodine (I_2) and methyl (CH_4). $^{127}I_2$ and $^{129}I_2$ have a large number of powerful absorption lines in the region of visible light; so for the optimum saturated absorption molecules for frequency stabilization of He-Ne lasers of 633 nm and 621 nm and Ar^+ lasers of 515 nm, their frequency stability can reach 10^{-12} and above. The 3.39-μm He-Ne laser that adopts methyl (CH_4) for stabilization obtained a frequency stability that can be compared with the microwave frequency standard compared for the first time. The CO_2 frequency stabilizing laser in its early years of development adopted the CO_2 saturated fluorescent light for frequency stabilization, but owing to the contradiction between pressure widening and signal-to-noise ratio, the stability was merely 10^{-12} and the reproducibility 1.5×10^{-10}. Later, experiments were made on the saturated absorption molecules of SF_6, with a frequency stability of 5×10^{-14} ($\tau = 10$ s) obtained. But SF_6 possesses a superfine structure an ordinary system is incapable of distinguishing. Therefore, the reproducibility is but 10^{-10}. The development of frequency stabilizing lasers since the 1960s is shown in Tab. 6.1-1.

Tab. 6.1-1 A list of the development of frequency stabilizing lasers

Frequency stabilizing laser	Wavelength	Reference frequency standard	Frequency stability	Reproducibility	Date
He-Ne	633 nm	Gain curve center			1963
He-Ne	1.15 μm	Lamb dip	10^{-9}~10^{-10}	10^{-7}	1965
He-Ne	633 nm	Lamb dip	10^{-9}~10^{-10}	10^{-7}	1965
He-Ne:Ne	633 nm	Saturated absorption			1967
He-Ne:CH_4	3.39 μm	Saturated absorption	$10^{-13}\sqrt{\tau}$	3×10^{-12}	1969
CO_2:CO_2	all branch lines	Saturated absorption	10^{-12}	1.5×10^{-10}	1970
He-Ne:I_2	633 nm	Saturated absorption	5×10^{-13}	10^{-10}	1972
Dye		Passive cavity			1973
Ar^+ :I_2	515 nm	Saturated absorption	5×10^{-14}	1.5×10^{-12}	1977
He-Ne:I_2	612 nm	Saturated absorption	5×10^{-13}	10^{-12}	1979
CO_2:SF_6	10.6 μm	Saturated absorption	5×10^{-14}	10^{-10}	1980
He-Ne:I_2	640 nm	Saturated absorption			1983

A comparison of frequency stabilizing laser with the microwave frequency standard shows that the former may obtain a narrower relative line width. Therefore, it may have even a higher frequency stability. Laser possesses an absolute frequency four orders higher than that of the cesium (Cs) frequency standard. So, in an identical time interval, the accuracy of length measurement can be greatly improved. Currently, the stability of the frequency stabilizing laser is already equivalent to even better than the Cs frequency standard. Laser frequency measurement is expanded to the region of visible light and, with the emergence of the phase-locked frequency chain, capable of transmitting the radio frequency on the laser frequency band. Some phase-locked systems can operate for hours on end. All this shows is that the day is not far off when the laser is applied to the optical frequency standard, thereby replacing the currently available cesium beam frequency standard to become the standard for frequency.

The application of the optical frequency standard will make it possible to unify the measuring standard for time (frequency) and length. The three characteristic parameters of the monochromatic electromagnetic wave, viz. frequency ν, wavelength λ, and propagating velocity c, can be associated by $c = \lambda\nu$. If c is regarded as a constant and given an accurate value, then the wavelength can be found just from the frequency of the monochromatic wave. Hence, if we have a standard for the optical frequency, we can obtain a standard for length (wavelength). With the development of the laser frequency stabilizing technology, the stability and reproducibility of the laser frequency have already far surpassed the accuracy of the ^{86}Kr wavelength criterion. Some scientists have measured the wavelength and frequency of the 3.39 μm line of methyl stabilized He-Ne laser adopting independent criteria. The values of light velocity found based on the results of measurement are in very good agreement with each other. Therefore, setting independent length criteria is no longer of any significance. At the 15th International Metrological Conference 1975, a light velocity value $c = 299, 792, 458$ m/s in vacuum was recommended, whose inaccuracy is $\pm 4 \times 10^{-9}$. If this numerical value is regarded as the defined value of light velocity, and optical frequency measurement is established with the optical frequency standard as the basis, then the criterion for length can be obtained from the criteria for time and frequency. At the present time, the accuracy of frequency (time) measurement has already reached the 10^{-14} order, the accuracy of length obtained from which can be greatly increased. Thus, the standards for length and time measurement will be unified.

With what had been achieved in the frequency stabilizing laser and optical frequency measurement, the new definition of the "meter" was passed at the 17th International Conference on Measures and Weights, 1983. It states that the "meter" is the length of path traversed by light in vacuum in a time interval of 1/299, 792, 458 s. This is the first time in metrology that a basic physical constant was used to define a basic unit and also the third time the "meter" was defined. This shows that the unit of length has become a derived unit and it is only owing to the importance of the unit of length in the international system of units that it is still retained as a basic unit.

6.1.4 Requirements on the reference standard frequency (reference spectral line)

Of the two kinds of methods for active frequency stabilization, since one consists in frequency stabilization by making use of the central frequency of the atomic spectral line as the standard while the other uses the external reference frequency as the standard, there are the following requirements on the reference standard spectral line:

(1) The stability and reproducibility of the spectral line central frequency should be good as frequency stability and reproducibility of the frequency stabilizing laser are dependent on those of the reference spectral line. As a frequency standard, it is required to have a long-term stability better than 10^{-13}. Hence the reference spectral line frequency shift (a result of the atomic collision, the Stark and Zeeman effect, and an unstable power) should be controlled.

(2) The line width should be narrow; mainly effort should be made to eliminate the Doppler broadening and collision broadening.

(3) There should be an adequate signal-to-noise ratio as the signal intensity is often in contradiction with the line width. For instance, resonance spectral lines of a great transition probability are very strong, but the energy level lifetime is short and natural broadening

great. Increasing the atomic density will strengthen the signal, but collision broadening will ensue. So it is required that, under the condition of a definite signal-to-noise ratio, the width of the spectral line should be reduced as much as possible.

(4) The frequency of the spectral line should match that of the controlled laser; that is, the frequency of the reference spectral line should fall near the peak value of the controlled laser's gain curve.

6.2 Lamb dip frequency stabilization

6.2.1 Lamb dip

In "The Principle of Laser" there is a description of the "hole" burning effect of the gain curve of an inhomogeneously broadened lineshape. And it is pointed out that, in the inhomogeneously broadened lineshape generated by the Doppler effect, an oscillation frequency can burn two "holes" (symmetric with the central frequency ν_0) in its gain curve. The one that appears on the oscillation frequency ν is called the "original hole" and that in the location symmetric with the gain curve is called the "image hole", as shown in Fig. 6.2-1(a). The areas of the two holes are proportional to the inverted population in the active medium that takes part in the stimulated radiation and have made contributions. The larger the area is, the more apparent it is implied that the greater the number of particles that take part in the stimulated radiation and the stronger the laser output power (light intensity) will be. Suppose there is a single mode laser, whose oscillation frequency is made to vary by changing the length of its resonator. As the small signal gains at different oscillation frequencies are different, the depth of the burnt "hole" on the gain curves, too, is different. When continuously changing the laser oscillation frequency, the area of the burnt hole far from the center of the spectral line is small. As the oscillation frequency moves toward the central frequency, the area of the burnt hole will enlarge and the depth increase, while the interval between the two "holes" is reduced. When the oscillation frequency is located at the central frequency, the two "holes" will become one, the area of which is smaller than the sum of the areas of the two holes that deviate not far from the center of the spectral line, showing that the number of particles that have made contributions has decreased. So the output power reaches the minimum value. There appears a dip at ν_0 of the curve, as shown in Fig. 6.2-1(b), which is what is spoken of as the Lamb dip, the mechanism of the generation of which is described in "The Principle of Laser".

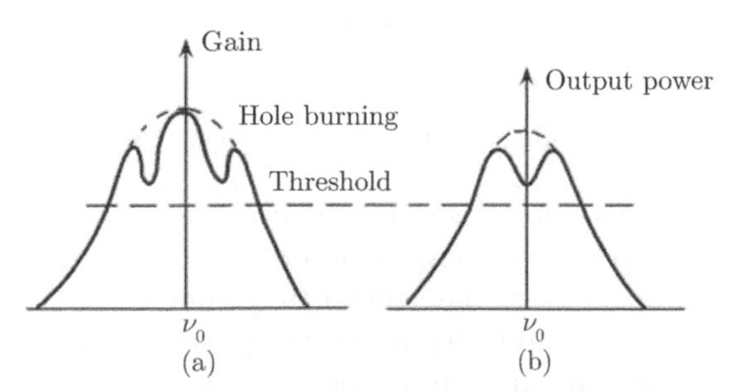

Fig. 6.2-1 (a) The hole burning effect of gain curve and (b) The Lamb dip

6.2.2 The principle of frequency stabilization with the Lamb dip

The Lamb dip frequency stabilizing method consists in controlling the laser cavity length by driving the piezoelectric ceramic ring with an electronic servo-system using the gain curve central frequency ν_0 as the reference standard frequency. It can make the frequency stabilized at ν_0, whose frequency stabilizing device is as shown in Fig. 6.2-2. The laser tube is made of quartz with a very small expansion coefficient and of the extracavity structure. The two reflective mirrors of the resonator are mounted on an invar support, one of which is pasted on the ceramic ring, the length of which is about several centimeters and the ring's inner and outer surfaces are each connected to an electrode with a modulating voltage of frequency f applied on both surfaces. When the voltage on the outer surface is positive and that on the inner is negative, the ceramic ring stretches whereas it contracts when the contrary is the case. By changing the voltage on the ceramic ring, we shall be able to adjust the length of the resonator to compensate for the change in the cavity length brought about by external factors. In general, the silicon phototriode is adopted as the photoelectrical receiver, which is capable of converting the optical signal into the corresponding electrical signal. The frequency selection amplifier simply does selective amplification and output of the signal of a certain specific frequency f. What a phase-sensitive detector does is compare the phase of the signal voltage after frequency selection amplification with that of the reference signal voltage. When the frequency selection amplifying signal is zero, the phase-sensitive output is zero. When the frequency selection amplifying signal is in phase with the reference signal, the DC voltage of phase-sensitive output is positive while it is negative when the contrary is true. Apart from providing the phase-sensitive detector with the reference signal voltage, the audio oscillator gives a sinusoidally modulated signal of frequency f to be applied on the ceramic ring to modulate the cavity length.

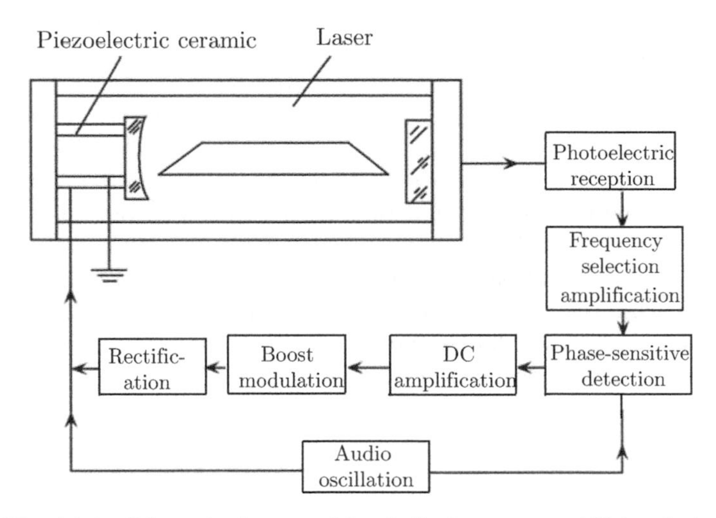

Fig. 6.2-2 Schematic diagram of Lamb dip frequency stabilizing device

Below we shall have a discussion on how to make use of the principle of Lamb dip for frequency stabilization. Figure 6.2-3 shows a laser output power-frequency curve. The output power has a minimum value at the central frequency ν_0 of the atomic spectral line, which is selected as the frequency stabilizing point, the process of frequency stabilizing being as follows: two kinds of voltage are applied on the piezoelectric ceramic, one is a dc voltage (0~300 V) for controlling the laser operation frequency ν and the other, the modulating voltage with frequency f (e.g., 1 kHz) for modulating the cavity length L, or the laser oscillation frequency ν, thereby giving the laser power P the corresponding modulation. If it happens that the laser oscillation frequency overlaps the central frequency of the spectral line ($\nu = \nu_0$), then the modulating voltage would make the oscillation frequency vary near ν_0 at frequency f (at point C in the figure). Hence the laser output power will periodically vary at frequency $2f$ (near point C). As the frequency selecting amplifier operates at the specified frequency f, it cannot pass through the frequency selecting amplifier, and the servo-system has no output signal to send to the piezoelectric ceramic while the laser continues to operate at ν_0. If the laser is disturbed by outside factors so that the laser oscillation frequency deviates from ν_0, for example, $\nu > \nu_0$ (at point D in the figure), then the laser power will vary at frequency f (as f_D in the figure) and the amplitude δ_p of variation is the error signal of the discriminator. Its phase is the same as that of the voltage of the modulating signal. This optical signal is transformed into the corresponding electrical signal by the photoelectric receiver and is, after frequency selection and amplification, sent into the phase-sensitive detector for phase comparison with the modulating signal of frequency f input from the audio oscillator to obtain a dc voltage. The magnitude of this voltage is proportional to the error signal, whether it's positive or negative depending on the phase relationship between the error signal and modulating signal. Now, as the two are in phase, a negative dc voltage is output from the phase-sensitive detector which, after dc amplification, modulation and step up and rectification, will be fed to the piezoelectric ceramic. This voltage will make the piezoelectric ceramic shorten and the cavity length stretch, causing the laser oscillation frequency to return to ν_0. Similarly, if the laser frequency $\nu < \nu_0$ (at point B in the figure), the phase of the output power is opposite to that of the modulating signal although it varies still at frequency f (as f_B in the figure) and the amplitude is still δ_p. Now, if a positive dc voltage is from the phase-sensitive detector, which makes the piezoelectric ceramic ring stretch, then the cavity length will shorten and so the laser oscillation frequency will automatically return to ν_0.

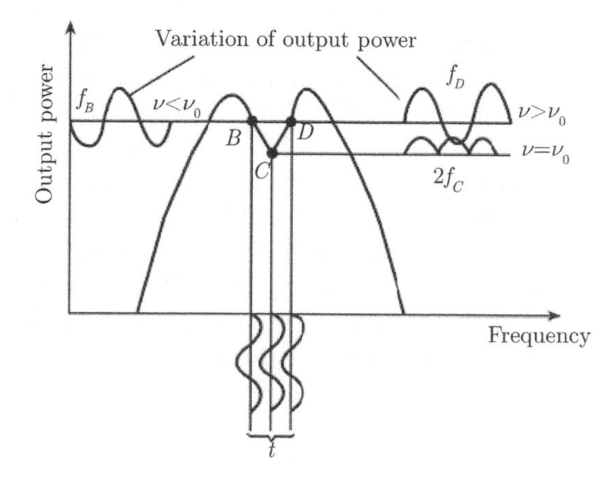

Fig. 6.2-3 The principle of frequency stabilization by Lamb dip

In a word, the essence of frequency stabilization by Lamb dip is: with the central frequency ν_0 of the spectral line as the reference standard, when the laser oscillation frequency deviates from ν_0, output an error signal→discriminate the magnitude and direction of frequency deviation via the servo-system, output a dc voltage to regulate the expansion and contraction of piezoelectric ceramic to control cavity length→have the laser oscillation frequency automatically locked at the center of Lamb dip.

6.2.3 Problems for attention when stabilizing frequency using Lamb dip

(1) The frequency stabilizing laser is required to be not only of single transverse mode, but also of single longitudinal mode. So the frequency stabilizing He-Ne laser generally selects a short cavity, for example, the cavity length is 230 mm, the longitudinal mode interval is about 650 MHz. Thus, when a certain longitudinal mode is in the center of the Lamb dip,

the adjacent longitudinal modes on the two sides can be shifted beyond the net gain curve so as to ensure that the output will be the single longitudinal mode. In order to obtain the single transverse mode, in general, a half confocal resonator made up of flat concave mirrors should be used. The curvature radius of the concave mirror should be rather large while the diameter of the capillary tube should be rather thin. In this way, single transverse mode can be realized by properly regulating the reflective mirror.

(2) Based on the above discussion, it can be seen that the frequency stability is related to the gradient of the two sides of the Lamb dip center. The greater the gradient, the greater the error signal will be. Hence the high sensitivity and even better stability. To obtain a fairly high frequency stability, the Lamb dip should preferably be both narrow and deep. The depth of Lamb dip is proportional to the excitation parameter β, which can be increased by regulating the laser's discharge current and the parameters of the laser tube and reducing the loss of the resonator, thus increasing the depth of the Lamb dip. If the frequency stability is required to be better than 10^{-9}, the Lamb dip depth should be about 1/8 of the output power. The total line width of laser output depends on Doppler broadening while the width of the Lamb dip is determined by the homogeneous broadening and is proportional to the atmospheric pressure. So by appropriately reducing the atmospheric pressure, the dip can become narrower. But if the atmospheric pressure is too low, the laser output power will be reduced.

(3) The symmetry of the Lamb dip line shape also affects the stability of frequency. When the Lamb dip is asymmetric, the curve gradient of the two sides of ν_0 will be different and so will the error signal. The error signal obtained by the side of the small gradient is very small and so the sensitivity is very poor and very hard to be accurately adjusted to the center of the dip. It has been found by experiment that if a He-Ne laser is charged with pure Ne isotopes (Ne^{20} or Ne^{22}), then the Lamb dip line shape obtained is symmetric while if natural Ne is charged, as it contains (Ne^{20}or Ne^{22}) isotopes, the difference in the central frequency between the spectral lines of the two kinds of isotopes is $\nu_{\text{Ne}}^{22} - \nu_{\text{Ne}}^{20} \cong 890$ MHz. Therefore, the Lamb dip obtained will be asymmetric and not sufficiently sharp. Figure 6.2-4 shows the line shapes of pure Ne^{20} and natural Ne. It can be seen from the figure that when the laser output power is very small, the dip is very shallow or no dip but a bell shape is seen; when the laser power is very great, the dip of Ne^{20} is deep and perfectly symmetric. That is why all He-Ne laser tubes actually used for frequency stabilization are all charged with single isotope Ne^{20} or Ne^{22}as the active medium.

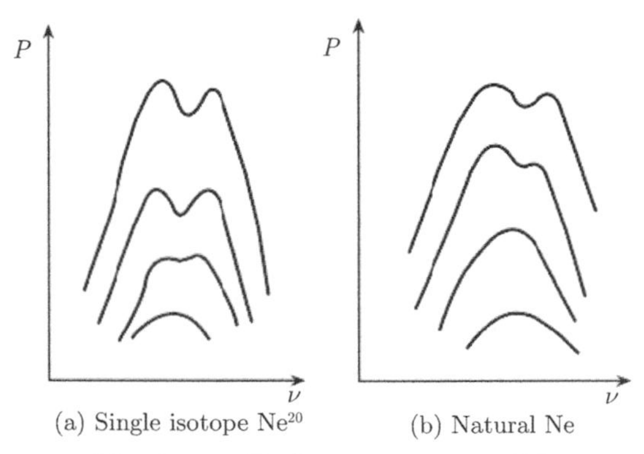

Fig. 6.2-4 The Lamb dip curves of Ne

(4) It is the central frequency ν_0 of the atomic transition spectral line that is used as the reference standard for frequency stabilization by Lamb dip, so the drift of ν_0 itself will directly affect the long-term stability and reproducibility of the frequency. The displacement

of ν_0 is mainly due to the disturbances of physical processes in the laser tube as described below. First, is the frequency shift brought about by the atmospheric pressure, especially the diffusion of He in the laser tube from the plasma tube to the outside which, after a rather long period of time, will cause appreciable changes so that the atmospheric pressure in the tube cannot remain constant. The pressure displacement and broadening include the atomic interaction and the effect of the atomic electric field during the interaction. For instance, when a certain atom emits photons, there will be a second atom that moves close to it and so the light-emitting atom, influenced by the second electric field of atoms, makes the energy level that emits photons experience energy displacement. Therefore, the energy level difference is not constant in the whole period of interaction but varies with the distance between atoms, with the transient photon frequency following suit and the ultimate spectral line distribution exhibiting asymmetric displacement and broadening. In a closed-type laser tube, the atmospheric pressure in the tube and the mixing ratio vary with time and the tube wall and electrodes will discharge gases with impurities in the course of operation while the pressure displacement varies as a result. For a 633 nm He-Ne laser, if an accuracy of the 10^{-10} order is desired and the error source is required to be one induced only by pressure displacement, then it is indispensable to have the variation of atmospheric pressure and gases with impurities controlled below 0.13 Pa, which is very hard to realize for commonly used lasers. Hence the displacement of the central frequency ν_0 of atomic transition will affect the long-term stability and reproducibility using the Lamb dip frequency stabilizing method. Second, it is the Stark effect that brings about the frequency shift. Perform dc discharge in an isotope tube and there will be induced the electrified particle streams from the two poles. The movement of different electrons and ions will form a radial two-pole electric field, which makes the laser transition energy level of the Ne atoms generate the second-order Stark effect displacement and broadening, thus causing displacement of the spectral line frequency. The influence of the above-mentioned factors sets the displacement of the Lamb dip central frequency at about an order of 10^{-7}. None of these disturbances can be adjusted with the servo-system while what can be done is simply reduce their effects as much as possible.

6.3 The Zeeman effect frequency stabilization

6.3.1 The Zeeman effect

When a light-emitting atomic system is placed in a magnetic field, the atomic spectral line will split under the action of the magnetic field. This phenomenon is referred to as the Zeeman effect. If a He-Ne laser oscillates in a single longitudinal mode, when the spectral line central frequency is consistent with the cavity's resonance frequency, there is no frequency pulling effect. The laser output frequency is ν_o (e.g., the 632.8 nm laser). If a longitudinal magnetic field is applied along the direction of the light beam, then we can observe along the direction of the magnetic field that one spectral line is symmetrically split into two. One is a left rotating circularly polarized light, whose frequency is higher than that of the spectral line ($\nu_0 + \Delta\nu$) when no magnetic field is applied, and the other is a right rotating circularly polarized light, whose frequency is lower than that of the spectral line ($\nu_0 - \Delta\nu$) when no magnetic field is applied. The light intensities of the two are equal and their sum is equal to the light intensity of the original spectral line (as shown in Fig. 6.3-1). The frequency difference between the two spectral lines is

$$\Delta\nu = 2\frac{g\mu_{\text{B}}H}{h} \tag{6.3-1}$$

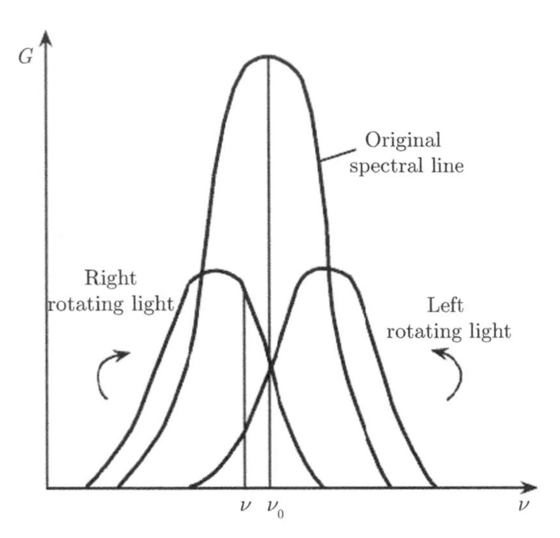

Fig. 6.3-1 The Zeeman effect

where g is the Landau factor, μ_B the Bohr magneton, h the Planck constant, and H the strength of the magnetic field. The intersection of the two split spectral lines is exactly the central frequency of the original spectral line. This is the longitudinal Zeeman effect.

The reason why the Zeeman effect is generated is because of the splitting of the atomic level under the action of the externally applied magnetic field, as shown in Fig. 6.3-2. When no magnetic field is applied ($H = 0$), during transition from a high level to a low one, the atom emits a streak of light of frequency ν_0; after the magnetic field is applied, the two levels will split up, as shown by the right side of Fig. 6.3-2. When changing from a high level to a low one in accordance with the rule of selection, the atoms will emit polarized light of three frequencies ($\nu_1 = \nu_0 + \Delta\nu$, ν_0, $\nu_2 = \nu_0 - \Delta\nu$).

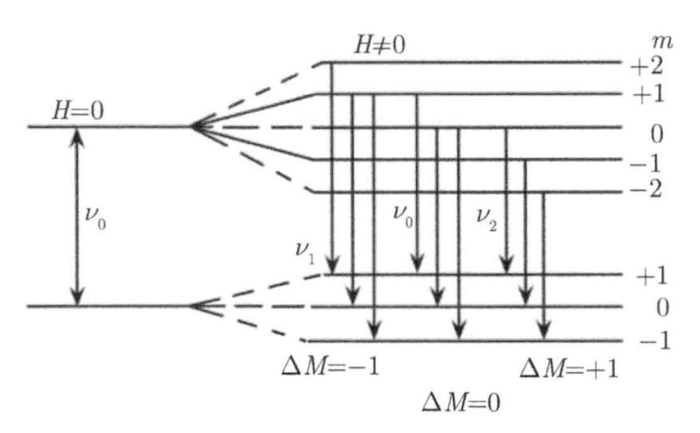

Fig. 6.3-2 The splitting of atomic energy level in a magnetic field

It can be seen from Fig. 6.3-1 that, when the laser oscillation frequency is at ν_0, the light intensity of the left rotating circularly polarized light is equal to that of the right rotating circularly polarized light. If the laser oscillation frequency has deviated from ν_0 (say at ν), then the light intensity (I_R) of the right rotating light is greater than that (I_L) of the left rotating light. Otherwise, there is $I_R < I_L$. Given the difference between the light intensities of the two circularly polarized lights output by the laser, we shall be able to find the direction and magnitude of deviation of the laser oscillation frequency from the central frequency. Thus effort can be made to form a control signal to regulate the resonator to have it stabilized at the central frequency of the spectral line. The curve at the intersection of the two spectral lines split by the longitudinal Zeeman effect has a fairly steep gradient, which can serve as a very sensitive frequency stabilizing reference point. So both the stability and reproducibility of the frequency are rather high.

6.3.2 The Zeeman effect double frequency stabilizing laser

1. The structure of the double frequency stabilizing laser

The double frequency stabilizing laser is made up of the double frequency laser tube with a longitudinally uniform magnetic field applied on it, an electro-optic modulator, and an electronic servo-system, as shown in Fig. 6.3-3.

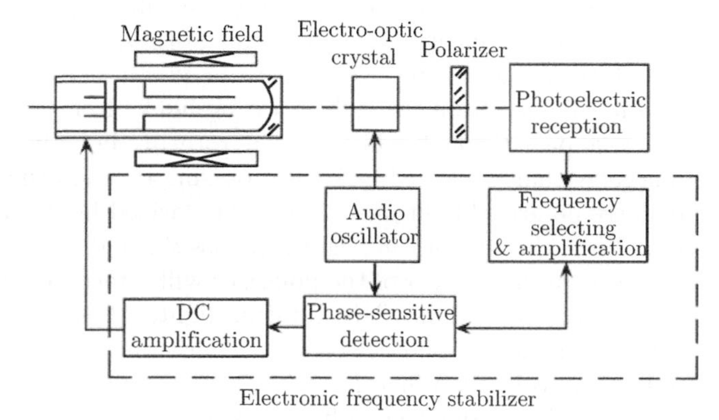

Fig. 6.3-3 Schematic diagram of double frequency stabilizer

The double frequency laser tube is an internal cavity tube using the piezoelectric ceramic to control the cavity length, with the shell of the tube made of quartz glass. The cavity mirror is made up of plane-concave reflective mirrors, in which the plane mirror is glued together with the piezoelectric ceramic ring, the laser tube is charged with highly purified helium and neon gases, He^3:Ne^{20} being about 7:1, and the atmospheric pressure charged about 400 Pa. Either too high an atmospheric pressure charged or a content of other gases will increase laser noise. In the discharge region a homogeneous longitudinal magnetic field of an intensity of 300 G is applied. The magnetic field is generated by a permanent magnetic ring or an electromagnetic coil, the structure of whose laser tube is as shown in Fig. 6.3-4.

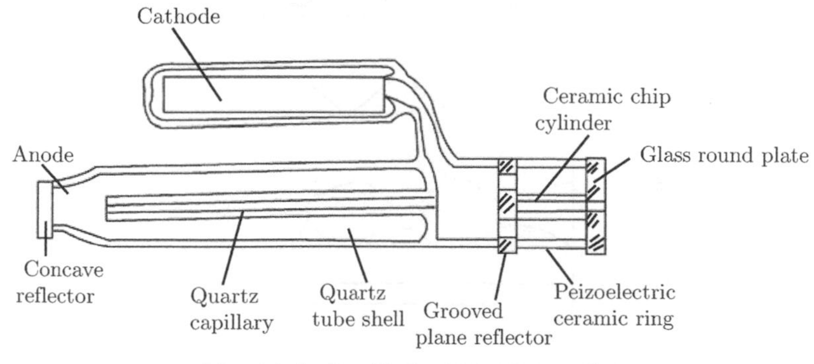

Fig. 6.3-4 Double frequency laser tube

As a frequency stabilizing laser tube, single mode output is required. For instance, for a He-Ne laser tube of output 632.8 nm, if only the cavity length is below 100 mm, the single longitudinal mode output can be ensured. If single transverse mode output is desired, then it is advisable to appropriately choose the diameter of the capillary, the curvature radius, and reflectivity of the cavity mirror to attain the goal.

The electro-optic modulator is composed of the electro-optic crystal and polarizers. When passing through a crystal with a voltage ($V_{\lambda/4}$) of 1/4 wavelength applied on it, the circularly polarized light will become linearly polarized light while the polarizer allows only the light parallel to the polarizing axis to pass through. So, by combining the two and using $\pm V_{\lambda/4}$ to make the left rotating and right rotating circularly polarized light alternately pass through the polarizer, we shall be able to complete frequency discrimination after determining the light intensities of the left and right rotating light by comparison.

The principle of frequency discrimination can be explained as follows: when the left rotating and right rotating circularly polarized light output by the double frequency laser enters the electro-optic crystal (on which an alternately varying $V_{\lambda/4}$ of frequency f is applied), it becomes two streaks of linearly polarized light normal to each other. Properly set the direction of the polarizer's polarizing axis. When the voltage is a positive half cycle, the linearly polarized light changed from the right rotating light after passing through the electro-optic crystal can just pass through while the left rotating light just cannot. When the voltage becomes a negative half cycle ($-V_{\lambda/4}$), after passing through the crystal, the left and right rotating circularly polarized light has a linearly polarized light direction opposite to that described above. While the left rotating light can pass through, the right one cannot. Hence, the photoelectric receiver in the rear of the polarizer will alternately receive the light intensity signals $I_{\nu_{\rm L}}$ and $I_{\nu_{\rm R}}$ of the left and right rotating light, whose variation frequency is f. When $I_{\nu_{\rm L}} > I_{\nu_{\rm R}}$, the phase of the photoelectric receiver's output signal voltage is the same as that of the modulating voltage. When $I_{\nu_{\rm L}} < I_{\nu_{\rm R}}$, then the phase of the output signal voltage is opposite to that of the modulating voltage. When $I_{\nu_{\rm L}} = I_{\nu_{\rm R}}$, then the output signal is a dc voltage, whose operation principle is as shown in Fig. 6.3-5.

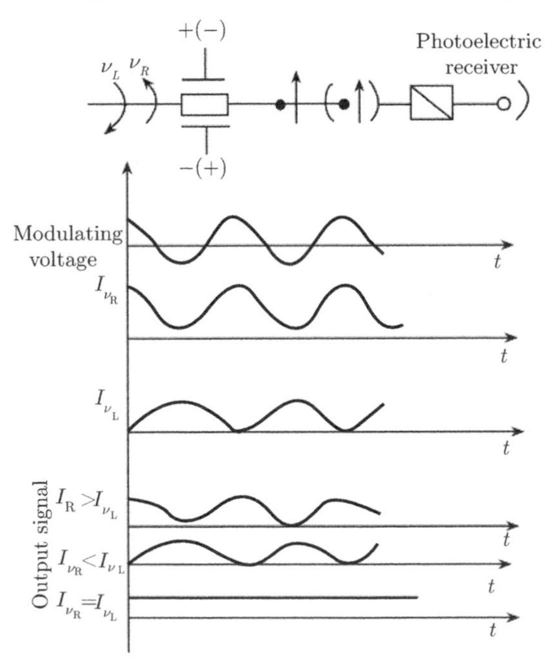

Fig. 6.3-5 Schematic diagram of frequency discrimination using an electro-optic modulator

The electronic servo-system consists of a 1 kHz audio oscillator, a frequency selecting amplifier, a phase-sensitive detector, and a dc amplifier.

2. The operation principle of the double frequency stabilizing laser

For a single mode laser with an oscillating frequency $\nu = q\dfrac{c}{2nL}$, when the laser begins to oscillate, for the particles in the active medium, under the action of powerful light, the refractive index n will vary; the variation amount Δn at the spectral line center ν_0 is zero. When the oscillation frequency $\nu > \nu_0$, then Δn is an increment, or an increase of the effective refractive index. Conversely, when $\nu < \nu_0$, the Δn is a decrement, or a decrease of the effective refractive index. Both cases have the tendency to pull the oscillation frequency toward the center of the spectral line, which is the frequency pulling effect.

After a longitudinal magnetic field is applied, because of the Zeeman effect, the spectral line is split into two spectral lines located at the two sides of ν_0 and equidistant from the central frequency; the central frequency of the former is a left rotating light with central frequency $\nu_{0\mathrm{L}}(>\nu_0)$ while the latter is a right rotating light with central frequency $\nu_{0\mathrm{R}}(<\nu_0)$, their gain curves $G(\nu_\mathrm{L})$ and $G(\nu_\mathrm{R})$ being as shown in Fig. 6.3-6. The frequency pulling effect can make the frequencies of the two streaks of circularly polarized light each move toward the maximum value of their respective gain curves, or ν_L moves toward ν_{0L} and ν_R toward ν_{0R}.

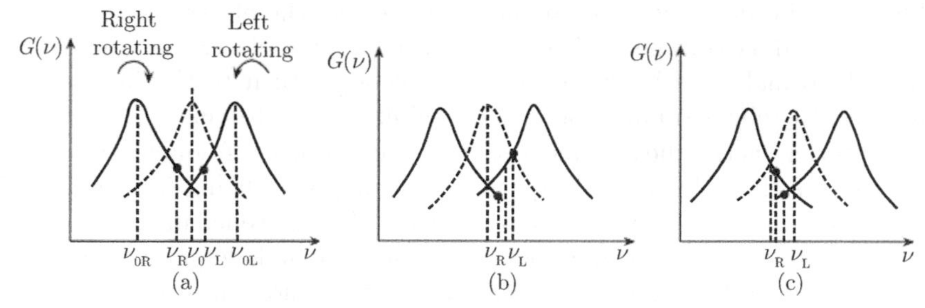

Fig. 6.3-6 The gain curves of a double frequency laser

For the left rotating circularly polarized light,

$$\nu_\mathrm{L} = q\frac{c}{2nL(1+\Delta n_\mathrm{L})} \quad or\ (1+\Delta n_\mathrm{L})\nu_\mathrm{L} = q\frac{c}{2nL} = \nu \tag{6.3-2}$$

As $\nu_\mathrm{L} < \nu_{0\mathrm{L}}$, then $\Delta n_\mathrm{L} < 0$ and so $\nu_\mathrm{L} > \nu$; that is, the frequency of the left rotating light is higher than the oscillation frequency with no magnetic field applied.

$$\nu_\mathrm{L} - \nu = -\Delta n_\mathrm{L}\nu_\mathrm{L} \tag{6.3-3}$$

Similarly, the frequency of the right rotating circularly polarized light is

$$\nu_R = q\frac{c}{2nL(1+\Delta n_\mathrm{R})} \tag{6.3-4}$$

As $\nu_\mathrm{R} > \nu_{0\mathrm{R}}$, then $\Delta n_\mathrm{R} > 0$ and so $\nu_\mathrm{R} < \nu$. The frequency of the right rotating light is lower than the oscillation frequency with no magnetic field applied, that is,

$$\nu - \nu_\mathrm{R} = \Delta n_\mathrm{R}\nu_\mathrm{R} \tag{6.3-5}$$

To sum up, after a longitudinal magnetic field is applied on a laser tube, the original laser with a single oscillating frequency will split into two streaks of laser with different frequencies, that is, the left rotating light with a rather high frequency and the right rotating light with a low frequency. So such a laser is spoken of as a double frequency laser. The difference in frequency between the two streaks of circularly polarized light is

$$\Delta\nu = \nu_\mathrm{L} - \nu_\mathrm{R} = \sqrt{\frac{\ln 2}{\pi^3}\cdot\frac{4\nu_0}{hQ}\left(\frac{g\mu_\mathrm{B}H}{\Delta\nu_\mathrm{D}}\right)}$$

where Q is the quality factor of the cavity and $\Delta\nu_{\mathrm{D}}$ is the Doppler line width. It's clear that for the difference in frequency, apart from the generation of Zeeman splitting because a longitudinal magnetic field is applied, it is related to the quality factor of the cavity.

Below we shall briefly discuss the principle of frequency stabilization for the double frequency laser. The frequency stabilizing reference point of the double frequency laser is the intersection of the left and right rotating light curves, as shown in Fig. 6.3-6(a). If the laser oscillation frequency $\nu = \nu_0$, it can be seen from the figure that the gain of the left rotating light is equal to that of the right rotating light, or $G_{\mathrm{L}} = G_{\mathrm{R}}$. So the power (light intensity) output is equal ($I_{\nu_{\mathrm{L}}} = I_{\nu_{\mathrm{R}}}$). Now the photoelectric receiver outputs a DC signal while the electronic servo-system has no signal to output. So the laser frequency remains unchanged. If an outside disturbance makes the laser frequency drift ($\nu > \nu_0$), as shown in Fig. 6.3-6 (b), then $G_{\mathrm{L}} > G_{\mathrm{R}}$. Therefore, $I\nu_{\mathrm{L}} > I\nu_{\mathrm{R}}$. Now the phase of the error signal output by the photoelectric receiver is opposite to that of the modulating voltage. Conversely, if $\nu < \nu_0$, as shown in Fig. 6.3-6 (c), then $G_{\mathrm{L}} < G_{\mathrm{R}}$, $I_{\nu_{\mathrm{L}}} < I_{\nu_{\mathrm{R}}}$. Then the phase of the error signal output by the photoelectric receiver will be the same as that of the modulating voltage. After amplification this error signal by frequency selection, the electronic servo-system will output the corresponding voltage, which can control the piezoelectric ceramic to modulate the cavity length to make the laser oscillation frequency return to the intersection of the two spectral lines, thereby attaining the goal of stabilizing the frequency.

In precision interferometric measurement, the double frequency frequency stabilizing laser has more advantages than the single frequency frequency stabilizing laser mainly because the former uses the beat frequency method to measure the difference in frequency. So it has a stronger anti-disturbance ability and does not make very high requirements on the operating conditions (temperature, humidity, cleanliness, etc.) and can work long continuously under the condition of no constant temperature. This is especially beneficial to industrial interferometers.

6.3.3 The Zeeman effect absorption frequency stabilization

The device of Zeeman effect absorption frequency stabilization is as shown in Fig. 6.3-7, which consists in installing a Zeeman absorptive tube on the optical path outside the cavity of a single frequency laser, in which is charged gas Ne of low atmospheric pressure (in contrast to a tube for charging the mixed gas He-Ne, one in which only Ne is charged has a smaller pressure displacement with the spectral line) and definite electric current is energized in the tube. Some of the excited Ne atoms can absorb the laser injected into the tube, but as the absorption spectral line is rather wide, it is inadvisable to use it as the reference frequency. If a longitudinal magnetic field is applied on the Ne tube, because of the Zeeman effect, the absorbing line of the Ne atom relative to the atom's initial central line will split into two streaks of overlapping absorbing lines, as shown in Fig. 6.3-8. Therefore, Ne absorption becomes bi-directional dispersion in character; that is, it possesses different absorption coefficients for the left and right rotating circularly polarized light of identical frequency but in opposite directions. The difference in absorption depends on how much the laser oscillation frequency deviates from the central frequency and it is only in the center ν_0 of the spectral line that the absorption of the two streaks of circularly polarized light is equal. It can be seen from the figure that the two absorbing lines cross each other at C where the gradient is steepest. With this point as the reference point for frequency stabilization, we can obtain sensitive frequency discriminating accuracy.

The principle of absorption frequency stabilization by the Zeeman effect: The linearly polarized light output from the single mode He-Ne laser, in passing through the electro-optic crystal with a positive-negative alternating $V_{\lambda/4}$ rectangular voltage applied on it, becomes two streaks of alternately changing left rotating and right rotating circularly polarized light.

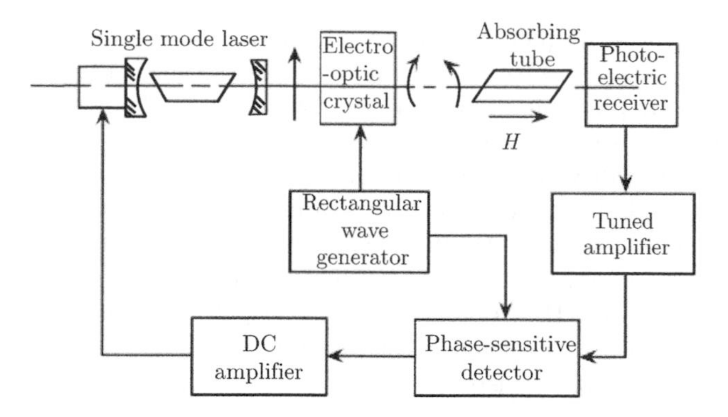

Fig. 6.3-7 Zeeman effect absorption frequency stabilizing device

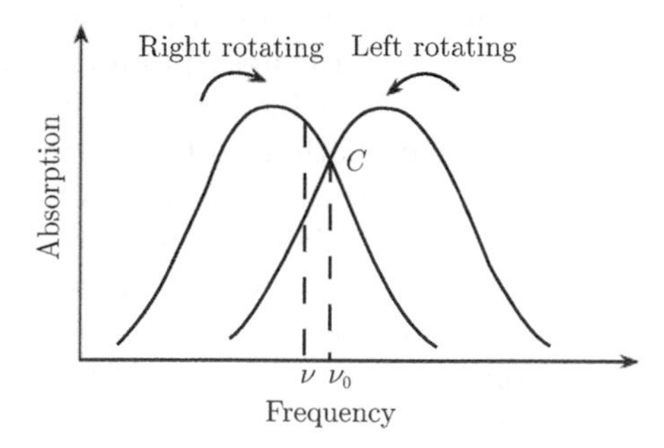

Fig. 6.3-8 Zeeman split Ne absorption line

Then, in passing through the Ne absorbing tube with a longitudinal magnetic field applied on it, the two streaks of alternating circularly polarized light will get modulated in the absorbing tube, resulting in the formation of the error signal. The amplitude of this error signal is proportional to the magnitude of the frequency difference in deviation and its phase is related to the direction of deviation. This error signal will be received by the opto-electric receiver and, after amplification and through the servo-system, will go to control the expansion and contraction of the cavity length, ensuring that the laser oscillation frequency will be stabilized at ν_0.

6.4 Saturated absorption frequency stabilization (anti-Lamb dip frequency stabilization)

It is known from the above discussions on the methods of Lamb dip frequency stabilization, double frequency frequency stabilization, etc. that the key to enhancing the stability and reproducibility of frequency is the selection of a stable and the narrowest possible reference frequency. In all the above-mentioned methods of frequency stabilization, it is the atomic transition central frequency of the laser itself that is used as the reference frequency, but the central frequency of atomic transition is apt to vary owing to its vulnerability to the influence of the discharge condition and some other factors, which limit its stability and reproducibility. To remedy such a situation, usually an outside reference frequency standard is adopted for frequency stabilization; for example, saturated absorption is used for this purpose. This is done by placing an absorption tube filled with low atmospheric pressure gas atoms (or molecules) in the resonator. The tube has absorption lines very well coordinated with the laser oscillation frequency and, as the atmospheric pressure in the tube is very low, the collision broadening is very small and can be neglected. The pressure displacement of the absorption line's central frequency, too, is very small. The absorption tube is in general free from the discharge effect and so the central frequency of the spectral line is fairly stable, forming a dip in a steady position and of very narrow width in the center of the absorption line. With this dip as the frequency stabilizing reference point, the frequency stability and the accuracy of reproducibility can be greatly improved.

The device of saturated absorption frequency stabilization is as shown in Fig. 6.4-1. The laser tube is connected in series with the absorption tube and placed in an extracavity resonator. It has a very strong absorption peak with respect to where the laser oscillation frequency is. For instance, for He-Ne laser with wavelength 632.8 nm, filling the tube with Ne or I_2, and for a laser with wavelength 3.39 μm, filling methyl (CH_4) can both yield the result of the absorption line in consistency with the oscillation frequency. The atmospheric pressure inside the absorption tube is in general only 0.13~1.3 Pa, so the influence from the atmospheric pressure and the discharge condition is insignificant, making it possible to obtain a frequency discriminating accuracy higher than that obtained by the previously discussed frequency discriminator. The mechanism of the generation of absorption dip in the center of the absorption line by the gas in the absorption tube is similar to that for the Lamb dip. For the light with $\nu = \nu_0$, the light intensities of its two columns of traveling wave propagating forward and backward are both absorbed by the $\upsilon_z = 0$ molecules, or the two columns of light intensity act on one and the same group of molecules. Hence the ease with which to reach saturation. For the $\nu \neq \nu_0$ light, the light intensities of the two columns of forward and backward propagating light are absorbed by two groups of molecules with longitudinal velocities $+\upsilon_z$ and $-\upsilon_z$, respectively. So it's not easy for the absorption to reach saturation, with the absorption dip appearing at ν_0 of the absorption line, as shown in Fig. 6.4-2 (b). The dip in the center of the absorption line implies that the absorption is minimum, so there appears a spike at ν_0 of the laser output power (light intensity), which is usually referred to as the anti-Lamb dip, as shown in Fig. 6.4-2(c). The anti-Lamb dip can be used as a very good reference point for frequency stabilization, whose frequency stabilizing procedure is similar to that of the Lamb dip frequency stabilization and will not be repeated here.

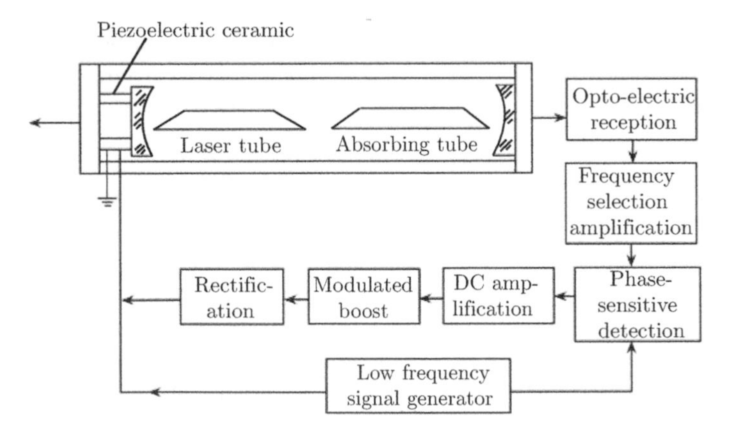

Fig. 6.4-1 The device of frequency stabilization with saturated absorption

Originally the frequency stabilization by saturated absorption used the Ne atomic gas as the absorption medium, but its effect is not ideal since the lifetime of the lower energy level of Ne is short, the absorption rather weak, and the anti-Lamb dip not very obvious. Currently, molecular gases are often adopted to serve as gases for frequency stabilization using saturated absorption. For instance, the electronic transition of $^{127}I_2$ molecular steam ground state [R(127) of band 11-5, P(33) of band 6-3, R(477) of band 9-2, etc.] exactly falls within the gain curves of the He-Ne laser's 612-nm, 633-nm, etc. oscillating lines. These absorption lines have rather stable electrical features and magnetic characteristics with the width of the anti-Lamb dip of the MHz order of magnitude. So it can be used as a standard. It happens that the $^{127}I_2$ steam ground state electronic absorption line [P(13) of band 43-0] is in agreement with the 514.5 nm oscillation line of the Ar^+ laser. So it can be used to stabilize the argon ion laser. The branch P(7) of the asymmetric vibration band ν_3 of the methane (CH_4) has 6 spectral lines in the neighborhood of 3.39 μm of He-Ne laser, of which the component $F_2^{(2)}$ happens to be located within the bandwidth of the laser spectral line of 3.39 μm He-Ne ($3S_2$-$3P_4$) laser, SF_6 has a strong absorbing line near 10.6 μm of CO_2 laser, and so on.

Using the absorption line of molecular gases as the reference frequency standard has the following merits:

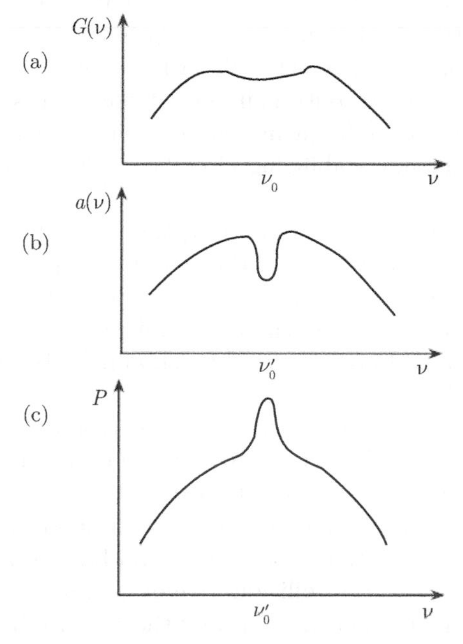

Fig. 6.4-2 The anti-Lamb dip

(1) The vibration transition lifetime of molecules is longer than the lifetime of the Ne atomic energy level, reaching up to the $10^{-2}\sim10^{-3}$ s order of magnitude. Therefore, the natural width of the molecular spectral line is much narrower than the atomic spectral line. The molecular absorption line generates the transition between the ground state and the vibration energy level and the absorption tube has very strong absorption without stimulation by discharging, avoiding the disturbance by discharge.

(2) The absorption tube atmospheric pressure is rather low, the spectral line broadening induced by molecular collision is very small, the width of its anti-Lamb dip having only extremely narrow spectral lines below 10^5 Hz.

(3) The molecule's ground state dipole moment is zero (e.g., the methane molecule is a typical spherically symmetric molecule). So both the Stark effect and Zeeman effect are very small, the frequency shift and broadening induced from which can be neglected. Thus, by virtue of its simple and compact construction, the intracavity absorption saturated frequency stabilizing laser has found wide application.

As both the tunable range of the laser spectral line and absorption line are very narrow, it is very important to select suitable gas molecules as the medium for saturated absorption to enable the absorption spectral line to be in agreement with the laser spectral line. For instance, the methane molecule has a strong absorption line in the vicinity of the 3.39-μm laser, but in the center of the CH_4 absorption line, the frequency is about 100 MHz higher than that in the center of the He-Ne's 3.39-μm laser spectral line. In order to make the central frequencies of the two overlap, He of high atmospheric pressure can be filled in the He-Ne laser tube to force the laser spectral line center to shift to the center of the absorption line by means of pressure displacement. The pressure displacements of He and Ne are (0.2 ± 0.004)MHz/Pa and (0.09 ± 0.03)MHz/Pa, respectively. If a mixed gas of He^3:Ne^{20} = 24:1 in a He-Ne laser tube (capillary length 300 mm, diameter 3 mm, discharge current 5 mA), the total atmospheric pressure is 667 Pa, and the methane absorption tube atmospheric pressure is 1.33 Pa. Now, the laser spectral line center and the methane absorption line center basically overlap.

In addition, as the peak value of the absorption spectral line does not overlap that of the laser gain curve, the saturated absorption peak is often located in the inclined background of the Lamb dip in the laser output power curve. So the primary derivative signal has a rather large fundamental wave background, which will bring about control errors in frequency stabilization. Therefore, in the servo-control circuits of such lasers, the spectral line's third derivative signal is often used as the frequency discriminating signal (that is, the adoption of a third harmonic as a frequency stabilizing circuit) to eliminate the influence from the background.

As the frequency stability of the saturated absorption frequency stabilizing laser is ultimately dependent on the frequency stability of the absorption spectral line while also related to the width of the spectral line and the signal-to-noise ratio, it is very important to select an ideal absorbing medium, which should satisfy the following conditions:

(1) The frequency of the absorbing spectral line should be basically in agreement with that of the laser gain spectral line.

(2) The absorption coefficient should be large. For this reason, the low energy is preferably the ground state. As the atomic absorption line wavelength is most often located in the waveband of the visible light and ultraviolet light, it is not easy to coordinate with most gas lasers. So very often the molecular absorption line and molecular vibration-rotation spectral line are used for enrichment. In addition, it is easy to find lines to match laser spectral lines while the polarizing rate is low and the collision frequency shift is smaller than with atoms.

(3) The excited state has a rather long lifetime and the spectral line a small natural width.

(4) The atmospheric pressure is low, the spectral line collision broadened, and the frequency shift is small.

(5) The molecular structure should be stable. Make every effort to do without the intrinsic electric moment and magnetic moment so as to reduce collision, Stark and Zeeman frequency shift, and broadening.

Tab. 6.4-1 A list of several saturated absorption operation systems

Laser	Operation wavelength/μm	Absorption gas
Ar^+	0.515	$^{127}I_2$
He-Ne	0.633	^{20}Ne, $^{127}I_2$, $^{129}I_2$
He-Ne	3.39	CH_4
CO_2	10.6	CO_2, SF_6, OsO_4
Dye	0.657	Ca

6.5 Other frequency stabilizing lasers

What is described above is the adoption of the laser atomic spectral line itself and the outside saturated absorption spectral line for the He-Ne laser as the frequency discriminating standard for frequency stabilization. In actual application, the frequently used CO_2 laser, pulse operated laser, semiconductor laser, etc. also, need to have their frequency stabilized. Correspondingly, a number of unique frequency stabilizing methods have emerged.

6.5.1 Frequency stabilization for the CO_2 laser

The frequency stabilizing CO_2 laser has found application in metrology, the optical frequency measuring chain and lidar. It has many oscillation lines within a range of 9~ 11 μm. Certain media capable of saturated absorption can be used to make the frequency of certain oscillation spectral lines of the CO_2 laser stable, including mainly SF_6, OsO_4, etc., as well as the gas CO_2 itself as the absorptive medium. By using the spontaneous transition fluorescence spectral line of wavelength 4.3 μm, we can perform frequency stabilizing for any one oscillation spectral line of the CO_2 laser, which is called the fluorescence frequency stabilizing method. With this method, a very good signal-to-noise ratio can be obtained, the stability reaching 10^{-12} and reproducibility being 1.5×10^{-10}.

The fluorescence frequency stabilizing system of the CO_2 laser is as shown in Fig. 6.5-1, which consists of a tunable single frequency CO_2 laser, the CO_2 absorption cell, the InSb fluorescence detector and the electronic servo-system. One mirror of the resonator is glued to the piezoelectric ceramic, on which are applied two kinds of driving voltage, whose function is the same as that of the previously discussed Lamb dip method.

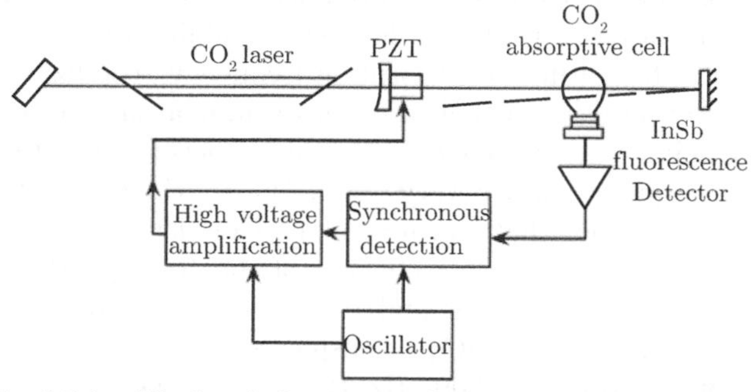

Fig. 6.5-1 CO_2 laser's fluorescence frequency stabilizing system

All the laser transitions of CO_2 molecules and the spontaneous transitions of wavelength 4.3 μm possess a common upper vibration energy level, so if an absorption cell filled with gas CO_2 is placed in the resonator, when it is illuminated with the CO_2 laser, the molecules on the ground state energy level (00°0) will transit onto the energy level (00°1) after absorbing photons. Colliding with one another, the stimulated molecules emit fluorescence of wavelength 4.3 μm during their transition to the ground state, the Doppler width of whose spectral line is about dozens of MHz as shown in Fig. 6.5-2. Owing to the Doppler effect, the molecules in thermal motion along the direction of the light's travel will become stimulated after absorbing laser of a certain specific frequency. The molecules without a thermal motion velocity ($v_z = 0$) in the direction of light's travel will absorb laser of frequency ν from two directions at the same time, thus experiencing a narrow dip upon attaining saturated absorption first. This narrow fluorescence saturated absorption curve can be utilized as the reference standard for CO_2 laser frequency stabilization. During the transition from the upper energy level (00°1) to the vibration ground state (00°0), CO_2 fluorescence emits a 4.3 μm fluorescence. The upper energy level has a very long lifetime of about 2.2 ms, so the fluorescence spectral line is very narrow, which is often checked with InSb (under liquid nitrogen). In the early stage the absorption cell is placed inside the cavity, then outside it to change the signal-to-noise ratio. The contrast to the 4.3-μm fluorescence, saturated absorption dip is up to 16% and the concave width is 0.9 MHz. The stability at sampling time 0.1~100 s is 6×10^{-12}, repeatability is 10^{-10}.

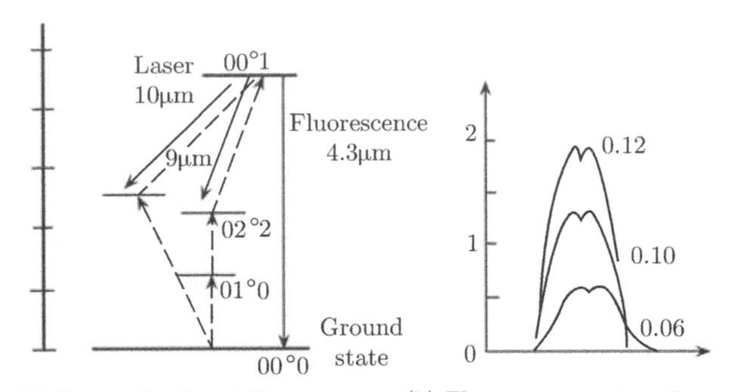

(a) Energy level and fluorescence (b) Fluorescence saturation curve

Fig. 6.5-2 Energy level and spectral line of CO_2 fluorescence

Apart from the above, the saturated absorption with the molecules of SF_6, OsO_4, etc. is used for CO_2 laser frequency stabilization. All the devices adopted are controlled using the extracavity absorption quasi-traveling wave. SF_6 has very strong absorption lines in the neighborhood of 10.5 μm and 10.6 μm, the wavelength of which overlapping the laser spectral line of P(16) and P(18) of CO_2. Borde and others in France observed and measured the absorption lines of SF_6 10.5 μm under several mTorr atmospheric pressure using a 6-m absorption chamber. The lines consist of two parts, one of doublets and the other of triplets. By locking the laser frequency at a peak of a doublet, a stability of 3×10^{-13}/s is obtained.

It's better to use OsO_4 for saturated absorption frequency stabilization than SF_6. The even number isotopes ^{190}Os and ^{192}Os nucleus of the osmium atom does not have a magnetic moment and the spectral line does not have a superfine structure; the OsO_4 molecule is very heavy and the Doppler frequency shift is very small. Within the range 10.2~10.6 μm, OsO_4 has many strong absorption lines that can match the CO_2 laser wavelength, of which the $^{192}OsO_4$ absorption line near the gain curve center of the 10.53-μmP(14) is strongest. By using a 250-cm long absorption chamber and telescope beam expanding laser (diameter d = 2.5-cm), under the condition of OsO_4 atmospheric pressure 40 mTorr, the saturated absorption line width obtained is 250~300 kHz, the laser frequency stability is 3×10^{-12}/ 100 s, and the repeatability is 2×10^{-11}.

6.5.2 The Ar^+ laser (using $^{127}I_2$ saturated absorption) for frequency stabilization

The argon ion laser is a gas laser that yields the maximum output power in continuous operation in the region of visible light. In general but several watts, the continuous power has rich laser spectral lines, of which the 515 nm (green line) emitted by $Ar^+ 3p^4 4p^4 D^0_{5/2} - 3p^4 4s^2 P_{3/2}$ transition is most intense. This line can agree with the wavelength of lines 43-0, P (13), and R(15) ($\nu'' = 0$, $J'' = 13$ to $\nu' = 43$, $J' = 12$ and $\nu'' = 0$, $J'' = 15$ to $\nu' = 43$, $J' = 16$, respectively) of the iodine molecule electronic transition $X' \sum^{+}_{og} \rightarrow B^3 \prod^{+}_{ou}$, respectively, and can constitute the Ar^+/I_2 saturated absorption system. For its location in the waveband of the visible light and very high power and rather narrow spectral line, it is very suitable for use as a standard for length and can find important applications in such fields as verification of the theory of relativity, geodesic survey, radar communication, etc. The 43-0, P(13) line superfine structure component a of $Ar^+/^{127}I_2$ has been internationally recommended to be used as the secondary standard for wavelength, the wavelength in vacuum being

$$\lambda = 514673466.2 \times 10^{-15} \text{ m}$$

and the corresponding frequency is 582490603.6 MHz.

The earliest 515 nm iodine steady argon ionic laser adopted the scheme for checking the linear absorption of fluorescence using the iodine molecular beam. As the lower energy level of the above-mentioned two iodine absorption lines are both in the vibration ground state ($\nu'' = 0$), the population is large, so the absorption coefficient is rather large. Their upper energy level has a rather long lifetime, and the natural line width is about 70 kHz. But the short-term stability of the argon ion is not good (because of the heavy discharge current and cooling by flowing water). So, first the F-P reference cavity is used for frequency stabilization to raise its short-term stability by one or two orders of magnitude. Such a method of stabilizing the frequency with molecular beam linear absorption has yielded a stability of $7 \times 10^{-14}/1000$ s.

Borde and Spieweck made investigations on frequency stabilization with the ion laser using outer cavity saturated absorption; the former placed the iodine absorption chamber in a ring-shaped interferometer, the latter using a device as shown in Fig. 6.5-3. Both of them had the laser frequency pre-stabilized on the inclined side of the F-P reference cavity's transmission curve to improve the short-term frequency stability of laser. Then the reference cavity frequency was locked on the saturated absorption peak. Both used the beam expanded laser and the low-pressure absorption chamber. According to reports, Borde and others have already obtained a line width close to the natural width 70 kHz, the stability for $\tau = 100$ s is 5×10^{-14}, and repeatability is 1.5×10^{-12}. The stability obtained by Spieweck is 1×10^{-11}/s, $5 \times 10^{-13}/100$ s when the iodine atmospheric pressure is 2.5 Pa (1Pa $\approx$ 7.5 mTorr). The pressure frequency shift of the spectral line is linear, being -5k Hz/Pa. At the peak modulating amplitude $\Delta\nu_{\rm m}$when 5 MHz $< \Delta\nu_{\rm m} <$ 15 MHz, there is no modulated frequency shift. The repeatability obtained by appropriately controlling the cooling temperature is 2×10^{-11}. He carefully studied different factors of the generation of spectral line frequency shift, the main ones (that brought about $\sim 10^{-12}$ error) being the iodine atmospheric pressure frequency shift, iodine impurity, absorption spectral lines asymmetry (influence of recoil splitting), the wave-front curvature, and light beam adjustment not good, etc. The total amount of frequency shift is about 1×10^{-11}.

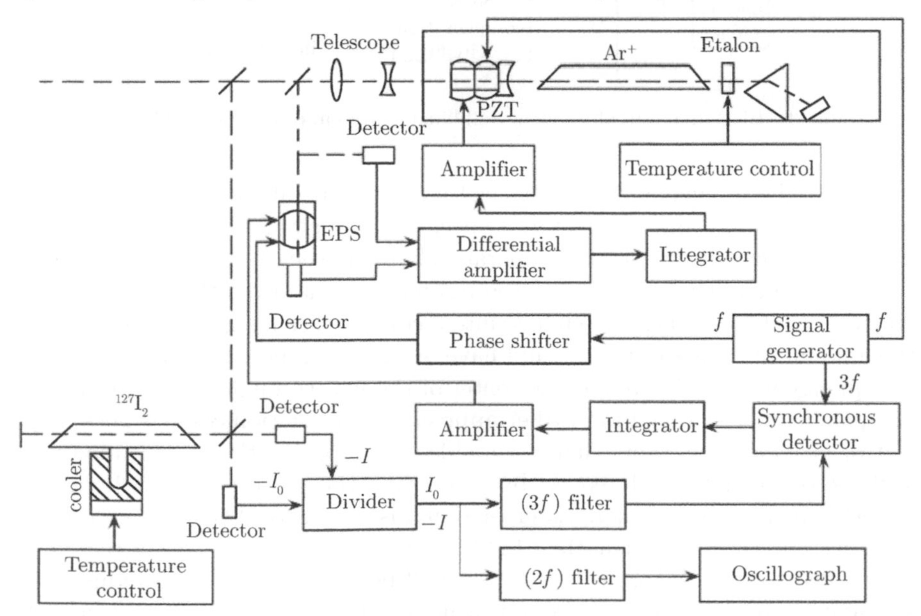

Fig. 6.5-3 Principle of the frequency stabilizing system using Ar^+/I_2 extracavity saturated absorption

6.5.3 Frequency stabilization for the pulsed laser

The frequency stability of the pulsed laser has all along been a hard nut to crack. In recent years, with the development of the pulsed laser in high resolution spectroscopy, laser chemistry, and lidar applications, people are compelled to explore methods for stabilizing the pulsed optical frequency, of which more seems to have been done on frequency stabilization for the pulsed solid lasers with better results obtained.

During stimulation, the solid pulsed laser needs to have rather great stimulating energy, so the variation of both the voltage and temperature in the system is very intensive. According to reports abroad, fairly good effects have been obtained in frequency stabilization using the method of compensation and locking for the cavity body and other parameters in the course of formation of the pulse. The frequency stabilizing device used is as shown in Fig. 6.5-4, which is an Nd:YAG pulsed laser with a repeat frequency of 10 Hz and a duration of 5 ms. Since the solid laser has a rather wide bandwidth plus the existence of the spatial hole burning effect, the laser just has to output in single mode (longitudinal or transverse). For this reason, an F-P fused quartz etalon should be inserted in the cavity. In order to eliminate the spatial hole burning effect, a 1/4 wave plate each is mounted at the two ends of the YAG rod while the cavity length is regulated through the PZT ceramic on a cavity mirror and the $LiNbO_3$ crystal. With the F-P confocal interferometer on the optical path outside the cavity as standard for the optical frequency, the whole laser system is cooled with thermostatic water flow.

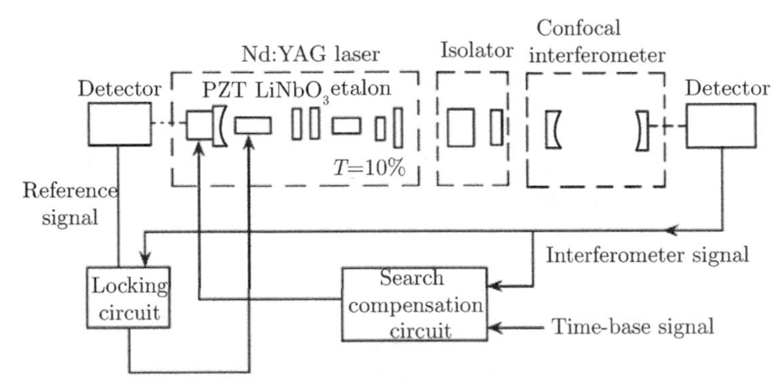

Fig. 6.5-4 Block diagram of frequency stabilizing system of the YAG pulsed laser

To ensure that the frequency of each optical pulse will be aligned with the transmission peak value of the confocal interferometer, a special search circuit is adopted which, the instant the optical pulse appears, starts to operate, changing the cavity length with a scanning voltage. Once the photodetector receives the resonance signal from the F-P interferometer, the search circuit will automatically change into a compensation signal voltage. The magnitude of the amount of compensation should have been determined in advance to ensure that the laser frequency will be directed at the center of the F-P throughout the pulse cycle. But now the frequency of laser may have rather appreciable fluctuation, so it is necessary to use another set of fast locking system that is capable of comparing the magnitude of the laser power on detectors 1 and 2. Moreover, by comparing the pulse power with the power of the F-P interferometer transmission, an error signal is obtained. Then, through the locking circuit servo-system, the optical length of $LiNbO_3$ crystal, or the oscillation frequency of the laser, is controlled so that the optical frequency will be stabilized at the maximum gradient of the F-P cavity transmission curve, constituting a fast frequency stabilizing circuit. After simultaneous operation of the two sets of systems, the frequency fluctuation of the optical pulse can be smaller than 200 kHz.

6.5.4 Frequency stabilization for the semiconductor laser

With its merits of small type, reliability, and long lifetime, the semiconductor laser would be of great significance for its application in such fields as superheterodyne optical communication, precision measurement, etc. if appropriate frequency stabilizing methods can be found to improve its frequency stability.

Previously, it was mainly the frequency stabilizing system as shown in Fig. 6.5-5 that was adopted for performing frequency stabilization for the semiconductor laser, with the F-P confocal interferometer used as the reference for the frequency. An audio scanning voltage is applied on the interferometer. When the laser oscillation frequency deviates from the central frequency of the scanning interferometer, a change in the transmission light intensity will take place so that an error signal will be obtained and then applied on the laser's thermostat controller via the servo-system. Finally, by regulating the laser's operation temperature, the laser frequency will be stabilized. As the frequency stabilizing system is constructed by using the amplitude characteristic of the optical resonator, the gradient of the two sides of the frequency discriminating curve quickly approximates to zero. The anti-disturbance ability of the system is very poor. Even a small hop of the laser frequency will lead to loss of locking, which, worse still, is not easily detected. So effective frequency stabilization cannot be realized.

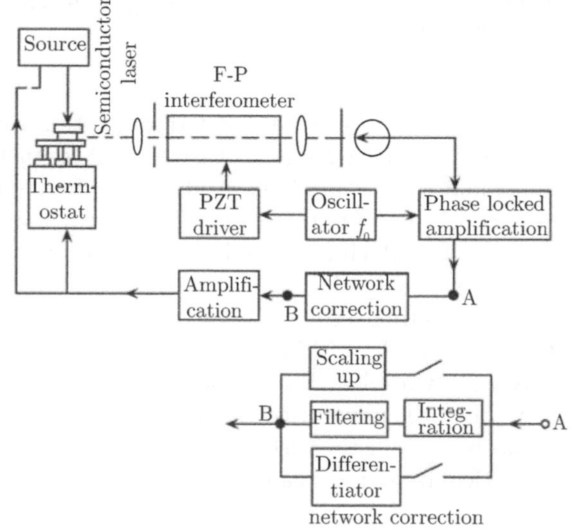

Fig. 6.5-5 The frequency stabilizing system of the semiconductor laser

In the early 1980s, a method of using "the cesium atoms saturated absorption for frequency stabilization" was reported abroad, the frequency stabilizing system of which is as shown in Fig. 6.5-6. The semiconductor laser is installed in a thermostat constructed using the Bohr

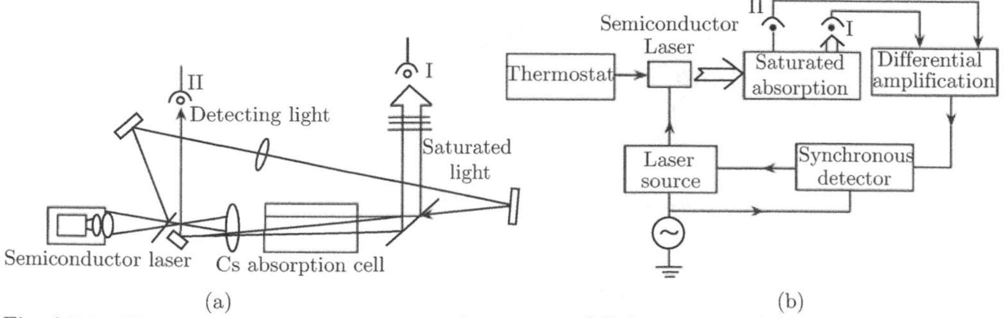

Fig. 6.5-6 The cesium saturated absorption frequency stabilizing system of the semiconductor laser

effect, the thermostatic accuracy reaching 10^{-3}~10^{-4} K. The frequency stabilizing system consists of the Cs saturated absorption device and the electronic servo-control system.

The energy level distribution of the cesium atom ground state $6^2S_{1/2}$ and the first excited state $6^2S_{3/2}$ is as shown in Fig. 6.5-7. The interval is 852.1 nm; the superfine structure of the ground state and excited state is as shown in the figure. The Doppler width of the D_2 line is about 370 MHz at ordinary temperature. So the absorption line proceeding from ground states 3 and 4 is easy to split. The lifetime of the excited state is very short ($\tau = 10^{-8}$ s) while the relaxation time between the two superfine energy levels of ground states 3 and 4 is rather long. Apart from the transition of ground state 3 to excited state $F = 2$ and ground state 4 to excited state $F = 5$, the remaining transition lines are all easy to get saturated. This is because the atoms that very quickly drop from the excited state fall on the two superfine energy levels of the ground state in the same probability. Such a pump effect between superfine energy levels makes even the semiconductor laser with rather small output power sufficiently capable of enabling them to generate the saturated absorption effect.

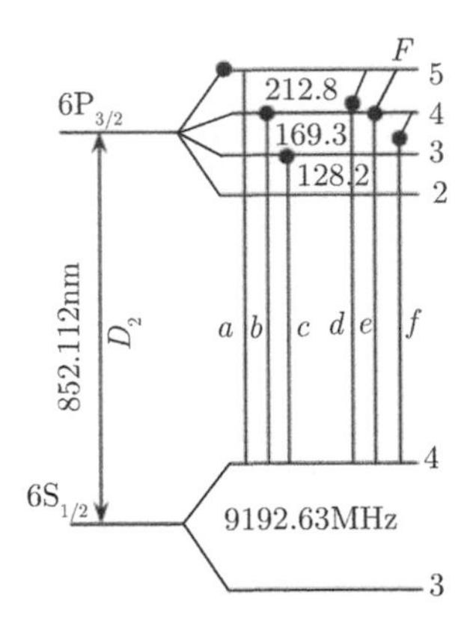

Fig. 6.5-7 The energy level graph of cesium atoms

The principle of the semiconductor laser Cs-saturated absorption frequency stabilization consists in, first, applying the modulating signal of frequency f_0 on the electrodes of the semiconductor laser. Scan (modulate) around line D_2 the laser frequency to detect the signals on the light beam in Fig. 6.5-8(a). On the base of line D_2 there are 6 lines of saturated absorption A, B, C, D, E, and F, of which A, B, and C are Lamb dips in the transition from ground state energy level 4 to energy level $F = 5$, 4, 3 of excited state $6^2P_{3/2}$, respectively, while D, E, and F are the intersecting resonant Lamb dips from ground state energy level 4 to the excited state $F = 4$, 5 and 3, 4. Line A is rather weak as the particles now can only return to ground state energy level 4. The frequency difference between lines B and E is smaller than 20 MHz. As both the absorption line itself and laser spectral line each have a definite width, there is no way to distinguish one from the other. To remove the effect of the base of line D_2, an identical base signal can be taken out from the saturated light beam.

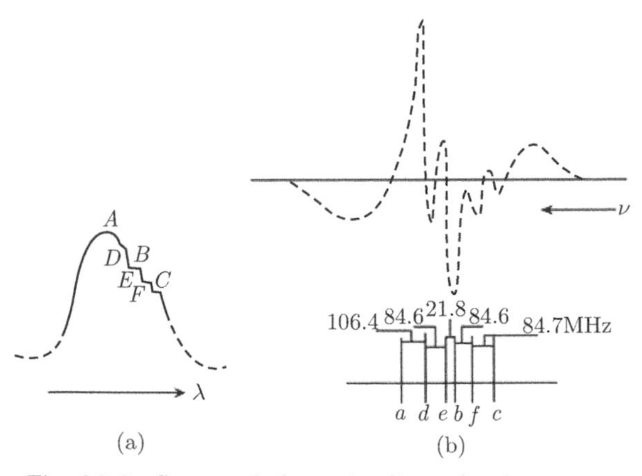

Fig. 6.5-8 Saturated absorption lines of cesium atoms

Choose a suitable amplitude to eliminate the base elements in the detecting light beam by differentiation. The final differential signal is as shown in Fig. 6.5-8(b). With the system locked on line D of a rather strong amplitude, the stability of the semiconductor laser can be estimated from the variation of the feedback voltage of the control system and is expressed by the Allan variance. When the average time is 0.2~1 s, the stability can reach 9×10^{-12}.

6.6 Measurement of frequency stability and reproducibility

As the frequency of laser is extremely high, it is very difficult to directly measure the stability of laser. Commonly used electronic instruments are equally helpless in displaying the variation of the extremely high optical frequency. So usually the beat frequency method is used for relative measurement, which is similar to the differential frequency in electronic technology. By performing frequency mixing of two columns of light waves, the signal of the differential frequency is of the radio frequency order. Both the frequency spectral analyzer and frequency meter can respond to this beat frequency signal. For the frequency spectral analyzer, its output potential level is proportional to the spectral density of the beat frequency, while what is measured by the frequency meter is the frequency of the beat frequency, the data of which will be processed with Allan variance.

6.6.1 The principle of beat frequency

Laser has good coherence. When two beams of light are superimposed one over the other, the difference value of the initial phase is temporarily stable or slowly varying. Hence the phenomenon of interference. The coherence availability between two beams of light waves has provided a method for measuring the stability of the light wave frequency, i.e., the beat frequency measuring method.

Suppose the transient frequencies of two beams of light wave of very little difference in frequency are $\nu_1(t)$ and ν_2 (t), respectively, and their optical fields are, respectively,

$$\begin{aligned} E_1(\nu_1) &= A_{c1}\cos(2\pi\nu_1 t) \\ E_2(\nu_2) &= A_{c2}\cos(2\pi\nu_2 t) \end{aligned} \tag{6.6-1}$$

where A_{c1} and A_{c2} are the amplitudes of the two light waves. When the two beams of light (whose propagation directions are parallel and overlap) are vertically injected onto the photodetector, the synthetic vibration output is proportional to the light intensity (the square of the optical field); that is, the photocurrent output is

$$\begin{aligned} i_p \approx I = E^2(t) &= [E_1(\nu_1) + E_2(\nu_2)]^2 \\ &= A_{c1}^2\cos^2(2\pi\nu_1 t) + A_{c2}^2\cos^2(2\pi\nu_2 t) + A_{c1}A_{c2}\cos[2\pi(\nu_1+\nu_2)t] \\ &\quad + A_{c1}A_{c2}\cos[2\pi(\nu_1-\nu_2)t] \end{aligned} \tag{6.6-2}$$

where the average value of the first and second terms, or the average value of the cosine function, is equal to 1/2, while the frequency of the third term (sum frequency term) is so high that optical detectors currently available are incapable of responding to it. Their average value is zero. The fourth term (difference frequency term) is much slower relative to the optical frequency. When the difference frequency signal is lower than the cut-off frequency of the photodetector, there is photocurrent output

$$i_p = A_{c1}A_{c2}\cos[2\pi(\nu_1-\nu_2)t] \tag{6.6-3}$$

It can be seen from this equation that the frequency of the difference frequency signal current $(\nu_1 - \nu_2)$ varies proportionally with the frequencies ν_1 and ν_2 of the two beams of light. If laser I has a very high frequency stability relative to laser II, it can be considered that $\nu_1 \cong \nu_0$ (as the reference frequency) and that $\Delta\nu = \nu_1 - \nu_2 = \nu_0 - \nu_2$. Hence the variation of the frequency value of the beat frequency is mainly induced by the action of frequency drift of laser II. The frequency stability of laser II relative to laser I is $\Delta\nu/\nu_0$.

The principle of the beat frequency method is as shown in Fig. 6.6-1. The difference frequency ν_B between the two signals has a rather low numerical value. The length of its M periods measured with a counter is τ, from the fluctuation of which the stability of the two signal sources can be calculated. When the stability of laser I as the reference signal is one order of magnitude higher than that of laser II to be measured, it can be considered that the results obtained are all generated by the laser as the measured signal. Below we shall derive the calculating formula:

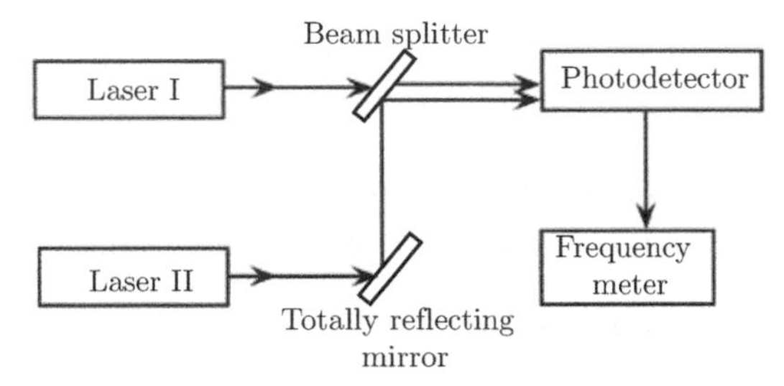

Fig. 6.6-1 The principle of frequency analyzing method

$$\nu_\text{B} = \frac{1}{T_1} = \frac{1}{\tau/M} \tag{6.6-4}$$

where $\tau = MT_1$ and T_1 is the period of the difference beat ν_B. In consideration of the instability of frequency, there is

$$T_1 = T_0 + \Delta T_1 \tag{6.6-5}$$

where T_0 is a constant value, which corresponds to

$$\nu_\text{B} = \nu_{\text{B}0} + \Delta\nu_\text{B} \quad \text{and} \quad \tau = \tau_0 + \Delta\tau \tag{6.6-6}$$

where $\nu_{\text{B}0}$ and τ_0 represent its constant values. According to the theory of errors, there is

$$\frac{\Delta T_1}{T_1} = \frac{\Delta\nu_\text{B}}{\nu_\text{B}}, \qquad \Delta\nu_\text{B} \cong \frac{\Delta T_1}{T_0^2}$$

But

$$T_0 = \frac{\tau_0}{M}, \qquad \Delta T_1 = \frac{\Delta\tau}{M}$$

so

$$\Delta\nu_\text{B} = \frac{M\Delta\tau}{\tau_0^2} \tag{6.6-7}$$

At moment t, the frequency relative fluctuation under the sampling length condition is

$$y_{t,\tau} = \frac{\Delta\nu_\text{B}}{\nu_\text{B}} = \frac{M\Delta\tau}{\nu_0\tau_0^2} \tag{6.6-8}$$

6.6.2 Frequency stability and reproducibility measured with beat frequency technology

As the frequency stability and reproducibility of one and the same batch types of He-Ne lasers are different, it is necessary to make an assessment of the relative frequency stability and reproducibility using the beat frequency technology. First, the interference comparison of all the lasers should be made with the Kr^{86} reference wavelength to obtain the wavelength value measured. The common reference wavelength found should be used as the comparison criterion for the frequency stability. Then the beat frequency of all lasers is measured using the reference laser to find the relative frequency stability.

The beat frequency experimenting device is as shown in Fig. 6.6-2, in which SL_1 and SL_2 are two lasers for frequency stabilization, each stabilized (locked) at the center of the reference criterion via the frequency stabilizer. The light output by the two lasers completely overlap after getting reflected by the total reflector R and partial reflecting and transmitting mirror S and injected onto the photodetector to obtain a difference frequency electric signal. Then it is amplified and input to the beat frequency measuring and processing system and can be directly observed by means of a frequency spectral analyzer. And the frequency values of beat frequency can be read out with a frequency counter. Suppose the frequency of SL_1 serving as the reference frequency $\nu_0 = 4.74 \times 10^{14}$ Hz ($\lambda_0 = 632.99142$ nm) and SL_2 represents the lasers to be measured. Under the same operation condition, the frequencies of the beat frequencies of the lasers relative to the reference laser are measured one after another. If the frequency of a certain laser relative to that of the reference laser experiences $\Delta\nu_1$, $\Delta\nu_2, \cdots, \Delta\nu_N$, N is the number of times of sampling measurement. As the fluctuation in frequency is random, the frequency stability is often processed using the Allan variance in statistics. This is because within the average time of observation, significant variation of frequency may occur anytime. If the maximum frequency variation value is expressed with an arithmetic average value, then very many errors will be generated. Hence in general the statistic method is adopted. The double sampling Allan variance of laser frequency deviation where ν_{2i} and ν_{2i-1} are the frequencies of two adjacent difference frequency signals continuously measured within the sampling average time τ. The double sampling Allan variance of the laser frequency stability

$$\sigma^2(2,\tau) = \frac{1}{N}\sum_{i=1}^{N}\left(\frac{\nu_{2i} - \nu_{2i-1}}{2}\right)^2$$

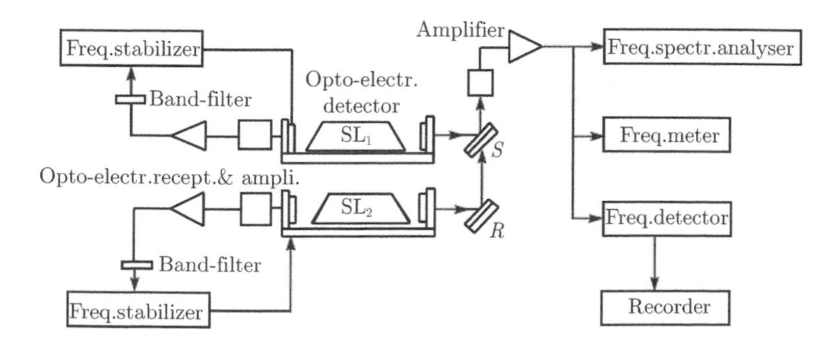

Fig. 6.6-2 Beat frequency experimenting device

$$S_\nu = \frac{\sqrt{\sigma^2(2,\tau)}}{\overline{\nu}} = \frac{1}{\overline{\nu}}\sqrt{\frac{\frac{1}{2}\sum_{i=1}^{N}(\nu_{2i}-\nu_{2i-1})^2}{2N}} \tag{6.6-9}$$

where $\overline{\nu}$ is the average frequency of laser, and $1/2$ is the assumption that each laser has a factor of identical action on the fluctuation of the frequency of difference frequency. Therefore, simply by finding the N groups of frequency sequence of adjacent difference frequencies on the digital frequency meter, the frequency stability of each laser within the sampling time τ can be found by calculation using the above equation. In addition, descriptions of determining the stability of frequency using the method of phase difference can be found in relevant literature. For instance, by adopting the conventional double light beam interference method to measure how much two laser interference fringes have moved, the accuracy of measuring frequency can reach the order of magnitude 10^{-13} under the second-order sampling condition.

The relationships between phase and frequency and frequency stability are

$$\nu(t) = \frac{1}{2\pi}\frac{\mathrm{d}}{\mathrm{d}t}[\varphi(t)]$$

$$y(t) = \frac{1}{2\pi\nu_0}\frac{\mathrm{d}}{\mathrm{d}t}[\varphi(t)] = \dot{\varphi}/(2\pi\nu_0) = \frac{1}{2\pi\nu_0}\left(\frac{\varphi(t+\tau)-\varphi(t)}{\tau}\right) \tag{6.6-10}$$

$$\sigma_y^2(\tau) = \left(\frac{(\overline{y}_{k+1}-\overline{y}_k)^2}{2}\right) \tag{6.6-11}$$

Phase is generally expressed by (rad) or (°), but such a unit makes it impossible for phase and the amount of its change to directly reflect the variation of frequency, so it has been recommended that the following value be taken to characterize phase:

$$x(t) = \varphi(t)/2\pi\nu_0 \tag{6.6-12}$$

That is, the phase characterization unrelated to the signal carrier frequency is taken. In reality, it represents the relative phase variation of light wave carrier frequency in a period of oscillation, and its unit is the "second" just as that of time. As it has taken into consideration the influence from the carrier frequency, after finding $x(t)$ at different frequencies, the results can be compared to measure the variation of frequency. The phase change by 2π rad in a period is true of any frequency, but when characterizing $x(t)$ with the phase of an independent carrier frequency, for the phase change in the same period, different values are taken at different carrier frequencies.

Suppose there are two signals

$$\begin{aligned} V_1(t) &= V_1\sin(2\pi\nu_0 t+\varphi_1) \\ V_2(t) &= V_2\sin(2\pi\nu_0 t+\varphi_2) \end{aligned} \tag{6.6-13}$$

Here we assume that the frequencies of the two signals are equal and the instability of frequency is represented by the phase angles $\varphi_1(t)$ and $\varphi_2(t)$. It can be seen from Fig. 6.6-3 that, at time t_1,

$$x_1(t) = \left.\frac{\varphi_1-\varphi_2}{2\pi\nu_0}\right|_{t_1} = \Delta t_1$$

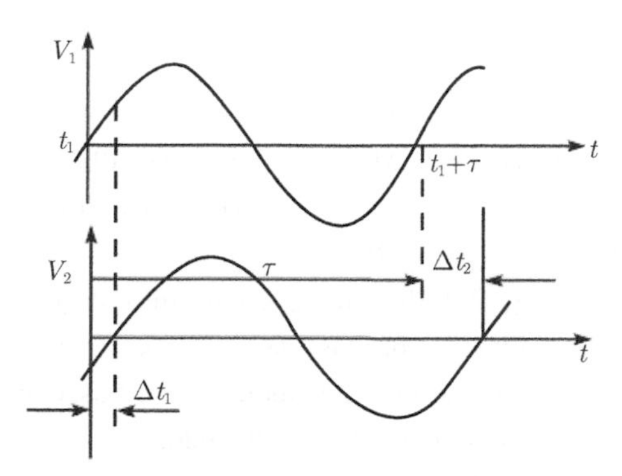

Fig. 6.6-3 Phase change diagram

After a period τ,

$$x_2(t) = \left.\frac{\varphi_1 - \varphi_2}{2\pi\nu_0}\right|_{t_1+\tau} = \Delta t_2$$

The relative value of the average frequency difference of the two signals within time τ is

$$\overline{y}(t) = \frac{1}{\tau}\int_{t_1}^{t_1+\tau} y(t)\mathrm{d}t = \frac{x_2(t) - x_1(t)}{\tau} \tag{6.6-14}$$

Repeat the measurement a number of times. Within the sampling interval $t_{\mathrm{k}} - t_{\mathrm{k+1}}$ for the kth sampling, there is

$$\sigma_\gamma(\tau) = \left\langle \frac{(\overline{y}_{k+1} - \overline{y}_k)^2}{2} \right\rangle = \left| \frac{1}{2\tau^2 M} \sum_{j=1}^{M} (x_{j-2} - 2x_{j+1} + x_j)^2 \right|^{1/2} \tag{6.6-15}$$

The symbol $\langle\ \ \rangle$ represents infinite time averaging. The above equation shows that the double sampling variance of frequency stability is the secondary differentiation of the phase value and that three adjacent phase values should be taken every time a calculation is to be made. In other words, when calculating two neighboring average frequencies, one and the same phase value will have to be used. If the phase difference between two signals can be accurately recorded for a long time, then the frequency stability for an arbitrary sampling length can be obtained.

Exercises and problems for consideration

1. Compare the similarity and dissimilarity between Lamb dip and anti-Lamb dip.
2. Analyze the operation principle of the phase-sensitive detector in the frequency stabilizing servo circuit and its role.
3. In a He-Ne laser, the Ne atoms spectral line width $\Delta\nu_D = 1.5 \times 10^9$ Hz, its spectral line's central frequency $\nu_0 = 4.7 \times 10^{14}$Hz. If no frequency stabilizing measure is taken, what is the stability of frequency of such a laser?
4. There is a frequency stabilized CO_2 laser. Its cavity length is regulated with a ring-shaped piezoelectric ceramic PZT. Its length L is 1 cm, sensitivity $m = 2.5 \times 10^{-4}/\mu\mathrm{m}/(\mathrm{V} \cdot \mathrm{cm})$. It has been found by measurement that the greatest variation (i.e., the maximum range of cavity length regulation) ΔL is 0.1 μm. In order to enable the frequency stabilizing system to operate normally, how high a voltage should the error signal be amplified to? [Note that $m = \Delta L/(V \cdot L)$].
5. We have two CO_2 lasers of identical structure and dimensions but different materials, one of quartz glass and the other of hard glass. If no account is taken of the influence of other factors, when temperature varies by 0.5°C, compare the stability of the two. (It is known the linear expansion coefficient of quartz glass $\alpha = 6 \times 10^{-7}/°\mathrm{C}$, and that of hard glass is $\alpha = 10^{-5}/°\mathrm{C}$.)

References

[1] Application of Laser in Metrological Technology (a collection of translated literature), Beijing, Intelligence Office, Chinese Institute of Metrologic Science, 1973.

[2] Characteristics and Application of He-Ne Lasers (a collection of translated literature), Beijing, Chinese Institute of Metrologic Science, 1974.

[3] Tu Shigu, Frequency Stabilizing and Measuring Techniques, Beijing, Science Press, 1986.

[4] G. Birnbaum. Frequency Stabilization of Gas Lasers. *Proc. IEEE*, 1967, 55(6), 1015-1025.

[5] A. J. Wallard. Frequency Stabilization of the Heliun-Neon Laser by Saturated Absorption on Iodine Vapour. *J. Phys. ESci. Instrum.*, 1972, 5(9), 926.

[6] P. Cerez and J. Bennett. Helium-Neon Laser Stabilized by Saturated Absorption in Iodine at 612 nm. *Appl. Opt.*, 1979, 18(7), 1079.

[7] Zhang Peilin et al., A Study on Frequency Stabilization of Bifrequency Lasers, Laser, 1978, 5(3), pp. 1 8.

[8] Y. L. Sun and R. L. Byer. Submegahertz Frequency-Stabilized Nd:YAG Oscillator. *Optics Lett.*, 1982, 7(9), 408-410.

[9] J. L. Hall. External Dye-Laser Frequency Stabilizer. *Optics Lett.*, 1984, 9(11), 502-504.

[10] P. H. Lee and M. L. Skolnick. Saturated Neon Absorption Inside a 6328 Å Laser. *Appl. Phys. Lett.*, 1967, 10(11), 303.

[11] Wang Yiqiu et al., The Principles of the Quantum Frequency Scale, Beijing, Science Press, 1986.

CHAPTER 7

The Nonlinear Optical Technology

7.1 Overview

7.1.1 The nonlinear optical effect

The emergence of laser has led to the discovery of the nonlinear effect in the light frequency range. Nonlinear optics has broken through the limitations of light wave linear superposition and independent transmission in conventional optics, revealing the process of variation between light wave fields in the medium in energy exchange, phase correlation, mutual coupling, and growth and decline. For instance, when propagating in a nonlinear optical medium, laser with a certain frequency may generate coherent radiation at a new frequency. If two light waves of different frequencies are input, under different conditions, a light wave output of a frequency that is the sum of or a difference between the frequencies of the two can be obtained. This, in a definite sense, is tantamount to a realization of the "synthesis" and "splitting" of photons and continuous tuning within the range of a certain wavelength from the infrared to ultraviolet. For a beam of "completely unrecognized" wave due to wave array distortion, its phase disturbance in the process of propagation can be corrected using the spatial phase conjugative effect to make its phase factor invert, that is, the light wave is inverted by time, completely compensating for the phase distortion. Hence the jocular similes "the time's river is reversed" and "become youthful in one's old age". Such nonlinear optical effects correspond to laser frequency doubling, sum frequency, difference frequency, optical parametric amplification and oscillation, stimulated scattering, and optical phase conjugation, respectively. In a certain sense, nonlinear optics belongs in the category of the interaction between intense light and material. Laser is an indispensable means of and tool for studying nonlinear optics. Following the adoption of Q switching, mode locking, amplification, and mode selecting technologies described in the previous chapters, the peak power and monochromatic directional brightness of laser have been greatly enhanced so that appreciable nonlinear optic effects can be generated. Nonlinear optics has deepened our understanding of the mechanism of interaction between light and material and enriched the content of laser technology, providing investigations on individual laser techniques with new contents and methods. In particular, it has scored great success in finding wide application in frequency spectral tuning of laser, wavelength transformation, and light beam quality improvement. This chapter will be devoted to a discussion of the above-mentioned nonlinear optical technologies that are rather mature and in extensive use.

7.1.2 Nonlinear polarization

Under the action of an external optical wave electric field E, polarization will be induced within a medium, the response being expressed by the intensity vector P of the electric polarization. Nonlinear optics implies that the response of material to an external optoelectric field is not a linear function of its amplitude while for the greater part of nonlinear optical effects, P can be denoted using the E vector power series expanded expression. Taking into account the finite but not zero response time, the nth-order polarization intensity is

$$P^{(n)}(r,t) = \varepsilon_0 \int_{-\infty}^{\infty} \mathrm{d}r_1 \cdots \int_{-\infty}^{\infty} \mathrm{d}r_n R^{(n)}(t-\tau_1, \cdots, t-\tau_n) | E(r,\tau_1) \cdots E(r,\tau_n) \qquad (7.1\text{-}1)$$

where ε_0 is the vacuum dielectric constant, $R^{(n)}$ is the nth-order response function, and the vertical line represents the dot product. This is an expression for time convolution multiple integral under spatial nonlocal domain field approximation. But usually the description of the properties of a medium does not have to use a response function because the response function is involved in discussion of the relationship between the electric field and medium polarization in the time domain while in general the opto-electric field applied always has its specific frequency. So, for a discussion on the above-mentioned problem in the frequency domain, the physical quantity tensor of polarizability is introduced. For this purpose, by adopting the Fourier transform in convolution operation, the expression for medium-induced total polarization strength is obtained as follows:

$$P(r,\omega) = \varepsilon_0[\chi^{(1)} \cdot E + \chi^{(2)} : EE + \chi^{(3)} \vdots EEE + \cdots] \tag{7.1-2}$$

χ represents the electric polarizability of the medium, of which $\chi^{(1)}$ is called the linear polarizability. $\chi^{(2)}$, $\chi^{(3)}$, $\cdots$ in the second term and above are called the nonlinear polarizability. Adopting the classical anharmonic oscillator model, we can give distinct physical images and the relevant properties.

1. The anharmonic oscillator model

Suppose the medium is composed of N classical anharmonic oscillators per unit volume. The electron's response to the externally driving electric field can be described as its motion in the anharmonic potential well, its equation of motion being

$$\ddot{r} + \Gamma \dot{r} + \frac{1}{m}\frac{\partial}{\partial r}U(r) = -\frac{e}{m}E(t) \tag{7.1-3}$$

where Γ is the damping coefficient, e and m are the electron's charge and mass, and r is the displacement the electron leaves the balanced position. Suppose the potential energy function within the medium is

$$U(r) = \frac{1}{2}m\omega_0^2 r^2 + \frac{1}{3}mAr^3 \tag{7.1-4}$$

where the first term is the harmonic term, ω_0 the natural resonant frequency; the second term is the anharmonic term which, substituted into Eq. (7.1-3), yields

$$\ddot{r} + \Gamma \dot{r} + \omega_0^2 r + Ar^2 = -\frac{e}{m}E(t) \tag{7.1-5a}$$

The above equation represents that the electrons, driven by the external electric field, do damped anharmonic motion. The anharmonic term in the equation is generally very small. Therefore, an approximate solution to the equation can be found level by level using the perturbation method, that is, by assuming the form of the solution to be an approximate superposition of one level over another.

$$r = r_1 + r_2 + \cdots + r_k \tag{7.1-5b}$$

Substituting Eq. (7.1-5b) into Eq. (7.1-5a), after operation and collation, we obtain the following simultaneous equations:

$$\ddot{\gamma}_1 + \Gamma \dot{\gamma}_1 + \omega_0^2 \gamma_1 = -\frac{e}{m}E(t) \tag{7.1-6a}$$

$$\ddot{r}_2 + \Gamma \dot{r}_2 + \omega_0^2 r_2 = -Ar_1^2 \tag{7.1.6b}$$

Suppose there are two kinds of Fourier components of frequency ω_1 and frequency ω_2, whose optical field is

$$E(t) = \frac{1}{2}[E_1 e^{-i\omega_1 t} + E_2 e^{-i\omega_2 t}] + c.c. \tag{7.1-7}$$

in which *c.c.* represents the complex conjugation term. Using Fourier transform, we can find the solutions for Eq. (7.1-6) to be

$$r_1(t) = \frac{-e}{2m}[E_1 L(\omega_1) e^{-i\omega_1 t} + E_2 L(\omega_1) e^{-i\omega_2 t}] + c.c. \tag{7.1-8}$$

$$\begin{aligned} r_2(t) = \frac{-e^2 A}{2m^2}\Big[&\frac{1}{2}E_1^2 L(2\omega_1)L^2(\omega_1)e^{-i2\omega_1 t} + \frac{1}{2}E_2^2 L(2\omega_2)L^2(\omega_{2'})e^{-i2\omega_2 t} \\ &+ |L(\omega_1)|^2 |E_1|^2 + |L(\omega_2)|^2 |E_2|^2 \\ &+ L(\omega_1+\omega_2)L(\omega_1)L(\omega_2)E_1E_2 e^{-i(\omega_1+\omega_2)t} \\ &+ L(\omega_1-\omega_2)L(\omega_1)L^*(\omega_2)E_1E_2^* e^{-i(\omega_1-\omega_2)t}\Big] + c.c. \end{aligned} \tag{7.1-9}$$

where $L(\omega_n) = \dfrac{1}{(\omega_0^2 - \omega_n^2) - i\omega_n\Gamma}$ is Lorentzian lineshape function. It can be seen from the above-mentioned solutions that the characteristic of nonlinear response is, for an optical field of frequencies ω_1 and ω_2, the intensity of electric polarization induced in a nonlinear medium has not only components of ω_1 and ω_2, but also components of frequencies $2\omega_1$, $2\omega_2$, and $\omega_1 \pm \omega_2$. These polarization intensity components will radiate electromagnetic waves of corresponding frequencies. These are optical effects of frequency multiplication, sum frequency, and difference frequency in nonlinear optics.

2. The properties of the polarizability tensor

Under electric dipole moment approximation, the intensity of electric polarization of an electrical charge density N is defined as

$$P = -Ner \tag{7.1-10}$$

For the linear term $P_L = -Ner$, by means of the relationship between $\chi^{(1)}$ and P, the first-order polarizability can be found to be

$$\chi^{(1)}(\omega) = \frac{Ne^2}{\varepsilon_0 m} L(\omega) \tag{7.1-11}$$

or the linear polarizability that usually represents dispersion and absorption of a medium.

A comparison with the second-order nonlinear term $P_{NL} = -Ner_2$ yields

$$\chi^{(2)}(2\omega) = \frac{Ne^3 A}{\varepsilon_0 m^2} L(2\omega)L^2(\omega) \tag{7.1-12}$$

The above equation represents the polarizability of frequency doubling. Similarly, the polarizability of sum frequency and difference frequency can be obtained.

$$\chi^{(2)}(\omega_1+\omega_2) = \frac{Ne^2}{\varepsilon_0 m^2} AL(\omega_1+\omega_2)L(\omega_1)L(\omega_2) \tag{7.1-13}$$

$$\chi^{(2)}(\omega_1-\omega_2) = \frac{Ne^2}{\varepsilon_0 m^2} AL(\omega_1-\omega_2)L(\omega_1)L*(\omega_2) \tag{7.1-14}$$

Apart from being related to frequency, the polarizability tensor is determined by its physical symmetry in terms of its principal properties:

(1) The first term $P = \varepsilon_0\chi^{(1)}E$ in Eq. (7.1-2) represents linear polarizability. For an isotropic medium, the linear polarizability is a scalar unrelated to direction while for an anisotropic medium, an optical wave field in a certain direction not only leads to the polarization in that direction, but also that in other directions. $\chi^{(1)}$ is no longer a scalar, but a second-order tensor that associates the two vectors P and E. Similarly, the second-order nonlinear polarizability $\chi^{(2)}$ is a third-order tensor that associates the three vectors P, E, and E_k. So the relationship of the second-order polarization intensity P with the electric field intensity E should be written as

$$P_i(\omega_1, \omega_2) = \varepsilon_0 \sum_{jk} \chi_{ijk}(\omega_1, \omega_2) E_j(\omega_1) E_k(\omega_2) \tag{7.1-15}$$

To sum up, the corresponding polarizabilities of all orders of nonlinear polarization are higher-order tensors one after another. Equation (7.1-2) can be written as

$$\begin{aligned}\frac{1}{\varepsilon_0} P_i(\omega) = &\sum_j \chi_{ij}(\omega) E_j(\omega) + \sum_{jk} \chi_{ijk}(\omega_1, \omega_3) E_j(\omega_1) E_k(\omega_2) \\ &+ \sum_{jkl} \chi_{ijkl}(\omega_1, \omega_2, \omega_3) E_j(\omega_1) E_k(\omega_2) E_l(\omega_3) + \cdots \end{aligned} \tag{7.1-16}$$

where all orders of the polarizability tensors weaken one after another with a difference of several orders of magnitude. For instance, χ_{ijk} $(\omega_1,\ \omega_2)$ is in general seven to eight orders of magnitude lower than χ_{ij} (ω). As we are mainly investigating the second-order nonlinear optical effects, we shall only discuss the third-order polarizability tensors. Strictly speaking, $\chi_{ijk}(\omega_1,\ \omega_2)$ is a quantity varying with frequency, or a quantity with dispersion. When the relevant frequencies taking part in the nonlinear optical effect are all in one and the same transparent wave range, their dispersion can be neglected. (When the relevant frequency is in the visible light and near infrared range, the dispersion of $\chi_{ijk}(\omega_1,\omega_2)$ does not exceed 10%.) So, the polarizability tensors $\chi_{ijk}(2\omega)$, $\chi_{ijk}(-\omega_3,\ \omega_1,\ \omega_2)$, and $\chi_{ijk}(-\omega_3,\ \omega_1,\ -\omega_2)$ of the second-order nonlinear optical effects (including frequency multiplication, sum frequency, difference frequency, parametric oscillation, etc.) are identical. According to the definition of the third-order tensor, in a Cartesian coordinate system, the third-order polarizability tensor has 27 tensor elements in all. Because of their permutation symmetry, or $\chi_{ijk}(-\omega_3,\ \omega_1,\ \omega_2) = \chi_{ikj}(-\omega_3,\ \omega_2,\ \omega_1)$, the tensor elements can be reduced to 18.

(2) Far from where ion resonant frequency is, polarization is simply brought about by electronic displacement while the contribution by the ions can be neglected (e.g., the near infrared, middle infrared, and visible light wave range). In addition, if the nonlinear medium is loss-free (that is, the frequencies of all the fields taking part in the nonlinear process are lower than the electronic absorption band), then it can be proved that the nonlinear polarizability tensor possesses overall permutation symmetry. Kleinman was the first to point out this property of χ_{ijk}. Obviously, by using the overall permutation symmetry, the independent tensors of χ_{ijk} can be further reduced.

(3) As the polarizability tensor is used for describing the characteristics of a medium's response to the optical field, the spatial symmetry of the structure of a medium itself will limit the number of independent components of nonlinear polarizability tensors. For example, it can be proved that, for 11 kinds of crystals with a centrosymmetric structure, all the independent components of their third-order nonlinear polarizability tensors are zero. For the other 21 kinds of crystals without centrosymmetry, restrained by spatial symmetry, some of their independent components are zero while others are equal to one another or are equal in numerical value but opposite in sign. They have even less independent tensor components.

7.1.3 The equation of wave coupling in a nonlinear medium

The wave equation for optical wave propagation in a non-ferromagnetic nonlinear medium without free electrical charges can be derived from the Maxwell equation set:

$$\nabla \times \nabla \times E(r,t) + \varepsilon_0\mu_0 \frac{\partial^2 E(r,t)}{\partial t^2} + \mu_0\sigma \frac{\partial E(r,t)}{\partial t} = -\mu_0 \frac{\partial^2 P(r,t)}{\partial t^2} \quad (7.1\text{-}17)$$

where μ_0 is the permeability of vacuum, σ represents electric conductivity related to optical loss, and $P(r,\ t) = P_L(r,\ t)\ +\ P_{NL}\ (r,\ t)$ is the total polarization intensity including the linear part P_L and the nonlinear part P_{NL} while playing the role of the oscillation source. Taking the transverse field condition $\nabla \cdot E = 0$ and neglecting the loss, for the optical field propagating along the z-axis, we have

$$\frac{\partial^2 E(z,t)}{\partial z^2} - \mu_0 \frac{\partial^2 D(z,t)}{\partial t^2} = \mu_0 \frac{\partial^2 P_{NL}(z,t)}{\partial t^2} \quad (7.1\text{-}18)$$

where $D(z,\ t) = \varepsilon E(z,t)$ is the electric displacement vector and $\varepsilon = \varepsilon_0[1 + x^{(1)}\ (\omega)]$ is the dielectric tensor of the medium. Assume the interacting optical waves are quasi-monochromatic plane waves. Now the light wave electric field and polarization intensity can be expressed as

$$E(z,t) = \frac{1}{2}\sum_m E_m(z,\omega_m)\mathrm{e}^{\mathrm{i}(k_m z - \omega_m t)} + c.c. \quad (7.1\text{-}19)$$

$$P_{NL}(z,t) = \frac{1}{2}\sum_m P_{NLm}(z,\omega_m)\mathrm{e}^{\mathrm{i}\omega_m t} + c.c. \quad (7.1\text{-}20)$$

Substituting the above expanded expression into Eq. (7.1-18) and taking the amplitude envelope's slowly varying approximation while neglecting the second-order derivative term of time, we obtain the equation for the amplitude component of the optical wave electric field:

$$\left(\frac{\partial}{\partial z} + \frac{1}{v_{gm}}\frac{\partial}{\partial t}\right) E_m(z,\omega_m) = \frac{\mu_0}{2\mathrm{i}k_m}\omega_m^2 P_{NLm}(z,\omega_m)\mathrm{e}^{-\mathrm{i}(k_m z - \omega_m t)} \quad (7.1\text{-}21)$$

in which $v_{gm} = \left(\frac{\partial k}{\partial \omega}|_{\omega_m}\right)^{-1}$ is the group velocity while at $P_{NLm} = 0, E_m = f(z - v_g t)$ represents the forward traveling wave independently propagating in the medium. P_{NLm} plays the role of the nonlinear oscillation source. As the above equation has taken into account the variation of the field amplitude with time, it is called the transient state wave coupling equation, which is used for processing the nonlinear optical effect of the mode-locked laser pulses. When the pulse width $\Delta t \gg \frac{L}{c/n}$ (L stands for the length of the medium), then the variation of the field amplitude with time is not obvious in the medium. It can be considered that the opto-electric field and amplitude component of the polarization intensity are not related to t. Neglecting the quantities related to t, we obtain the steady-state wave coupling equation usually suited to the Q-switched laser pulse in the medium.

$$\frac{\mathrm{d}E_m(z,\omega_m)}{\mathrm{d}z} = \mathrm{i}\frac{\mu_0\omega_m^2}{2k_m} P_{NLm}(z,\omega_m)\mathrm{e}^{-\mathrm{i}k_m z} \quad (7.1\text{-}22)$$

m takes $1, 2, \cdots, n+1$, respectively. Using its conjugative form, we find that there are $2\ (n + 1)$ equations in all, constituting $n + 1$ waves in the interacting wave coupling equation set in nonlinear optics. Each specific nonlinear optical effect is mainly determined by the nonlinear relationship between the nonlinear polarization intensity and the light wave electric field. Each polarization that determines the frequency component can be expressed as

$$P_{NLm}(\omega_m) = \varepsilon_0 K \chi_{m\alpha_1\cdots\alpha_n}(-\omega_m; \omega_1, \cdots, \omega_n) E_{\alpha_1}(z, \omega_1) \cdots E_{\alpha_n}(z, \omega_n) \tag{7.1-23}$$

where K is the numerical factor determined by intrinsic conversion symmetry.

$$K = \frac{1}{2^{n-1}} \frac{n!}{\gamma!} \tag{7.1-24}$$

γ represents that, of the n frequencies, γ are identical; ω_m represents the algebraic sum of n frequencies, of which if the frequency is a negative value, then its corresponding electric field takes the conjugative form.

For the second-order nonlinear optical effect, there are three waves that interact with one another. Suppose the frequency relationship is $\omega_3 = \omega_1 + \omega_2$. The corresponding polarization components are obtained, respectively, as follows:

$$P_{NL1}(z, \omega_1) = \varepsilon_0 \chi(-\omega_1; -\omega_2, \omega_3) : E_2^*(\omega_2) \mathrm{e}^{-\mathrm{i}k_2 z} E_3(\omega_3) \mathrm{e}^{\mathrm{i}k_3 z} \tag{7.1-25a}$$

$$P_{NL2}(z, \omega_2) = \varepsilon_0 \chi(-\omega_2; -\omega_1, \omega_3) : E_1^*(\omega_1) \mathrm{e}^{-\mathrm{i}k_1 z} E_3(\omega_3) \mathrm{e}^{\mathrm{i}k_3 z} \tag{7.1-25b}$$

$$P_{NL3}(z, \omega_3) = \varepsilon_0 \chi(-\omega_3; -\omega_1, \omega_2) : E_1(\omega_1) \mathrm{e}^{-\mathrm{i}k_1 z} E_2(\omega_2) \mathrm{e}^{\mathrm{i}k_2 z} \tag{7.1-25c}$$

which, substituted into Eq.(7.1-22), respectively, yield the three equations of wave coupling:

$$\frac{\mathrm{d}E_1}{\mathrm{d}z} = \frac{\mathrm{i}\omega_1}{n_1 c} d_{\text{eff}} E_2^* E_3 \mathrm{e}^{-\mathrm{i}\Delta k z} \tag{7.1-26a}$$

$$\frac{\mathrm{d}E_2}{\mathrm{d}z} = \frac{\mathrm{i}\omega_2}{n_2 c} d_{\text{eff}} E_1^* E_3 \mathrm{e}^{-\mathrm{i}\Delta k z} \tag{7.1-26b}$$

$$\frac{\mathrm{d}E_3}{\mathrm{d}z} = \frac{\mathrm{i}\omega_3}{n_3 c} d_{\text{eff}} E_1 E_2 \mathrm{e}^{\mathrm{i}\Delta k z} \tag{7.1-26c}$$

where $d_{\text{eff}} = \frac{\chi_{\text{eff}_i}}{2}$ is called the effective nonlinear coefficient with the Kleinman approximation already taken, n_i is the refractive index corresponding to ω_i, and $\Delta k = k_1 + k_2 - k_3$ are the phase factors. In the equation, the spatial variation of each optical wave electric field has the intervention by the other optical wave electric fields, showing that there is energy transfer and exchange between the light waves in the medium. Such energy transfer is coupled through the effective nonlinear coefficient d_{eff} of the nonlinear medium.

Using the conjugative form of Eq. (7.1-26) and the relation $\frac{\mathrm{d}I_i}{\mathrm{d}z} = \frac{1}{2} n_i c \varepsilon_0 \frac{\mathrm{d}(E_i E_i^*)}{\mathrm{d}z}$ of light intensity I_i with E_i, we obtain

$$\frac{1}{\omega_1} \frac{\mathrm{d}I_1}{\mathrm{d}z} = \frac{1}{\omega_2} \frac{\mathrm{d}I_2}{\mathrm{d}z} = -\frac{1}{\omega_3} \frac{\mathrm{d}I_3}{\mathrm{d}z} \tag{7.1-27}$$

which is called the Manley-Rowe relation. Letting $I_i \propto \rho_i h\omega_i, \rho_i$ being the number of photons flowing through a unit cross section in unit time, we obtain from Eq. (7.1-27):

$$\frac{\mathrm{d}\rho_1}{\mathrm{d}z} = \frac{\mathrm{d}\rho_2}{\mathrm{d}z} = -\frac{\mathrm{d}\rho_3}{\mathrm{d}z} \tag{7.1-28}$$

The above equation shows that the amount of increase of the ω_1 photon flow density along the direction of the z-axis is equal to that of the ω_2 photon flow density while also equal to the amount of decrease of the ω_3 photon flow density. Put another way, every time a ω_3

photon is decreased, then an ω_2 and ω_3 photon each is increased, respectively, and vice versa. The three equations for wave coupling have given the relation of "splitting" and "synthesis" of photons, including the sum frequency ($\omega_3 = \omega_1 + \omega_2$), frequency doubling ($2\omega = \omega + \omega$), and difference frequency ($\omega_3 = \omega_1 - \omega_2$). From $\omega_3 = \omega_1 + \omega_2$ and Eq. (7.1-27), it is not hard to find

$$\frac{\mathrm{d}I_1(z)}{\mathrm{d}z} + \frac{\mathrm{d}I_2(z)}{\mathrm{d}z} + \frac{\mathrm{d}I_3(z)}{\mathrm{d}z} = 0 \tag{7.1-29}$$

That is, $I_1(\mathrm{z}) + I_2(\mathrm{z}) + I_3(\mathrm{z}) =$ constant. Despite the energy exchange between the optical waves, the total energy is conservative.

7.2 The laser frequency doubling (SHG) technology

The laser frequency doubling technology is also called the second harmonic generation (SHG) technology and is a nonlinear optical effect discovered the earliest in experiment. The experiment on ruby laser SHG performed in 1961 by Franken and others marked the beginning of extensive experiment and theoretical research on nonlinear optics. Laser SHG is the principal method for transforming laser in the direction of short wavelength and has already reached the stage of very wide application with commercially available components and devices.

7.2.1 The wave coupling equation of SHG and its solution

In terms of the frequency relation, SHG is equivalent to the case of $\omega_1 = \omega_2 = \omega$, $\omega_3 = 2\omega$ in the sum frequency generation (SFG). Noting $P_{NL}(2\omega) = \frac{1}{2}P_{NL}(\omega_3)$ and from Eqs. (7.1-26), we can obtain the wave coupling equations for the base frequency E_ω and the second harmonic electric field $E_{2\omega}$:

$$\frac{\mathrm{d}E_\omega}{\mathrm{d}z} = \frac{\mathrm{i}\omega}{cn_\omega} d_{\mathrm{eff}} E_\omega^* E_{2\omega} \mathrm{e}^{-\mathrm{i}\Delta kz} \tag{7.2-1a}$$

$$\frac{\mathrm{d}E_{2\omega}}{\mathrm{d}z} = \frac{\mathrm{i}\omega}{cn_{2\omega}} d_{\mathrm{eff}} E_\omega E_\omega \mathrm{e}^{\mathrm{i}\Delta kz} \tag{7.2-1b}$$

1. The non-depletion approximation

When the second harmonic light generated is of small signal approximation, the corresponding incident base frequency light loss can be neglected, i.e., $\frac{\mathrm{d}E_\omega}{\mathrm{d}z} \approx 0$. Eq. (7.2-1b) can be directly integrated as

$$\begin{aligned} E_{2\omega}(L) &= \frac{\mathrm{i}\omega}{cn_{2\omega}} d_{\mathrm{eff}} E_\omega^2 \int_0^L \mathrm{e}^{\mathrm{i}\Delta kz}\mathrm{d}z \\ &= \mathrm{i}\frac{2\pi L d_{\mathrm{eff}}}{n_{2\omega}\lambda_\omega} E_\omega^2 \mathrm{sinc}\left(\frac{\Delta kL}{2}\right) \mathrm{e}^{\mathrm{i}\Delta kL/2} \end{aligned} \tag{7.2-2}$$

where $\mathrm{sinc}\left(\frac{\Delta kL}{2}\right) = \frac{\sin(\Delta kL/2)}{\Delta kL/2}$, L is the length of the crystal, and λ_ω is the vacuum wavelength of the fundamental frequency light. As the per unit area's optical power or light intensity $I = \frac{1}{2}nc\varepsilon_0 |E|^2$, Eq. (7.2-2) can be expressed by the second harmonic intensity:

$$I_{2\omega} = \frac{8\pi^2 L^2 d_{\mathrm{eff}}^2}{\varepsilon_0 n_\omega^2 n_{2\omega} \lambda_\omega^2 c} I_\omega^2 \mathrm{sinc}^2\left(\frac{\Delta kL}{2}\right) \tag{7.2-3}$$

The ratio of the output frequency second harmonic intensity $I_{2\omega}$ to the base frequency light intensity I_ω is used to characterize its conversion efficiency, called the SHG efficiency; that is,

$$\eta_{\text{SHG}} = \frac{I_{2\omega}}{I_\omega} = \frac{8\pi^2 L^2 d_{\text{eff}}^2}{\varepsilon_0 n_\omega^2 n_{2\omega} \lambda_\omega^2 c} I_\omega \text{sinc}^2\left(\frac{\Delta k L}{2}\right) = \eta_{\max}\text{sinc}^2\left(\frac{\Delta k L}{2}\right) \tag{7.2-4}$$

It is known from the properties of sinc function that η_{SHG} takes the maximum value when $\Delta kL = 0$ as shown in Fig. 7.2-1, but $L \neq 0$. So $\Delta k = 0$ is the crucial factor that ensures that a highly efficient SHG can be obtained, called the phase matching condition. This concept is of paramount importance in nonlinear optics. In a certain sense, the phase matching condition determines the mode and efficiency of energy exchange between optical waves.

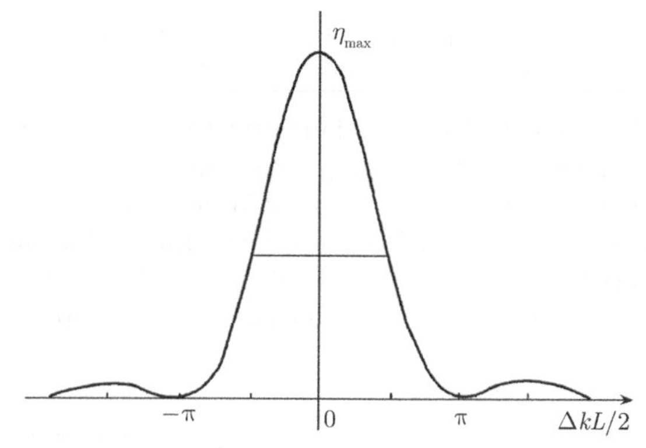

Fig. 7.2-1 The curve of relationship between frequency and phase factor in SHG

2. A physical analysis of phase matching

When $\Delta k = 0$, the base frequency light is converted into second harmonic light generated in large quantities, the non-depletion approximation becomes ineffective, and the wave coupling equations can be changed to

$$\frac{\mathrm{d}\varepsilon_\omega}{\mathrm{d}z} = \frac{\omega d_{\text{eff}}}{cn}\varepsilon_\omega\varepsilon_{2\omega} \tag{7.2-5a}$$

$$\frac{\mathrm{d}\varepsilon_{2\omega}}{\mathrm{d}z} = \frac{\omega d_{\text{eff}}}{cn}\varepsilon_\omega^2 \tag{7.2-5b}$$

where $E_\omega = \varepsilon_\omega \mathrm{e}^{\mathrm{i}\varphi_\omega}$ and $E_{2\omega} = \varepsilon_{2\omega}\mathrm{e}^{\mathrm{i}\varphi_{2\omega}}$. ε is the mode (real number) of E, or $\varepsilon^2 = |E|^2$. The phase relation is specified as $2\varphi_\omega - \varphi_{2\omega} = \frac{3}{2}\pi$ to ensure, under the phase matching condition, that the energy of the fundamental frequency light will steadily transfer to SHG $\left(\frac{\mathrm{d}\varepsilon_\omega}{\mathrm{d}z} < 0, \frac{\mathrm{d}\varepsilon_{2\omega}}{\mathrm{d}z} > 0\right)$ and there is $n_\omega = n_{2\omega} = n$ at that.

From the conservation of energy relation $\frac{\mathrm{d}}{\mathrm{d}z}[\varepsilon_\omega^2(z) + \varepsilon_{2\omega}^2(z)] = 0$, we obtain the boundary condition $\varepsilon_\omega^2(z) + \varepsilon_{2\omega}^2(z) = \varepsilon_\omega^2(0)$, which, substituted into Eq. (7.2-5b), yields

$$\frac{\mathrm{d}\varepsilon_{2\omega}(z)}{\mathrm{d}z} = \frac{\omega d_{\text{eff}}}{cn}[\varepsilon_\omega^2(0) - \varepsilon_{2\omega}^2(z)] \tag{7.2-6}$$

After integration, rigorous solutions under the phase matching condition are obtained:

$$\varepsilon_{2\omega}(z) = \varepsilon_\omega(0)\tanh\left[\frac{\omega d_{\text{eff}}}{cn} z\varepsilon_\omega(0)\right] \tag{7.2-7}$$

$$I_{2\omega}(z) = I_\omega(0)\tanh^2\left[\frac{\omega d_{\text{eff}}}{cn} z \left(\frac{2I_\omega(0)}{nc\varepsilon_0}\right)^{\frac{1}{2}}\right] \tag{7.2-8a}$$

$$I_\omega(z) = I_\omega(0) - I_{2\omega}(z) = I_\omega(0)\text{sech}^2\left[\frac{\omega d_{\text{eff}}}{cn} z \left(\frac{2I_\omega(0)}{nc\varepsilon_0}\right)^{\frac{1}{2}}\right] \tag{7.2-8b}$$

From Fig. 7.2-2 we can easily see the process of "growth and decline" of the fundamental frequency light and second harmonic light generated in a crystal as well as the graphic representation of the energy transfer of the optical wave. The characteristic length of a crystal is defined as:

$$L_{\text{SHG}} = \frac{cn}{\omega d_{\text{eff}}}\left(\frac{cn\varepsilon_0}{2I_\omega(0)}\right)^{\frac{1}{2}} = \frac{cn}{\omega d_{\text{eff}}}|E_\omega(0)|^{-1} \tag{7.2-9}$$

When $z = L_{\text{SHG}}$, over 50 percent of the fundamental frequency light has been converted into second harmonic light generated, and L_{SHG} can be used as a reference standard for choosing the length of an SHG crystal. For a rather large effective nonlinear coefficient d_{eff} and rather strong fundamental frequency light intensity input, adopting a rather short SHG crystal we can obtain a very high second harmonic intensity. Under the $\Delta k = 0$ condition, for a sufficiently long crystal, the SHG conversion efficiency η_{SHG} approximates to 100 percent.

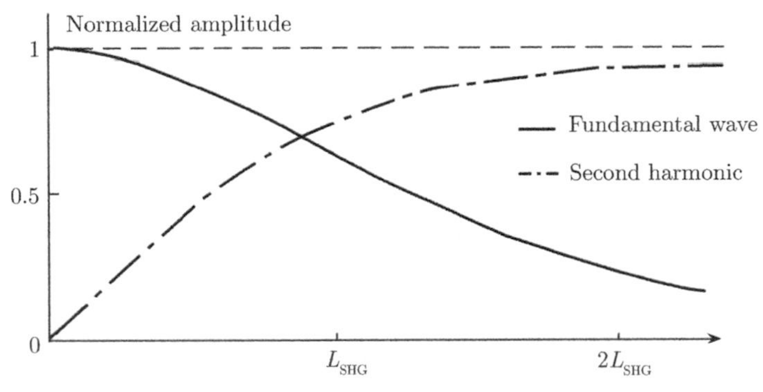

Fig. 7.2-2 The "growth and decline" relation between fundamental wave and second harmonic under the phase matched condition

7.2.2 The meaning and method of phase matching

Phase matching is one of the most important concepts in nonlinear optics. The phase matching condition for SHG is $\Delta k = 0$, or $2k_\omega = \boldsymbol{k}_{2\omega}$. From the relationship between photon momentum P and the wave vector k, we can derive $P_\omega + P_\omega = P_{2\omega}$. In the process of SHG, the frequency relation shows the conservation of energy while phase matching shows the conservation of momentum under the high conversion efficiency condition, ensuring that energy transfer will steadily proceed one way from fundamental frequency light to second harmonic light. The phase matching condition controls the direction of energy transfer between optical waves. In addition, from the relation of the wave vector with the phase velocity $k = \dfrac{\omega}{c/n} = \dfrac{\omega}{v_p}$, we can find that when $2k_\omega = k_{2\omega}$, there is $v_{p_\omega} = v_{p_{2\omega}}$, showing that, in a crystal, the equiphase surfaces of the fundamental frequency light wave and the second harmonic light wave possess identical velocity ensuring that the phase relationship $2\varphi_\omega = \dfrac{3}{2}\pi + \varphi_{2\omega}$ will remain unchanged throughout the entire process of movement. It is a kind of coherence process that is unrelated to the spatial coordinates and with a constant phase difference. The SHG optical wave obtained under this condition will be synchronously superimposed and its interference strengthened. Specific details can be dealt with from the perspective of optical parameters. From $v_{p_\omega} = \dfrac{c}{n_\omega} = v_{p_{2\omega}} = \dfrac{c}{n_{2\omega}}$, the phase matching condition requires that $n_\omega = n_{2\omega}$. For the purpose of compensating for the dispersion effect bound to exist in a medium, the corresponding angular phase matching method has been invented.

1. The angular phase matching

Have the fundamental frequency light injected into a SHG crystal at a particular angle and in a polarization state. We can use the birefringence effect possessed by the SHG crystal itself to offset the dispersion effect so as to meet the requirement for phase matching. Angular matching is the most frequently used, most important, and highly efficient method of generating second harmonic light radiation. According to the mode of configuration of the fundamental frequency light electric field polarizing state, it is divided into the parallel and orthogonal type, the corresponding angular matching being spoken of as type I and type II angular phase matching. Under the normal dispersion condition, for a uniaxial crystal, the analytical expression for the matching angle of type I and type II can be obtained.

(1) The negative uniaxial crystal type I matching mode ($o^\omega + o^\omega \to e^{2\omega}$)

The fundamental frequency light electric field takes the direction the o light is polarized. Under the condition of satisfying definite incident angle $\theta^I_{m\text{neg}}$, angular matching makes the SHG electric field polarize along the direction of the e light. Substitute the expression $n^\omega_\text{o} = n^{2\omega}_\text{e}(\theta^I_{m\text{neg}})$ into the refractive index curvature equation of the SHG e light:

$$\frac{1}{[n^{2\omega}_\text{e}(\theta)]^2} = \frac{\cos^2\theta}{n^{2\omega}_\text{o}} + \frac{\sin^2\theta}{n^{2\omega}_\text{e}} \tag{7.2-10}$$

where $n^{2\omega}_\text{o}$ and $n^{2\omega}_\text{e}$are the principal axis refractive indices of given SHG. The analytical solution for the type I matching angle of the negative uniaxial crystal is

$$\theta^1_{m\text{neg}} = \arcsin\left[\frac{(n^\omega_\text{o})^{-2} - (n^{2\omega}_\text{o})^{-2}}{(n^{2\omega}_\text{e})^{-2} - (n^{2\omega}_\text{o})^{-2}}\right]^{1/2} \quad . \quad (n^\omega_\text{o} \geqslant n^{2\omega}_\text{e}) \tag{7.2-11}$$

The above equation has a definite requirement on the principal axis refractive index of the negative uniaxial crystal, namely, $n^\omega_\text{o} \geqslant n^{2\omega}_\text{e}$; otherwise, $\theta^1_{m\text{neg}}$does not exist, showing that at $n^\omega_\text{o} < n^{2\omega}_\text{e}$, the dispersion with the crystal is so serious that birefringence is insufficient to compensate for the phase mismatch caused by dispersion. This should be given careful consideration when selecting the SHG crystal.

(2) The negative uniaxial crystal type II matching mode ($o^\omega + e^\omega \to e^{2\omega}$)

The fundamental frequency light electric field takes the o light and e light as two polarization states, $2k_\omega = \dfrac{\omega}{c}[n^\omega_\text{o} + n^\omega_\text{e}(\theta)]$. The SHG electric field is in the direction of the polarized e light, $k_{2\omega} = \dfrac{2\omega}{c} n^{2\omega}_\text{e}(\theta)$. The phase matching condition $2k_\omega = k_{2\omega}$ is turned into a relation of the refractive index:

$$n^{2\omega}_\text{e}(\theta^{\text{II}}_{m\text{neg}}) = \frac{1}{2}[n^\omega_\text{o} + n^\omega_\text{e}(\theta^{\text{II}}_{m\text{neg}})] \tag{7.2-12}$$

Finding the solution simultaneously with the refractive index curvature equations of the fundamental frequency e light and SHG e light, we have

$$2\left[\frac{\cos^2\theta^{\text{II}}_{m\text{neg}}}{(n^{2\omega}_\text{o})^2} + \frac{\sin^2\theta^{\text{II}}_{m\text{neg}}}{(n^{2\omega}_\text{e})^2}\right]^{1/2} = n^\omega_\text{o} + \left[\frac{\cos^2\theta^{\text{II}}_{m\text{neg}}}{(n^\omega_\text{o})^2} + \frac{\sin^2\theta^{\text{II}}_{m\text{neg}}}{(n^\omega_\text{e})^2}\right]^{1/2} \quad (n^\omega_\text{o} \geqslant n^{2\omega}_\text{e}) \tag{7.2-13}$$

which is the solution for $\theta^{\mathrm{II}}_{m\mathrm{neg}}$.

(3) The positive uniaxial crystal type I matching mode ($\mathrm{e}^{\omega} + \mathrm{e}^{\omega} \to \mathrm{o}^{2\omega}$)

The fundamental frequency light is in the polarizing direction of the e light and the SHG light is in the polarizing direction of the o light. By substituting the equality $n^{\omega}_{\mathrm{e}}(\theta^{\mathrm{I}}_{m\mathrm{pos}}) = n^{2\omega}_{\mathrm{o}}$ into the refractive index equation of the fundamental frequency e light, we can obtain the corresponding matching angle:

$$\theta^{\mathrm{I}}_{m\mathrm{pos}} = \arcsin\left[\frac{(n^{\omega}_{\mathrm{o}})^{-2} - (n^{2\omega}_{\mathrm{o}})^{-2}}{(n^{\omega}_{\mathrm{o}})^{-2} - (n^{\omega}_{\mathrm{e}})^{-2}}\right]^{1/2} \quad (n^{\omega}_{\mathrm{e}} \geqslant n^{2\omega}_{\mathrm{o}}) \tag{7.2-14}$$

(4) The positive uniaxial crystal type II matching mode ($\mathrm{e}^{\omega} + \mathrm{o}^{\omega} \to \mathrm{o}^{2\omega}$)

The fundamental frequency light electric field takes two polarization states of the o light and e light, $2k_{\omega} = \dfrac{\omega}{c}[n^{\omega}_{\mathrm{o}} + n^{\omega}_{\mathrm{e}}(\theta)]$. The SHG electric field is of the o light polarizing direction, $k_{2\omega} = \dfrac{2\omega}{c} n^{2\omega}_{\mathrm{o}}(\theta)$. The phase matching condition $2k_{\omega} = k_{2\omega}$ is transformed into the relation of the refractive index:

$$n^{2\omega}_{\mathrm{o}} = \frac{1}{2}[n^{\omega}_{\mathrm{o}} + n^{\omega}_{\mathrm{e}}(\theta^{II}_{m\mathrm{pos}})] \tag{7.2-15}$$

which, substituted into the equation of the refractive index curvature, can solve the corresponding matching angle as

$$\theta^{\mathrm{II}}_{m\mathrm{pos}} = \arcsin\left[\frac{1 - \left(\dfrac{n^{\omega}_{\mathrm{o}}}{2n^{2\omega}_{\mathrm{o}} - n^{\omega}_{\mathrm{o}}}\right)^2}{1 - \left(\dfrac{n^{\omega}_{\mathrm{o}}}{n^{\omega}_{\mathrm{e}}}\right)^2}\right]^{1/2} \quad (n^{\omega}_{\mathrm{e}} \geqslant n^{2\omega}_{\mathrm{o}}) \tag{7.2-16}$$

With respect to the type I and type II matching modes of positive and negative uniaxial crystal, there is geometric graphic representation in the cross section of the corresponding refractive index curved surfaces. The requirements on the refractive index of the principal axis can be clearly seen from Fig. 7.2-3(a) and (b). The SHG light always appears in the

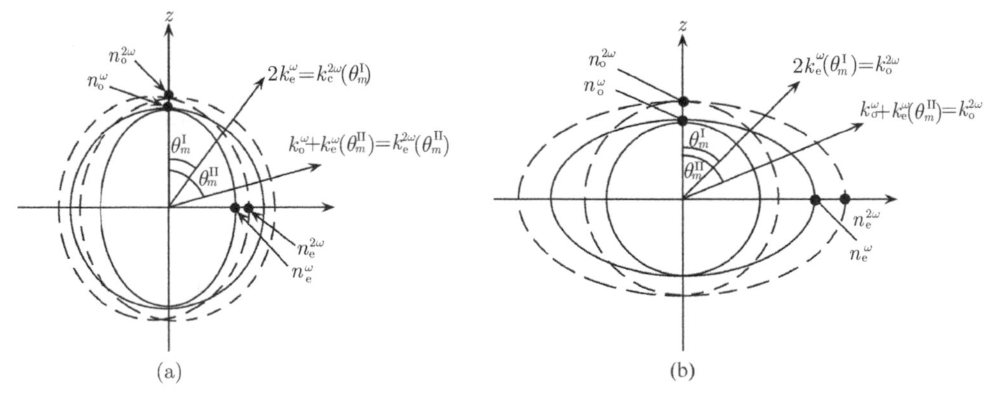

Fig. 7.2-3
(a) Type I and type II angular matching for negative uniaxial crystal SHG;
(b) Type I and type II angular matching for positive uniaxial crystal SHG

corresponding polarization state of a refractive index as low as possible, while the fundamental frequency light cannot take the corresponding polarization state of a high refractive index by itself. Compensation is made for the phase mismatch brought about by normal dispersion in the birefringence effect introduced in the above mode, thus meeting the requirement for phase matching.

(5) The angular matching for the biaxial crystal

Under ordinary conditions, the lower the symmetry of a crystal, the greater its nonlinear polarization coefficient will be. In the biaxial crystal, there exist two optical axes whose included angles with the z-axis are equal and are not zero, the symmetry lower than that of the uniaxial crystal. Therefore, most crystals (KTP, BNN, LBO, etc.) with a rather high SHG conversion efficiency are biaxial crystals. For such crystals, their angular matching of SHG is also divided into type I and type II (the parallel and orthogonal) modes. As the refractive index curved surface of the biaxial crystal is no longer a rotary curved surface with the z-axis as the symmetric axis, the spatial trajectory of the phase matching condition is a complex double-blade curved surface, with the direction of phase matching related to not only the matching angle θ, but also the azimuthal angle φ. The biaxial crystal index surface is a double-layer double-blade curved surface. Apart from two optical axial directions, it gives two refractive indices, n' (θ, φ) and n'' (θ, φ), to a wave vector of definite frequency in the $K(\theta, \varphi)$ direction. In the SHG, it gives four refractive indices to the fundamental frequency light and frequency doubled light. Under the constraint of satisfying type I type II angular matching condition, the four refractive indices can determine from the above equation the spatial trajectories for phase matching. Under normal dispersion conditions, these spatial trajectories have 13 topological structures. For every specific frequency doubled biaxial crystal, the corresponding numerical solution can be found on a computer.

2. The optical aperture effect and the noncritical phase matching (NCPM)

In an SHG crystal, what takes part in the nonlinear interaction is a light beam of a light spot diameter A. No matter whether it is type I or type II phase matching, the polarization state of the SHG is always different from that of the fundamental frequency light. During type I phase matching, the fundamental frequency light itself is in two different polarization states. So it cannot be ensured that the energy flow direction of the fundamental frequency light is consistent with that of the SHG light; that is, the e light and o light are "walking off" in direction. With the increase of the distance of propagation, the distance between the energy flow directions of the e light and o light increases as well. Such a phenomenon is called the "optical aperture effect", with the overlap seen only within the aperture length L_a. In the negative uniaxial crystal in Fig. 7.2-4, consider the type I phase matching. The included angle between the energy flow directions of the e light and o light is

$$\tan\alpha = \frac{1}{2}(n_{\mathrm{o}}^{\omega})^2[(n_{\mathrm{e}}^{2\omega})^{-2} - (n_{\mathrm{o}}^{2\omega})^2]\sin(2\theta_{\mathrm{m}}) \tag{7.2-17}$$

The aperture length

$$L_a = A/\tan\alpha \tag{7.2-18}$$

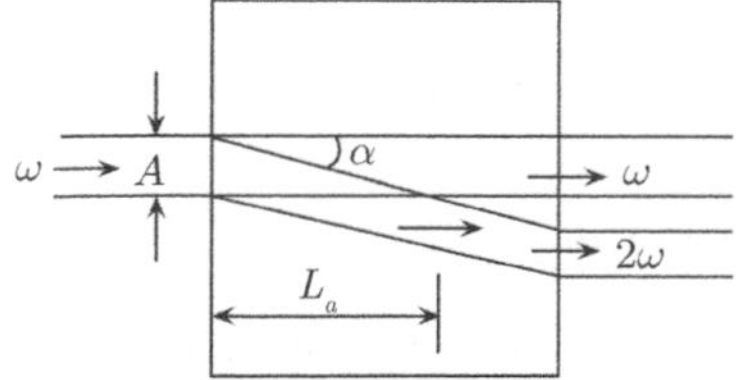

Fig. 7.2-4 The optical aperture effect and aperture length

When $L > L_a$, the energy of the fundamental frequency light no longer overlaps that of the second harmonic light generated, leading to a reduction in the optical power density and a bigger spot. In order to increase the aperture length L_a in the hope of obtaining a large fundamental frequency light aperture A, it is necessary to expand the beam. But the SHG conversion efficiency is proportional to the light intensity I_ω while inversely proportional to A^2, $\eta_{\text{SHG}} \propto \dfrac{1}{A^2}$. When A becomes larger, η_{SHG} will become smaller. Therefore, after the beam of the fundamental frequency light is expanded, the SHG light may not necessarily increase. In addition, the angle of divergence $\Delta\theta$ will lead to a deviation from the phase matching angle while also affecting the efficiency of SHG. The phase factor

$$\Delta kL/2 = [n_{\text{e}}^{2\omega}(\theta) - n_{\text{o}}^{\omega})]\omega L/c \tag{7.2-19}$$

where $\theta = \theta_{\text{m}} + \Delta\theta$; $n_{\text{e}}^{2\omega}(\theta)$is expressed by Eq. (7.2-10). Performing series expansion of the above equation around θ_{m} and taking first-order approximation, we find the relation of Δk with the angle of divergence $\Delta\theta$:

$$\Delta k \approx \frac{\omega}{c}\left[\frac{(n_{\text{e}}^{2\omega})^{-2} - (n_{\text{o}}^{2\omega})^{-2}}{(n_{\text{o}}^{\omega})^{-3}}\right]\sin(2\theta_{\text{m}})\Delta\theta \tag{7.2-20}$$

Phase mismatch is proportional to the divergence angle of the fundamental frequency light, and the influence of $\Delta\theta$ on η_{SHG} is fairly great. Therefore, angular phase match is in general unstable, which is also referred to as the critical phase match.

It is known from Eqs. (7.2-17) and (7.2-20) that both $\tan\alpha$ and Δk are proportional to $\sin(2\theta_{\text{m}})$. As long as $\sin(2\theta_{\text{m}}) = 0$, then both α and Δk will be zero, or it is required that either $\theta_{\text{m}} = 0$ or $\theta_{\text{m}} = \dfrac{\pi}{2}$. But it is impossible for θ_{m}=0 to satisfy the phase matching condition and it is necessary to select $\theta_{\text{m}} = \dfrac{\pi}{2}$, that is, the fundamental frequency light is injected normal to the optical axis. After the geometric parameter θ_{m} is determined to be $\dfrac{\pi}{2}$, phase match can be attained by changing the temperature. Generally speaking, the index of refraction of the e light is considerably influenced by temperature. Take type I matching mode as an example. At $\theta_{\text{m}} \equiv \dfrac{\pi}{2}$ and $T = T_{\text{m}}$, if $n_{\text{e}}^{2\omega}\left(\dfrac{\pi}{2}, T_{\text{m}}\right) = n_{\text{o}}^{\omega}(T_{\text{m}})$, the ellipse is "tangent to" the circular curvature of the refractive index in the direction normal to the optical axis, not only are the values of the refractive indices made equal, but also the first-order derivatives are equal, fully satisfying the condition for phase matching. $\theta_{\text{m}} \equiv \dfrac{\pi}{2}$ makes $L_\alpha \to \infty$, eliminating the optical aperture effect, making full use of the spatial length of the SHG crystal, helping raise the SHG conversion efficiency while greatly reducing the unfavorable influence of the divergence angle $\Delta\theta$ on the phase matching condition. Therefore, temperature matching is in reality a special case of angular matching with $\theta_{\text{m}} = \dfrac{\pi}{2}$, usually also spoken of as the noncritical phase matching (NCPM).

3. The effective nonlinear coefficient d_{eff}

In angular phase matching, the fundamental frequency light and the second harmonic light generated must take specified polarizing directions, that is, either the o polarization or the e polarization, without a doubt. But there should be different combinations to realize phase matching. So, for crystals of the same type, the effective nonlinear coefficient d_{eff} is different for different modes of matching. It is a function of θ and φ. In $d_{\text{eff}} = \tilde{a}_{\text{i}} \cdot \dfrac{\chi^{(2)}}{2} : a_j a_k$, the unit vector a should be consistent with the specified polarization state. The uniaxial crystals $a_{i,j,k}$ correspond to the unit matrices of the o light or e light:

$$a_\mathrm{o} = \begin{bmatrix} \sin\varphi \\ -\cos\varphi \\ 0 \end{bmatrix}, \quad a_\mathrm{e} = \begin{bmatrix} -\cos\theta\cos\varphi \\ -\cos\theta\sin\varphi \\ \sin\theta \end{bmatrix} \tag{7.2-21}$$

where φ is the azimuthal angle on the X-Y plane. Both the optical wave electric field and the polarization intensity vector can be expressed with a_o or a_e. For the negative uniaxial crystal type I matching mode,

$$P_{2\omega} = a_\mathrm{e} P_{2\omega}^\mathrm{e} = \frac{\varepsilon_\mathrm{o}}{2}[d_{il}] : a_\mathrm{o} a_\mathrm{o} E_\omega^o E_\omega^o \tag{7.2-22}$$

where $[d_{il}]$ is the SHG crystal's quadratic polarization matrix, i = 1, 2, 3; l= 1, 2, 3, 4, 5, 6. From the orthonormality of a_e, we have

$$P_{2\omega}^\mathrm{e} = \frac{\varepsilon_\mathrm{o}}{2}\tilde{a}_\mathrm{e} \cdot [d_{il}] : a_\mathrm{o} a_\mathrm{o} E_\omega^o E_\omega^o \tag{7.2-23}$$

The effective nonlinear coefficient of the negative uniaxial crystal type I matching mode is obtained:

$$d_\mathrm{eff\ neg}^\mathrm{I} = \tilde{a}_\mathrm{e} \cdot [d_{il}] : a_\mathrm{o} a_\mathrm{o} \tag{7.2-24a}$$

Similarly, we can obtain

$$d_\mathrm{eff\ pos}^\mathrm{I} = \tilde{a}_\mathrm{o} \cdot [d_{il}] : a_\mathrm{e} a_\mathrm{e} \tag{7.2-24b}$$

$$d_\mathrm{eff\ neg}^\mathrm{II} = \tilde{a}_\mathrm{e} \cdot [d_{il}] : a_\mathrm{e} a_\mathrm{o} \tag{7.2-24c}$$

$$d_\mathrm{eff\ pos}^\mathrm{II} = \tilde{a}_\mathrm{o} \cdot [d_{il}] : a_\mathrm{e} a_\mathrm{o} \tag{7.2-24d}$$

in which

$$a_\mathrm{o} a_\mathrm{o} = \begin{bmatrix} \sin^2\varphi \\ \cos^2\varphi \\ 0 \\ 0 \\ 0 \\ -\sin 2\varphi \end{bmatrix} \tag{7.2-25a}$$

$$a_\mathrm{e} a_\mathrm{e} = \begin{bmatrix} \cos^2\theta\cos^2\varphi \\ \cos^2\theta\sin^2\varphi \\ \sin^2\theta \\ -\sin 2\theta\sin\varphi \\ -\sin 2\theta\cos\varphi \\ \cos^2\theta\sin 2\varphi \end{bmatrix} \tag{7.2-25b}$$

$$a_\mathrm{e} a_\mathrm{o} = \begin{bmatrix} -\dfrac{1}{2}\cos\theta\sin 2\varphi \\ \dfrac{1}{2}\cos\theta\sin 2\varphi \\ 0 \\ -\sin\theta\cos\varphi \\ \sin\theta\sin\varphi \\ \cos\theta\cos 2\varphi \end{bmatrix} \tag{7.2-25c}$$

For a specific SHG uniaxial crystal, in the corresponding matching mode, the corresponding effective nonlinear coefficient $d_{\text{eff}}(\theta_{\text{m}}, \varphi)$ can be found using Eq. (7.2-24), in which θ_{m} is the matching angle, the azimuthal angle φ being determined by $|d_{\text{eff}}|_{\text{max}}$.

For example, for a $\bar{4}2m$ class negative uniaxial crystal,

$$[d_{il}] = \begin{bmatrix} 0 & 0 & 0 & d_{14} & 0 & 0 \\ 0 & 0 & 0 & 0 & d_{25} & 0 \\ 0 & 0 & 0 & 0 & 0 & d_{36} \end{bmatrix} \tag{7.2-26}$$

which, substituted into Eq. (7.2-24a), gives the type I matching mode effective nonlinear coefficient:

$$d_{\text{eff neg}}^{\text{I}} = (-\cos\theta\cos\varphi - \cos\theta\sin\varphi, \sin\theta)\begin{bmatrix} 0 \\ 0 \\ -d_{36}\sin 2\varphi \end{bmatrix} = -d_{36}\sin\theta\sin 2\varphi \tag{7.2-27}$$

When $\varphi = \pm\dfrac{\pi}{4}$, $\left|d_{\text{eff neg}}^{\text{I}}\right|$ takes the maximum value. Therefore, $d_{\text{eff}}(\theta, \varphi)$ can determine what value the spatial azimuthal angle φ of the fundamental frequency optical wave vector should take. As for the calculation of d_{eff} for the biaxial crystal, the reader is referred to the relevant literature and monographs.

7.2.3 The quasi-phase matching method (QPM)

1. The coherent length L_{c}

First let's make an analysis of the spatial variation of the SHG light intensity $I_{2\omega}$ within the crystal under the condition of not satisfying $\Delta k = 0$. By adopting non-depletion approximation, from Eq. (7.2-3), when Δk is never zero, there is

$$I_{2\omega}(z) = \frac{8\omega^2 d_{eff}^2}{\varepsilon_0 n_\omega^2 n_{2\omega}\lambda_\omega^2 c^3(\Delta k)^2} I_\omega^2 \sin^2\left(\frac{\Delta k z}{2}\right) \tag{7.2-28}$$

The above equation shows that the SHG light intensity appears to vary periodically in the z direction. When $0 \leqslant z \leqslant \dfrac{\pi}{|\Delta k|}$, $I_{2\omega}$ exhibits a tendency to rise, showing that the energy exchange process is mainly from a transfer of fundamental frequency light to SHG light. At $z = \dfrac{\pi}{|\Delta k|}$, $I_{2\omega}$ reaches maximum, as shown in Fig. 7.2-5. Thus the coherent length L_{c} of SHG is defined as:

$$L_{\text{c}} = \frac{\pi}{|\Delta k|} = \frac{\lambda_\omega}{4\,|n_\omega - n_{2\omega}|} \tag{7.2-29}$$

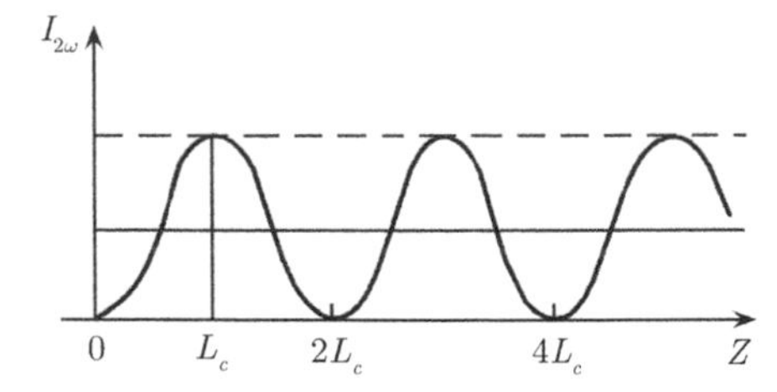

Fig. 7.2-5 Spatial variation of SHG intensity under $\Delta k \neq 0$ condition

Obviously, when $\Delta k = 0$, there is $L_{\text{c}} \to \infty$. Within the entire crystal length L, the fundamental frequency light is always transferring energy to the SHG light. But when $\Delta k \neq 0$, in the spatial range of $L_{\text{c}} \leqslant z < 2L_{\text{c}}$, $I_{2\omega}$ tends to decrease, showing that the energy exchanging process is mainly manifested as the SHG light "spitting back" to the fundamental frequency light wave what has been transferred to it. Therefore, the increase in the length of the crystal has not led to an intensified SHG light while how to make the SHG light increase monotonically in a medium under the condition of limited coherent length has led to the emergence of the method of quasi-phase matching (QPM).

2. The spatial modulating $d_{\text{eff}}(z)$

The two processes of the rise and fall of the SHG light intensity correspond to z belonging to the odd-numbered L_c and even-numbered L_c region, respectively. From an analysis of the wave coupling Eq. (7.2-1b), we find

$$\left.\frac{\mathrm{d}E_{2\omega}}{\mathrm{d}z}\right|_{z=2nL_c} = \mathrm{e}^{-i\pi}\left.\frac{\mathrm{d}E_{2\omega}}{\mathrm{d}z}\right|_{z=2(n+1)L_c} \tag{7.2-30}$$

where $n = 0, 1, 2, \cdots, 0 \leqslant z < L$; L is the length of the medium. This is the reversed phase process of the two phases difference π. Bloembergen et al. first pointed out that, after going through the coherent length L_c, by having the SHG effect's corresponding third-order polarizability tensor χ_{ijk} change sign, the tendency of the SHG light in the even-numbered L_c to fall can be reversed, which in reality is spatial modulation of d_{eff}.

With L_c as spatial spacing, have the adjacent signs reversed and $\frac{\mathrm{d}E_{2\omega}}{\mathrm{d}z}$ in Eq. (7.2-30) of the same sign and the goal of monotonic rise of the SHG light intensity is attained. Let

$$d_{\text{eff}} = \begin{cases} +d_{\text{eff}}, & z \in (2n+1)L_c \\ -d_{\text{eff}}, & z \in 2nL_c \end{cases} \tag{7.2-31}$$

The medium's length $L = NL_c$, N being a positive integer. Under this condition, the electric field of the outgoing SHG light is

$$\begin{aligned} E_{2\omega}(L) &= \frac{\mathrm{i}\omega}{n_{2\omega}c}E_\omega^2\int_0^{NL_c} d_{\text{eff}}(z)\mathrm{e}^{\mathrm{i}\Delta kz}\mathrm{d}z \\ &= \frac{\mathrm{i}\omega d_{\text{eff}}}{n_{2\omega}c}\left[\left(\int_0^{L_c} - \int_{L_c}^{2L_c} + \int_{2L_c}^{3L_c} - \cdots + (-1)^{N-1}\int_{(N-1)L_c}^{NL_c}\right)\mathrm{e}^{i\Delta kz}\mathrm{d}z\right] \\ &= \frac{4\omega d_{\text{eff}}}{n_{2\omega}c\pi}E_\omega^2 NL_c \end{aligned} \tag{7.2-32}$$

3. The characteristics of quasi-phase matching

(1) Quasi-phase matching is different from the method of angular phase matching for compensating for the dispersion effect. As it does not adopt the birefringence effect, it makes special requirements on the state of polarization of the optical wave, which is equivalent to an increase in the availability of the fundamental frequency light. There does not exist the problem of serious optical aperture effect and phase mismatch.

(2) Quasi-phase matching is not divided into type I and type II, nor is its selection of the effective nonlinear coefficient d_{eff} restricted by θ_{m} and φ. It can choose from the nonlinear polarization matrix the large component d_{il} to receive the fundamental frequency light along the spatial direction to which it corresponds so as to obtain high SHG light intensity.

(3) For a material with a large nonlinear coefficient within a certain wavelength range but very small birefringence, when unable to realize angular phase matching, QPM can be adopted for SHG to broaden the range of materials to choose from and the corresponding SHG waveband.

(4) The material adopted by quasi-phase matching is the "ferroelectric crystal" of "poly-plate multi-domain" with its adjacent nonlinear polarizability tensors opposite in sign, as

shown in Fig. 7.2-6 while its corresponding second-order polarizability tensors remain unchanged. Hence the linear optical properties do not change either.

Fig. 7.2-6 The superlattice material for QPM; the widths of white and black regions are both L_c while effective nonlinear coefficients are of opposite signs

7.2.4 The way of frequency doubling (SHG)

Generally speaking, an SHG crystal can be placed either outside or inside a laser resonator, the two ways being known as extracavity and intracavity SHG, respectively. However, when the fundamental frequency light has a high repeatability but rather low peak power, the intracavity SHG way is in general adopted, which has a rather high conversion efficiency.

1. The extracavity SHG

For extracavity SHG, the pulsed Q-switching laser is in general adopted as the fundamental frequency light source. In order to obtain a rather high conversion efficiency, sometimes multiple stage amplification is adopted to increase the peak value power of the fundamental frequency light so as to obtain a rather high SHG light intensity. In experiments on the extracavity SHG, the focusing method is also sometimes adopted to increase the optical power density of the fundamental frequency light in the SHG crystal. In Fig. 7.2-7, the Gauss light beam injected onto the crystal is represented with the confocal parameter z_0, within which the divergence of the optical beam is rather small. z_0 is also called the collimation length. The relation of z_0 with the beam waist radius w_0 is

$$z_0 = \frac{1}{2}kw_0^2 = \pi w_0^2 n_\omega / \lambda_\omega \tag{7.2-33}$$

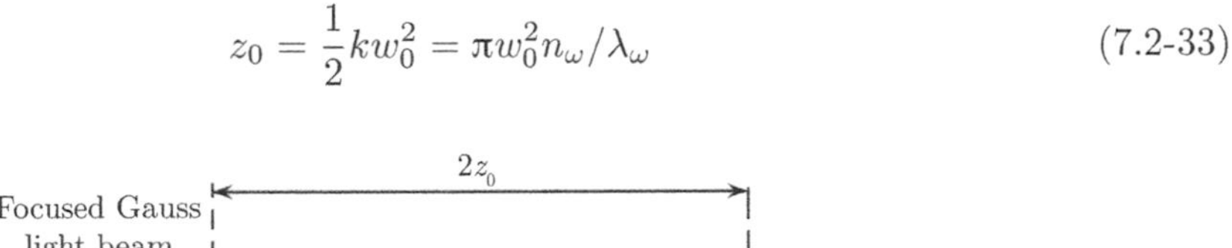

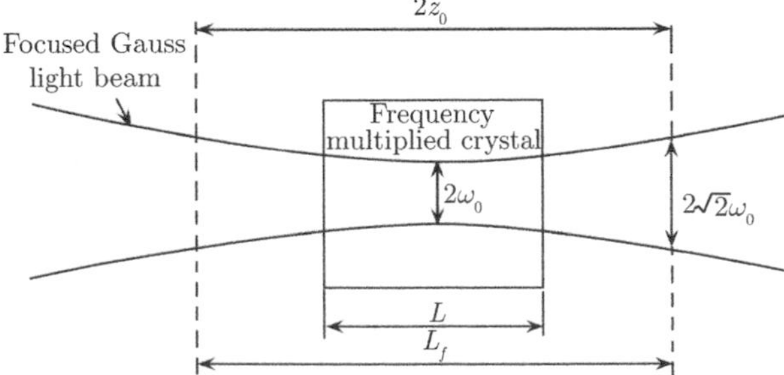

Fig. 7.2-7 The confocal focusing condition for Gauss beam SHG

If twice the collimation length $2z_0 = L_f$ is chosen as the length of the crystal, then this condition is the condition for confocal focusing.

Under this condition, the SHG efficiency $\eta_{\mathrm{SHG}} \propto L_f$. The efficiency of extracavity SHG is usually lower than that of intracavity SHG.

2. Intracavity SHG

Intracavity SHG consists in placing the nonlinear SHG crystal in the laser resonator to make the fundamental frequency light in the cavity pass through the SHG crystal back and forth. Under appropriate conditions, a rather high conversion efficiency can be obtained. Suppose the reflectivity of the fundamental frequency light by the laser output mirror is R. Then the power of the fundamental frequency light in the cavity is $\dfrac{1}{1-R}$ times greater than that outside the cavity. If $R \approx 1$, then the efficiency of intracavity SHG will be considerably spectacular. For continuously operating and quasi-continuously operating lasers with Q switched at a high repeatability, the intracavity SHG way is adopted as a rule, as shown in Fig. 7.2-8 (a). Intracavity SHG makes high requirements on the optical homogeneity and transparency of the SHG crystal. On occasions of the use of high average power, rather high requirements are also made on the thermal conductivity of the SHG crystal, with the temperature control adopted where necessary. Shown in Fig. 7.2-8 (b) is the formation of double path SHG by inserting an acousto-optic Q switch and a harmonic reflective mirror in the resonator of the continuous laser, the merits of which are simple construction and high frequency doubling efficiency but the thermal lens effect of the laser is rather appreciable and the harmonic reflective mirror has increased the cavity loss.

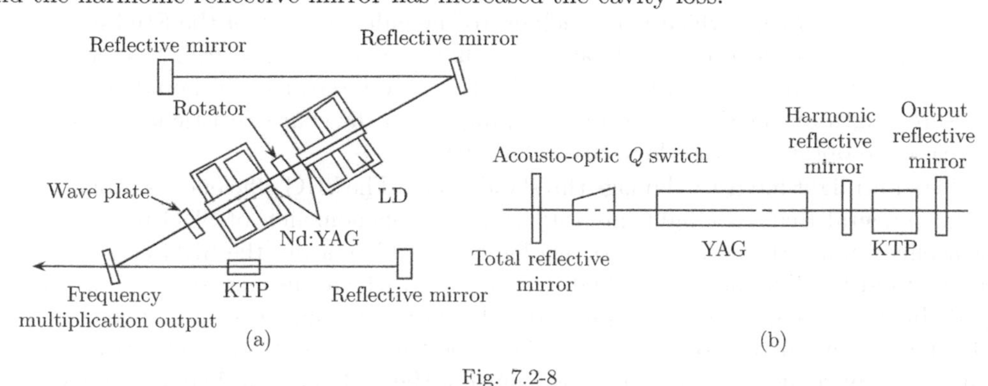

Fig. 7.2-8
(a) Intracavity CW-SHG device; (b) A straight cavity mode quasi-continuous SHG laser

In addition, no optical device in the cavity is allowed to cause the SHG absorption loss. Like the extracavity case, the intracavity SHG also has its confocal focusing condition; that is, the SHG crystal should be located at the beam waist of the Gauss light beam in the cavity as much as possible. $2z_0$ can also be given an overall consideration as a reference value for the length of the SHG crystal. For the multiple longitudinal continuous laser SHG in the cavity, owing to the sum frequency generation (SFG) effect crossing the longitudinal modes and the knowledge from Eqs. (7.2-12) and (7.2-13) that, when the difference between ω_1 and ω_2 is of the order of longitudinal mode spacing, $\chi^{(2)}(2\omega_{1,2})$ is almost equal to $\chi^{(2)}(\omega_1+\omega_2)$. Therefore, the SFG competes with the SHG, leading to the SHG laser output becoming very unstable, or the cropping up of the "green light problem". Theoretical analysis and experimental method in this regard can be found in relevant literature.

7.2.5 The SHG crystal

SHG crystal materials are the physical foundation of realizing laser frequency doubling technology. In accordance with the requirement on spatial symmetry, they are chosen from among 21 kinds of point groups without an inversion symmetry center. The SHG crystal is mainly constructed of inorganic oxides and semiconductor materials; the former are often used in the near infrared, visible light, and ultraviolet waveband, while the latter, in the medium and far infrared waveband. An SHG crystal of practical value should satisfy the following requirements:

(1) A wide transparent waveband. From the fundamental frequency to SHG waveband, there is a high transmissivity with extremely small absorption loss. In practice, different wavebands should have different SHG crystals chosen for coordinated use.

(2) A large nonlinear coefficient d_{il}. Use the nonlinear coefficient d_{36} of KDP as a standard for comparison. The nonlinear coefficient of the currently frequently used KTP crystal is 21.5 times the standard value.

(3) A suitable birefringence. In angular phase matching, the birefringence effect possessed by the SHG crystal itself is utilized to overcome the dispersion effect to attain the goal of phase matching. For a crystal with too small a birefringence, although the phenomenon of SHG can be observed, it cannot satisfy the condition for phase matching. So a high SHG efficiency cannot be obtained. For instance, for the quartz positive uniaxial crystal with which the SHG efficiency is first observed from ruby laser, owing to too weak a birefringence effect, the refractive index curvatures of the e light of the fundamental frequency and the o light of SHG do not intersect. Angular phase matching cannot be realized. However, for a crystal with too great a birefringence, although phase matching can be realized, there exists a serious optical aperture effect that makes the aperture length L_{a} far smaller than the characteristic length L_{SHG}, which is not helpful to the enhancement of the SHG efficiency, either. An ideal birefringence is preferably one that can reach $\theta_{\mathrm{m}} = 90°$, that is, to the effect of temperature matching. In actual application, the birefringence should ensure that the aperture length of the SHG crystal is no smaller than the characteristic length so as to make it easier to obtain a rather high SHG efficiency.

(4) A high anti-light intensity damage threshold value. The SHG efficiency is proportional to the fundamental frequency light intensity while the phenomenon of SHG is exactly the nonlinear effect under the action of a strong light. For this reason, the SHG crystal has to be able to stand the illumination by high power laser without the phenomenon of getting damaged. In the sense of practical application, the optical damage threshold is required to be at least no lower than 1 $\mathrm{MW/cm^2}$. At the same time, a great fundamental frequency light intensity can reduce the characteristic length of the SHG crystal, helping increase the SHG light intensity.

(5) Stable physico-chemical performance. An SHG crystal of practical application value should possess such characteristics as good optical homogeneilty, freedom from deliquescence, and rather small variation of the various aspects of its performance with environmental factors.

(6) A large receiving angle and receiving bandwidth. This is beneficial to reducing the divergence of the fundamental frequency light and the unfavorable influence of non-ideal monochromaticity on phase matching.

The requirements on the SHG crystal as discussed above are in fact also applicable to other second-order nonlinear effects. Research on nonlinear optical materials such as the SHG crystal has fundamentally raised the nonlinear optical efficiency and expanded the range of application. In particular, the significance of the theory of "anionic base group" put forward by the Chinese scientist Chen Chuangtian has far surpassed that of the Miller coefficient. The highly efficient, wide spectral band, and extensively-applied crystals such as β-BBO, LBO, and SBBO have successfully been developed by the Fujian Institute for Material Structures, the Chinese Academy of Sciences as never before in the rest of the world. The development of micrometer supercrytal material using the quasi-phase matching technique for implementing SHG is equally eye-catching. Owing to the disuse of the conventional method of angular phase matching, the selection of material for SHG can be extended to ferroelectric crystals such as those of weak birefringence or optically isotropic (cubic crystalline system) so that the incident angle of the optical wave and polarization state can be selected in the direction in which d_{il} is maximum. To adapt to SHG of different wavelengths, the thickness of the plate domain can be changed. The adoption of the Fibnacci sequence makes it possible for the variable periodic structure to realize SHG for different fundamental frequency light, optical parametric oscillation, or third harmonic generation (THG) at the same time free from the optical aperture effect at that while applicable to direct SHG for semiconductor laser, etc. The photon super-lattice materials polysynthetic multi-domain periodically polarized lithium niobate (PPLN) and lithium tantalate (PPLT) developed by the National Key Solid Laser Microstructure Laboratory of Nanjing University are at the internationally advanced level.

7.3 Optical parametric oscillation technology

Optical parametric oscillation (OPO) technology is capable of expanding laser to the low frequency range to generate tunable coherent radiation. Generally speaking, the output frequency of a laser is as a rule discrete while numerous scientific research and technological applications require high brightness radiation sources of frequencies that are continuously tunable. The adoption of the optical parametric oscillation method with the second-order nonlinear optical effect as a foundation makes it possible to realize continuous tuning within a considerably large frequency range. By virtue of its high conversion efficiency, simple device construction, and stable performance, this method has found wide application.

In general, a definite phase matching mode must be adopted to realize the optical parametric process. At the same time, in the parametric process, the role of the medium is embodied in d_{eff}. As a means of energy exchange between optical waves, the medium itself does not take part in the exchange of energy. The interacting optical wave frequencies are not related to the medium's eigenenergy level. The average energy flow from the optical field to the medium is zero, neither giving the optical field a gain nor causing it any loss. The optical parametric oscillation effect provides gain through the pump wave E_{p}. The nonlinear crystal, as the medium of energy transfer, has energy coupled to the signal optical wave E_{s} to make it amplified while generating a new idler optical wave E_{i}. This process is completely a process of parametric interaction, whose frequency tuning mode is constrained by the phase matching condition. With different optical wave frequencies and the corresponding nonlinear crystals selected, the tuning range of the optical parametric oscillation covers the medium, near infrared, and visible wavebands.

7.3.1 Optical parametric amplification

In the process of optical parametric amplification (OPA), there are two pumping wave inputs. Suppose E_{p} and E_{s} are collinearly injected into a nonlinear crystal, of which E_{p} is strong pump wave $\left(\dfrac{\mathrm{d}E_{\text{p}}}{\mathrm{d}z} \approx 0\right)$ while E_{s} is weak signal wave $\left(\dfrac{dE_{\text{s}}}{dz} \neq 0\right)$. From Eq. (7.1-26) we obtain the corresponding wave coupling set of equations:

$$\frac{\mathrm{d}E_{\text{i}}}{\mathrm{d}z} = \frac{\mathrm{j}\omega_{\text{i}}}{n_{\text{i}}c} d_{\text{eff}} E_{\text{s}}^{*} E_{\text{p}} \mathrm{e}^{-\mathrm{j}\Delta kz} \tag{7.3-1a}$$

$$\frac{\mathrm{d}E_{\text{s}}}{\mathrm{d}z} = \frac{\mathrm{j}\omega_{\text{s}}}{n_{\text{s}}c} d_{\text{eff}} E_{\text{i}}^{*} E_{\text{p}} \mathrm{e}^{-\mathrm{j}\Delta kz} \tag{7.3-1b}$$

whose frequency relation is $\omega_{\text{p}} - \omega_{\text{s}} = \omega_{\text{i}}$, equivalent to a difference frequency process with a weak signal input, whose boundary conditions are $E_{\text{i}}(0) = 0$, $E_{\text{s}}(0) \neq 0$. In the presence of phase mismatch $\Delta k \neq 0$, the solutions are

$$E_{\rm s}(z) = E_{\rm s}(0)\left[\cosh(bz) - \frac{{\rm j}\Delta k}{2b}\sinh(bz)\right]{\rm e}^{-{\rm j}\Delta kz/2} \quad (7.3\text{-}2a)$$

$$E_{\rm i}(z) = {\rm j}\frac{\omega_{\rm i}d_{\rm eff}/n_{\rm i}c}{b}E_{\rm p}E_{\rm s}^{*}(0)\sinh(bz){\rm e}^{-{\rm j}\Delta kz/2} \quad (7.3\text{-}2b)$$

where

$$b = \left[\Gamma^2 - \left(\frac{\Delta k}{2}\right)^2\right]^{1/2}, \ \Gamma^2 = \frac{\omega_{\rm i}\omega_{\rm s}d_{\rm eff}^2\left|E_{\rm p}\right|^2}{n_{\rm i}n_{\rm s}c^2} \quad (7.3\text{-}3)$$

b is the net gain coefficient, and $E_{\rm i}$ the electric field of the idler wave. When $\Delta k > 2\Gamma$ and b is an imaginary number, the hyperbolic sinusoidal function becomes the sinusoidal function. The light intensity appears in the form of fluctuation in the nonlinear crystal and is unable to obtain sustained increase. Therefore, phase matching is the necessary condition for performing optical parametric amplification. Only by satisfying the idler wave for phase matching will it be possible to obtain gain in the nonlinear crystal. The solutions of satisfying phase matching $\Delta k = 0$ ($b = \Gamma$) are, respectively,

$$E_{\rm s}(z) = E_{\rm s}(0)\cosh(\Gamma z) \quad (7.3\text{-}4)$$

$$E_{\rm i}(z) = {\rm j}\left(\frac{\omega_{\rm i}n_{\rm s}}{\omega_{\rm s}n_{\rm i}}\right)^{1/2}E_{\rm s}^{*}(0)\sinh(\Gamma z) \quad (7.3\text{-}5)$$

Here the signal wave $E_{\rm s}$ gets amplified, and the idler wave comes into being from nothing. With an increase in the distance of propagation, both waves rise monotonically in the nonlinear crystal, approximately an exponential-type increase, obtaining net gain from the consumption of the pump wave $E_{\rm p}$. The idler wave $E_{\rm i}$ and the signal wave $E_{\rm s}$ are equivalent in the wave coupling equations and it is only because different boundary values have been taken that the expressions of the solutions are different. In the process of amplification, according to the Manley-Rowe relation, photons of equal quantities are obtained from the pump wave while for amplification, by taking into account the positive feedback conditions, self-excited oscillation can be formed. In fact, there is no need for the signal light $\omega_{\rm s}$ to be provided from the outside, but its role is played by the spontaneous radiation (noise). In the resonator, the pump light is made use of to overcome loss and obtain an output. Therefore, the nonlinear crystal can be placed in the corresponding resonator. So long as the high frequency pump wave $\omega_{\rm p}$ is input to make $\omega_{\rm i}$ and $\omega_{\rm s}$ in background radiation amplified in the cavity, if reflective mirrors M_1 and M_2 are placed at the two ends of the nonlinear crystal, then optical parametric amplification will become an optical parametric oscillator, as shown in Fig. 7.3-1. When the parametric gain is greater than the loss, we obtain parametric oscillation output. Increasing the effective length of the nonlinear crystal will enhance the availability of the pump light.

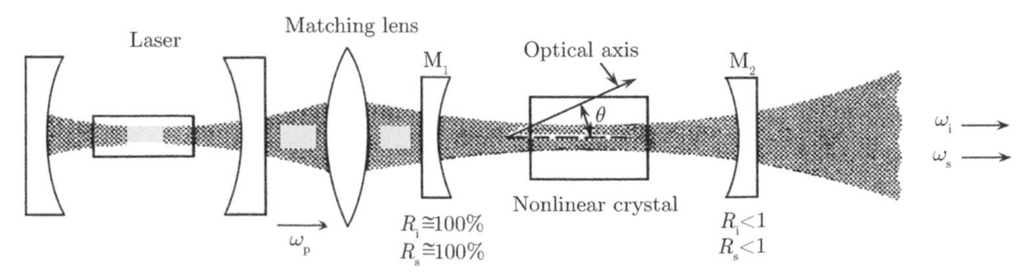

Fig. 7.3-1 The optical parametric oscillation device

7.3.2 Optical parametric oscillation

Optical parametric oscillation "splits" a pump photon ω_p into a signal photon ω_s and an idler photon ω_i:

$$\omega_p = \omega_s + \omega_i \tag{7.3-6}$$

in conformity with the condition for the conservation of photon energy while the signal wave and idler wave that obtained a gain during optical parametric oscillation must satisfy the condition for collinear phase matching (conservation of momentum):

$$\boldsymbol{k}_p = \boldsymbol{k}_s + \boldsymbol{k}_i \tag{7.3-7a}$$

where $\boldsymbol{k}_p$, $\boldsymbol{k}_s$, and $\boldsymbol{k}_i$ are the wave vectors of three kinds of light, respectively. Using the $k = \omega n/c$ relation, we can obtain the relation between the given frequency and refractive index when the three waves act in the same direction:

$$n_p\omega_p = n_s\omega_s + n_i\omega_i \tag{7.3-7b}$$

1. The longitudinal mode condition

The optical parametric oscillator possesses the structure and properties of a commonly used optical resonator and obeys the corresponding resonator longitudinal mode condition. For the standing wave cavity, the oscillation signal and idler optical wave should each satisfy the self-reconstructing condition for a round trip:

$$\varphi_s + \frac{\omega_s}{c} \cdot 2L_s = 2s\pi \tag{7.3-8a}$$

$$\varphi_i + \frac{\omega_i}{c} \cdot 2L_i = 2i\pi \tag{7.3-8b}$$

where $L_{s,i} = L + (n_{s,i} - 1)\, l$ are the corresponding optical cavity lengths of ω_s and ω_i, respectively. L is the geometric cavity length; l is the geometric length of the nonlinear crystal; s and i are positive integers; and $\varphi_{s,i}$ are the phase shifts of ω_s and ω_i led to by the reflective mirrors.

For a ring-shaped traveling wave cavity, the self-reconstructing conditions for a cycle should be satisfied:

$$\varphi_s + \frac{\omega_s}{c} \cdot L_s = 2s\pi \tag{7.3-9a}$$

$$\varphi_i + \frac{\omega_i}{c} \cdot L_i = 2i\pi \tag{7.3-9b}$$

The above two equations are spoken of as the double resonance conditions (DRC).

2. The threshold value condition

In an optical parametric oscillator, the pump wave E_p provides the single pass gain and the optical loss is mainly the output loss. Like ordinary oscillators, the gain being equal to the loss is the threshold value condition while the output is the part that has exceeded the threshold value. To simplify the process of analysis, non-depletion approximation is still taken for the pump light while (at $\Delta k = 0$), for the coupled wave equations of the signal wave and the idler wave in the cavity, under the condition of oscillation, the back-and-forth loss parameters can be assumed to be α_s and α_i, respectively:

$$\frac{dE_i}{dz} = -\alpha_i E_i + \frac{j\omega_i}{n_i c} d_{eff} E_p E_s^* \tag{7.3-10a}$$

$$\frac{\mathrm{d}E_{\mathrm{s}}^*}{\mathrm{d}z} = -\alpha_{\mathrm{s}}E_{\mathrm{s}}^* + \frac{-\mathrm{j}\omega_{\mathrm{s}}}{n_{\mathrm{s}}c}d_{\mathrm{eff}}E_{\mathrm{p}}^*E_{\mathrm{s}} \tag{7.3-10b}$$

where the relation of $\alpha_{\mathrm{i,s}}$ with σ that represents optical loss in Eq. (7.1-17) is $\alpha_{\mathrm{i,s}} = \dfrac{\mu_0 c\sigma_{\mathrm{i,s}}}{2n_{\mathrm{i,s}}}$.

Suppose the reflectivity of the optical parametric oscillation cavity mirrors is denoted by R_{i} and R_{s}, respectively:

$$\alpha_{\mathrm{i}} = \frac{2(1-R_{\mathrm{i}})}{L}, \quad \alpha_{\mathrm{s}} = \frac{2(1-R_{\mathrm{s}})}{L} \tag{7.3-11}$$

From the steady-state condition

$$\frac{\mathrm{d}E_{\mathrm{i}}}{\mathrm{d}z} = 0 = \frac{\mathrm{d}E_{\mathrm{s}}^*}{\mathrm{d}z} \tag{7.3-12}$$

it is found from the equation set (7.3-10) that E_{i} and E_{s} have a relation of non-zero solution:

$$\det\begin{bmatrix} -\alpha_{\mathrm{i}} & \dfrac{\mathrm{j}\omega_{\mathrm{i}}}{n_{\mathrm{i}}c}d_{\mathrm{eff}}E_{\mathrm{p}} \\ \dfrac{-\mathrm{j}\omega_{\mathrm{s}}}{n_{\mathrm{s}}c}d_{\mathrm{eff}}E_{\mathrm{p}}^* & -\alpha_{\mathrm{s}} \end{bmatrix} = 0 \tag{7.3-13}$$

The pump threshold condition is obtained as

$$\Gamma^2 = \frac{\omega_{\mathrm{s}}\omega_{\mathrm{i}}}{n_{\mathrm{s}}n_{\mathrm{i}}c^2}d_{\mathrm{eff}}^2\left|E_{\mathrm{p}}\right|^2 = \alpha_{\mathrm{i}}\alpha_{\mathrm{s}} \tag{7.3-14}$$

Γ is the single pass gain coefficient given by Eq. (7.3.3), the corresponding threshold pump intensity being

$$I_{\mathrm{p_{th}}} = \frac{n_{\mathrm{p}}n_{\mathrm{s}}n_{\mathrm{i}}c^3\varepsilon_0\alpha_{\mathrm{i}}\alpha_{\mathrm{s}}}{2\omega_{\mathrm{s}}\omega_{\mathrm{i}}d_{\mathrm{eff}}^2} = \frac{2n_{\mathrm{p}}n_{\mathrm{s}}n_{\mathrm{i}}c^3\varepsilon_0}{\omega_{\mathrm{s}}\omega_{\mathrm{i}}d_{\mathrm{eff}}^2}\frac{(1-R_{\mathrm{i}})(1-R_{\mathrm{s}})}{L^2} \tag{7.3-15}$$

Equations (7.3-14) and (7.3-15) give the threshold condition and pump intensity of the double resonance mode, respectively.

For the single resonance mode (SRM), the reflective mirror reflects only one kind of optical wave ω_{s}. The single pass gain coefficient must be equal to the back-and-forth loss coefficient:

$$\Gamma_{\mathrm{s}} = \frac{\sqrt{2(1-R_{\mathrm{s}})}}{L} \tag{7.3-16}$$

A comparison with the double resonance power gain gives

$$\frac{(\Gamma_{\mathrm{s}}L)^2}{(\Gamma L)^2} = \frac{I_{\mathrm{sp_{th}}}}{I_{\mathrm{p_{th}}}} = \frac{2}{1-R_{\mathrm{i}}} \tag{7.3-17}$$

When R_{i} in the double resonance approaches 1, the threshold pump power of the single resonance will be increased many times.

7.3.3 The operation mode of optical parametric oscillation

The optical parametric oscillator has a form of structure similar to that of the laser resonator. The pump light emitted by the laser is injected into the optical parametric oscillator providing the signal wave and idler wave with gain in the nonlinear crystal. Once the gain exceeds the threshold value or the pump intensity is stronger than the threshold value intensity in Eq. (7.3-15), the optical parametric oscillator will have signal wave and idler wave output. The effect of the optical parametric oscillator's resonator on the transverse mode of the signal wave and idler wave and the judgment of the stability can be analyzed by adopting a method similar to that adopted for the laser resonator. What is worth notice is the pump wave's Gauss light beam mode must match the eigen-Gauss light beam mode of the optical parametric resonator. There are usually two modes of operation of the optical parametric oscillator, the double resonance operation (DRO) and the single resonance operation (SRO). By double resonance is meant that the two reflective mirrors M_1 and M_2 of the optical parametric oscillator are both of high reflectivity to ω_s and ω_i (e.g., $R_i \approx 1$, $R_s \approx 1$). If high reflectivity is only for ω_i, then the resonance is called single resonance. For the double resonance mode, it is required that, in a resonator, the different longitudinal mode conditions of the signal wave and idler wave should each be satisfied at the same time. It is known from Eqs. (7.3-8) and (7.3-9) that the relation of variation between the signal wave frequency and idler wave frequency due to a change in the geometric length of the cavity length is

$$-\Delta L \propto \Delta\omega_s = \Delta\omega_i \tag{7.3-18}$$

while we have from Eq. (7.3-6)

$$\Delta\omega_s = -\Delta\omega_i \tag{7.3-19}$$

In the course of tuning, as Eqs. (7.3-18) and (7.3-19) are contradictory to each other, the double resonance operating mode will cause instability of the frequency, leading to the fluctuation of the output energy. For this reason, for the double resonance mode, it is very important to strictly control the structure of the cavity to minimize the effect of ΔL. The single resonance mode is able to bypass the above-mentioned contradiction. In Eq. (7.3-18) there is only one frequency that varies, which is proof of good stability that makes it a fairly popular optical parametric operating mode. Optical parametric oscillation is capable of continuously operating with pulses in two states. The magnitude of the threshold pump intensity of the double resonance mode is expressed by Eq. (7.3-15). When the transmissivity is very small, $T_i = T_s \approx 10^{-2}$, the threshold value pump intensity is about the kW/cm^2 order. Adoption of the Q-switched laser pumping can satisfy the threshold value condition. With respect to the pulsed operation mode, in order to shorten the time for setting up optical parametric oscillation and obtain a rather high conversion efficiency, the length of the resonator should be reduced as much as possible. In addition, for the one-way traveling wave pump mode, the signal wave and idler wave (reversed wave) reflected by the output mirror will generate the SFG to return the energy to the pump wave, seriously affecting the improvement of the conversion efficiency. Therefore, for the double resonance mode, it is advisable to adopt the ring-shaped traveling wave cavity.

7.3.4 The frequency tuning method for the optical parametric oscillator

The principal use of optical parametric oscillation is to perform frequency tuning for laser. It can be seen from Eq. (7.3-6) of conservation of photon energy that the pump wave frequency ω_p is fixed while ω_s and ω_i are variables, which cannot be determined by the condition of conservation of energy alone. The condition for the conservation of momentum that makes the gain maximum should also be considered, which can be given by Eq. (7.3-7) in the collinear case. By writing the two equations simultaneously and eliminating the fixed frequency ω_p, we have

$$\frac{\omega_s}{\omega_i} = \frac{n_i - n_p}{n_p - n_s} \tag{7.3-20}$$

The above equation has determined the mutual relation between ω_s and ω_i under the condition that the conservation of energy is unified with the conservation of momentum. But, under normal dispersion condition $\frac{dn}{d\omega} > 0$, there is $n_p > n_i$, $n_p > n_s$ for a normal dispersion medium, leading to $\frac{\omega_s}{\omega_i} < 0$. For this reason, it is necessary to adopt the birefringence effect of the anisotropic crystal to offset the dispersion effect. On the premise of satisfying the condition of conservation of energy and momentum, through the variation of the refractive index, ω_s and ω_i should be made to vary correspondingly under the constraint by Eq. (7.3-20) to attain the goal of frequency tuning. There are two modes of frequency tuning with birefringence as the basis, the angular tuning and temperature tuning.

(1) Angular tuning

The mode of tuning that leads to a variation of the optical wave frequency due to the change in the included angle θ relative to the crystal axis formed by the incident pump optical wave is called angular tuning. As ω_p is a high frequency, according to the law of compensating for normal dispersion, for a negative uniaxial crystal, the e light polarization will be taken for the polarization state of the pump light while both ω_s and ω_i that are generated polarize along the direction of the o light, which is the matching mode of type I. If one of them polarizes in the direction of the e light, then it is the type II matching mode. Take the negative uniaxial crystal type I matching mode as an example. The pump wave takes the polarization state of the e light while the signal wave and idler wave take the polarization state of the o light. When the included angle between k_p and the optical axis is θ, under the condition $\Delta k = 0$, there is

$$\omega_p n_p(\theta) = \omega_s n_s + \omega_i + n_i \tag{7.3-21}$$

When changing the angle $\Delta\theta$, $\Delta k = 0$ should still be made to hold. On the premise of guaranteeing that the gain will be maximum, there is bound to be a pair of new frequencies ω_s and ω_i to satisfy the phase matching condition of Eq. (7.3-21).

$$\omega_p(n_p + \Delta n_p) = (\omega_s + \Delta\omega_s)(n_s + \Delta n_s) + (\omega_i + \Delta\omega_i)(n_i + \Delta n_i)$$

Noting that ω_p does not change while $\Delta\omega_i = -\Delta\omega_s$, expanding the above equation, and performing collation, we obtain

$$\Delta\omega_s = \frac{\omega_p \Delta n_p - \omega_s \Delta n_s - \omega_i \Delta n_i}{n_s - n_i} \tag{7.3-22}$$

As the pump wave ω_p is of the e light, whose refractive index n_p is a function of θ while the signal wave and idler wave are not related to the direction but dependent only on the frequency, that is,

$$\Delta n_p = \left.\frac{\partial n}{\partial\theta}\right|_{\theta=\theta_0} \Delta\theta, \quad \Delta n_s = \left.\frac{\partial n_s}{\partial\omega_s}\right|_{\omega=\omega_s} \Delta\omega_s, \quad \Delta n_i = \left.\frac{\partial n_i}{\partial\omega_i}\right|_{\omega=\omega_i} \Delta\omega_i$$

substituting the above into Eq. (7.3-22), we have the variation rate of the oscillation frequency relative to angle θ.

$$\frac{\partial\omega_s}{\partial\theta} = \frac{\omega_p\left(\frac{\partial n_p}{\partial\theta}\right)}{(n_s - n_i) + \left[\omega_s\left(\frac{\partial n_s}{\partial\omega_s}\right) - \omega_i\left(\frac{\partial n_i}{\partial\omega_i}\right)\right]} \tag{7.3-23}$$

where $\frac{\partial n_p}{\partial\theta}$ can be found from Eq. (7.2-10). The above equation shows the relationship of variation of the signal oscillation frequency with the orientation of the crystal, which is the expression for type I angular frequency tuning of the negative uniaxial crystal. The expression for type II angular frequency tuning can be found in a similar manner. However, it should be noted that when the signal wave or idler wave takes the e light polarization state, apart from a consideration of the dispersion relation $\frac{\partial n}{\partial \omega}$, attention should also be given to $\frac{\partial n_e}{\partial \theta}$, that is, Δn is a full differential expression. It can be seen from Eq. (7.3-23) that, as the pumping wave frequency ω_p does not change, if the pump light takes the o light polarization state, then $\frac{\partial n_p}{\partial \theta} = 0$, leading to ω_s not changing either and angular tuning becoming meaningless. So the positive uniaxial crystal is in general not adopted as an angle tuning crystal. For the non-collinear type angular tuning, if only the projection relation is adopted, the method of tackling the problem is similar to the above-mentioned, the key being ensuring that, for the relation between ω_p, ω_s, and ω_i, apart from satisfying the conservation of energy, conservation of momentum, too, should be satisfied. Only for the process $\Delta k = 0$, will there be maximum gain in the optical parametric oscillation. In terms of the geometric graphs of the refractive index curved surface, when k_p and the crystal rotate $\Delta\theta$ relative to each other, the rotation is in fact equivalent to moving n_p's length in $k_s + k_i = k_p$'s corresponding $n_p\omega_p = n_s\omega_s + n_i\omega_i$ to make the spatial direction of the maximum gain in optical parametric oscillation change. In order to adapt to this change, ω_s and ω_i, too, have to change to make $\Delta k = 0$ remain unchanged. In this way, there will appear, in the "noise" provided by spontaneous radiation, new refractive index curved surfaces that ω_s and ω_i correspond to that will satisfy the phase matching conditions and yield the maximum gain while the dispersion effect $\frac{\partial n}{\partial \omega} \neq 0$, in this sense, has become a beneficial factor to optical parametric oscillation frequency tuning, as shown in Fig. 7.3-2.

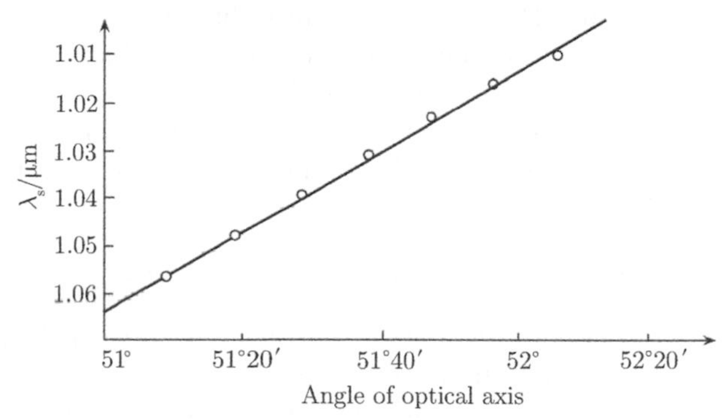

Fig. 7.3-2 The angle tuning curve of $LiNbO_3$ crystal

(2) Temperature tuning

The variation of temperature will change the refractive index of a crystal. In optical parametric oscillation, the variation of the refractive index is the direct factor that causes the signal wave and idler wave frequencies to change. Therefore, tuning can be performed by changing the temperature of the nonlinear crystal itself.

From

$$\Delta n(T, \omega) = \frac{\partial n}{\partial T}\Delta T + \frac{\partial n}{\partial \omega}\Delta\omega \tag{7.3-24}$$

we have

$$\frac{\Delta\omega_s}{\Delta T} = \frac{\omega_p \frac{\partial n_p}{\partial T} - \omega_s \frac{\partial n_s}{\partial T} - \omega_i \frac{\partial n_i}{\partial T}}{(n_s - n_i) + \left[\omega_s \left(\frac{\partial n_s}{\partial \omega_s}\right) - \omega_i \left(\frac{\partial n_i}{\partial \omega_i}\right)\right]} \tag{7.3-25}$$

For types I and II matching modes, from the refractive index surface equation and the relation between specific refractive indices and temperature and the dispersion curve, we can obtain the relation of variation of ω_s with temperature. Usually nonlinear crystals with their refractive index very sensitive to temperature variation should be chosen and an incident angle $\theta = \frac{\pi}{2}$ be taken to reduce the influence of the oplical aperture effect, as shown in Fig. 7.3-3.

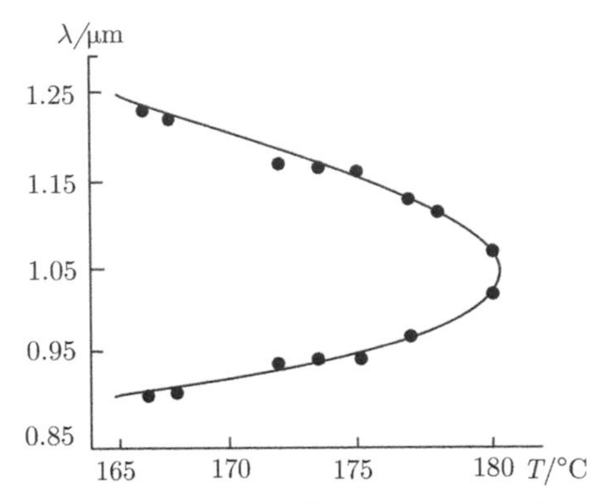

Fig. 7.3-3 The LBO crystal temperature tuning curve

In addition, adopting the electro-optic effect that changes the refractive index can also realize tuning. Its responding velocity is fast, but the tuning range is rather small. By using the PPLN material and the quasi-phase matching method, we shall be able to convert the frequency of the pump wave into a certain fixed frequency. A change in temperature can also realize frequency tuning. By adopting pump light sources of different wavelengths and nonlinear crystals, the optical parametric oscillation technology, apart from providing tunable coherent light sources from visible to infrared waveband, owing to the special correlation existing between the phase and polarization state of the signal wave and idler wave, is already applied in quantum optics, optical field compressed state technique, and the photon entangled pair as well as research fields of optical quantum communication, etc.

7.4 Stimulated Raman scattering

Raman scattering is a kind of non-elastic light scattering with its incident light frequency not equal to that of the scattered light frequency. In the frequency spectrum of the scattered light, information about the energy level transition of the medium, molecular vibration, rotation, and microscopic-movement information on various elements excitation is contained. Raman scattering has all along been an important method for spectral analysis, the study of the microstructure, and the dynamic process of materials. The discovery of the effect of stimulated Raman scattering (SRS) under the action of ruby laser has not only provided experiment on Raman scattering with an ideal light source so that research on spectroscopy has made significant progress, but also it has enabled the laser tuning technique to make appreciable advancement, providing a strong coherent light source in the new spectral waveband, leading to the appearance of different kinds of Raman frequency shift lasers based on stimulated Raman scattering and the generation of Stokes and anti-Stokes radiation of different levels and orders, making it possible to expand the range of coherent radiation tuning in the infrared and ultra-violet frequency direction, respectively.

7.4.1 The Raman scattering effect

In the course of Raman scattering, the particle absorbs a pump photon and emits a scattering photon, the particle's initial and terminal states at different energy levels. Its frequency shift ν_R is the difference between the upper and lower energy levels divided by the Planck constant, called the Raman eigen-frequency (Raman mode) of the medium. Figure 7.4-1 shows the probability of obtaining the scattering photon adopting the quantum mechanical calculation of the second-order double photon process.

$$W_s \propto (1 + \overline{n_R})\overline{n_L} \tag{7.4-1}$$

where $\overline{n_R}$ and $\overline{n_L}$ are the photon degeneracy of the Raman scattering light and the incident pump light, respectively.

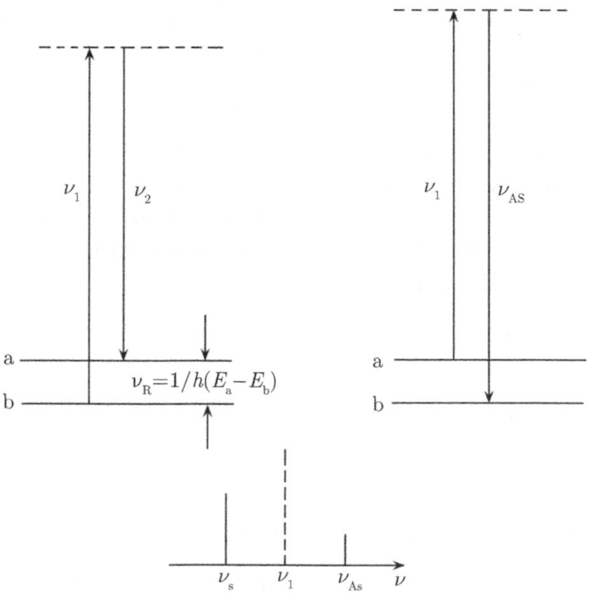

Fig. 7.4-1 Raman scattering energy level transition

An ordinary light source is used as the incident pump light, whose photon degeneracy $\overline{n_L} \ll 1$; correspondingly, $\overline{n_R} \ll 1$, making $\overline{n_R n_L}$ negligible. In W_s only the factor term "1" in parentheses is retained, which is equivalent to the case of spontaneous radiation, called the spontaneous Raman scattering. The scattered light generated is characterized by incoherence. What is particularly apparent is that the intensity I_{AS} of the anti-Stokes (AS) line is far weaker than the intensity I_s of the Stokes (S) line. This is because when the medium is in the thermal equilibrium state, the upper energy level population $N_a \ll N_b$ (population at lower energy level), making the scatter photon number that generates the anti-Stokes spectral line far smaller than the photon number that generates the Stokes spectral line and there are basically no higher-order Stokes lines and anti-Stokes lines that make their appearance.

The adoption of laser with a high photon degeneracy ($\overline{n_L} \gg 1$) as the pump light gives Raman scattering the properties of stimulated radiation. Experiment shows that stimulated Raman scattering mainly has the following characteristics:

(1) An obvious threshold value character. Stimulated scattering will not appear until the pump light has reached or exceeded a definite intensity.

(2) The Stokes lines are distributed as diffuse lines and are strongest when in the direction of spatially overlapping the pump light. The high frequency anti-Stokes lines are not consistent with the pump light in direction and are distributed as ring-shaped sharp lines in a definite spatial solid angle direction.

(3) Under the action of strong pump light, the intensity of the anti-Stokes lines of an identical order is close to that of the Stokes lines. On the two sides of the pumping light frequency, there are multiple equally spaced spectral line distributions; the number of orders of the side frequency line can exceed ten orders.

(4) Unrestricted by the macro-symmetric character of a medium, stimulated Raman scattering can invariably occur in the solid, liquid, and gas media.

The relationship between the first-order Stokes line's frequency ω_{S}, the anti-Stokes line's frequency ω_{AS}, and the pump frequency ω_{L} is

$$\omega_{\mathrm{L}} - \omega_{\mathrm{s}} = \omega_{\mathrm{R}} = \omega_{\mathrm{AS}} - \omega_{\mathrm{L}} \tag{7.4-2}$$

whose frequency difference is the frequency of the optical branch phonon and has a "resonance" relation with the upper and lower energy levels of the Raman medium. Hence

$$\chi_{\mathrm{R}}(\omega) \propto \frac{1}{\omega_{\mathrm{R}}^2 - (\omega_{\mathrm{L}} - \omega)^2 + \mathrm{j}(\omega_{\mathrm{L}} - \omega)\Delta\omega_{\mathrm{R}}} \tag{7.4-3}$$

The corresponding Raman polarizability $\chi_{\mathrm{R}}(\omega)$ should be a complex number, or $\chi_{\mathrm{R}} = \chi'_{\mathrm{R}} + \mathrm{i}\chi''_{\mathrm{R}}$, its imaginary part being represented as gain ($\chi'_{\mathrm{R}} < 0$) or absorption ($\chi''_{\mathrm{R}} > 0$), showing that there is energy exchange existing between the Raman medium and the external optical field. For the stimulated Raman Stokes scattering, the case can be handled as a non-parametric process, the polarization intensity being

$$P_{\mathrm{s}}(\omega_{\mathrm{s}}) = \frac{3\mathrm{j}}{4}\varepsilon_0\chi_{R_{\mathrm{s}}}\left|E_{\mathrm{L}}\right|^2 E_{\mathrm{s}}\mathrm{e}^{\mathrm{j}(k_{\mathrm{L}}-k_{\mathrm{L}}+k_{\mathrm{S}})\cdot z} \tag{7.4-4}$$

During resonance, $\chi_{R_{\mathrm{s}}} \approx i\chi''_{R_{\mathrm{s}}}$ while neglecting the optical Kerr effect ($\chi'_{R_{\mathrm{s}}} \approx 0$):

$$P_{\mathrm{s}}(\omega_{\mathrm{s}}) = \frac{3!}{2}\varepsilon_0\chi''_{R_{\mathrm{s}}}\left|E_{\mathrm{L}}\right|^2 E_{\mathrm{s}}\mathrm{e}^{\mathrm{j}(k_{\mathrm{L}}-k_{\mathrm{L}}+k_{\mathrm{s}})\cdot z} \tag{7.4-5}$$

The amplitude equation of the Stokes wave is obtained:

$$\frac{\mathrm{d}E_{\mathrm{s}}}{\mathrm{d}z} = \frac{3\omega_{\mathrm{s}}}{4cn_{\mathrm{s}}}\left|\chi''_{R_{\mathrm{s}}}\right|\left|E_{\mathrm{L}}\right|^2 E_{\mathrm{s}}\mathrm{e}^{\mathrm{j}(k_{\mathrm{L}}-k_{\mathrm{L}}+k_{\mathrm{s}})\cdot z}\mathrm{e}^{-\mathrm{j}k_{\mathrm{s}}\cdot z} \tag{7.4-6}$$

where the phase factor $\Delta \boldsymbol{k}_{\mathrm{L}} = \boldsymbol{k}_{\mathrm{L}} - \boldsymbol{k}_{\mathrm{L}} + \boldsymbol{k}_{\mathrm{s}} - \boldsymbol{k}_{\mathrm{s}} \equiv 0$, showing that the non-parametric process is not related to Δk. The spatial directionality of the S line is not strong, but is mainly concentrated on the direction of the pump light and is distributed as a "diffuse line". From the above equation we obtain the small signal power gain coefficient of the Raman medium

$$G_{\mathrm{R}} = \frac{3\omega_{\mathrm{s}}}{\varepsilon_0 c^2 n_{\mathrm{s}} n_{\mathrm{L}}}\left|\chi''_{R_{\mathrm{s}}}\right| I_{\mathrm{L}} \tag{7.4-7}$$

where I_L is the pump light intensity.

$$\chi''_{R_{\mathrm{s}}} = -\frac{(4\pi)^2 n_{\mathrm{L}} c^4 \varepsilon_0 N\left(\dfrac{\mathrm{d}\sigma}{\mathrm{d}\Omega}\right)}{3n_{\mathrm{s}}\omega_{\mathrm{L}}\omega_{\mathrm{s}}^3 h\Delta\omega_{\mathrm{R}}} \tag{7.4-8}$$

where $\Delta\omega_R$ is the Raman gain line width, N the density of the ground state population, and $\dfrac{\mathrm{d}\sigma}{\mathrm{d}\Omega}$ the medium's differential Raman scattering cross section.

The Raman gain coefficient of the unit pump intensity

$$g_{\mathrm{R}} = \frac{G_{\mathrm{g}}}{I_{\mathrm{L}}} = \frac{3\omega_{\mathrm{s}}}{\varepsilon_0 c^2 n_{\mathrm{s}} n_{\mathrm{L}}}\left|\chi''_{R_{\mathrm{s}}}\right| = \frac{16\pi^2 c^2 N\left(\dfrac{\mathrm{d}\sigma}{\mathrm{d}\Omega}\right)}{n_{\mathrm{s}}^2\omega_{\mathrm{L}}\omega_{\mathrm{s}}^2 \mathrm{h}\Delta\omega_{\mathrm{R}}} \tag{7.4-9}$$

In a resonator with single pass loss coefficient α and cavity mirror reflectivity R_1 and R_2, respectively, the Raman scattering oscillation threshold value condition is

$$R_1 R_2 \exp[2(G_{\mathrm{R}} - a)L] \geqslant 1 \tag{7.4-10}$$

from which the corresponding threshold value pump intensity can be found.

For the anti-Stokes wave, during thermal equilibrium, the population of the upper energy level is far smaller than that of the lower energy level, and it is known from Eq. (7-4-3) that $\chi'_{R_{AS}}(\omega_{AS}) > 0$. So, in the stage of formation of the S wave, the AS wave is unable to directly obtain a gain from the Raman medium via the pump light. The wave coupling theory must be adopted for analysis.

7.4.2 Wave coupling analysis

In the stimulated Raman scattering effect, multiple equally spaced spectral line features are seen to appear aside from the phenomenon of the AS photons appearing only in a particular direction. The equation of coupling for wave interaction in the third-order nonlinear optical effect must be adopted for an explanation since, under the condition that there do not exist the equally spaced eigen-energy levels in the Raman optical medium, the multiple level equally spaced Stokes lines and anti-Stokes lines that have appeared do not correspond to the transition of the multiple equally spaced energy levels. The operation mechanism of the Raman optical medium in the state of thermal equilibrium is different from that of the ordinary lasing material. Gain is not provided by the inverted population but it is the pump wave that provides the stimulated Raman scattering optical wave with gain while the multiple equally spaced spectral lines are a result of the process of optical wave coupling, energy exchange, and the parametric action related to Δk.

1. Stimulated anti-Stokes radiation

The double photon process cannot explain the phenomenon of stimulated anti-Stokes radiation. Experimentally, have the Raman medium cooled to an extremely low temperature. The upper energy level population is extremely small but still capable of generating an anti-Stokes wave of an intensity close to that of the Stokes wave. This shows that the anti-Stokes wave obtains its energy mainly from wave coupling. Furthermore, the "sharp line" distribution possessed by the anti-Stokes wave shows that its directionality is very strong while also implying that the energy exchange process is a parametric action related to $\Delta k = 0$.

From Eq. (7.4-2) we obtain the frequency relation between the pump wave, the first-order Stokes optical wave, and anti-Stokes optical wave:

$$\omega_{\mathrm{L}} + \omega_{\mathrm{L}} = \omega_{\mathrm{s}} + \omega_{\mathrm{AS}} \tag{7.4-11}$$

which belongs in the third-order interaction between four waves. Adopting the resonance non-depletion approximation and letting $A_{\mathrm{i}} = \sqrt{\frac{n_{\mathrm{i}}}{\omega_{\mathrm{i}}}} E_{\mathrm{i}}$, we obtain the wave coupling equation for the Stokes wave and that for the anti-Stokes wave:

$$\frac{\mathrm{d}A_{\mathrm{s}}}{\mathrm{d}z} = \alpha_{\mathrm{s}} A_{\mathrm{s}} + \mathrm{j} b A_{\mathrm{AS}}^{*} \mathrm{e}^{-\mathrm{j}\Delta k \cdot z} \tag{7.4-12a}$$

$$\frac{\mathrm{d}A_{AS}}{\mathrm{d}z} = \alpha_{\mathrm{AS}} A_{\mathrm{AS}} + \mathrm{j} b A_{\mathrm{s}}^{*} \mathrm{e}^{-\mathrm{j}\Delta k \cdot z} \tag{7.4-12b}$$

where

$$\alpha_{\mathrm{s}} = \frac{3\chi''_{\mathrm{R}}(\omega_{\mathrm{s}})\omega_{\mathrm{L}}}{4cn_{\mathrm{L}}} |A_{\mathrm{L}}|^2 \sqrt{\frac{\omega_{\mathrm{s}}}{n_{\mathrm{s}}}}, \quad \alpha_{\mathrm{AS}} = \frac{3\chi''_{\mathrm{R}}(\omega_{\mathrm{AS}})\omega_{\mathrm{L}}}{4cn_{\mathrm{L}}} |A_{\mathrm{L}}|^2 \sqrt{\frac{\omega_{\mathrm{AS}}}{n_{\mathrm{AS}}}} \tag{7.4-13}$$

The parametric coupling coefficient

$$b = \frac{3\chi^{(3)}\omega_{\mathrm{L}}}{8cn} A_{\mathrm{L}}^2 \sqrt{\frac{\omega_{\mathrm{s}}\omega_{\mathrm{AS}}}{n_{\mathrm{s}} n_{\mathrm{AS}}}} \tag{7.4-14}$$

The first term on the right side of Eq. (7.4-12) is an expression of the non-parametric action. As to whether it is absorption or gain, this is to be determined by whether the imaginary part of Raman polarizability is positive or negative. Further analysis shows that, under the above-mentioned wave coupling condition, even with $\chi''_{\mathrm{R}}(\omega_{\mathrm{S}}) < 0$, it's impossible for the Stokes wave to steadily increase exponentially. When the Stokes optical wave reaches a definite intensity, owing to the consumption of the ground state population and the action of energy conversion of wave coupling, the gain of the Stokes line appears to be saturated. Now the non-parametric process does not play the dominant role. The increase in the pump intensity no longer directly leads to an increase in the Stokes intensity. Rather, an anti-Stokes photon, that is, $\omega_{\mathrm{AS}} = \omega_{\mathrm{R}} + \omega_{\mathrm{L}}$, is "synthesized" by the optical branch phonon $\omega_{\mathrm{R}} = \omega_{\mathrm{L}} - \omega_{\mathrm{s}}$ generated by the difference frequency resonance and a pump photon. Usually the Raman mode phonon has a very short lifetime and is a fast relaxation variable. Under the steady-state condition, it is not necessary to consider the wave coupling equation of the phonon. In the thermal equilibrium state, there is $\chi''_{\mathrm{R}}(\omega_{\mathrm{AS}}) > 0$, where α_{AS} is expressed as absorption. In the course of stimulated Raman scattering, the population distribution of the Raman medium is in a state of abrupt changes. As the process of ground state particles emitting pump photons by absorbing anti-Stokes photons ground state particles can emit pump photons by absorbing anti-Stokes photons and the second-order process of the upper energy level particle emitting AS photons by absorbing pump photons are offset by each other, α_{AS} may become smaller. But, on the premise of the conservation of photon energy, the ground state particle cannot emit an AS photon by directly absorbing a pump photon. So, in a Raman medium with the population not in inverted distribution, it's possible for α_{AS} to be zero. The gain of the anti-Stokes wave will mainly rely on the second term on the right side of the equation, with the parametric process related to phase matching Δk playing the principal role. The numerical solution is given by Bleombergen and Shen. It can be seen from Fig. 7.4-2 that the maximum gain of the anti-Stokes wave occurs in a region near $\Delta k = 0$, manifested as "sharp line" distribution. The conclusions drawn from the analysis of wave coupling are in agreement with the experimental results of stimulated Raman scattering. $\Delta k = 0$ is the vector phase matching condition for the anti-Stokes wave. For an isotropic medium, the dispersion effect makes it impossible for the anti-Stokes wave and

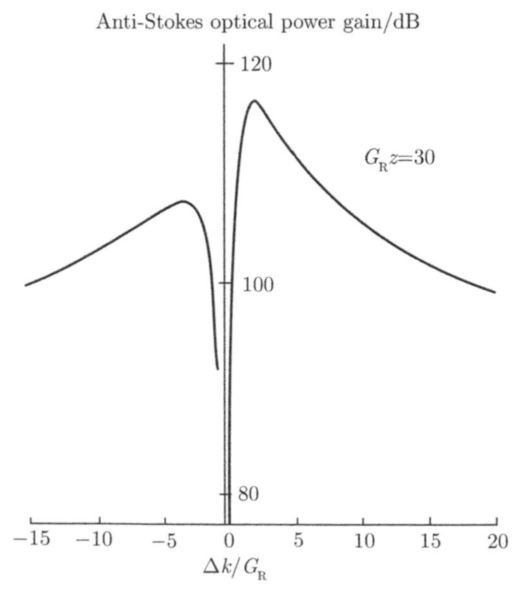

Fig. 7.4-2 Curve of relation between anti-Stokes optical power gain and Δk

Stokes wave to be generated collinearly. The condition for the conservation of its momentum is expressed as

$$2\boldsymbol{k}_{\mathrm{L}} = \boldsymbol{k}_{\mathrm{S}} + \boldsymbol{k}_{\mathrm{AS}} \quad \text{or} \quad \boldsymbol{k}_{\mathrm{L}} + (\boldsymbol{k}_{\mathrm{L}} - \boldsymbol{k}_{\mathrm{S}}) = \boldsymbol{k}_{\mathrm{L}} + \boldsymbol{k}_{\mathrm{R}} = \boldsymbol{k}_{\mathrm{AS}} \tag{7.4-15}$$

while $\boldsymbol{k}_{\mathrm{R}}$ is the wave vector of the coherent optical branch phonon. With this equation, we can determine the direction of the anti-Stokes optical wave's wave vector $\boldsymbol{k}_{\mathrm{AS}}$ relative to wave vector $\boldsymbol{k}_{\mathrm{L}}$ of the pumping optical wave. The corresponding wave vector of the Stokes optical wave is "automatically" selected from the wave vectors of the Stokes optical wave in the "diffuse lines". The Stokes optical wave and anti-Stokes optical wave in definite directions have the coherence character in the sense of satisfying phase matching conditions. As shown in Fig. 7.4-3, this property of coherence has many important roles in the coherent anti-Stokes Raman scattering (CARS).

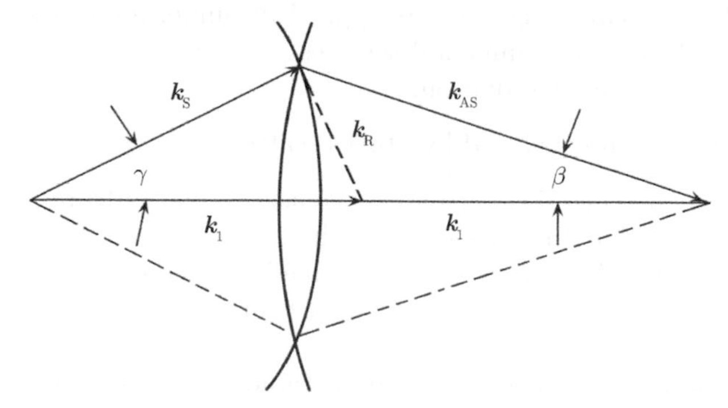

Fig. 7.4-3 Graphical representation of non-collinear phase matching of anti-Stokes line

2. Higher-order stimulated Raman scattering

As the Raman mode optical branch phonon of frequency ω_{R} serves as the mesomeric particle for energy exchange, multilevel radiation spectral lines occur in stimulated Raman scattering. If the pump light is regarded as a zeroth-order spectral line, then the radiation frequency relations between the S wave and AS wave of the second-order and above are

$$\omega_{\mathrm{S}_{n+1}} + \omega_{\mathrm{S}_{n-1}} = 2\omega_{\mathrm{S}_n} \tag{7.4-16a}$$

$$\omega_{\mathrm{AS}_{n+1}} + \omega_{\mathrm{AS}_{n-1}} = 2\omega_{\mathrm{AS}_n} \tag{7.4-16b}$$

This implies that the radiation wave of every order can serve as the pump wave of the next order. When the medium's length and pump power are increased, the power of the first-order Stokes line increases to the maximum. When stable, the energy is transferred to the second-order Stokes line, and then to the third-order S line, and so on. In reality, it is a "staircase" transfer of the energy of the incident pump light till $I_{\mathrm{L}}(\omega_{\mathrm{L}})$ is expended to below the threshold value. The coupling equation for the higher-order multi-level Stokes wave is expressed by intensity. The numerical solution after correction of the Gaussian beam distribution is in fairly good agreement with experimental results. The directionality of the higher-order Stokes wave is manifested as a mixing of the diffuse line spot and sharp line halo, showing that both the non-parametric and parametric factors are playing a part. The diffuse line spot has distinct threshold value character. The flare angle of the sharp line halo can be found by calculation using the relation between the phase matching condition and the dispersion of the medium. For the higher-order anti-Stokes radiation, it is mainly the parametric action that counts. The level by level energy conversion is constrained by the condition of conservation of momentum. Influenced by the dispersion, the anti-Stokes photon made its appearance in a particular spatial direction in the form of a sharp line halo, its phase matching condition being

$$\boldsymbol{k}_{\mathrm{AS}_{n+1}} + \boldsymbol{k}_{\mathrm{AS}_{n-1}} = 2\boldsymbol{k}_{\mathrm{AS}_n} \tag{7.4-17}$$

Under ordinary conditions, as $n_{\mathrm{AS}_{n+1}} \neq n_{\mathrm{AS}_{n-1}} \neq n_{\mathrm{AS}_n}$, it's impossible for it to be generated collinearly. With an increase in the number of levels of the anti-Stokes wave, the order is raised, the frequency "violet shifted", and the spatial flare angle of the corresponding halo, becomes increasingly large. In the visible light waveband, there will appear a pattern of a concentric circular ring distribution of "red, orange, yellow, green, blue, indigo and violet" from interior to exterior.

7.4.3 Application of stimulated Raman scattering in tuning

The stimulated Raman scattering effect is applied in numerous fields including physics, chemistry, biology, medicine, environmental sciences, etc. Our focus is mainly on its application in laser tuning of coherent radiation.

1. Tunable infrared radiation generated by Stokes scattering

The stimulated Raman scattering media that generate such infrared tuning are mainly metallic steam and gas media. In order to make the corresponding Raman scattering cross section of the eigen-energy level difference in a medium increase and the tuning range broaden, usually different lasers are used at different tuning wavebands to provide a suitable pump light source.

By inputting dye laser of dozens of kW of peak value power in alkaline metal steam, we can obtain a tuning range of 10^2 cm^{-1} order of magnitude, the characteristic being that the frequency of the pumping light and a certain allowable upward transition of the ground state approach resonance, so that it is possible to strengthen the Raman scattering cross section and the corresponding third-order nonlinear polarizability, thereby reducing the threshold value pump intensity. For example, when using laser from the blue to ultraviolet waveband to pump the cesium steam, the Stokes output will be continuous band by band within 2.5~ 15 μm. In addition, using nitrogen molecule laser to pump potassium atom steam and dye laser to pump the rubidium atom steam, etc. can all yield tunable infrared coherent radiation with a peak power of several kW in the infrared region.

Hydrogen molecule and nitrogen molecule gases have very strong Raman modes. By means of a pump light of the MW/cm^2 order of magnitude, a strong Raman output can be obtained in a gas cell tens of centimeters long and of several atmospheres. TEACO_2 laser can also be used as a Raman pumping light source of molecular gases. In the application of using laser to separate the uranium isotope, when adopting the molecular method, 16 μm infrared radiation is required. By using a high-power CO_2 laser to pump CH_4 to generate a 16 μm Stokes output, the conversion efficiency is 10%, or a 16 μm radiation can be obtained from stimulated Raman scattering generated by the revolving transition of the parahydrogen molecule.

2. Tunable ultraviolet radiation generated by anti-Stokes scattering

Although it is possible in theory and by experiment to estimate and observe the emergence of higher-order Stokes scattering light, it is emphatically necessary to greatly increase the power density of the pump light before the anti-Stokes line can reach the level of practical significance of being used. But this will bring about some disadvantageous factors. To begin with, there is the very high requirement on the anti-damage ability of the optical medium itself. Then, there will be other nonlinear optical phenomena such as the additional "self-focus", multi-photon absorption ionization, etc. that are all in the way of stimulated Raman scattering. To remedy such a situation, it is proposed that the non-parametric method be adopted to generate the anti-Stokes line so that the Raman medium can be made to be in a state of population inverted distribution. Thus, when the first-order scattering occurs, the gain of the anti-Stokes line is positive while that of the Stokes line is negative. Our previous discussion on the Stokes line can now be applied to the anti-Stokes line. This assumption has been verified. We can adopt the method of photodissociation to form the inverted distribution of the meta-stable state and ground state. Then we can obtain the stimulated anti-Stokes radiation of the said meta-stable state. For instance, after thallium chloride is dissociated with ArF laser, the thallium generated is in the meta-stable state. and it should be pumped with SHG of the Q-switched or the YAG laser of THG. Thallium will, from $6P^2P^0_{3/2}$ to the ground state $6P^2P^0_{1/2}$, generate stimulated anti-Stokes radiation. Such a radiation can be tuned in the vacuum ultraviolet waveband, the efficiency being about 10%.

3. Stimulated spin inverted Raman scattering tuning

When acted on by an external magnetic field H, the electrons will be in quantized circumferential motion within the plane perpendicular to the magnetic field and the corresponding energy level is called the Landau level $(n + 1/2)\hbar\omega_{\mathrm{L}}$, in which ω_{L} is the Larmor precession frequency. Owing to the Zeeman effect, every Landau level can be split into two sublevels that correspond to the two spin states of the electrons, whose interval is

$$\Delta E_{\mathrm{s}} = g_{\mathrm{ef}}\mu_{\mathrm{B}}H \tag{7.4-18}$$

where μ_{B} is the Bohr magneton and g_{ef} is the effective factor. For free electrons $g_{\mathrm{ef}} = 2$.

The corresponding frequency difference is $\Delta\nu_{\mathrm{s}} = \Delta E_{\mathrm{s}}/h$, i.e.,

$$\Delta\nu_{\mathrm{s}} = \frac{\mu_B}{h} g_{\mathrm{ef}} H \tag{7.4-19}$$

In principle, when injecting the pumping light, we can observe the Raman scattering of the spectral line $h\ (\nu_{\mathrm{p}} \pm \Delta\nu_{\mathrm{s}})$. But as g_{ef} is not great, the effect is not obvious. In a solid medium, $g_{\mathrm{ef}} \approx 50$ for certain materials (e.g., InSb), and the S line can be observed. In Eq. (7.4-19), the change in the magnetic field intensity H can regulate the frequency difference so that the frequency of the S line gets corresponding tuning. Such a method is also called magnetic tuning, which is currently widely used in the n type semiconductor indium antimonide. Such lasers have become spin inverted Raman (SIR) lasers. As at low temperatures (about 4K) the concentration of electrons in the conduction band of InSb is rather low, its Raman gain is very small, and the Stokes line is rather weak. For this reason, generally InSb alone is not output as Raman amplification tuning but is output in the difference frequency mode, with $\nu_{\mathrm{p}} - \Delta\nu_{\mathrm{s}} = \nu_{\mathrm{F}}$ as the frequency condition, $\Delta\nu_{\mathrm{s}}$ continuously varying till the continuous tuned output of ν_F. When $\Delta\nu_{\mathrm{s}} = \nu_{\mathrm{F}}$, it is the spin inverted resonance (degenerated frequency mixing) condition. In addition, it is necessary to satisfy the phase matching condition under the action of the magnetic dipole torque which, at collineation, is $n_{\mathrm{p}}\omega_{\mathrm{p}} - n_{\mathrm{s}}\omega_{\mathrm{s}} = n_{\mathrm{F}}\omega_{\mathrm{F}}$ ($\omega_{\mathrm{s}} = 2\pi\Delta\nu_{\mathrm{s}}$). Adopting the frequency mixing effect we can convert more pumping light into tuned coherence for output, the tuning range depending on the intensity of the magnetic field. When a pulsed CO_2 laser is adopted for pumping, the peak power of the tuned output is 200 W. When the tuning range reaches 2000 cm^{-1} (if the superconductivity condition is used to augment H), $\Delta\nu_{\mathrm{s}}$ may be even greater. Furthermore, if CO laser is to be used for pumping InSb, as 5.3 μm is close to the energy gap of InSb, its Raman scattering cross section will become larger while reducing the threshold light intensity $I_{\mathrm{p_t}}$, and it is possible for the operation to be in the continuous wave mode.

7.4.4 Raman frequency shift laser devices

The Stokes wave and anti-Stokes wave in the stimulated Raman scattering possess the coherence of stimulated radiation. In terms of the frequency domain, performing "bidirectional" tuning of pump laser with Raman mode integer times frequency shift makes it possible to provide a high brightness coherent light source in certain wavebands that current laser oscillation is incapable of reaching. As stimulated Raman scattering belongs in the third-order nonlinear optical effect, its operation material is not so rigorously restricted by the point group symmetry as with the second-order effect. In the numerous solid, liquid, and gas media, stimulated Raman scattering has invariably been realized. In particular, speaking of the gas medium, it is easy to obtain Raman operation material of large volume, long optical path, and good optical homogeneity. What's more, it can stand exceedingly great light intensity and has a high optical damage threshold value and a very quick recovery time. Hence its wide application.

Raman laser devices are roughly divided into the following kinds:

1. The single path traveling wave Raman laser device

The pumping light passes through the Raman medium in the mode of the traveling wave and the stimulated scattering wave, too, is output in the form of single path amplification. For the first-order Stokes radiation, the pumping light direction mainly generated by the interacting paths of the largest overlap is mainly forward scattering of weak directionality. For the anti-Stokes radiation, according to the vector phase matching condition $\Delta k = 0$, halo can be observed in specified spatial direction. When the pumping light is intensified and the path of interaction lengthened, corresponding higher-order Raman scattering can occur in the direction of different spatial angles. But, at the same time, it is possible for the emergence of competition by other third-order nonlinear optical effects unfavorable to stimulated Raman scattering, thereby reducing the conversion efficiency.

2. The Raman laser resonator

Incorporate a corresponding reflective mirror each at the two ends of the Raman medium to form a rather strong spatial restriction condition while increasing the optical path of equivalent interaction at the same time. Provide the first-order Stokes wavelength with suitable forward feedback and the corresponding threshold value condition can be determined from Eq. (7.4-10). Generally speaking, its threshold value pump intensity is lower than that of the amplification device of the single path traveling wave. Because of the non-directionality of the first-order Stokes wave, the two reflective mirrors of the Raman resonator can be placed in some direction not collinear with the pumping light, while for the anti-Stokes wave of strong directionality, the axial line of the resonator must take the spatial direction determined according to the vector phase condition $\Delta k = 0$ as standard to make the spatial width of the pumping laser pulse greater than the length of the Raman medium, which will yield fairly good resonant effect. For instance, in the application of laser ranging, gas methane is adopted as the Raman frequency shift medium. Pumped by the Q switch Nd:YAG laser, the first-order Stokes wave radiation is generated and laser of 1.54 μm, which is harmless to the human eye, is obtained. Under appropriate atmospheric pressure, the adoption of the resonator structure gives an energy conversion efficiency of up to 30% and fairly good beam quality as well.

3. The Raman waveguide, fiber optic laser devices

If the medium is of the waveguide structure, the nonlinear effect can be observed at a rather low pump power level. The reason lies in that the waveguide inner wall provides the total reflection of light with a long path of interaction while the tiny aperture can increase the power density of light. Therefore, even with a low pump power, it is still possible to obtain a high stimulated Raman scattering efficiency. For gas and liquid Raman media, slender glass tubes are adopted for packaging to form the waveguide structure, with suitable focusing lens selected by referring to the corresponding numerical aperture to have the pump light coupled into the waveguide. This device is applied on occasions where the stimulated scattering cross section is rather small or where pumping is quasi-continuous or continuous while in the waveguide type solid Raman medium, the stimulated Raman effect of optical fibers is most noteworthy. Apart from a very long path of interaction and a thin and small chip diameter, optical fibers suffer very low losses in the near infrared and visible light waveband, providing the stimulated Raman scattering with a great enhancement factor. Under definite conditions, adopting a pump light of 1~10 W orders of magnitude makes it possible to reach the threshold value pump power needed by the stimulated Raman scattering, thus enabling the continuous laser to also be used as the pump source of stimulated Raman scattering. The fiber optic Raman laser is capable of tuning within the 1~1.6 μm range, which is of great significance for the study and application of fiber-optic communication and fiber-optic soliton lasers.

7.5 Optical phase conjugation

Optical phase conjugation is in essence a representation of the wave nature of light. The optical phase conjugation technology generated from the nonlinear optical effect can be applied in the real-time optical holography, adaptive optics, laser beam quality improvement, elimination of phase distortion, and such fields as optical signal and image processing, high resolution and nonlinear laser spectroscopy, etc.

7.5.1 The concept of optical phase conjugation

There are two forms of spatial conjugation for the optical wave phase plane. One is the forward phase conjugate wave, whose direction of propagation is the same as that of the initial optical wave, with the spatial distributions of the equiphase surfaces all in mirror symmetry with the planes whose initial optical waves are in the vertically propagating direction of the plane. Restricted by the phase matching condition, the forward phase conjugate wave can only occur at the very small spatial angle or a very short length of interaction, its use value and range of application greatly constrained. Therefore, this section will mainly discuss the second kind of phase conjugation, that is, the backward phase conjugation effect, whose direction of propagation is opposite to that of the initial optical wave. However, the equiphase surface distributions overlap, and the phase matching condition is not restricted in an arbitrary incident direction but is automatically satisfied.

1. The significance of phase conjugation

Suppose the expression of a monochromatic optical wave electric field of angular frequency ω forward propagating along the z axis is

$$\begin{aligned} E(r,t) &= \frac{1}{2}A(x,y)\exp[\mathrm{i}(\omega t - kz)] + c.c. \\ &= \frac{1}{2}[\Psi(r)\exp(\mathrm{i}\omega t) + \Psi^*(r)\exp(-\mathrm{j}\omega t)] \end{aligned} \tag{7.5-1}$$

The optical wave electric field at the same frequency as that of $E(r,\ t)$ but propagating in the opposite direction along the z-axis and with the complex amplitude spatially conjugate with it is

$$E_c(r,t) = \frac{1}{2}A_c(x,y)\exp[j(\omega t - k_c z)] + c.c.$$
$$= \frac{1}{2}[\Psi_c(r)\exp(j\omega t) + \Psi_c^*(r)\exp(-j\omega t)] \quad (7.5\text{-}2)$$

As $A_c = A^*$, $\psi_c = \psi^*$, and $k_c = -k$, E_c can be expressed as

$$E_c(r,t) = \frac{1}{2}A^*(x,y)\exp(kz)\exp(j\omega t) + c.c.$$
$$= \frac{1}{2}\{\Psi^*(r)\exp[-j\omega(-t)] + \Psi(r)\exp[j\omega(-t)]\} \quad (7.5\text{-}3)$$

Comparing the above equation with Eq. (7.5-1), we obtain the relation

$$E_c(r,t) = E(r,-t) \quad (7.5\text{-}4)$$

If the moment of conjugation is taken as $t = 0$, then the backward phase conjugate optical wave $E_c(r,\, t)$ at moment t and the forward incident optical wave $E(r,\, -t)$ at moment $-t$ are completely identical, or the backward spatial phase conjugation is equivalent to time reversion. In this sense, the device for realizing spatial phase conjugation is called the phase conjugation mirror (PCM).

2. Characteristics of the phase conjugation mirror

As shown in Fig. 7.5-1, for an ordinary plane reflective mirror, according to the law of linear optics, if it is desired to maintain the continuity of the tangent component of the incident optical wave vector $\boldsymbol{k}_i$ on the reflective surface, the only way to make this possible is for the normal component of the reflective optical wave vector $\boldsymbol{k}_r$ to be of opposite sign relative to $\boldsymbol{k}_i$. From this the law of reflection is derived, which states that the incident angle is equal to the reflective angle. For the divergent spherical wave, after getting reflected by a plane mirror, it still continues to keep its property of divergence. If the phase conjugation mirror (PCM) is adopted to make both the tangent component and normal component of the incident optical wave vector $\boldsymbol{k}_i$ on the surface of the mirror opposite in sign, that is, there is $\boldsymbol{k}_c = -\boldsymbol{k}_i$, then wave vector inversion is realized that makes an optical wave injected from any direction return along the original optical path. For the divergent spherical wave, after being reflected by the phase conjugation mirror, it becomes a convergent spherical wave opposite in the direction of propagation but identical in the shape of wave array. From the point of view of time reversion, the phase conjugation mirror has realized "reflection" of the incident optical wave unrelated to the incident angle and in the complete sense of reflection, that is, the wave vector inversion with the wave array surfaces overlapping. By using this characteristic of the PCM, the phase distortion of the optical wave can be restored in a compensating way. For example, suppose a plane wave propagating along the z-axis is $A_i e^{jkz}$ + c.c., in which A_i is a real number which, after passing through a distorting medium with its refractive index non-uniformly distributed, becomes $A_i e^{jkz} e^{j\varphi(x,y)} + c.c.$ distorted wave, whose equiphase surface appears in the form of irregular distribution. According to an analysis of the spatial angular spectrum of the wave models, all distorted waves can be decomposed into a number of planar subwaves of different directions. The PCM is capable of performing wave vector inversion of these subwaves, which is equivalent to taking conjugation for the phase factor $e^{jkz}\, e^{j\varphi(x,y)}$. After the action by the PCM, the optical wave propagating along the $-z$ direction is $rA_i e^{-j(-k)z} e^{-j\varphi(x,y)} + c.c$, in which r is the amplitude reflection coefficient of the PCM. When passing through the distorting medium once again, the passage is equivalent to being multiplied by the phase factor $e^{j\varphi(x,y)}$, just offset by its conjugate wave $e^{-j\varphi(x,y)}$. The optical wave that has returned to its initial position is $rA_i e^{jkz}$, still a planar wave, with phase distortion repaired. In reality, it can directly be observed from the time reversion characteristic of the PCM that, no matter how circuitous the optical path for the optical wave to reach the PCM may be, the equiphase surface of the optical wave that returns to the initial position through the same optical path after having been acted on by the PCM is still the shape as when setting out on its trip. In this sense, the PCM has made the "rejuvenation" of optical wave conversion a reality. For the case of A_i being a complex number, such as a Gauss beam, this conclusion still holds. Because of this characteristic, the PCM is of important use value in improving the laser beam quality.

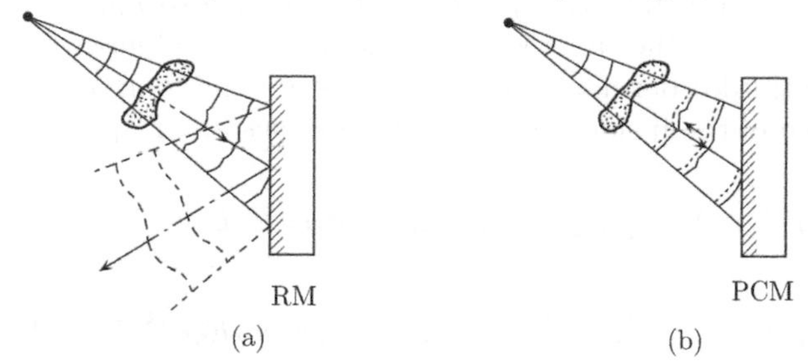

Fig. 7.5-1 An ordinary reflective mirror (RM):
(a) Mirror in conjugation with phase (PCM); (b) Comparison of action with incident optical wave

7.5.2 Method and application of optical phase conjugation

It should be pointed out that the phenomenon of optical phase conjugation itself is not in the category of the optical nonlinear effect. For instance, in the recording process of holographic photography in optical holography, the object wave E_3 and the reference wave E_1 generate interference on the holographic plate to form a grating containing "all the information" on the object. Then, by utilizing the reproduced optical wave E_2 opposite in direction to the wave vector of E_1 to illuminate the holographic photograph, it will be possible to observe the real imager of the said object in the spatial position of the original object. But the diffraction optical wave E_4 that has formed this real image is none other than the phase conjugate wave $E_4 = E_3^*$ of the object wave E_3. However it is impossible for the process of ordinary holographic recording and that of reproduced readout to proceed at the same time, there is no real-time character. And so there is no time reversion characteristic of the PCM either, making it impossible to remove phase distortion dynamically and in real time. By making use of the conjugate phase generated by the nonlinear optical effect, we shall be able to realize real-time holography. At present, the most important and most frequently used nonlinear optical method is the four-wave mixing (FWM) of the third-order nonlinear optical effect and the stimulated backward Brillouin scattering (SBBS), which will be dealt with respectively below.

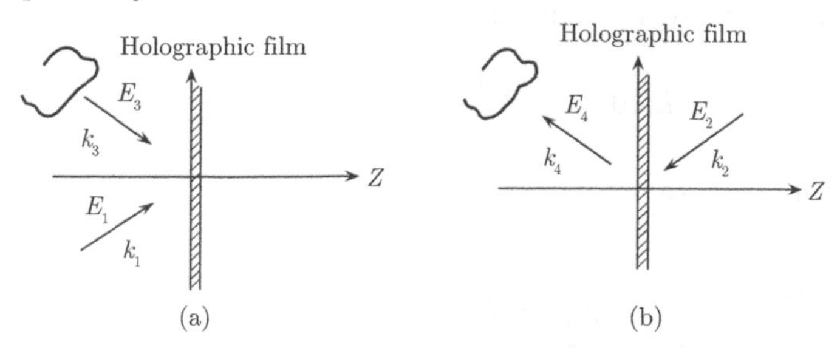

Fig. 7.5-2 The non-real-time optical holographic recording and reproducing process:
(a) Recording; (b) Reproduction

1. The four-wave frequency mixing method

The superposition of the schematic diagrams of the two steps of holographic recording and reproduction as shown in Fig. 7.5-2 is highly enlightening for experimental optical path configuration for the generation of optical phase conjugation effect. First of all, the pump wave vectors $\boldsymbol{k}_1$ and $\boldsymbol{k}_2$ are opposite in direction, which correspond to the reference wave vector and the readout wave vector, respectively. For a wave vector $\boldsymbol{k}_3$ of an incident object wave in a certain direction, there will be an inverse wave vector $\boldsymbol{k}_4$ that satisfies the phase matching condition making its appearance in a direction opposite to that of $\boldsymbol{k}_3$ that generates the conjugate optical wave $E_4 \propto E_3^*$ corresponding to E_3. A case in which all the optical wave frequencies are the same is called the degenerate four-wave mixing (DFWM), which belongs in the third-order nonlinear optical effect. The components of the third-order nonlinear polarization of the incident and reflective waves are, respectively,

$$P_{NL3}(z,\omega) = \frac{3}{4}\varepsilon_0\chi^{(3)}(-\omega,\omega,\omega,-\omega)E_1E_2E_4^*e^{j(k_1+k_2-k_4)\cdot z} \tag{7.5-5a}$$

$$P_{NL4}(z,\omega) = \frac{3}{4}\varepsilon_0\chi^{(3)}(-\omega,\omega,\omega,-\omega)E_1E_2E_3^*e^{j(k_1+k_2-k_3)\cdot z} \tag{7.5-5b}$$

where χ_{eff}^3 of the transparent loss-free medium is a real number. Under the strong pump condition, adopt the non-depletion approximation, noting that the opposite direction propagation condition between pump waves $\boldsymbol{k}_1 = -\ \boldsymbol{k}_2$ and, during phase matching, $\Delta\boldsymbol{k} = \boldsymbol{k}_1 + \boldsymbol{k}_2 - (\boldsymbol{k}_3 + \boldsymbol{k}_4) = 0$, the energy transfer by the pump wave to the incident optical wave and its conjugate optical wave is most effective. Here there is inevitably $\boldsymbol{k}_4 = \boldsymbol{k}_\text{c} \equiv -\ \boldsymbol{k}_3 = -\ \boldsymbol{k}_\text{p}$. Thus we obtain the steady-state wave coupling equation between the incident optical wave $E_\text{p} = E_3$ and the backward conjugate wave $E_\text{c} = E_4$:

$$\frac{dE_\text{p}}{dz} = -j\kappa^* E_\text{c}^*, \quad \frac{dE_\text{c}}{dz} = j\kappa^* E_p^* \tag{7.5-6}$$

where $\kappa^* = \dfrac{3\omega}{4nc}\chi_{\text{eff}}^{(3)}E_1E_2$ is the coupling coefficient.

From the boundary conditions E_p ($z = 0$) $= E_\text{p}$ (0) and E_c ($z = L$) $= 0$, we obtain the solutions for the wave coupling equation:

$$E_\text{c}(z) = j\frac{\kappa^*}{|\kappa|}\frac{\sin|\kappa|(z-L)}{\cos|\kappa|L}E_\text{p}^*(0) \tag{7.5-7a}$$

$$E_\text{p}(z) = \frac{\cos|\kappa|(z-L)}{\cos|\kappa|L}E_\text{p}(0) \tag{7.5-7b}$$

The backward conjugate wave at the incident surface $z = 0$ is

$$E_\text{c}(0) = -j\left(\frac{\kappa^*}{|\kappa|}\tan|\kappa|L\right)E_\text{p}^*(0) \tag{7.5-8a}$$

The incident wave at $z = L$ is

$$E_\text{p}(L) = E_\text{p}(0)\sec|\kappa|L \tag{7.5-8b}$$

When $\dfrac{\pi}{4} < |\kappa|L < \dfrac{3\pi}{4}$, there is $|E_\text{c}(0)| > |E_p(0)|$, showing that the intensity of the backward conjugate wave has exceeded that of the incident wave. Now the nonlinear medium is like a conjugate reflective amplifier. In particular, when $|\kappa|L = \dfrac{\pi}{2}$, there is $\dfrac{E_\text{c}(0)}{E_\text{p}^*(0)} = \infty =$

$\frac{E_p(L)}{E_p(0)}$, which is tantamount to the appearance of "self-excited oscillation". For the incident detecting wave $E_p(0)$ input that is non-zero, the corresponding system reflectivity R and transmissivity T of DFWM can be obtained from Eq. (7.5-8):

$$R = \frac{|E_c(0)|^2}{|E_p(0)|^2} \tan^2 |\kappa| L, \quad T = \frac{|E_p(L)|^2}{|E_p(0)|^2} = \sec^2 |\kappa| L \tag{7.5-9}$$

Owing to the energy intervention by the pump wave, there is the case of occurrence of both R and T being greater than 100%. DFWM is capable of performing the role of parametric amplification of the forward incident optical wave and the backward conjugate optical wave. Furthermore, as the phase matching condition $\Delta k = 0$ here is satisfied by $\boldsymbol{k}_1 + \boldsymbol{k}_2 = 0$ and $\boldsymbol{k}_p + \boldsymbol{k}_c = 0$, respectively, for $\boldsymbol{k}_c$ injected from different directions, there will automatically occur $\boldsymbol{k}_c = -\boldsymbol{k}_p$ in the opposite direction, and without affecting the relation $\boldsymbol{k}_1 = -\boldsymbol{k}_2$ at that. Therefore, the DFWM shown in Fig. 7.5-3 is applicable to the compensation for phase distortion in the wide angle large visual field. A more deep-going analysis shows that, in DFWM, if it is desired to highly efficiently generate the backward conjugate wave, then the two opposite pump light intensities $\boldsymbol{I}_1$ and $\boldsymbol{I}_2$ should be equal.

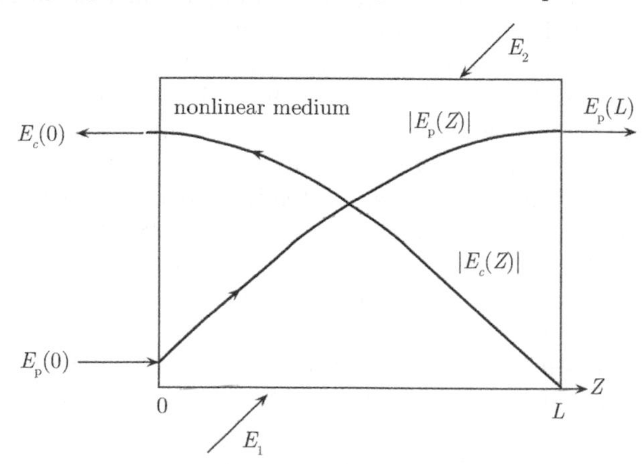

Fig. 7.5-3 Parametric amplification of degenerate four-wave mixing (DFWM)

2. The method of stimulated Brillouin scattering (SBS)

Apart from the previously discussed stimulated Raman scattering (SRS) as a kind of non-elastic optical scattering, there is the coherent light scattering of the acoustic branch frequency shift as a result of the coupling action between the coherent density wave inside a medium generated by the electrostrictive effect of the opto-electric field and the incident optical wave, called the stimulated Brillouin scattering (SBS). Both SRS and SBS are capable of realizing the optical phase conjugation effect, but the frequency shift of the former is rather large and phase mismatch is rather serious, which leads to a low efficiency of generating the phase conjugate wave and so is not adopted very often while for the latter, the phase conjugate wave generated by the Brillouin backward scattering (SBBS) is a phenomenon of nonlinear optical phase conjugation discovered experimentally the earliest and is already in wide application. To sum up, the phase conjugative effect generated by SBS has the following characteristics:

(1) SBS is a kind of third-order nonlinear optical effect of non-parametric action with a threshold value. Only when the pump photon intensity has exceeded the threshold value will it be possible to observe the backward phase conjugate wave. In general, the threshold

value intensity is of the 10 M$\sim$GW/cm^2 order. By properly reducing the optical beam cross section and increasing the length of interaction, a low threshold optical power can be obtained.

(2) In SBS, the path of the overlapping action is longest and the gain greatest, hence the phase conjugate wave is mainly generated by the stimulated backward Brillouin scattering (SBBS). The optical frequency is the first-order S wave frequency and the frequency shift is the shift amount of the forward moving grating's Doppler effect, which is of the GHz order, far smaller than the frequency shift amount of SRS. The phase mismatch that results is not obvious and a fairly good phase matching effect can be obtained.

(3) The generation of the phase conjugation effect by SBBS does not need additional input of pump light. The refractive index grating formed by the coherent density wave of the incident optical wave itself, acting on the optical wave, forms the phase conjugate optical wave. By integrating with each other the pump light and incident light have the self-conjugate character, which makes it conveniently applied in the laser resonator.

(4) It has been proved by theoretical analysis and experimental investigation that the phase conjugation effect of SBBS is positively correlated with the degree of distortion of the incident optical wave array surface within a definite range. The inhomogeneous optical field distribution can contribute to the emergence of a highly efficient phase conjugate wave, which is after all a characteristic of the nonlinear effect.

(5) As there is no action by any other pump light, from the point of view of conservation of energy, the PCM power reflectivity R adopting the SBBS method is bound to be smaller than 100%. The typical value is 40%$\sim$70%, leading to the inevitable residual losses.

(6) A comparison of the lifetime τ of a stimulated phonon with the incident laser pulse width Δt shows that, if $\tau \ll \Delta t$, the process is a steady-state one, otherwise, a transient one. And τ is inversely proportional to the gain line width of the SBS. Besides, while experiencing phase conjugation, SBBS is also able to generate a pulse width compressing effect at the same time to make the pulse width of the backward phase conjugate rays distinctly smaller than that of the incident light.

In a word, owing to the simple structure of the device, stable performance, and practical and convenient operation, SBBS has found extensive application in high power and high quality beam lasers.

In addition to DFWM and SBBS, there are other methods for generating optical phase conjugation such as the photon echo, photorefractive effect, etc.

3. The laser phase conjugate resonator

In a laser resonator, under the action of pumping, the laser medium will, apart from providing gain, inevitably generate irregular phase distortion, seriously affecting the laser beam quality. A laser resonator that adopts the PCM, because of the wave vector inverting performance of phase conjugation, is capable of eliminating the distortion to enhance the quality of the beam. First, based on the meaning of backward phase conjugate optical wave and the matrix optical method, we obtain the transform matrix of the PCM. Suppose the near axis ray vector injected onto the PCM is $\begin{bmatrix} r_{\mathrm{i}} \\ \theta_{\mathrm{i}} \end{bmatrix}$. According to the meaning of the backward phase conjugation, the PCM converts the incident ray to $\begin{bmatrix} r_{\mathrm{i}} \\ -\theta_{\mathrm{i}} \end{bmatrix}$; that is, there is the following equation that holds:

$$\begin{pmatrix} r_{\mathrm{r}} \\ \theta_{\mathrm{r}} \end{pmatrix} = \begin{pmatrix} r_{\mathrm{i}} \\ -\theta_{\mathrm{i}} \end{pmatrix} = \begin{bmatrix} A & B \\ C & D \end{bmatrix} \begin{pmatrix} r_{\mathrm{i}} \\ \theta_{\mathrm{i}} \end{pmatrix} \tag{7.5-10}$$

After multiplication between the matrices two relations are obtained:

$$Ar_i + B\theta_i = r_i \tag{7.5-11a}$$

$$Cr_i + D\theta_i = -r_i \tag{7.5-11b}$$

The transform matrix is

$$\begin{bmatrix} A & B \\ C & D \end{bmatrix}_{\text{PCM}} = \begin{bmatrix} 1 & 0 \\ 0 & -1 \end{bmatrix} \tag{7.5-12}$$

From this the PCM's property of eliminating phase distortion can be proved. Suppose the intracavity total transmission matrix including the action of phase distortion medium is $T_L = \begin{bmatrix} a & b \\ c & d \end{bmatrix}$, and det $T_L = 1$, the initial ray vector $\begin{pmatrix} r_0 \\ \theta_0 \end{pmatrix}$. After action by the PCM followed by a round trip cycle, there is

$$\begin{pmatrix} r_1 \\ \theta_1 \end{pmatrix} = \begin{pmatrix} d & b \\ c & a \end{pmatrix} \begin{pmatrix} 1 & 0 \\ 0 & -1 \end{pmatrix} \begin{pmatrix} a & b \\ c & d \end{pmatrix} \begin{pmatrix} r_0 \\ \theta_0 \end{pmatrix} \tag{7.5-13}$$

From det $T_L = ad - bc = 1$, there is

$$\begin{pmatrix} r_1 \\ \theta_1 \end{pmatrix} = \begin{pmatrix} 1 & 0 \\ 0 & -1 \end{pmatrix} \begin{pmatrix} r_0 \\ \theta_0 \end{pmatrix} \tag{7.5-14}$$

That is, a conjugate wave vector of opposite directions but identical wave array surfaces has been obtained. In particular, under the condition of the initial plane wave $\theta_0 = 0$, there is $r_1 \equiv r_0$. When passing through the distorting medium back and forth, there is still the plane wave that has been obtained in the originally investigated spatial position. The PCM has removed the phase distortion. It is like a slave (passive) distorting mirror always in agreement with the incident wave array surface that performs conversion by inversion of the incident wave vectors from all directions. For the Gauss wave, according to the PCM's wave vector inversion character, parameter q has the conversion relation of opposite sign conjugation:

$$q_r = -q_i^* \tag{7.5-15}$$

reflecting that the wavefront curvature radius R represented by parameter q's real part is opposite in sign, while the beam width represented by the imaginary part remains unchanged. Equations (7.5-15) and (7.5-12) will be transformed in accordance with the following theorem of ABCD:

$$q_r = \frac{Aq_i^* + B}{Cq_i^* + D} \tag{7.5-16}$$

For the optical phase conjugate cavity (Fig. 7.5-4), the above equation holds in the sense of round-trip self-reproduction and it can thus be inferred that there may exist infinitely many Gauss modes in the laser phase conjugate cavity that are not constrained by the geometric parameters of the cavity and the transform matrix of the optical devices in the cavity, transcending the stability condition of the traditional cavity types $0 < G_1G_2 < 1$ and possessing a stability in the broad sense. In a practical experimenting device, an SBS medium with the property of self-pumped conjugation is usually adopted as the PCM and is often used at the amplification level for output following a back-and-forth traverse. The PCM is able to eliminate the phase distortion inflicted on the optical path of transmission and amplification to yield high power laser output of the same quality as that of the incident seed beam. Moreover, optical phase conjugation technology can be used for light beam purification, coherent beam synthesis, adaptive optics, automatic collimated focusing, optical tracing and addressing, lens-free imaging, dynamic interferometer, real-time optical image, and spatial correlated convolution as well as optical field compressing state in quantum optics, and so on. Owing to its accurate time reversion, return strictly along the original optical path, precise phase compensation, and narrow band filtering as well as the real-time holographic characteristics with respect to the incident light, the optical phase conjugation effect has broad prospects for application in such fields as laser technology, optoelectronics, information optics, etc.

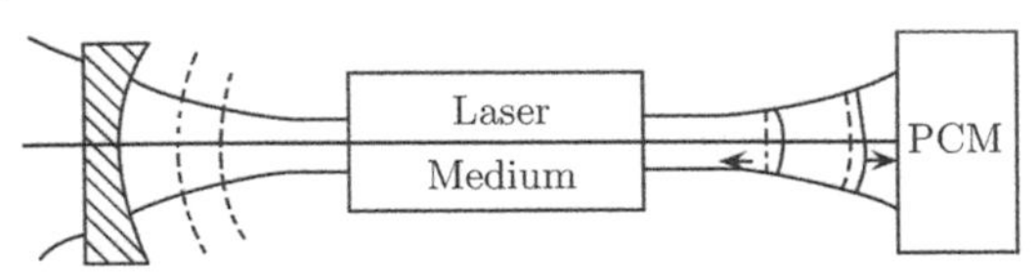

Fig. 7.5-4 The laser phase conjugation resonator

Exercises and problems for consideration

1. Adopt the steady-state wave coupling equation to infer the Manley Rowe relation of the sum frequency and difference frequency effects in the interaction between three waves and discuss its physical meaning.

2. Give examples to show the significance and role of the phase matching condition in the nonlinear optical effects.

3. It is known that the nonzero second-order nonlinear polarizing matrix elements of a certain negative uniaxial crystal are $d_{15} = d_{24} = d_{31} = d_{32} = d_{33}$. Prove that the crystal does not have type II angular phase matching effect and find the effective nonlinear coefficient of its type I angular phase matching.

4. Derive the expression for SHG light intensity $I_{2\omega}$ under the quasi-phase matching condition and plot the curve of relation between $I_{2\omega}$ and the crystal length $L = NLc$ (N is a positive integer).

5. Give the expression of relation of the optical parametric oscillation tuning frequency with angular variation using the equation of the uniaxial crystal refractive index surface.

6. Derive the expression for the scattering angle β of order 1 anti-Stokes line in the stimulated Raman scattering from the non-collinear phase matching and normal dispersion condition.

7. Prove that the two-round-trips matrix of the laser phase conjugate resonator is a unit matrix and discuss its significance.

References

[1] Laser Physics, compiled by Group of Laser Physics, Shanghai, Shanghai People's Press, 1975.

[2] N. Bloembergen. Nonlinear Optics. London: W. A. Benjamin Inc., 1977.

[3] Y. R. Shen. The Principles of Nonlinear Optics. New York: John Wiley & Sons, Inc., 1984.

[4] A. Yariv. Quantum Electronics (3rd Ed.). New York: John Wiley & Sons, Inc, 1989.

[5] Liu Songhao, He Guangsheng, High Light Optics and Its Application, Guangzhou, Guangdong Science and Technology Press, 1995.

[6] Yao Jianquan, Nonlinear Optical Frequency Conversion and Laser Spectrum Modulating Technique, Beijing, Science Press, 1995.

[7] Chen Chuangtian, The Anion Group Theory of the Crystal Electrooptic and Nonlinear Optic Effects, Acta Physica Sinica, 1976, 25:148.

[8] Lu Ya Lin, Mao Lun, and Ming Nai-Ben. Green and Ultraviolet Light Generati on in $LiNbO_3$ Optical Superlattice through Quasiphase Matching. *Appl. Phys. Lett.*, 1994, 64:3092-3094.

[9] Lu Baida, Laser Optics (2nd Ed), Chengdu, Sichuan University Press, 1992.

[10] Chen Jun, Optic Phase Conjugation and Its Application, Beijing, Science Press, 1999.

[11] Jun-Ichi Sakai. Phase Conjugate Optics. New York: McGraw-Hill, Inc., 1992.

[12] Tetsuo Kojima, Shuichi Fujikawa, and Koji Yasui. Stabilization of a High-Power Diode-Side Pumped Intracavity-Frequency-Doubled CW Nd:YAG Laser by Compensating for Thermal Lensing a KTP Crystal and Nd:YAG Rods. *IEEE J. Quantum Electron*, Vol. 35, pp. 377-380, 1999.

CHAPTER 8

The Laser Transmission Technology

Laser transmission is a branch of technology concerned with the study of the interaction between the laser beam and transmission medium, the principal task of which is to reveal the transmission characteristics and laws of the laser beam. The laser transmission medium can be divided into natural mediums (e.g., atmosphere, water) and artificial mediums (e.g., various kinds of optical waveguides, optical fibers, etc.). The laser information system in the earth's surface atmosphere involves laser atmospheric transmission technology; laser underwater detection involves laser underwater transmission technology; while the fiber-optic communication network involves fiber-optic transmission technology.

8.1 An overview of optical fibers

8.1.1 The fiber-optic waveguide structure and weak guide characteristics

The optical fiber is a kind of medium waveguide capable of transmitting optical frequency electromagnetic waves. The typical structure of an optical fiber is as shown in Fig. 8.1-1, which consists of a fiber core, the cladding, and the protective sleeve. The fiber core and cladding, constitute the light-transmitting waveguide structure while the protective sleeve simply plays the role of protection. The property of waveguide is determined by the refractive index distribution. In the second row of the figure are the schematic diagrams of cross section $n(r)$ of two kinds of typical fiber core refractive index. In engineering practice, $\varDelta$ is defined as the relative refractive index difference between the fiber core and the cladding, i.e.,

$$\varDelta = \left[1 - \left(\frac{n_2}{n_1}\right)^2\right] \Big/ 2 \tag{8.1-1}$$

When $\varDelta < 0.01$, Eq. (8.1-1) is reduced to

$$\varDelta = \frac{n_1 - n_2}{n_1} \tag{8.1-2}$$

which is spoken of as the weak conductivity condition and cannot be taken lightly. In theory, the characteristic of weak conduction of optical fiber is one of the important differences between optical fiber and microwave medium waveguide. Actual optical fibers, especially single mode optical fibers, have a very low doping concentration, making the refractive index difference between the fiber core and cladding very small. So, the basic implication of weak conduction is that very small refractive index difference is sufficient to constitute a good fiber-optic waveguide structure while providing technological manufacture with convenience. Many scholars in their comments hold the view that the introduction of the concept of weak conduction is of pioneering significance. The characteristic of weak conduction of optical fibers is of inestimable importance for the development of the theory of optical fibers.

The refractive index distribution in a medium cross section can in general be expressed by exponential type distribution as

$$n(r) = n_1\left[1 - 2\varDelta\left(\frac{r}{a}\right)^{a}\right]^{1/2} \qquad (0 < r \leqslant a) \tag{8.1-3a}$$

$$n(r) = n_1(1 - 2\Delta)^{1/2} = n_2 \qquad (r \geqslant a) \tag{8.1-3b}$$

where a is the fiber core radius; n_1 the refractive index on the axial line of optical fiber; n_2 that of the cladding; and α is a constant. When $\alpha = \infty$, it is step optical fiber; when $\alpha = 2$, it is square law gradient optical fiber, as shown in Fig. 8.1-1. The two kinds of optical fiber are the most widely applied fiber-optic waveguide and the focus in the following discussion.

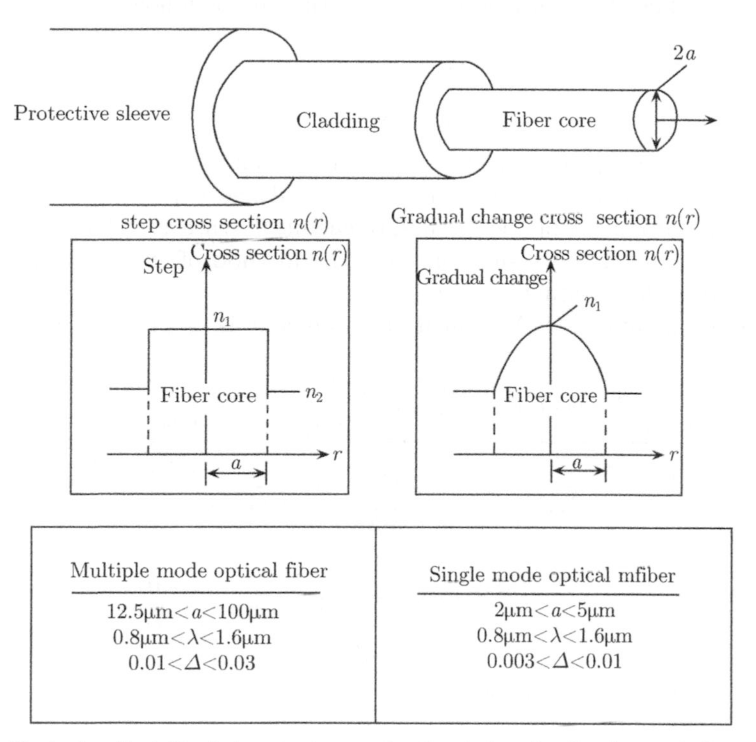

Fig. 8.1-1 Typical optical fiber's terminology, refractive index distribution, and dimension range: a is fiber core radius; λ is optical wave length in free space

The last row in Fig. 8.1-1 gives the structural parameters of the single mode optical fiber and multimode optical fiber. Integrate the fiber-optic core radius a, the relative refractive index difference Δ, and the operation optical wavelength λ while the structural parameter of optical fiber is expressed by the normalized frequency V of optical fiber, i.e.,

$$V \equiv k_0 a(n_1^2 - n_2^2)^{1/2} \tag{8.1-4}$$

where $k_0 = 2\pi/\lambda$ is the number of the optical wave in free space, under the weak conduction condition,

$$V \cong k_0 n_1 a\sqrt{2\Delta} \tag{8.1-5}$$

We shall prove in a later section that the single mode condition is

$$V < 2.045 \qquad \text{(step optical fiber)}$$

Otherwise it is a multiple mode optical fiber.

8.1.2 A brief introduction to manufacturing technology of optical fibers

The manufacturing technology of optical fibers consists of three principal processes: smelting, drawing, and moulding.

1. Smelting

The smelting process consists in synthesizing super-purity chemical raw material silicon tetrachloride and oxygen after high temperature chemical reaction into a low-loss quality quartz rod (called optical fiber prefabricated rod). During smelting, a small amount of impurities is doped to control the refractive index, such as the doping of germanium, phosphorous, boron, fluorine, etc. The chemical reactions of smelting are as follows:

$$SiCl_4 + O_2 \rightarrow SiO_2 + 2Cl_4 \uparrow$$

$$GeCl_4 + O_2 \rightarrow GeO_2 + 2Cl_4 \uparrow$$

in which SiO_2 is the very quartz and the raw material silicon tetrachloride is a gasifiable liquid. Liquids are in general easier to purify than solids, so it is inadvisable to have solid natural quartz purified but adopt the method of chemical synthesis. There are many kinds of smelting technologies. Here we shall only use the method of chemical vapor deposition (MCVD) to illustrate the smelting process.

A schematic diagram of the MCVD smelting technology is as shown in Fig. 8.1-2.

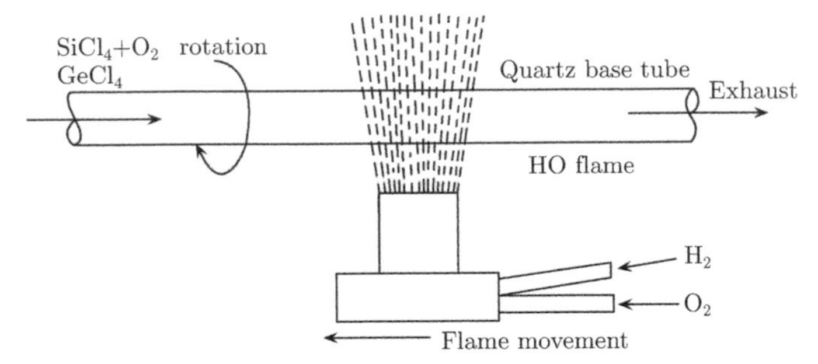

Fig. 8.1-2 Schematic diagram of the smelting technology

The chemical reaction is carried out in the quartz base tube so as to keep the super-purity chemical reaction from getting contaminated. The raw material gases are input through one end of the tube, including silicon tetrachloride, oxygen, and germanium tetrachloride and other dopants. The oxyhydrogen flame is used for heating outside the tube to 1400~1500°C to generate chemical reaction to produce quartz (SiO_2) and germanium glass (GeO_2) with chlorine exhausted from the other end as a waste gas. The synthesized SiO_2 and GeO_2 will be deposited on the tube wall in the form of powder and will melt into a layer of very thin transparent germanium containing quality quartz when given a high temperature. The flame moves back and forth and the tube rotates uniformly. Layer of layer of quality quartz deposits in the tube. When the deposited quartz reaches sufficient thickness, raise the flame temperature to 1700~2000°C and the quartz tube is softened. Because of its surface tensile force, the quartz tube automatically shrinks into a piece of solid quartz rod called the prefabricated rod that is capable of being drawn into optical fibers. The core of the prefabricated rod is high quality quartz that can transmit light, while the skin is ordinary quartz that simply plays the role of a protector.

2. Wire drawing

The process of wire drawing consists in stretching the rather thick prefabricated quartz rod into thin and long optical fibers, a schematic diagram of the device for this step being shown in Fig. 8.1-3.

Upon being led very slowly into the high temperature furnace (temperature in furnace about 2000°C), the prefabricated rod is softened and drawn into thin wires in turn by

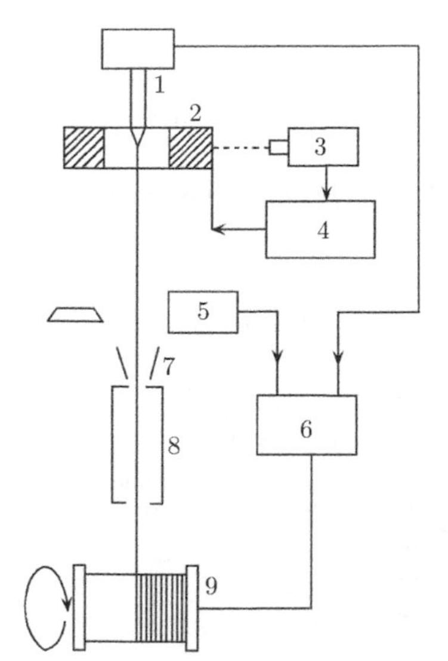

Fig. 8.1-3 Schematic diagram of the wire-drawing technology device: 1. Optical fiber base rod; 2. High temperature furnace; 3. Temperature measuring meter; 4. Furnace temperature control; 5. Non-contact caliber; 6. Speed regulator; 7. Optical fiber coater; 8. Solidifying furnace; 9. Wire-drawing wheel

the wire drawing wheel. To guarantee the accuracy of the optical fiber diameter, a laser diameter measuring meter is mounted that can monitor the outer diameter of optical fiber without contacting it while controlling the furnace temperature and speed of wire drawing according to the deviation signal so as to ensure that the outer diameter of the optical fiber will be within the allowable deviation range. To prevent the surface of the optical fiber from generating microcracks due to pollution from the outside, it is very important to coat a layer of protective paint after the optical fiber is formed and have the paint immediately solidified in the solidifying furnace until the fibers are finally wound round the sleeve of the wire-drawing wheel. Generally, for the sake of uniform wire drawing, the wire-withdrawing sleeve is separated from the wire-drawing wheel.

3. Sleeve moulding

In order to further protect optical fibers and enhance their mechanical strength, optical fibers with a layer of paint are in general coated with a layer of nylon. Usually there are two modes of sleeve moulding of optical fibers, loose sleeving and tight sleeving.

Loose sleeving of optical fibers is as shown in Fig. 8.1-4(a). In this case, the optical fibers are not too crowded in the nylon tube. The coating material of loosely sleeved optical fibers is in general epoxy resin with poor water resistance. To prevent moisture from getting into the tube, it is advisable to fill the tube with semifluid ointment.

Tight sleeved optical fibers are as shown in Fig. 8.1-4(b). During wire drawing, the material for the tight sleeved optical fibers is in general silicon rubber, in the exterior of which a layer of nylon is seamlessly sleeved to bring about tightly sleeved optical fibers, which cannot budge in the nylon tube.

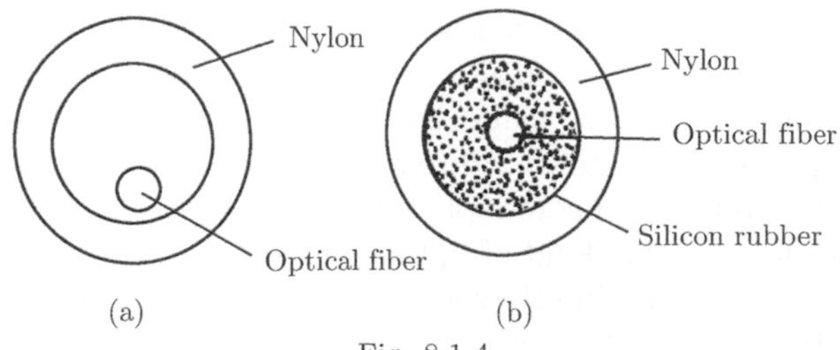

Fig. 8.1-4

(a) Loosely sleeved optical fibers; (b) Tightly sleeved optical fibers

The two kinds of sleeve moulding technology are similar to the conventional wire-making extrusion moulding technology and will not be dealt with here.

The loose sleeving and tight sleeving technologies each have their merits and demerits. For application in engineering, it is advisable to process and combine a number of optical fibers into optical cables.

8.1.3 Optical cables

For the sake of engineering application, it is necessary to strand a number of optical fibers into an optical cable, with its exterior in various protective sleeves to prevent possible damage done by all kinds of mechanical forces and during construction.

The structure of an optical cable is roughly divided into the layer strand type (a) and framework (groove) type (b), as shown in Fig. 8.1-5.

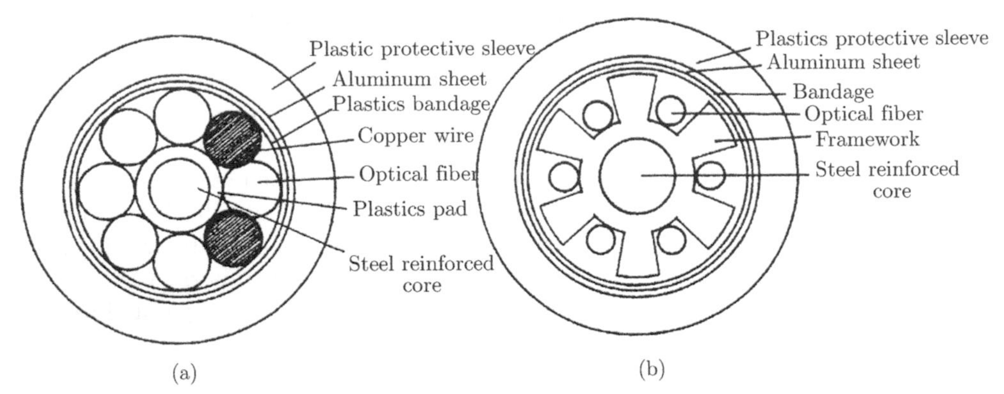

Fig. 8.1-5 Layer strand and framework type optical cables

The structure of the layer strand type of optical cable is quite similar to that of the electric cable and has a piece of steel reinforced core for increasing the tensile strength of the optical cable, especially for resisting the tensile force applied during construction.

Depending on need, a number of copper wires can be placed in the optical cable for remote supply of electric energy needed by the relay station or transmission of telemeter and monitoring signals.

The performance of the framework type of optical cable is fairly good, especially in its resistance to lateral pressure, for which it has a rather great strength. But its manufacturing technology is more complicated than that for the layer strand type cable, requiring special equipment while the layer strand type optical cable can be manufactured with the same traditional equipment for making electric cables and hence the lower cost.

In order to satisfy various applications, the optical cable has the following typical structures:

(1) The simple type optical cable: Its structure is as shown in Fig. 8.1-5 and is mainly used for piping but also for overhead transmission. The thin aluminum in it is used for preventing moisture rather than enhancing strength. Overhead application can also do without the layer of aluminum, whose main characteristics are lightness and low price.

(2) The metal-free optical cable: Its structure is the same as that of the simple type optical cable, only with the steel reinforced core replaced with glass steel rope or Kevlan rope and without the aluminum layer. This kind of optical cable is mainly used in communication lines that can resist strong electric interference or prevent thunderbolt.

(3) The armored cable: Its structure is as shown in Fig. 8.1-6. It is used in long distance trunk line and is of the direct-buried mode. To prevent water from seeping into the optical cable from the outside and flowing into the relay box, the optical cable is often filled with ointment.

(4) The submarine optical cable: Its structure is as shown in Fig. 8.1-7. It is used to cross rivers and sea and its exterior is wrapped with high strength steel wires and is highly pressure-resistant and water-proof. It has an extremely high tension-resistant strength mainly for preventing the powerful tensile force brought about by its own weight during

deep-sea construction that will cause the cable to fracture. Damage from the outside such as those done by ship anchors and the shark should also be prevented.

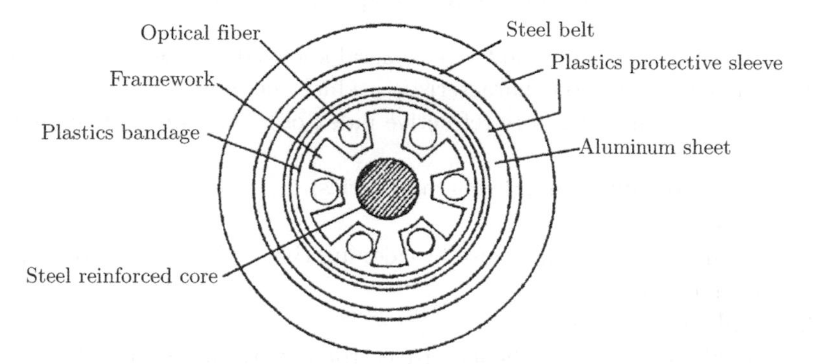

Fig. 8.1-6 The armored optical cable

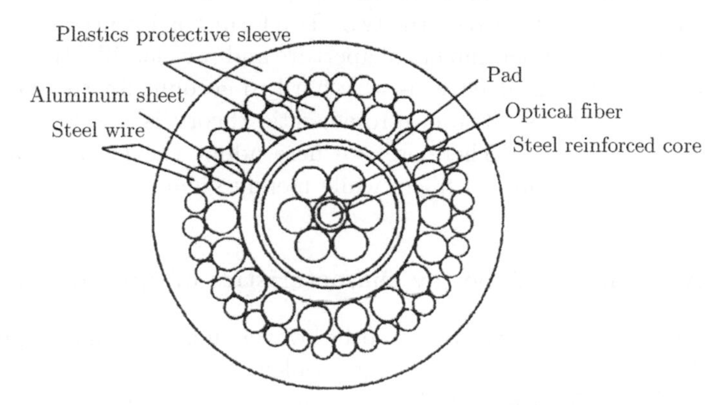

Fig. 8.1-7 Submarine optical cable

(5) The earth wire optical cable: Its structure is as shown in Fig. 8.1-8. It is used on the top of the electric power iron tower both as ground wire and as the optical cable and for communication or controlling and monitoring information. So there is no more need for installing electric poles for communication.

(6) The single core optical cable: Its structure is as shown in Fig. 8.1-9. It is used as the connecting line of equipment and has only one piece of optical fiber that is placed in the soft plastics pipe with the gap filled with Kevlan, which has a very great elastic modulus and so

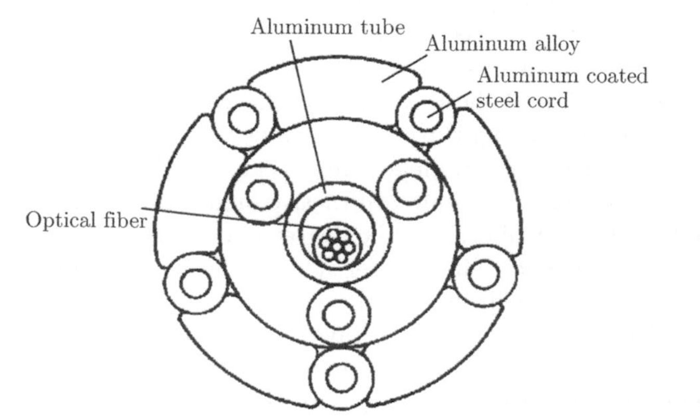

Fig. 8.1-8 The earth wire optical cable

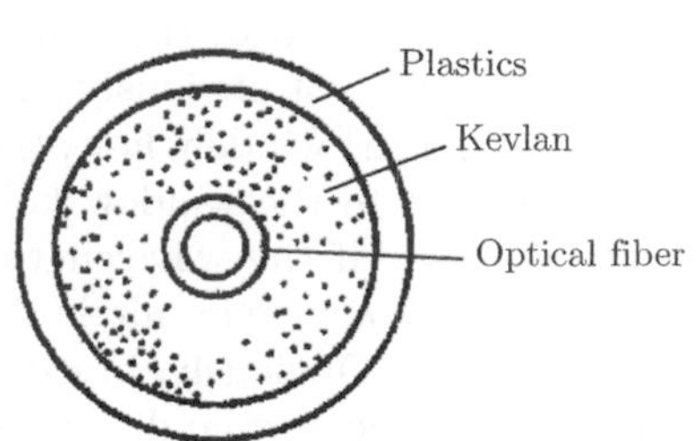

Fig. 8.1-9 The single core cable

very small flexibility.

8.1.4 The transmission characteristics of optical fibers

The previous was a brief introduction to the general knowledge about optical fibers. Below we shall discuss their transmission characteristics, which include:

- the relation between the fiber core refractive index cross section and the mechanism of light transmission;
- the single mode and multiple mode characteristics;
- the loss characteristic;
- the dispersion characteristic and pulse broadening (bandwidth characteristic);
- the polarizing characteristic;
- the nonlinear phenomenon, etc.

Analysis of optical rays and the theory of electromagnetic mode are the basic method of investigating the transmission characteristics of optical fibers and there is very strong mutual supplementary character between the two. The former is intuitive and practical, for instance, the concept of optical fiber numerical aperture and ray classification. But, owing to the approximation of the method itself, it is very hard to account for the concept of single and multiple mode optical fibers. The electromagnetic theory is a very rigorous method of analysis, but too complicated. Therefore, we here adopt the method of approximate analysis for the multiple mode optical fibers while merely giving the concept of the mode total amount and mode group.

8.2 An analysis of the ray characteristics of optical fibers

Compared with the electromagnetic wave in other wavebands, light has a special property of having a very short wavelength. In handling problems of light transmission, one of the frequently encountered problems is that the dimensions of optical components and devices and optical waveguides are much greater than the wavelength. So, usually it is always possible to adopt short wavelength approximation. This method of handling by approximation is called the ray method and the corresponding theory the theory of rays. Such a theory has light processed into optical rays (rays for short), with the phenomenon of light transmission in the waveguide explained using reflection and refraction in optics. Its advantages are it is intuitive and simple and, for certain problems, such as the stimulation of the optical waveguide, it would be very difficult to handle, while it's conveniently handled with the theory of rays. The proposition of the word "ray" is quite natural. We will think of the word "ray" when a very thin beam of light is propagating in space. In everyday life, we have such an experience, viz. in a dark room we can see a beam of light emitted into the room through the window. We can imagine that it has given the propagation direction of the optical radiation field. From the point of view of the wave theory, this can be regarded as a plane wave with part of it intercepted when passing through the window. If the hole for the window is diminished, the light beam will become thinner. But because of diffraction, the edge of light is not obviously sharp and there is a definite diffractive angle. Hence after passing through the window hole, the light beam will diffuse, the degree of diffusion related to the ratio of λ to the window hole diameter λ/a. The greater λ/a is, the more serious will diffusion be. If the wavelength $\lambda \rightarrow 0$, there will not be the phenomenon of marked diffraction even if a were infinitely diminished. So infinitely thin rays with sharp boundaries can be obtained. Obviously, in extreme cases such as this, the direction described by the rays is precisely the normal direction of the plane wave wavefront intercepted. In reality, so long as the condition $a \gg \lambda$ holds, the concept and method of rays can be applied, without having to emphasize $\lambda \rightarrow 0$.

The basis of the rays theory is the equation of rays:

$$\frac{\mathrm{d}}{\mathrm{d}s}\left[n(\boldsymbol{r})\frac{\mathrm{d}\boldsymbol{r}}{\mathrm{d}s}\right] = \nabla n(\boldsymbol{r}) \tag{8.2-1}$$

where $\boldsymbol{r}$ represents the position vector of a certain point on the ray in space, s is the path length from this point to the origin of the ray, and $n(\boldsymbol{r})$ the spatial distribution of the refractive index. By applying this equation and referring to the initial condition, it will be possible in principle to determine the trajectory of the ray in an arbitrary known refractive index distribution $n(\boldsymbol{r})$ medium.

8.2.1 Step optical fibers

For the step optical fiber shown in Fig. 8.1-1, the refractive indices in its core and cladding are uniformly distributed, the indices being denoted by n_1 and n_2, respectively; the fiber core and cladding form a reflection interface at the fiber core's radius a. Equation (8.2-1) shows that the trajectory of the ray in the homogeneous medium is a straight line. Obviously, the mechanism of light transmission of the optical fiber lies in the principle of the total reflection of a ray. The optical fiber can be regarded as a cylindrical waveguide, in which the trace of a ray can be in the principal section passing through the optical fiber's axial line as shown in Fig. 8.2-1(a) or not in the principal section passing through the optical fiber's axial line as shown in Fig. 8.2-1(b). To completely determine a ray, two parameters must be used, viz. the angle θ at which the ray is injected and the included angle φ between the ray and the optical fiber's axial line. There are two classes of rays in the cylindrical waveguide, the meridional ray and the oblique ray, as shown in Fig. 8.2-1.

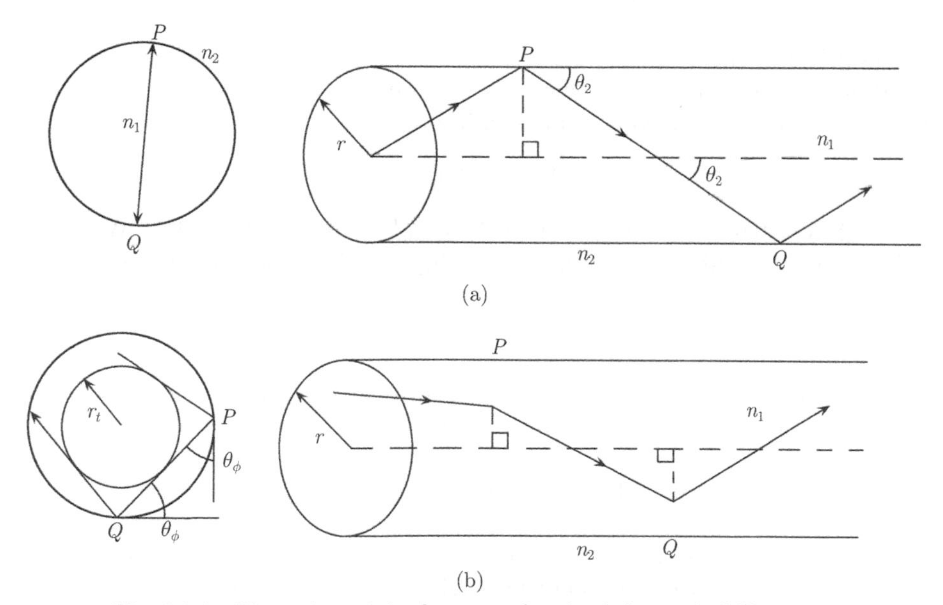

Fig. 8.2-1 The ray's path in the step refractive index optical fiber core: (a) Sawtooth path of meridional ray; (b) Oblique ray's spiral path and its projection on cross section of fiber core

1. The meridional ray

When an incident ray passes through the optical fiber's axial line and the incident angle θ_1 is greater than the interface critical angle $\theta = \sin^{-1}\frac{n^2}{n_1}$, the ray will steadily experience total reflection on the cylindrical interface, forming a zigzag optical path with the trajectory of the conducting ray in the optical fiber's principal section from beginning to end. Such a ray is referred to as the meridional ray and the plane that contains it is called the meridional plane.

Consider the optical fiber's meridional plane shown in Fig. 8.2-2. Suppose a ray is injected from a medium of refractive index n_0 via the waveguide end face central point A and propagates as a meridional ray after entering the waveguide. According to the law of refraction, there is

$$n_0 \sin\varphi_0 = n_1 \sin\varphi_1 = n_1 \cos\varphi_1 = n_1\sqrt{1-\sin^2\theta_1} \tag{8.2-2}$$

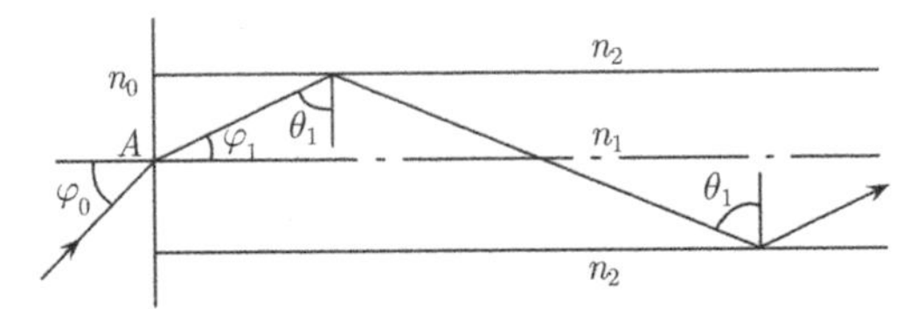

Fig. 8.2-2 Meridional ray in step optical fiber

When total reflection is generated, it is required that $\theta_1 > \theta_0$. So

$$\sin\varphi_0 \leqslant \frac{1}{n_0}(n_1^2 - n_2^2)^{1/2} \tag{8.2-3}$$

In general, $n_0 = 1$ (air), thus the meridional ray's corresponding largest incident angle is

$$\sin\varphi_{0\mathrm{m}}^{(m)} = (n_1^2 - n_2^2)^{1/2} = \mathrm{NA} \tag{8.2-4}$$

NA is called optical fiber's numerical aperture, which determines the maximum value $\varphi_{0\mathrm{m}}^{(m)}$ of the meridional ray half-aperture angle, which represents the collecting ray ability of the optical fibers. Under the weak conduction condition,

$$\mathrm{NA} \approx n_1(2\varDelta)^{1/2} \tag{8.2-5}$$

For instance, if $\varDelta = 0.01$, $n_1 = 1.5$, then $NA \approx \varphi_{0\mathrm{m}}^{(m)} \approx 0.21$ rad ($\approx$12°). What is worth notice is that the optical fiber's collecting ray ability is only determined by the distribution of the refractive index.

2. The oblique ray

When the incident ray does not pass through the optical fiber's axial line, the transmission ray will not be in the same plane. According to the spatial refractive propagation as shown in Fig. 8.2-3, such a ray is called the oblique ray. If it is projected onto the end cross section, it will be seen even more clearly. The transmission ray will completely be confined between two coaxial cylindrical planes, one of which is the core-cladding boundary, the other in the core, whose position depends on angles θ_1 and φ_2 (Fig. 8.2-3), is called the defocused plane. Obviously, with the growth of the incident angle θ_1, the inner defocused plane expands from inward to outward and approximates to the boundary interface. In the limit case, the incident surface of the ray on the end face of the optical fiber touches the cylindrical surface ($\theta_2 = 90°$). The ray being transmitted within the optical fiber evolves into a spiral line that is tangent to the cylindrical surface. The two defocused planes overlap as shown in Fig. 8.2-3(b).

Now let's analyze the maximum incident angle when the oblique ray satisfies the total reflection condition. According to Fig. 8.2-3(a), suppose the oblique ray is injected from point A on the end face and is then totally reflected at points B, C, etc. Plot a straight

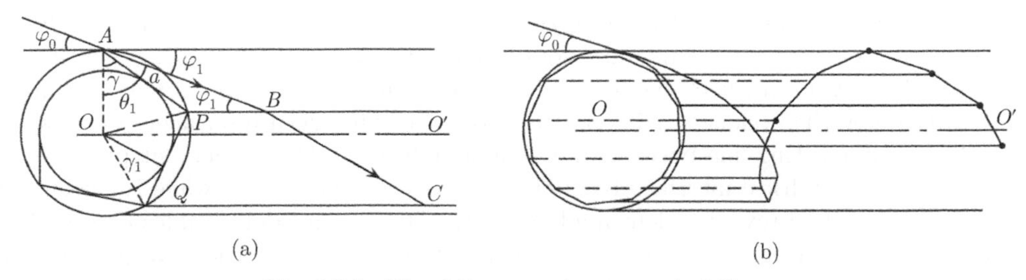

(a) (b)

Fig. 8.2-3 The oblique rays in step optical fibers

line through points B and C parallel to the axial line OO' and intersecting the end face circumference at points P and Q. Express the incident angle of the end face as φ_0 and the refractive angle as φ_1 (also called the axial line angle). $\alpha = \dfrac{\pi}{2} - \varphi_1$ represents the included angle between the refracted ray and the end face, γ is the included angle between the incident surface and the meridional plane AOO', and θ_1 is the incident angle of the refracted ray at the interface. As the respective planes where α and γ are located are perpendicular to each other, according to the principle of solid geometry,

$$\cos\theta_1 = \cos\alpha\cos\gamma = \sin\varphi_1\cos\gamma \tag{8.2-6}$$

When satisfying the total reflection condition $\sin\theta_1 \geqslant n_2/n_1$, it is required that

$$\cos\theta_1 = (1-\sin^2\theta_1)^{1/2} \leqslant \frac{1}{n_1}(n_1^2 - n_2^2)^{1/2} \tag{8.2-7}$$

The allowable maximum axial angle $\varphi_{0\mathrm{m}}^{(s)}$ in the waveguide obtained from Eq. (8.2-6) is

$$\sin\varphi_{1\mathrm{m}}^{(s)} = \frac{(n_1^2 - n_2^2)^{1/2}}{n_1\cos\gamma} = \frac{\sin\varphi_{0\mathrm{m}}^{(m)}}{n_1\cos\gamma} \tag{8.2-8}$$

By applying the refraction law, when $n_0 = n_1 = 1$ (air), the maximum incident angle is

$$\sin\varphi_{0\mathrm{m}}^{(s)} = \frac{\sin\varphi_{0\mathrm{m}}^{(m)}}{\cos\gamma} \tag{8.2-9}$$

where $\varphi_{0\mathrm{m}}^{(m)}$ is the maximum incident angle at which the meridional ray is transmitted. As $\cos\gamma < 1$, it can be seen that when satisfying the condition $\theta_1 > \theta_\mathrm{c}$, the incident angle of the oblique ray can take $\varphi_{0\mathrm{m}}^{(s)} > \varphi_{0\mathrm{m}}^{(m)}$. It can be seen from Eqs. (8.2-8) and (8.2-9) that, when $\gamma = 0$, $\varphi_{0\mathrm{m}}^{(s)}$ takes the minimum value and makes $\varphi_{0\mathrm{m}}^{(s)} = \varphi_{0\mathrm{m}}^{(m)}$. When $\cos\gamma = (1/n_1)\sin\varphi_{0\mathrm{m}}^{(m)}$, $\sin\varphi_{0\mathrm{m}}^{(s)} = \pi/2$, that is to say, the condition $\theta_1 > \theta_\mathrm{c}$ does not seem to have any restriction on φ_1. In fact, in the case of the cylindrical interface, is it quite certain that the oblique ray that satisfies the condition $\theta_1 > \theta_\mathrm{c}$ can transmit power loss-free by generating optical total reflection? For this question, it is necessary to analyze the influence of the axial ray angle φ_1. For the oblique ray, its longitudinal propagation constant should be

$$\beta = k_0 n_1 \cos\varphi_1 \tag{8.2-10}$$

taking into consideration the ray $\varphi_1 > \pi/2 - \theta_\mathrm{c}$, which, substituted into Eq. (8.2-10), yields

$$\beta < k_0 n_1 \cos\theta_\mathrm{c} = k_0 n_2 \tag{8.2-11}$$

It will be seen from an analysis of the electromagnetic theory that $\beta = k_0 n_2$ is exactly the cut-off condition for waveguide guidance mode, that is to say, once there is $\varphi_1 > \pi/2 - \theta_c$, even though $\theta_1 > \theta_c$ is satisfied, the guidance mode is cut off. To explain this, we might as well analyze how the oblique ray is reflected from the cylindrical interface. As shown in Fig. 8.2-4, the intersecting line between the oblique light wave at the incident plane of the interface and the cylindrical interface is an ellipse. Now the plane light wave is injected at point A and the central ray's incident angle $\theta_1 > \theta_c$. But the incident angles of rays 1, 2, and 3 are not identical, and $\theta_2 > \theta_1 > \theta_3$. Hence, it is possible for the rays on the right side of point A not to satisfy the condition $\theta_3 > \theta_c$, the degree of its deviation from the total reflection condition depending on the interface curvature, that is, the axial ray angle φ_1. This shows that, in the case of the bent interface, any optical wave with its axial ray angle reaching a definite extent can only be partially reflected even if the condition of total reflection of the plane boundary is satisfied. So there is an endless optical power leaking out of the core region.

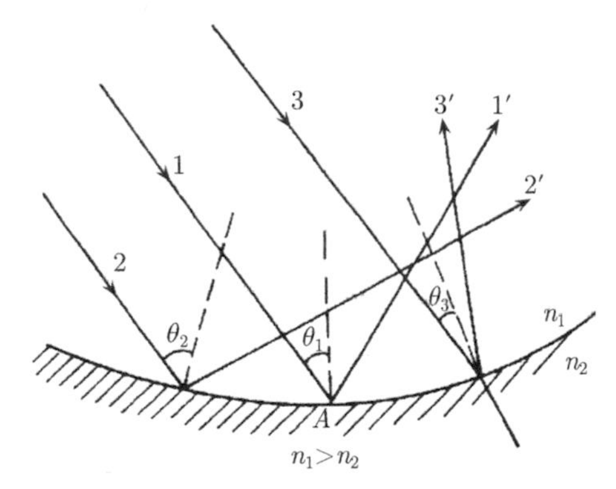

Fig. 8.2-4 Reflection of optical wave from the cylindrical interface

It can be seen from the above discussion that all likely incident rays at point A on the cylindrical interface can be resolved into three parts as shown in Fig. 8.2-5.

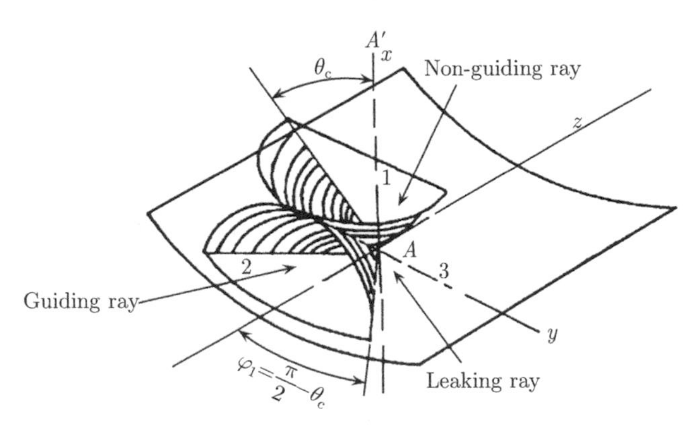

Fig. 8.2-5 The guiding ray, nonguiding ray, and leaking ray in step optical fiber

(1) Non-guiding rays (refracted rays). When $\theta_1 > \theta_c$, they correspond to the rays in the semi-circular cone (region 1) made with the interface normal passing through point A as the axial line and θ_c as the coning angle. It's obvious that none of this part of rays satisfies the total reflection condition, part of the rays having been refracted into the cladding.

(2) The guiding rays. When $\theta_1 > \theta_c$, $\varphi_1 < \pi/2 - \theta_c (\beta > k_0 n_2)$, they correspond to the rays in the semi-circular cone (region 2) made with the intersecting line AA' passing through

point A and touching the cylindrical surface as the axial line and with $\varphi_1 = \pi/2 - \theta_c$ as the coning angle. This part of rays will all be totally reflected at point A, so the optical power will be confined in the core and propagating loss-free. According to the refraction law, the incident angular aperture is found by calculation as

$$\sin\varphi_{0m} = n_1 \sin\varphi_{1m} = n_1 \cos\theta_c = (n_1^2 - n_2^2)^{1/2} = \text{NA} \tag{8.2-12}$$

This result is the same as Eq. (8.2-4), showing that the total incident numerical angular aperture including the meridional rays and oblique rays is exactly the numerical angular aperture of the meridional rays.

(3) The leaking rays (tunnel rays). If $\theta_1 > \theta_c$, $\varphi_1 > \pi/2 - \theta_c(\beta < k_0 n_2)$, they are the rays located in the regions outside the two semi-circular cones. Although this part of rays satisfies the condition $\theta_1 > \theta_c$, no total reflection of light ever occurs on the bent interface, that is, the optical power cannot be completely confined in the core, part of the power leaking outside of it.

3. Diffusion of optical pulses induced by different optical paths

It is known from the above discussion that, for the guiding rays that form different included angles with the fiber axis in the step optical fibers, the optical paths each have traversed when axially passing through the same distance are different. Therefore, if an optical pulse sets off guiding rays of different angles at the incident end, as the optical path traversed by a streak of ray is different from that of another, they will reach the terminal one after another, thereby causing the broadening of the optical pulse, which is spoken of as diffusion of the optical pulse.

Take the meridional ray as an example. Let's make an analysis of the time delay difference of different rays. For a ray that forms angle φ with the optical axis, its velocity along the axis is $v_s = v\cos\varphi = (c/n_1)\cos\varphi$. When $\varphi = 0$, the velocity is highest, when $\varphi = \theta_c$, the velocity is lowest. Therefore, the difference between the longest time and shortest time spent by a ray in traversing an axial distance L is

$$\Delta\tau = \frac{Ln_1^2}{cn_2} - \frac{Ln_1}{c} \cong \frac{Ln_1}{c} \cdot \Delta \tag{8.2-13}$$

It is clear that optical pulse diffusion is proportional to Δ; the smaller Δ is, the smaller $\Delta\tau$ will be.

8.2.2 The gradual varying refractive index optical fiber

We have seen in the step refractive index optical fiber that, for the rays that form different dip angles with the optical axis, their optical paths are different when passing through the same axial distance. A ray with a big dip angle has a long optical path whereas one with a small dip angle has a short optical path. It's only natural for people to think that, if the refractive index is made to decrease with an increase in the distance from the axis, then, although a ray that has deviated greatly from the optical axis has traversed a long path, as the refractive index encountered is small, the optical path of a ray with a big dip angle will be compensated for to a certain extent, thus reducing the maximum delay difference. Hence the generation of optical fibers with graded refractive index distribution.

We shall first proceed from the ray equation to find the trajectory of rays in optical fibers with generally axially symmetric refractive index distribution, followed by an in-depth discussion of the rays in optical fibers of the square law distribution and their maximum group delay difference.

1. The ray equation in the cylindrical coordinate system

It can be proved using the ray Eq. (8.2-1) that, in an isotropic inhomogeneous medium, the ray is always deflected toward the direction of a great refractive index. For the square law optical fibers, the rays getting away from the axial line will steadily bend toward the axial line, it being inevitable that there exists a turning point at which to leave the axis. This is a feature of the movement of rays in such optical fibers.

The ray's unit vector $\boldsymbol{i}_s(r, \varphi, z)$ in the cylindrical coordinate system can be expanded as

$$\boldsymbol{i}_s = \frac{d\boldsymbol{r}}{ds} = \frac{d\boldsymbol{r}}{ds}\boldsymbol{e}_r + r\frac{d\varphi}{ds}\boldsymbol{e}_\varphi + \frac{dz}{ds}\boldsymbol{e}_z \tag{8.2-14a}$$

and

$$\left.\begin{aligned} \frac{dz}{ds} &= \cos\theta_z \\ \frac{dr}{ds} &= \sin\theta_z \sin\theta_\varphi \\ r\frac{d\varphi}{ds} &= \sin\theta_z \cos\theta_\varphi \end{aligned}\right\} \tag{8.2-14b}$$

θ_z and θ_φ represent the included angles between $\boldsymbol{i}_s$ and $\boldsymbol{e}_z$ and $\boldsymbol{e}_\varphi$, respectively.

The ray vector $\boldsymbol{i}_s$ corresponds to the transmission direction vector of the local plane wave in wave optics. So the light transmission vector $\boldsymbol{\beta}$ is parallel to the ray vector $\boldsymbol{i}_s$. As the mode of $\boldsymbol{\beta}$ is nk_0,

$$\boldsymbol{\beta} = nk_0\boldsymbol{i}_s \tag{8.2-15}$$

The three corresponding components are

$$\beta_r = nk_0\frac{dr}{ds} \tag{8.2-16}$$

$$\beta_\varphi = nk_0 r\frac{d\varphi}{ds} \tag{8.2-17}$$

$$\beta_z = nk_0\frac{dz}{ds} \tag{8.2-18}$$

There is obviously

$$\beta_r^2 + \beta_\varphi^2 + \beta_z^2 = (nk_0)^2 \tag{8.2-19}$$

where k_0 is the wave number of the plane waves in vacuum.

The three component equations of the ray Eq. (8.2-1) in the cylindrical coordinate system are

$$\frac{d}{ds}\left(n\frac{dr}{ds}\right) - rn\left(\frac{d\varphi}{ds}\right)^2 = \frac{\partial n}{\partial r} \tag{8.2-20}$$

$$\frac{d}{ds}\left(r^2 n\frac{d\varphi}{ds}\right) = \frac{\partial n}{\partial r} \tag{8.2-21}$$

$$\frac{d}{ds}\left(n\frac{dz}{ds}\right) = \frac{\partial n}{\partial z} \tag{8.2-22}$$

We have designated the optical fiber refractive index to be axially symmetrically distributed, i.e., $n = n(r)$ is unrelated to z, φ.

$$\frac{\partial n}{\partial \varphi} = 0 \tag{8.2-23}$$

$$\frac{\partial n}{\partial z} = 0 \tag{8.2-24}$$

From Eqs. (8.2-22), (8.2-24), and (8.2-18), we find

$$nk_0 \frac{\mathrm{d}z}{\mathrm{d}s} = \text{constant} = \beta_z \equiv \beta \tag{8.2-25}$$

This shows that, for the local plane wave in propagation in optical fibers, the axial component β_z of the propagation constant is a constant denoted by β. In reality, it is precisely because β is a constant that it is possible for the mode to exist, that is, that it is possible for the guiding ray to exist.

From Eqs. (8.1-21), (8.1-23), and (8.1-17), we find

$$k_0 n(r) r^2 \frac{\mathrm{d}\varphi}{\mathrm{d}s} = \text{constant} = r\beta_\varphi \equiv \upsilon \tag{8.2-26}$$

This is the second angular oriented invariance in optical fibers with gradient refractive index distribution, denoted by υ.

To determine the trajectory of a ray, it is necessary to find the relation of function between r-z and φ-z (or φ-r). Substituting Eq. (8.2-26) into Eq. (8.2-20), we can obtain

$$\frac{\mathrm{d}}{\mathrm{d}s}\left(n\frac{\mathrm{d}r}{\mathrm{d}s}\right) - \frac{\upsilon^2}{k_0^2 r^3 n} = \frac{\partial n}{\partial r} \tag{8.2-27}$$

$\frac{\mathrm{d}}{\mathrm{d}s}$ can be expressed as

$$\frac{\mathrm{d}}{\mathrm{d}s} = \frac{\mathrm{d}}{\mathrm{d}z}\frac{\mathrm{d}z}{\mathrm{d}s} = \frac{\beta}{nk_0}\frac{\mathrm{d}}{\mathrm{d}z} \tag{8.2-28}$$

Hence Eq. (8.2-27) can be written as

$$\frac{1}{2}\frac{\mathrm{d}}{\mathrm{d}z}\left[\beta^2\left(\frac{\mathrm{d}r}{\mathrm{d}z}\right)^2 + \frac{\upsilon^2}{r^2} - n^2 k_0^2\right] = 0 \tag{8.2-29}$$

The solution for Eq. (8.2-29) is

$$\beta^2\left(\frac{\mathrm{d}r}{\mathrm{d}z}\right)^2 + \frac{\upsilon}{r^2} - n^2 k_0^2 = C(\text{constant}) \tag{8.2-30}$$

The to-be-determined constant C can be determined by the characteristic of the ray at the turning point. The turning point radius r_t is defined as the maximum radius at which the ray leaves the axial line. Obviously, when $r = r_\mathrm{t}$, $\mathrm{d}r/\mathrm{d}s = 0$, or it can probably be expressed as $\mathrm{d}r/\mathrm{d}z = 0$. From Eq. (8.2-30), there is

$$\frac{\upsilon^2}{r_t^2} - n^2(r_t)k_0^2 = C \tag{8.2-31}$$

From Eqs. (8.2-23), (8.2-25), and (8.2-26), we obtain

$$\beta_\varphi^2(r_t) = \frac{\upsilon^2}{r_\mathrm{t}^2} \tag{8.2-32}$$

$$\beta_r^2(r_t) = 0 \tag{8.2-33}$$

$$\beta_z^2(r_t) = \beta^2 \tag{8.2-34}$$

From Eq. (8.2-19), we obtain:

$$\frac{v^2}{r_t^2} + \beta^2 = n^2(r_t)k_0^2 \tag{8.2-35}$$

Introduce the mark $g(r)$ and let

$$g(r) = n^2(r)k_0^2 - \frac{v^2}{r^2} - \beta^2 \tag{8.2-36}$$

Then according to Eq. (8.2-35), the condition of the turning point radius r_t can be expressed by

$$g(r_t) = 0 \tag{8.2-37}$$

By comparing Eq. (8.2-31) with Eq. (8.2-35), the to-be-determined constant C is found to be

$$C = -\beta^2 \tag{8.2-38}$$

By substituting the constant back into the ray trajectory equation (8.2-30), we obtain

$$\beta^2\left(\frac{\mathrm{d}r}{\mathrm{d}z}\right)^2 + \frac{v^2}{r^2} - n^2k_0^2 = -\beta^2 \tag{8.2-39a}$$

which, reduced, becomes

$$\frac{\mathrm{d}r}{\mathrm{d}z} = \frac{1}{\beta}\sqrt{g(r)} \tag{8.2-39b}$$

Therefore, the ray trajectory is

$$z - z_0 = \beta\int_{r_0}^{r}\frac{\mathrm{d}r}{\sqrt{g(r)}} \tag{8.2-40}$$

When the initial condition $z = z_0$ is used in Eq. (8.2-40), $r = r_0$.

The relation of φ with z can be found from Eqs. (8.2-24) and (8.2-26). Dividing one with the other yields

$$r^2\beta\frac{\mathrm{d}\varphi}{\mathrm{d}z} = v \tag{8.2-41}$$

Therefore, there is

$$\varphi - \varphi_0 = \frac{v}{\beta}\int_{z_0}^{z}\frac{\mathrm{d}z}{r^2} \tag{8.2-42}$$

When the initial condition $z = z_0$ is used in the equation, $\varphi = \varphi_0$. After obtaining $r(z)$ from Eq. (8.2-40), substitution into the integral equation (8.2-42) yields $\varphi(z)$. The angular variation of the ray's trajectory can now be expressed using a form more similar to Eq. (8.2-40). For this purpose, using the relation

$$\frac{\mathrm{d}\varphi}{\mathrm{d}z} = \frac{\mathrm{d}\varphi}{\mathrm{d}r}\cdot\frac{\mathrm{d}r}{\mathrm{d}z} = \frac{\mathrm{d}\varphi}{\mathrm{d}r}\cdot\frac{1}{\beta}\sqrt{g(r)} \tag{8.2-43}$$

Equation (8.2-41) can be rewritten as

$$\frac{\mathrm{d}\varphi}{\mathrm{d}r} = \frac{v}{r^2\sqrt{g(r)}} \tag{8.2-44}$$

Therefore

$$\varphi - \varphi_0 = v\int_{r_0}^{r}\frac{\mathrm{d}r}{r^2\sqrt{g(r)}} \tag{8.2-45}$$

It can be seen from Eqs. (8.2-40) and (8.2-45) that, so long as the refractive index distribution $n(r)$ of a medium and the initial condition of the ray are known, the trajectory of an optical ray can be determined.

Similar to the rays in the step refractive index optical fibers, the rays in the gradually varying refractive index distribution optical fibers can also be divided into the meridional and oblique rays, as shown in Fig. 8.2-6.

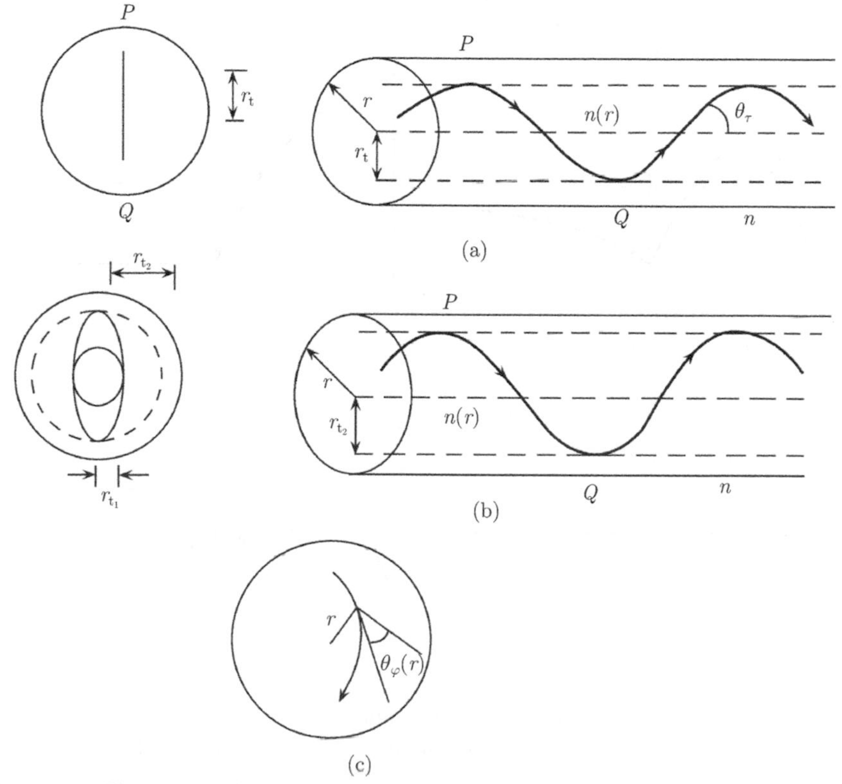

Fig. 8.2-6 The ray's path in the core of gradually varying cross section optical fiber and its projection on cross section of core:
(a) Path of meridional ray; (b) Path of oblique ray; (c) Included angle $\theta_\varphi(r)$

Substituting Eqs. (8.2-14b) and (8.2-17) into Eq. (8.2-26) yields $\upsilon = k_0 r n(r) \sin\theta_z \cos\theta_\varphi$. If the ray invariant $\upsilon = 0$, then $\theta_\varphi = \pi/2$. Therefore, it is the meridional ray passing through the axial line on the meridional plane back and forth. When the ray reaches the turning point $\mathrm{d}r/\mathrm{d}z = 0$, we obtain from Eq. (8.2-37) the turning point condition:

$$n^2(r_\mathrm{t})k_0^2 - \beta^2 = 0$$

Therefore, the turning point of the meridional ray satisfies the equation

$$n(r_\mathrm{t}) = \frac{\beta}{k_0} \tag{8.2-46}$$

When $\upsilon \neq 0$, the turning point radius Eq. (8.2-37) is

$$n^2(r_\mathrm{t})k_0^2 - \beta^2 - \frac{\upsilon^2}{r_\mathrm{t}^2} = 0 \tag{8.2-47}$$

This is a quadratic equation. So there exists the situation where the trajectories of two rays of both internal and external turning point radii r_{t_2} and r_{t_1} are alternately tangent to the two cylindrical planes of radii r_{t_2} and r_{t_1}, respectively. Such rays are oblique rays. In a special case, if $r_{t_1} = r_{t_2}$, the two cylindrical planes overlap. Now the oblique ray becomes a spiral line.

In the above discussion, we have up to now not mentioned the core radius of the optical fiber. Obviously, only when $r_{t_1} < a$ will the ray be completely confined in the optical fiber. If $r_{t_1} > a$, the ray will leave the optical fiber. But even if the ray satisfies $r_{t_1} < a$, the tunneling leaky mode may still exist, as shown in Fig. 8.2-7.

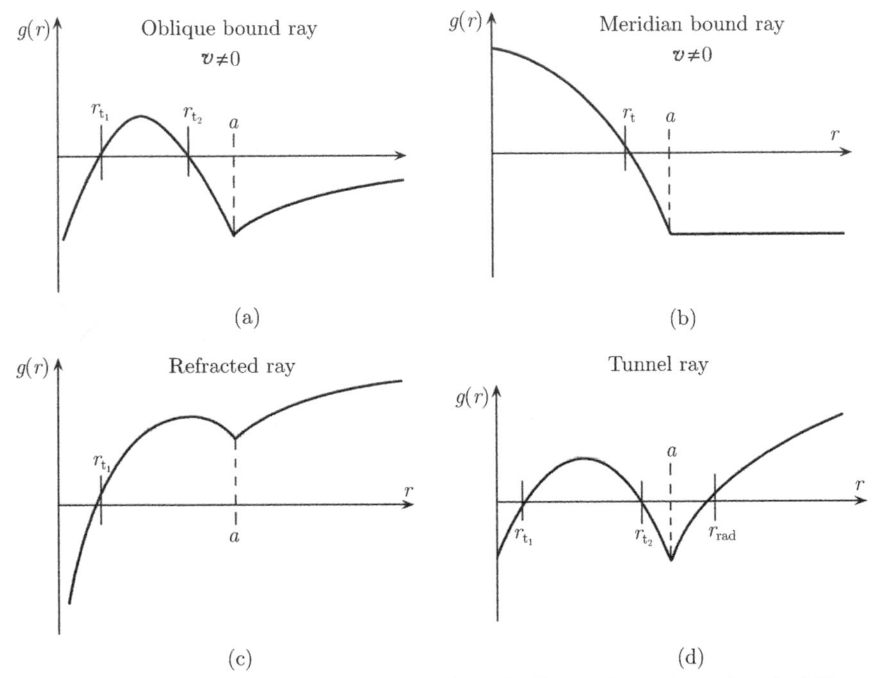

Fig. 8.2-7 A plot of the basic shapes of typical gradually varying refracted optical fiber $g(r)$ (Only when $g(r) > 0$ can the ray be transmitted as marked in thick line in the figure)

2. The ray's trajectory in square law gradient optical fibers

The $n(r)$ in the square law refractive index distribution optical fibers can be expressed by

$$n^2(r) = n_1^2\left[1 - 2\varDelta\left(\frac{r}{a}\right)^2\right] \tag{8.2-48}$$

Substituting Eq. (8.2-48) into Eq. (8.2-40), we can obtain the analytic expression of the ray's trajectory:

$$\begin{aligned} z &= \beta\int_{r_0}^{r}\frac{\mathrm{d}r}{\left[n_1^2k_0^2 - 2\varDelta\left(\frac{r}{a}\right)^2 n_1^2k_0^2 - \beta^2 - \frac{v^2}{r^2}\right]^{1/2}} \\ &= \beta\int_{r_0}^{r}\frac{r\mathrm{d}r}{(k^2r^2 - Ar^4 - v^2)^{1/2}} \\ &= \frac{\beta}{2\sqrt{A}}\left(\sin^{-1}\frac{2Ar^2 - k^2}{\sqrt{k^4 - 4Av^2}} - \sin^{-1}\frac{2Ar_0^2 - k^2}{\sqrt{k^4 - 4Av^2}}\right) \end{aligned} \tag{8.2-49}$$

where

$$k^2 = n_1^2k_0^2 - \beta^2 \tag{8.2-50}$$

$$A = 2\Delta\left(\frac{n_1 k_0}{a}\right)^2 \tag{8.2-51}$$

In the calculation, it has already been assumed that the $r = r_0$ at $z = 0$. The turning point radius r_t of the ray is given by Eq. (8.2-47). For the square law distribution medium expressed by Eq. (8.2-48), Eq. (8.2-47) becomes

$$g(r_t) = k^2 r_t^2 - A r_t^4 - \upsilon^2 = 0 \tag{8.2-52}$$

The solution for Eq. (8.2-52) is

$$r_{t_{1,2}} = \left(\frac{k^2 \pm \sqrt{k^4 - 4A\upsilon^2}}{2A}\right)^{1/2} \tag{8.2-53}$$

where the subscripts 1 and 2 of r_t correspond to taking "+" or "−" before the radical sign.

If the starting point of a ray is selected at a certain internal turning point, that is, assuming that $r_0 = r_{t_2}$ when $z = 0$, then there is

$$\sin^{-1}\frac{2Ar_0^2 - k^2}{\sqrt{k^4 - 4A\upsilon^2}} = \sin^{-1}\frac{2Ar_{t_2}^2 - k^2}{\sqrt{k^4 - 4A\upsilon^2}} = -\frac{\pi}{2}$$

Therefore, the Eq. (8.2-49) of ray trajectory can be reduced to

$$z = \frac{\beta}{2\sqrt{A}}\left(\sin^{-1}\frac{2Ar^2 - k^2}{\sqrt{A^4 - 4A\upsilon^2}} + \frac{\pi}{2}\right) \tag{8.2-54}$$

Letting $\sqrt{A}/\beta = \Omega$, we can transform Eq. (8.2-54) into

$$\sin^{-1}\frac{2Ar^2 - k^2}{\sqrt{k^4 - 4A\upsilon^2}} = 2\Omega z - \frac{\pi}{2}$$

or

$$\frac{2Ar^2 - k^2}{\sqrt{k^4 - 4A\upsilon^2}} = -\cos\Omega z \tag{8.2-55}$$

From Eq. (8.2-53), we obtain

$$r_{t_1}^2 - r_{t_2}^2 = \frac{1}{A}\sqrt{k^4 - 4A\upsilon^2} \tag{8.2-56}$$

By using Eq. (8.2-56), after simple operation, Eq. (8.2-55) can be further transformed into

$$r = [r_{t_1}^2 \sin^2(\Omega z) + r_{t_2}^2 \cos^2(\Omega z)]^{1/2} \tag{8.2-57}$$

The angular relationship of the ray's trajectory can be calculated with Eq. (8.2-42). Substituting Eq. (8.2-57) into the integral equation, we can obtain

$$\varphi = \tan^{-1}\left[\frac{r_{t_1}}{r_{t_2}}\tan(\Omega z)\right] \tag{8.2-58}$$

where $\varphi_0 = 0$ is already taken.

Equations (8.2-55) and (8.2-58) depict the ray's trajectory in the square law medium. When $\upsilon \neq 0$, the ray is an oblique ray, which has two turning point radii. When $\upsilon \neq 0$, the internal turning point radius $r_{t_2} = 0$. Now the ray's trajectory Eq. (8.2-55) can be reduced to

$$r = r_{t_1}\sin(\Omega z) \tag{8.2-59}$$

This equation shows that the meridional ray is of the sinusoidal oscillation type, whose spatial period Λ is

$$\Lambda = \frac{2\pi}{\Omega} \tag{8.2-60}$$

As shown in Fig. 8.2-8(a), it possesses the self-focusing property and is also called self-focusing optical fiber. A length of $\Lambda/4$ self-focusing optical fiber plays a role similar to that of an optical lens in being capable of focusing rays and imaging. The difference between the two lies in that one bends rays by relying on the refraction of the spherical surface while the other does so by relying on the variation of the gradient of the refractive index. The self-focusing lens is characterized by small dimensions, capability of obtaining an ultrashort focal distance, bending, and imaging, all these being very hard or even totally impossible for an ordinary lens to accomplish. It can be proved that the self-focusing lens' focal distance f (the distance from the focus to the main plane) is

$$f = \frac{1}{n(0)\Omega \sin(\Omega z)}$$

The variation of f with z is as shown in Fig. 8.2-8 (b), in which h is the distance of the main plane from the end face. When $z = \Lambda/4$, $f = f_{\min}$.

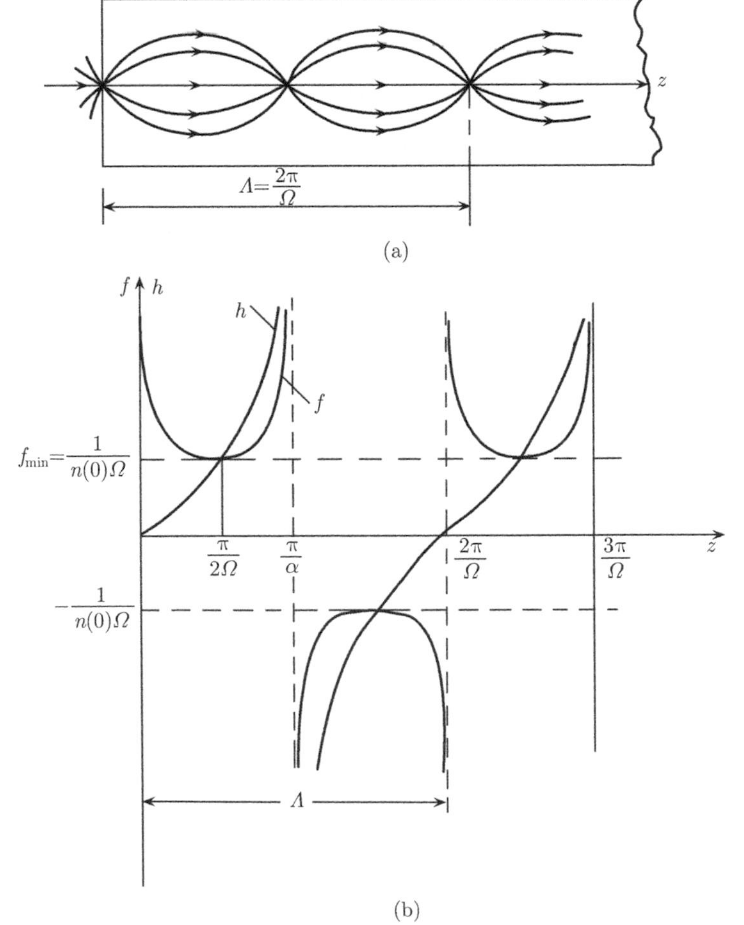

Fig. 8.2-8 The lens characteristic of self-focusing optical fibers:
(a) Meridional ray; (b) Periodic variation of f

If $r_{t_1} = r_{t_2}$, a special oblique ray will occur. From Eq. (8.2-57), there is

$$r = [r_{t_1}^2 \sin^2(\Omega z) + r_{t_2}^2 \cos(\Omega z)]^{1/2} = r_{t_1} = r_{t_2} \tag{8.2-61}$$

From Eq. (8.2-58), we have

$$\varphi = \tan^{-1}[\tan(\Omega z)] = \Omega z \tag{8.2-62}$$

It is thus clear that this is a kind of spiral line with a constant radius and rotating with spatial angular velocity Ω.

3. Group delay and maximum group delay difference of optical fibers in square law refractive index distribution optical fibers

The time a ray spends passing through a unit axial length is called the specific group delay $\overline{\tau}$, or the group delay of unit length. In an inhomogeneous medium, the trajectory of a ray is bent, and the time τ spent along the ray's trajectory over a distance s is

$$\tau = \frac{1}{c}\int_0^s n\mathrm{d}s \tag{8.2-63}$$

where c is the light velocity in vacuum, n the refractive index. This expression of the group delay has not taken material dispersion into account. We shall introduce material dispersion in the expression of the group delay later as a correction. From Eq. (8.2-25),

$$\mathrm{d}s = \frac{nk_0}{\beta}\mathrm{d}z \tag{8.2-64}$$

Hence Eq. (8.2-63) can be expressed by

$$\tau = \frac{k_0}{c\beta}\int_{z_0}^{z} n^2\mathrm{d}z \tag{8.2-65}$$

The refractive index n is in general a function of r. For the sake of the integral equation (8.2-65), it is necessary to transform integration over dz into integration with respect to dr. For this reason, making use of Eq. (8.2-39a), we obtain

$$\tau = \frac{k_0}{c}\int_{r_0}^{r} \frac{n^2(r)}{\sqrt{g(r)}}\mathrm{d}r \tag{8.2-66}$$

When light goes over distance s along the ray's trajectory, its axial advancing distance is l. It is known from Eq. (8.2-40) that

$$l = z - z_0 = \beta\int_{r_0}^{r} \frac{\mathrm{d}r}{\sqrt{g(r)}} \tag{8.2-67}$$

Therefore the specific group delay $\overline{\tau}$ is

$$\overline{\tau} = \frac{\tau}{l} = \frac{k_0}{c\beta} - \frac{\displaystyle\int_{r_0}^{r} \frac{n\mathrm{d}r}{\sqrt{g(r)}}}{\displaystyle\int_{r_0}^{r} \frac{\mathrm{d}r}{\sqrt{g(r)}}} \tag{8.2-68}$$

For the square law distribution medium, $g(r)$ is given by Eq (8.2-52), which, substituted into Eq. (8.2-68) and after integral operation, yields

$$\overline{\tau} = \frac{n_1^2k_0^2 + \beta^2}{2ck_0\beta} \tag{8.2-69}$$

Eq. (8.2-69) shows that the specific group delay is only related to the ray invariant β while unrelated to the angular modes number υ. That's why modes of identical β's all have identical specific group delays.

We know that, for the conduction mode, the magnitude of the propagation constant β is between n_2k_0 and n_1k_0. Therefore, the maximum group delay difference can be found from Eq. (8.2-69) to be

$$\Delta\tau = \tau_{\max} - \tau_{\min} = l(\overline{\tau}_{\max} - \overline{\tau}_{\min}) = \frac{n_1 l}{2c}\varDelta^2 = \tau_0\frac{\varDelta^2}{a} \tag{8.2-70}$$

where τ_0 is $(n_1l)/c$. It can be seen when compared with Eq. (8.2-13) that the group delay in the square law distribution optical fibers is only $\Delta/2$ of the step refractive index distribution optical fibers.

8.3 The attenuation and dispersion characteristics of optical fibers

8.3.1 Attenuation of optical fibers

Attenuation is an important index of optical fibers that shows the transmission loss of optical fibers with respect to optical energy and has decisive influence on the transmission distance for fiber optic communication.

The attenuation coefficient a is defined as the decibel of optical power attenuation over a fiber's unit length, that is,

$$a = \frac{10}{L}\lg\frac{P_{\text{i}}}{P_{\text{o}}} \quad (\text{dB/km}) \tag{8.3-1}$$

where P_{i} and P_{o} are input and output optical power of optical fibers, and L is the length of the optical fiber. The loss of optical fibers has two major sources, the absorption loss and scattering loss.

1. The absorption loss

The absorption loss is due to the absorption of optical energy by the fiber optic material and the harmful impurities in it. These impurities consume the optical energy in the optical fibers in the form of thermal energy.

The absorption loss of a material is a kind of inherent loss and is inevitable. We can only choose a material with minimum inherent loss for fabricating optical fibers. Quartz has the minimum absorption in the IR waveband and is an excellent material for optical fibers.

Absorption by harmful impurities is mainly due to the content of such ions as Fe, Co, Ni, Mn, Ca, V, Pt plus the OH ions in fiber material. As long as there are the above-mentioned impure substances of the ppm order, very great losses will be induced in optical fibers. In general the ppm ultra-purity chemical raw material is adopted to make low loss optical fibers. In the ultrapure raw materials adopted for modern optical fibers there are basically no metallic ions while the absorption loss of optical fibers is mainly caused by OH ions. The absorption of optical energy is mainly due to the resonance of the molecular valence bond. So the loss-wave length characteristic of optical fibers appears to be in the form of a peak, as shown in Fig. 8.3-1. The absorption by the OH ions is rather serious. The absorbing wavelength of its fundamental wave is located at 2.72 μm, the second and third harmonic absorption at 1.38 μm and 0.95 μm, respectively. At 1.24 μm, 1.3 μm, 0.88 μm and 2.22 μm are the combined resonance absorption peaks of OH ions and the fiber optic material SiO_2. The attenuation characteristics of optical fibers with an extremely low loss made of a material with an OH ion content smaller than 1 ppm are shown in Fig. 8.3-2. It can be seen from Fig. 8.3-1 that, when the wavelength is greater than 1.6 μm, the fiber-optic material "quartz" begins to exhibit an inherent loss. This implies that the inherent loss of quartz increases with a decrease of the optical frequency. So, although in principle the HE_{11} mode has no cut-off frequency, restricted by the optical frequency attenuation characteristic of the material, it is only possible to possess the low attenuation transmission characteristic in a certain frequency band. Since the single mode optical fiber has only one fundamental mode, its transmission optical path is shorter than the average optical path of higher-order modes of multimode optical fibers. Usually the attenuation of the single mode optical fiber is slightly smaller than that of the multimode optical fiber.

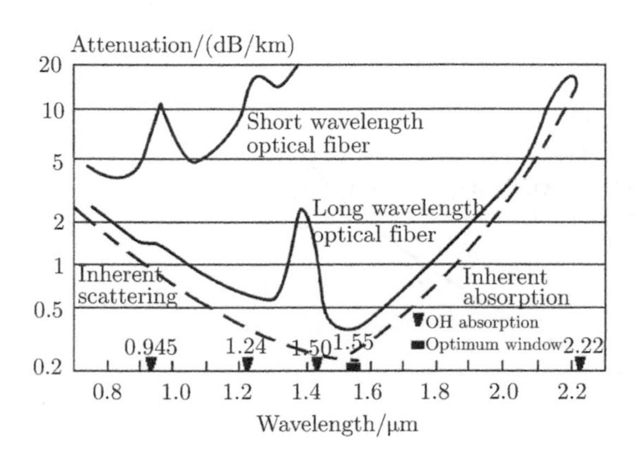

Fig. 8.3-1 Optical fiber attenuation-wavelength characteristic

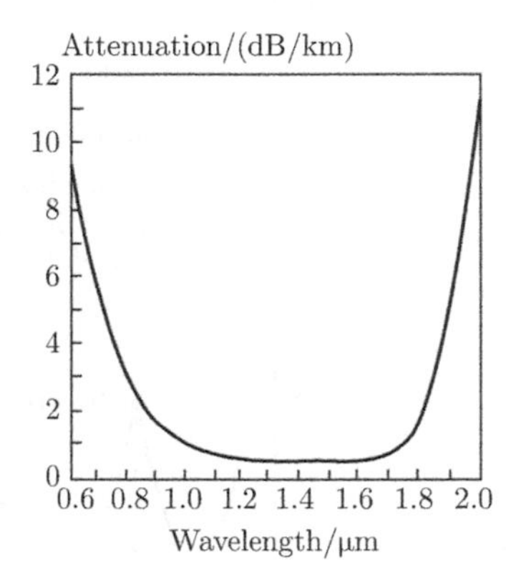

Fig. 8.3-2 OH radical-free low loss optical fiber

2. The scattering loss

Owing to the imperfection of the optical fiber fabricating technology, such as the presence of micro-bubbles, the inhomogeneity of the refractive index, and the internal stress, where there will be the scattering from the optical energy, which will increase the loss to optical fibers.

Another source of scattering loss is what is known as the Raleigh scattering, or the scattering occurring when particles with random fluctuation are encountered whose wavelength size is comparable. The Raleigh scattering loss is inversely proportional to the fourth power of the wavelength and can be reduced by adopting a rather long operation wavelength. If the raw material of the optical fiber is of many components and complicated structure, then the Raleigh scattering loss will be serious.

There also exist the so-called Brillouin and Raman scattering losses that are nonlinear scattering losses brought about by strong light in optical fibers. In general, in multimode optical fibers the density of optical energy is rather small, so this will not take place. But in single mode optical fibers, because of the very small core diameter, when the density of optical energy is sufficiently strong, the two kinds of loss may occur.

It can be seen from Fig. 8.3-1 that quartz optical fibers have several transmission windows with small attenuation. $\lambda = 0.8 \sim 0.9$ μm is a short wavelength window, which is generally used for the multimode optical fiber. When there is a rather small attenuation in the vicinity of $\lambda = 1.0$ μm, 1.3 μm, and 1.55 μm, the window is called the long wavelength window. The windows of the single mode optical fiber are 1.3 μm and 1.55 μm. At 1.55 μm, the attenuation is minimum, as low as 0.2 dB/km. The attenuation of pure SiO_2 optical fiber is as low as 0.16 dB/km. This is a low attenuation (dotted line) formed in the recession where the curves of Raleigh loss and material inherent loss intersect.

Making use of the attenuation-wavelength curve of optical fibers, we shall be able to distinguish between various loss factors to find the cause of trouble and improve technology.

According to the above-mentioned relation between loss and wavelength, we can obtain the following equation:

$$a = A/\lambda^4 + B + C(\lambda) \tag{8.3-2}$$

The optical fiber attenuation-wavelength curve thus plotted is shown in Fig. 8.3-3, from the dotted line in which the numerical values of A, B, and C are obtained, an inspection of which will show which of the losses is dominant. The slope A represents the Raleigh scattering loss, B the loss due to the imperfection of the waveguide, and C the extra loss due to OH ion absorption.

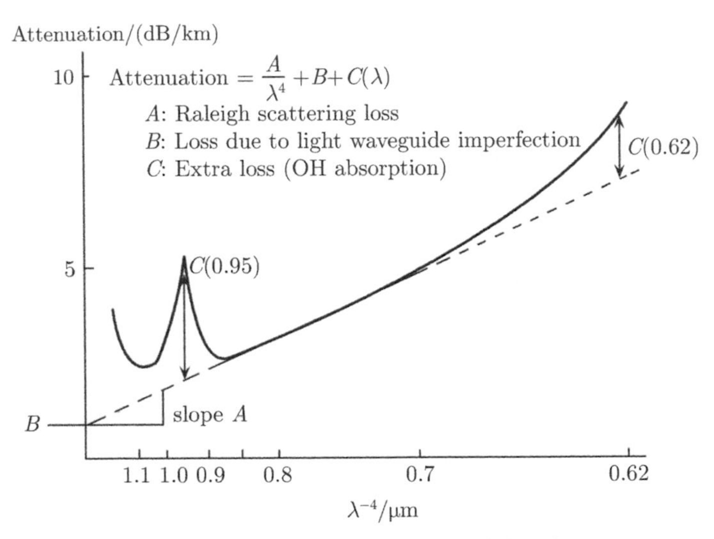

Fig. 8.3-3 Analysis of various optical fiber losses

8.3.2 Relationship between fiber-optic dispersion, bandwidth, and pulse broadening parameters

1. The dispersion of optical fibers

The dispersion of optical fibers will make pulse signals broaden, that is, restrict the bandwidth of the optical fiber or transmission capacity.

Generally speaking, there is the following relation between the single mode optical fiber's pulse broadening and dispersion:

$$\Delta\tau = d \cdot L \cdot \delta\lambda \tag{8.3-3}$$

where d is the total dispersion, L is the length of optical fiber, and $\delta\lambda$ is the spectral line width of the optical signal.

2. The bandwidth of optical fiber

There is a definite relationship between optical pulse broadening and the bandwidth of the optical fiber. It is shown by experiment that the frequency response $H(f)$ of the optical fiber is approximately of Gauss shape, as shown in Fig. 8.3-4.

$$H(f) = \frac{P(f)}{P(0)} = \mathrm{e}^{-(f/f_c)^2 \ln 2} \tag{8.3-4}$$

where $P(f)$ and $P(0)$ are the optical fiber's output AC optical power when the light intensity modulating frequency is f and 0, respectively, and f_c is the half power point frequency. There is obviously

$$10 \log H(f_c) = 10 \log \frac{P(f_c)}{P(0)} = -3 \text{ dB} \tag{8.3-5}$$

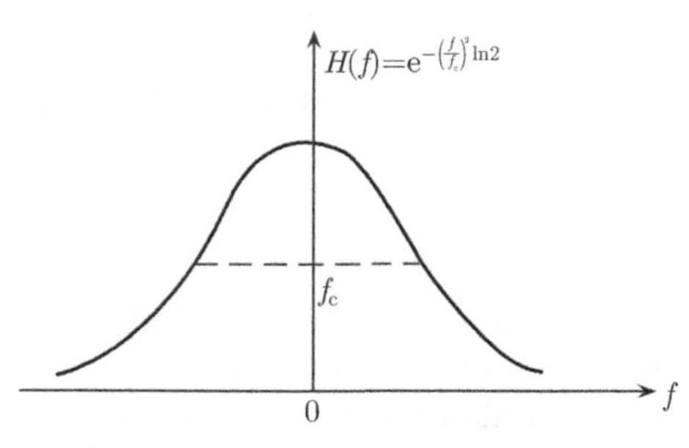

Fig. 8.3-4 The frequency response characteristic of optical fiber

f_c is called the 3-dB optical bandwidth of the optical fiber.

Generally, the optical energy is most often tested with optoelectronic devices while the output current of a tester is proportional to the optical power measured. For the above equation, there is

$$20\log\frac{I(f_c)}{I(0)}=-6\ \mathrm{d}B \tag{8.3-6}$$

So, f_c is also spoken of as the 6-dB electric bandwidth. In effect, the 3-dB optical bandwidth and the 6-dB electric bandwidth are equal f_c's.

3. The impact response of the optical fiber

To investigate the time domain characteristics of the optical fiber, input an infinitely narrow optical pulse in an optical fiber. The area of the pulse is 1, that is, the impact function $\delta(t)$. The optical fiber's output or the impact response

$$\begin{aligned} h(t) =& F^{-1}[H(f)] = F^{-1}\left[\mathrm{e}^{-\left(\frac{f}{f_{CL}}\right)^2\ln 2}\right] \\ =& \sqrt{\frac{\pi}{\ln 2}} f_{CL}\mathrm{e}^{-(\pi f_{CL}t)^2/\ln 2} \end{aligned} \tag{8.3-7}$$

is obviously of the Gauss shape, as shown in Fig. 8.3-5, in which f_{CL} is the bandwidth of the optical fiber of length L.

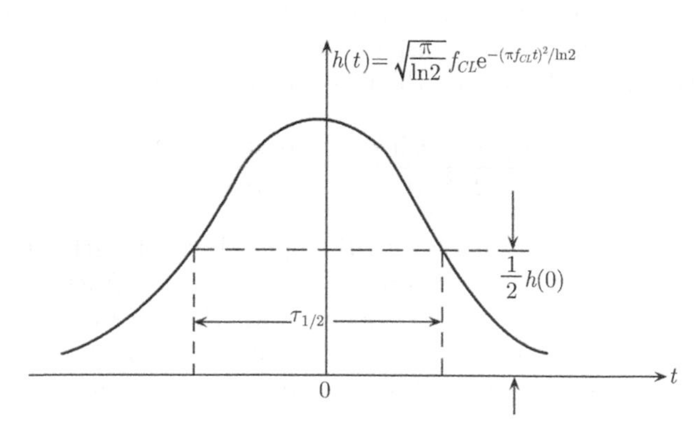

Fig. 8.3-5 The optical fiber's impact response and $\tau_{1/2}$

In engineering, for convenience in measurement, the pulse half-width $\tau_{1/2}$ is often adopted to measure the pulse width (see Fig. 8.3-5). The expression of Gauss pulse denoted by $\tau_{1/2}$ is

$$h(t)=\sqrt{\frac{\pi}{\ln 2}} f_{CL}\mathrm{e}^{-\left(\frac{t}{\tau_{1/2}}\right)^2\ln 2} \tag{8.3-8}$$

Comparing this with Eq. (8.3-7), we can obtain the equation of the relation between the bandwidth of the optical fiber and the pulse half-width as

$$f_{CL}=\frac{2\ln 2}{\pi\tau_{1/2}}\approx\frac{440(\mathrm{MHz})}{\tau_{1/2}(\mathrm{ns})} \tag{8.3-9}$$

This approximation equation is correct for the Gauss pulse. In general, in the case of a long distance, the pulse after transmission may approximate to the Gauss shape. For the non-Gauss shape pulse, Fourier transform should be adopted to show the relationship between $\tau_{1/2}$ and f_{CL}.

8.3.3 The dispersion characteristic of the optical fiber

There are mainly three kinds of origin of dispersion for the multimode optical fiber, the mode dispersion, material dispersion, and waveguide dispersion, of which the mode dispersion is the principal component. It is because of the difference in the transmission coefficients of the different transmission modes; that is, it is because the optical paths traversed by the various transmission modes are different that pulse broadening has been brought about.

The single mode optical fiber has only one transmission mode HE_1, so there is no mode dispersion. The dispersion of the single mode optical fiber arises mainly from the material dispersion, the waveguide dispersion, and the refractive index distribution dispersion, their meanings being as follows:

(1) The material dispersion: As the refractive index of the optical fiber material is not a constant with respect to the optical frequency, the propagation velocity of optical energy in the optical fiber differs from optical frequency to optical frequency. For a signal with a rather wide spectral line, after propagation, pulse broadening will occur. This is called material dispersion.

(2) The waveguide dispersion: Suppose an optical pulse with a definite spectral line width propagates in an ideal single mode optical fiber whose refractive index does not vary with the optical frequency. The result is still a phenomenon of pulse broadening. The reason is because, when the optical frequency changes, the propagation coefficient of the transmission mode changes with it bringing about dispersion, called waveguide dispersion.

(3) The refractive index distribution dispersion: The refractive index difference between the fiber core and cladding is defined by the following equation:

$$\varDelta = \frac{n_1^2(\lambda) - n_2^2(\lambda)}{2n_2^2(\lambda)} = \frac{n_1(\lambda) - n_2(\lambda)}{n_2(\lambda)} \tag{8.3-10}$$

Here suppose the refractive index varies with the optical wavelength. The dispersion that occurs when $\varDelta$ varies with the optical frequency is called the refractive index distribution dispersion. In general, the refractive index material for the fiber core and cladding varies with the optical frequency at an approximately identical ratio. So $\varDelta$ generally remains approximately unchanged. Usually the refractive index distribution dispersion is very small and can be neglected.

Consider a certain single mode component that makes up the optical pulse. Suppose it propagates in the waveguide with a group velocity $v_g = d\omega/d\beta$. Thus, the time needed for transmission over a distance L is

$$\tau_g = \frac{L}{v_g} = L\frac{d\beta}{d\omega} \tag{8.3-11}$$

τ_g is called the group delay time. The delay time over unit length is

$$\tau = \frac{1}{v_g} = \frac{d\beta}{d\omega} = \frac{1}{c}\frac{d\beta}{dk_0} \qquad \left(k_0 = \frac{\omega}{c}\right) \tag{8.3-12}$$

τ is called the specific group delay. Since the signal modulating bandwidth is much smaller than the optical carrier frequency, the group velocity of the optical pulse can be expressed by the group velocity at its carrier frequency. The propagation constant β of the guide mode is

$$\beta^2 = k_0^2[b(n_1^2 - n_2^2) + n_2^2] \tag{8.3-13}$$

Thus

$$\frac{d\beta}{dk_0} = \frac{\frac{1}{2}k_0(n_1^2 - n_2^2)\frac{db}{dk_0} + (n_1N_1 - n_2N_2)b + n_2N_2}{\sqrt{b(n_1^2 - n_2^2) + n_2^2}} \tag{8.3-14a}$$

where

$$N_1 = \frac{d(k_0n_1)}{dk_0}, \qquad N_2 = \frac{d(k_0n_2)}{dk_0} \tag{8.3-14b}$$

are group refractive indices. Using the normalized frequency $V = k_0a(n_1^2 - n_2^2)^{1/2}$, we write db/dk_0 as

$$\frac{db}{dk_0} = \frac{d}{dV}\left(\frac{dV}{dk_0}\right) = \frac{a}{\sqrt{n_1^2 - n_2^2}}(n_1N_1 - n_2N_2)\frac{db}{dV} \tag{8.3-15}$$

which, substituted into Eq. (8.3-4), under the weak guide condition, $n_1 \approx n_2$, yields

$$\frac{d\beta}{dk_0} \approx \left(b + \frac{V}{2}\frac{db}{dV}\right)(N_1 - N_2) + N_2 \approx \frac{d(bV)}{dV}(N_1 - N_2) + N_2 \tag{8.3-16}$$

The latter equality uses the approximation condition, that is, when db/dV is rather small, $b + (V/2)(db/dV) \approx d(bV)/dV$. Thus, the Eq. (8.3-12) of the specific group delay is written as

$$\tau \approx \frac{1}{c}\left[(N_1 - N_2)\frac{d(bV)}{dV} + N_2\right] \tag{8.3-17}$$

It can be seen that the specific group delay is a function of the frequency. Thus, the group delay difference or the pulse width generated by the various frequency components of the single mode optical pulse is

$$\Delta\tau_g = L \cdot \Delta\tau = L \cdot \frac{d\tau}{d\omega}\Delta\omega = L\frac{d^2\beta}{d\omega^2}\Delta\omega \tag{8.3-18}$$

or, expressed by the wavelength λ_0,

$$\Delta\tau = \frac{d\tau}{d\lambda_0}\Delta\lambda_0 \tag{8.3-19}$$

where $d\tau/d\omega$ (or $d\tau/d\lambda_0$) is exactly the optical fiber dispersion d in Eq. (8.3-3). The spectral width of most actual optical sources (such as the luminous diode or laser diode) is far greater than the Fourier spectral width of the signal pulse. Now the $\Delta\omega$ (or $\Delta\lambda_0$) in Eq. (8.3-18) or Eq. (8.3-19) should be replaced by the optical source spectral width. Using Eq. (8.3-17), considering that the refractive index dispersion characteristic of the core is not much different from that of the cladding, we take $dn_1/dk_0 \approx dn_2/dk_0$ and obtain

$$\begin{aligned}\Delta\tau &\approx \frac{1}{c\lambda_0}\left[\lambda_0\frac{dN_2}{d\lambda_0} - (N_1 - N_2)V\frac{d^2(bV)}{dV^2}\right]\Delta\lambda_0 \\ &\approx \frac{1}{c\lambda_0}\left[\lambda_0\frac{dN_2}{d\lambda_0} - (n_1 - n_2)V\frac{d^2(bV)}{dV^2}\right]\Delta\lambda_0\end{aligned} \tag{8.3-20}$$

where the first term is material dispersion, and the second is the waveguide dispersion (also called the in-mode dispersion). The corresponding pulse broadening is denoted by $\Delta\tau_n$ and $\Delta\tau_W$. For the multiple mode optical fiber, the group delays of the modes are different.

Taking the difference between the maximum and minimum values, we obtain the group delay difference, called the multiple mode group delay discretization. Therefore, the total group delay difference of the optical pulse in the multiple mode waveguide should be

$$\Delta\tau = \Delta\tau_m + \Delta\tau_n + \Delta\tau_W + \Delta\tau_{\text{others}} \tag{8.3-21}$$

where $\Delta\tau_m, \Delta\tau_n$, and $\Delta\tau_W$ represent the contributions by the multiple mode group delay discretization, material dispersion, and waveguide dispersion to pulse broadening, respectively.

In the following an estimate will be made of the magnitude of the three kinds of dispersion.

1. The material dispersion

Under the weak guide condition ($n_1 \approx n_2 \approx n$), and supposing the refractive index dispersion characteristic of the core is not much different from that of the cladding ($\mathrm{d}n_1/\mathrm{d}\lambda_0 \approx \mathrm{d}n_2/\mathrm{d}\lambda_0 \approx \mathrm{d}n/\mathrm{d}\lambda_0$), then the pulse broadening induced by material dispersion will be determined by the first term to the right side of Eq. (8.3-20), i.e.,

$$\Delta\tau_n \approx \frac{1}{c}\lambda_0 \frac{\mathrm{d}^2 n}{\mathrm{d}\lambda^2}\Delta\lambda_0 \tag{8.3-22}$$

It can be seen from classical optics that the dispersion characteristic of material can be written as

$$n^2 - 1 = \sum_i \frac{B_i \lambda_{0i}^2}{\lambda_0^2 - \lambda_{0i}^2} \tag{8.3-23}$$

where λ_0 is the optical wavelength, λ_{0i} is the inherent resonance wavelength of the vibrator, and B_i is material constant related to resonance wavelength. So we have

$$\frac{\mathrm{d}n}{\mathrm{d}\lambda_0} = \frac{1}{2n}\frac{\mathrm{d}(n^2-1)}{\mathrm{d}\lambda_0} = -\frac{\lambda_0}{n}\sum_i \frac{\lambda_{0i}^2 B_i}{(\lambda_0^2 - \lambda_{0i}^2)^2} \tag{8.3-24}$$

$$\frac{\mathrm{d}^2 n}{\mathrm{d}\lambda_0^2} = \frac{1}{n}\sum_i \frac{\lambda_{0i}^2(3\lambda_0^2 + \lambda_{0i}^2)B_i}{(\lambda_0^2 - \lambda_{0i}^2)^3} - \frac{1}{n}\left(\frac{\mathrm{d}n}{\mathrm{d}\lambda_0}\right)^2 \tag{8.3-25}$$

which, substituted into Eq. (8.3-22), gives

$$\Delta\tau_n \approx -\frac{\lambda_0}{cn}\left[\sum_i \frac{\lambda_{0i}^2(3\lambda_0^2 + \lambda_{0i}^2)B_i}{(\lambda_0^2 - \lambda_{0i}^2)^3} - \left(\frac{\mathrm{d}n}{\mathrm{d}\lambda_0}\right)^2\right]\Delta\lambda_0 \tag{8.3-26}$$

Take fused quart as an example. Adopting two models of resonance frequency, $\lambda_{01} = 0.1$ μm, $\lambda_{02} = 0.9$ μm, $B_1 = 1.0995$, $B_2 = 0.9$, $n = 1.4623$ ($\lambda_0 = 0.5$ μm), we then obtain from Eq. (8.3-23):

$$n^2 - 1 = \frac{1.0955\lambda_0^2}{\lambda_0^2 - (0.1)^2} + \frac{0.9\lambda_0^2}{\lambda_0^2 - (0.9)^2} \tag{8.3-27}$$

The dispersion characteristic calculated from Eq. (8.3-27) is as shown in Fig. 8.3-6. The curve $\Delta\tau/\Delta\lambda_0 - \lambda_0$ intersects the zero axis at $\lambda_0 = 1.27$ μm. This is an important characteristic of the material SiO_2. At this wavelength ($\lambda_0 = 1.27$ μm) of the wave, the optical fiber's transmission bandwidth is very great. Meanwhile, at the IR wavelength of zero material dispersion, the Raleigh scattering loss is already lower than 1 dB/km and the opto-electric detector can still operate. This model also shows that, if it is desired that the material dispersion should be zero, the material should at least have two resonance frequencies (one in the IR and the other in the ultraviolet band). In the optically guided fiber, doped SiO_2

is frequently used and the dispersion is slightly changed. But the zero dispersion point is still in the vicinity of $\lambda_0 = 1.3$ μm. If a GaAs semiconductor laser is adopted, the operation wavelength $\lambda_0 \approx 0.8$ μm, $\Delta\lambda_0 \approx 2 \times 10^{-3}$ mm(20 Å). It can be seen from Fig. 8.3-6 that $\Delta\tau_n \approx 0.3$ ns/km.

2. The waveguide dispersion

By considering the second term to the right of Eq. (8.3-20), the contribution of the wave guide dispersion to the pulse broadening is obtained:

$$\Delta\tau_W \approx -\frac{1}{c\lambda_0}(n_1 - n_2)V\frac{\mathrm{d}^2(bV)}{\mathrm{d}V^2}\Delta\lambda_0 \tag{8.3-28}$$

where the relation of $V(\mathrm{d}^2(\mathrm{bV})/\mathrm{d}V^2)$ with V is shown in Fig. 8.3-7. For the HE_{11} mode, take $V = 2 \sim 2.4$, $V(\mathrm{d}^2(bV)/\mathrm{d}V^2) \approx 0.2 \sim 0.1$, $(n_1 - n_2) \sim 0.015$, and the light source spectral width $\Delta\lambda_0/\lambda_0 \sim 0.0025$. From Eq. (8.3-38) it has been found by calculation that $\Delta\tau_W \approx 25$ ps/km. It can be seen that its influence is much smaller than that of the material dispersion.

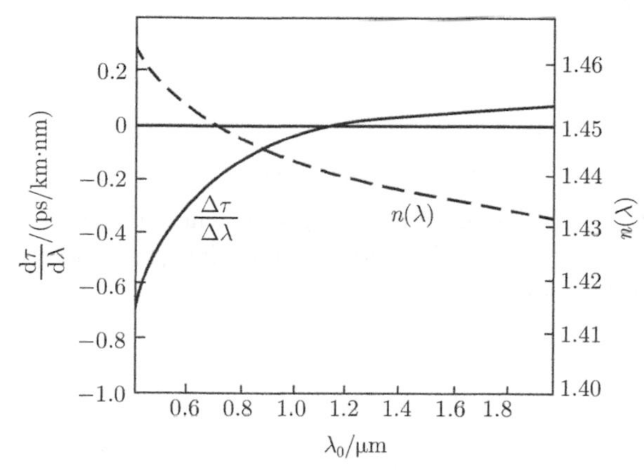

Fig. 8.3-6 The material dispersion characteristic of fused quartz

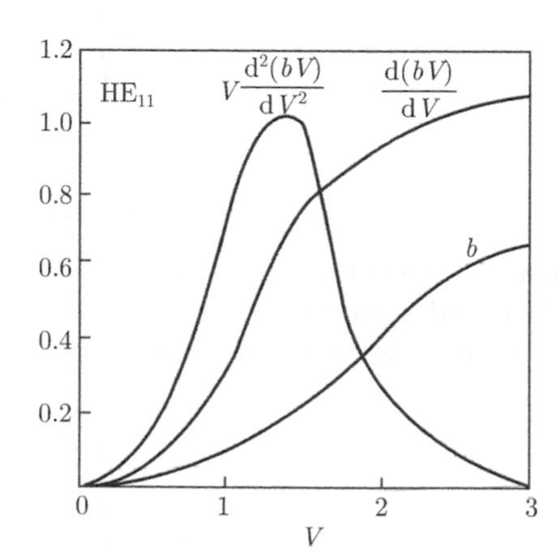

Fig. 8.3-7 The waveguide dispersion of single mode optical fiber

Fig. 8.3-8 is the dispersion characteristic of the single mode optical fiber plotted according to Eq. (8.3-30), in which the dashed line represents the pure material dispersion while the solid line represents the dispersion of the HE_{11} mode with three different core diameters a taken. When $V = 1.8$, the three curves intersect at the zero axis, respectively. It can be seen from the figure that, when a is rather large ($a > 5$ μm), the curves almost overlap the material dispersion, showing that the waveguide dispersion has become very small. When a rather small core diameter is taken, the zero dispersion point can be made to shift to $\lambda_0 = 1.55$ μm, which is exactly the least loss point of the Ge-doped fused quartz optical fiber.

3. The multiple mode group delay discretization

The afore-mentioned two kinds of dispersion are mainly concerned with the single mode optical fiber, in which the material dispersion plays the principal role. In the multiple mode optical fiber, although the delay of every mode is also affected by the material dispersion and waveguide dispersion, it's mainly influenced by the group delay between the modes.

Since the step optical fiber is but a special case in the gradual varying optical fiber, in order that the results of analysis will be of universal significance, we proceed from the group dispersion of the gradually varying optical fiber and obtain

$$\tau = \frac{N_1}{c}\left\{1 + \Delta\frac{a-2}{a+2}\left[\frac{M(\beta)}{M}\right]^{\frac{a}{a+2}} + \Delta^2\frac{3a-2}{2(a+2)}\left[\frac{M(\beta)}{M}\right]^{\frac{2a}{a+2}}\right\} \tag{8.3-29}$$

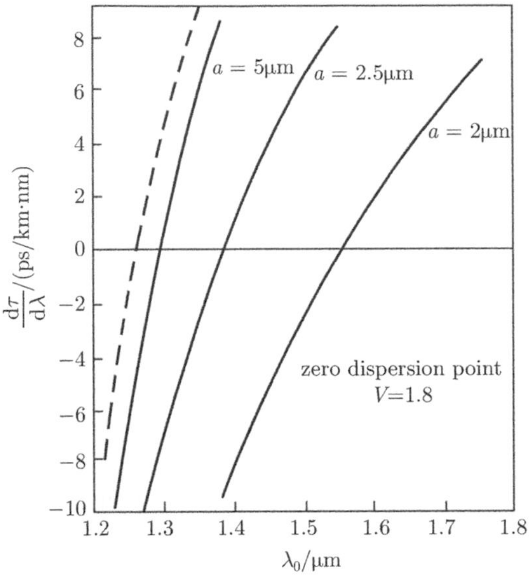

Fig. 8.3-8 The dispersion characteristic of single mode optical fiber

For the lowest order mode, obviously $M(\beta)/M \rightarrow 0$, thus $\tau \approx N_1/c$, whereas for the highest order mode, $M(\beta)/M \rightarrow 1$. From Eq. (8.3-29), it is found that the group delay difference between the highest and lowest order mode is

$$\Delta\tau = \begin{cases} \dfrac{N_1}{c}\dfrac{\Delta^2}{2} & (\alpha = 2) \\ \dfrac{N_1}{c}\left(\dfrac{\alpha-2}{\alpha+2}\right)\Delta & (\alpha \neq 2) \\ \dfrac{N_1}{c}\Delta & (\alpha \rightarrow \infty) \end{cases} \tag{8.3-30}$$

This equation shows that, when $\alpha = 2$, $\Delta\tau > 0$, making it clear that the higher order mode lags behind while at $\alpha < 2$, $\Delta\tau < 0$, the higher order mode is in the lead.

The above discussion shows that the magnitude of the specific group delay is closely related to the distribution of the index of refraction (α value). In order to minimize the group delay difference, the optimum distribution of index of refraction should make α approach 2 but not equal to 2. Now, neither the second nor the third term in Eq. (8.3-29) can be neglected, the corresponding group delay difference being

$$\Delta\tau = \frac{N_1}{c}\left[\Delta\frac{\alpha-2}{\alpha+2} + \Delta^2\frac{2\alpha-2}{2(\alpha+2)}\right] \tag{8.3-31}$$

If we let $\Delta\tau = 0$, solving the above equation, we obtain

$$\alpha \approx 2(1-\Delta) \tag{8.3-32}$$

That is, when the distribution of the index of refraction satisfies this condition, the highest order mode and the lowest order mode will be in synchronization. But the intermediate modes between the two are not necessarily in synchronization. It's not difficult to prove that, when $[M(\beta)/M]^{1/2} = [(\alpha-2)/(2-3\alpha)\Delta]^{(\alpha+2)/\alpha}$, the maximum group delay difference generated by these intermediate modes is

$$\Delta\tau_{\rm m} \approx -\frac{N_1}{c}\frac{\Delta^2}{8} \tag{8.3-33}$$

If the material dispersion is neglected, take $N_1 \approx n_1$. For the quartz optical fiber $n_1 = 1.5$, Δ is usually of the 0.01 order. From Eq. (8.3-30) the pulse broadening $\Delta\tau_{\rm m}$ due to the intramode group delay discretization in the multiple mode optical fiber is found to be about 50 ns/km, which is greater by one to two orders of magnitude than that induced by the material dispersion. If the square law gradual varying optical fiber is adopted, the pulse broadening will be reduced to $\Delta/2$ of the step optical fiber, or smaller by about two orders of magnitude, while when $\alpha = 2(1-\Delta)$, the pulse broadening will be 1/4 that of the square law gradually varying optical fiber. It is clear that it is of extremely great significance to adopt the gradually varying multiple mode optical fiber for increasing the transmission bandwidth of the optical fiber.

4. The refractive index distribution of single mode optical fiber and zero dispersion wavelength shift

In order to make single mode optical fibers with satisfactory performance, structures of different kinds of refractive index distributions have been proposed. The satisfactory performance as is here discussed mainly consists in:

(1) the core diameter should be as large as possible to facilitate coupling connection;

(2) the loss due to structure should be small, especially the bending loss;

(3) the optical fiber's internal stress should be small so as to protect the transmission performance and mechanical strength from being affected;

(4) the manufacturing technology is easy and production convenient;

(5) possessing excellent dispersion characteristics, in particular, the zero dispersion wavelength is able to be at small attenuation wavelength (1.55 μm).

The fundamental requirement on the single mode optical fiber is the guarantee of single mode transmission. Under the condition of ensuring single mode transmission, the core radius a should be increased as much as possible. According to the condition of single mode transmission,

$$V = \frac{2\pi}{\kappa} a n_1 \sqrt{\Delta} \leqslant 2.4$$

it is known that if too large an a is chosen, the refractive index difference Δ will become small while too small a Δ will lead to an increase in the bending loss. In general the core radius is chosen as $a = 4\sim8$ μm and Δ is chosen as 0.5% or thereabouts, the structure being as shown in Fig. 8.3-9(a), (b), and (c).

The structure of Fig. 8.3-9(a): The cladding is pure quartz. Too thick a pure quartz cladding has too high a melting point that makes smelting and wire drawing difficult. Usually germanium (Ge) is doped to increase the refractive index of the fiber core. If too much germanium is doped, the scattering loss will be increased. So phosphorus (P) can be used as a substitute that can also enhance the refractive index. But P has water absorbability that will shorten the lifetime of the optical fiber while slightly increasing attenuation at the wavelength of 1.55 μm.

For the structure shown in Fig. 8.3-9(b), as considerable Ge is doped, the attenuation is a little higher than usual.

Shown in Fig. 8.3-9(c) is the double clad concave type optical fiber with a fairly high performance. Its inner cladding is doped with fluorine (F) that makes the refractive index difference Δ^- a negative value. With a rather small amount of Ge doped in the fiber core,

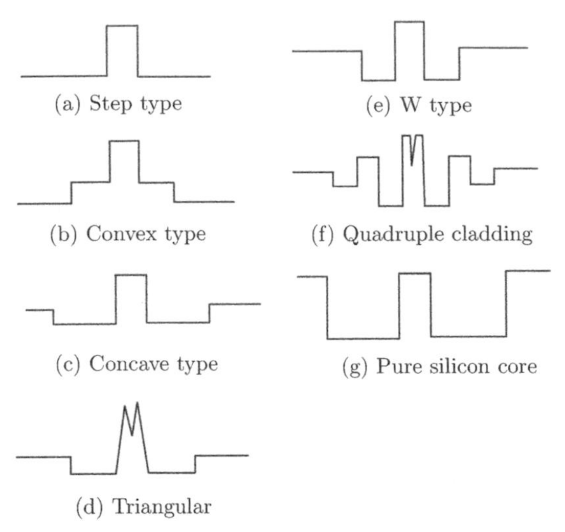

Fig. 8.3-9 The refractive index distribution of single mode optical fiber

a fairly great refractive index difference Δ^+ can be obtained. The total refraction difference $\Delta = \Delta^- + \Delta^+$, which basically satisfies the performance requirements 1~4. Currently, most factories produce single mode optical fibers with such a structure. The refractive index distribution of typical concave type single mode optical fibers is as shown in Fig. 8.3-10.

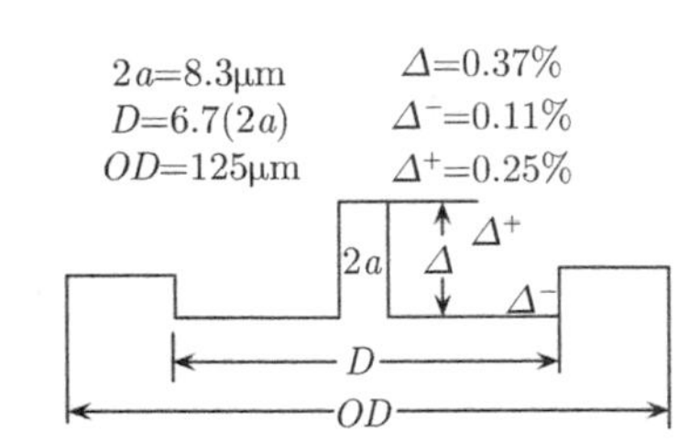

Fig. 8.3-10 The refractive index distribution of concave type single mode optical fiber

The above-mentioned optical fibers have zero dispersion at 1.3 μm or so, but attenuation there is not minimum. It is at 1.55 μm that attenuation is minimum. In order to realize long distance, large capacity fiber-optic communication, effort should be made to make zero dispersion wavelength displacement optical fibers.

The type W optical fiber shown in Fig. 8.3-9(e) can make the zero dispersion shift to where the wavelength is 1.55 μm, mainly utilizing the negative waveguide dispersion formed by the refractive index distribution to offset the positive material dispersion. Controlling the refractive index distribution can make the type W optical fiber possess two zero dispersion wavelengths (1.4 μm and 1.7 μm), while possessing a dispersion smaller than 2 ps/(km·nm) within the range $\lambda = 1.2\sim2.9$ μm, as shown in Fig. 8.3-11. The loss of the type W optical fiber is rather great, especially the bending loss. During zero dispersion wavelength displacement, it is necessary to ensure that attenuation will not increase or increase very little at wavelength 1.55 μm. Otherwise, nothing done can make up the loss.

For the triangular refractive index distribution optical fiber shown in Fig. 8.3-9(d), if its zero dispersion wavelength is shifted to 1.55 μm, attenuation will increase very little. At wavelength 1.55 μm, attenuation is merely 0.23 dB/km as shown in Fig. 8.3-12. Such optical fibers are convenient to manufacture and have yielded good results in actual use.

Figure 8.3-9(f) shows a quadruple cladding single mode optical fiber (QC), which possesses low dispersion within a large range and is zero dispersion at wavelengths 1.3 μm and 1.55 μm. But the production technology for such optical fibers makes rather high requirements and connection is difficult. In particular, it is inadvisable to use the method of electric arc fusion, but connection by gluing can be used. Its attenuation at $\lambda = 1.55$ μm is 0.4 dB/km, and is not used in practice as yet.

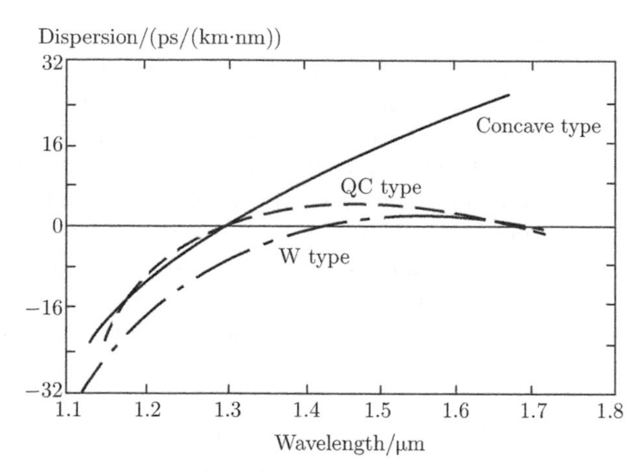

Fig. 8.3-11 The dispersion of several kinds of single mode optical fibers

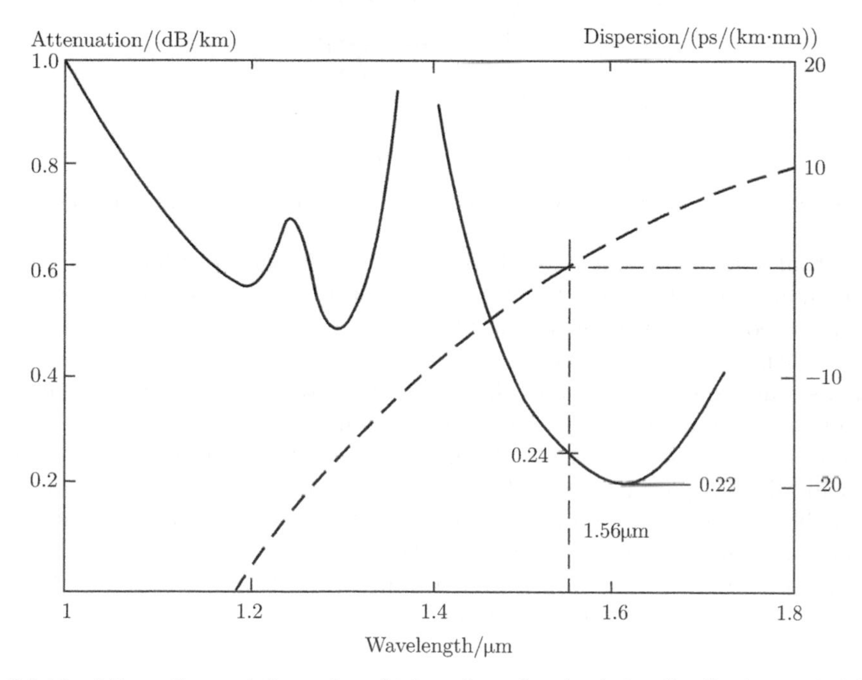

Fig. 8.3-12 Attenuation and dispersion of triangular refractive index distribution optical fiber

Figure 8.3-9(g) shows a pure silicon core optical fiber; that is, the fiber core is pure quartz SiO_2 without any dopant, but the cladding is doped with fluorine to make refraction decrease so as to constitute the refractive index difference Δ. The attenuation of such an optical fiber at 1.6 μm is as low as 0.16 dB/km, at present the optical fiber with the lowest attenuation. Its dispersion is similar to that of ordinary single mode optical fibers, with a zero dispersion wavelength of $\lambda_0 = 1.27$ μm.

8.4 Polarization and birefringence of the single mode optical fiber

8.4.1 The polarization characteristics of single mode optical fibers

The mode of an ideal single mode optical fiber is mode HE_{11}, which is linearly polarized and the polarizing direction is the radial direction of the optical fiber. After setting up an $x-y$ Cartesian coordinate on the optical fiber's cross section, the polarization in an arbitrary radial direction can be expressed by two independent polarizing components HE_{11}^x and HE_{11}^y. Under ideal conditions, the propagation constants of the two polarizing components are equal, that is, $\Delta\beta = \beta_x - \beta_y = 0$. During the propagation, the two components keep in phase from beginning to end and synthesize into the original radial polarizing state from beginning to end. That is to say, the HE_{11}^x and HE_{11}^y modes are degenerate, as shown in Fig. 8.4-1. Since the two modes are independent, they do not influence each other. For instance, when stimulating the HE_{11}^x mode only along the x-axis on the end face of the optical fiber, no HE_{11}^y mode will emerge in the optical fiber, and vice versa. If stimulating the HE_{11} mode along the direction between the x- and y-axes, then there will exist the HE_{11}^x and HE_{11}^y modes in the optical fiber from beginning to end, their amplitude value ratio remaining unchanged along the optical fiber.

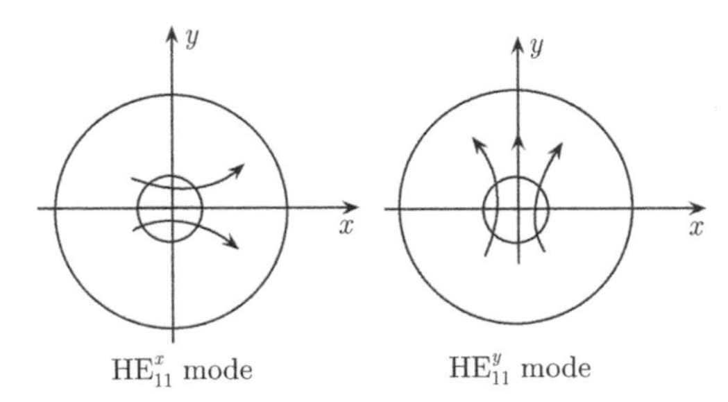

Fig. 8.4-1 Two polarizing states of fundamental mode in single mode optical fiber

For any actual optical fibers, it's very hard for the ideal conditions (circular cross section, perfectly straight, and defect-free) to exist. To facilitate explanation of the point we shall study a homogeneous and straight optical fiber with an elliptical cross section as shown in Fig. 8.4-2. In such a case, there will be two priority polarizing directions. One is the elliptical cross section's long axial direction and the other, the short axial direction, represented by the a-axis and b-axis in the figure, respectively. Obviously, apart from the linearly polarized light injected along the x-or y-axis, the linearly polarized light injected from other angular directions will all propagate in the HE_{11}^x and HE_{11}^y modes. Since $a \neq b$, the propagation constants of the two modes are no longer the same. In the process of propagation along z, they no longer remain in phase, thereby generating the variation of the polarizing state of the synthetic mode, that is, $\Delta\beta = \beta_x - \beta_y \neq 0$, generating the polarization evolution of the synthetic mode and making the degenerate state vanish.

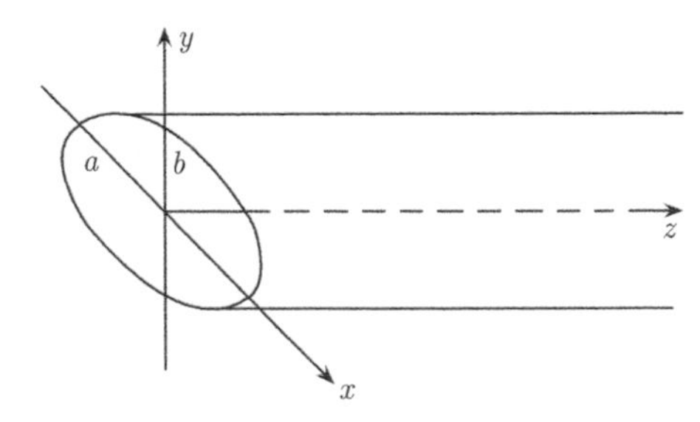

Fig. 8.4-2 The elliptical cross section optical fiber

It is known to all that the polarizing state synthesized by two orthogonal components is determined by their phase difference. HE_{11}^x and HE_{11}^y are exactly two orthogonal components. Clearly, the polarizing state of the synthetic mode (combined vibration) is determined by the propagation phase

$$\varphi = \Delta\beta \times l \tag{8.4-1}$$

When $\varphi = 0$, it is linearly polarized light while when $\varphi = \pi/2$ and the amplitudes of the two components are equal, it is circularly polarized light. When $\varphi = \pi$, it becomes linearly polarized light, but the polarizing direction has turned an angle of $\pi/2$; when $\varphi = 3\pi/2$, it becomes elliptically polarized light in the opposite direction. When $\varphi = 2\pi$, it will be restored to the original linearly polarized state, as shown in Fig. 8.4-3(a)-(e). When $\Delta\beta$ remains unchanged along the optical fiber, or under the homogeneous birefringence condition, the above-mentioned process of polarization evolution will repeat again and again periodically. Evidently, this period reflects the intrinsic characteristics of the elliptical cross-section optical fiber.

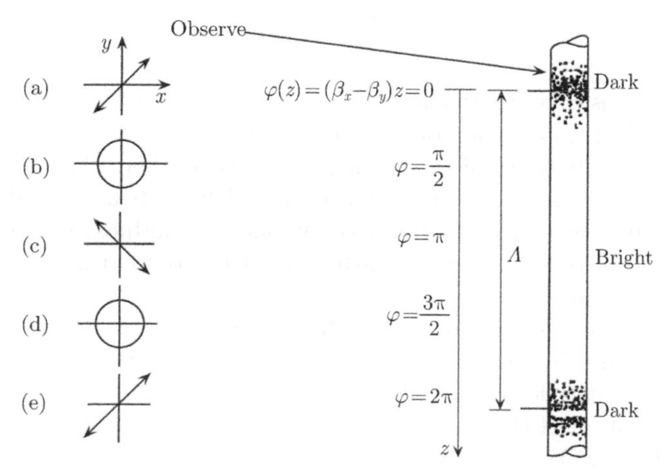

Fig. 8.4-3 The phenomenon of birefringence of polarized light in optical fiber

Within a repeated period of evolution of the polarization state, the distance Λ traversed by mode transmission is defined as the beat length of the single mode optical fiber, i.e.,

$$\Lambda \equiv 2\pi/\Delta\beta \tag{8.4-2}$$

It can be seen from the equation that the smaller $\Delta\beta$ is, the greater Λ will be. If the light intensity in the optical fiber is sufficiently strong, so strong that even in a dark room the scattered light of the fiber core-cladding can be seen with an unaided eye, then it is possible to observe the obvious periodic variation of the light intensity of this kind of scattered light along the length of the optical fiber, as shown in the graph to the right of Fig. 8.4-3. When $\varphi = 0$ or 2π, it is a dark region; when $\varphi = \pi$, it's a bright region and the distance between dark regions is the beat length Λ. The radiation graph of the electric dipole in the electromagnetic theory can account for this phenomenon. In the direction of polarization (equivalent to the direction of the electric vibration dipole), radiation is minimum while in the direction perpendicular to the direction of polarization, the radiation is strongest. So, it is darkest at $\varphi = 0$ or 2π while brightest at $\varphi = \pi$.

The above is the case of the evolution of polarization of single mode optical fiber when it is confirmed that $\Delta\beta$ is a constant. In reality, the problem is much more complicated. In particular, influenced by random perturbation factors, $\Delta\beta$ will vary randomly, thereby making the mode polarization experience random variation. In the following we shall discuss the causes of the generation of random variation of $\Delta\beta$. Then the method for controlling $\Delta\beta$ and the concept of special optical fibers will be given.

8.4.2 The birefringence of single mode optical fibers

Optical fiber polarization and birefringence of light are two kinds of manifestation of anisotropy in medium optics. Briefly, the birefringence of the single mode optical fiber refers to the phenomenon of difference emerging in the propagation constants of two originally degenerate modes. There are many methods of expressing birefringence, the simplest being the expression by means of $\Delta\beta$ or parameter B (normalized birefringence coefficient), i.e.,

$$B \equiv \frac{\Delta\beta}{\beta_{\text{av}}} = \frac{n_x - n_y}{n_{\text{av}}} \tag{8.4-3}$$

where n_{av} represents the average refractive index of the single mode optical fiber. If B is used to denote the beat length Δ, then

$$\Lambda = \frac{2\pi}{\Delta\beta} = \frac{2\pi}{\beta_{\text{av}} B} \tag{8.4-4}$$

Usually the beat length is between 10 cm~2 m, which makes actual application very difficult.

The main factors of the generation of birefringence are:

(1) The cross section ellipticity of the single mode optical fiber. Both the manufacturing technology and the two symmetric ends of the optical fiber being acted on by pressure will make the optical fiber's cross section exhibit out-of-roundness or ellipticity. Such a geometric distortion can be equivalent to distortion of the optical fiber's refractive index, thereby generating birefringence.

(2) The bending of the single mode optical fiber. All three effects upon bending of the optical fiber will generate birefringence: ① bending of the axial line; ② bending induces fiber core ellipticity; ③ the formation of the internal stress inside the optical fiber causes strain, leading to variation of the refractive index, or what is called the stress birefringence generated by the photoelastic effect.

(3) The torsion of the single mode optical fiber. The action of torsion not only causes geometric deformation of the optical fiber, but also generates the photoelastic effect.

(4) The electro-(magneto)-optic effect of the single mode optical fiber. Under the action of a strong electric (or magnetic) field, owing to the electro-(magneto)-optic effect, birefringence will also be generated.

It can be seen from the above discussion that there are many causes of the generation of birefringence. This is both helpful for using optical fibers in sensitive components but also harmful. For example, in the phase interferometer type sensor, it is desired that polarization will remain unchanged while not hoping for the generation of the birefringence effect. In a polarization type sensor, however, it is hoped that the birefringence effect will be apparent while not hoping that the birefringence in the optical fiber will generate disturbance. Hence the emergence of the polarization type single mode fiber-optic technology.

8.4.3 The polarization type single mode optical fiber

As is described above, in the fiber-optic sensing technology, according to the difference in the principle of an application, different requirements have been made on the polarization characteristics of the single mode optical fibers. Below we shall describe several polarization type single mode optical fibers with different characteristics, which will be of great use for choosing a proper optical fiber in application.

1. Low birefringence single mode optical fibers

The imperfection in the optical fiber manufacturing technology has brought about the ellipticity and the inner stress which cause the birefringence effect of the optical fiber. It's clear that if effort is made to improve the technological level so that both the fiber core ellipticity and inner stress are minimized, then both the stress birefringence and geometric birefringence will be reduced to a very low level and the corresponding beat length will reach above 100 meters. Such an optical fiber is called the low birefringence optical fiber. At present, two kinds of low birefringence optical fibers are available, the ideal circularly symmetric optical fiber and rotational optical fiber.

(1) The ideal circularly symmetric optical fiber. Research shows that the thermal expansion coefficient of ordinary optical fiber cladding is greater than that of the fiber core. What with the ellipticity of the core, the stresses in the directions of the two transverse coordinates on the cross section are different, thus generating stress (photoelastic effect) birefringence, which is much greater than the geometric birefringence of the degree of ellipticity. Based on such an understanding, people adopt the cladding and fiber core material with equal thermal expansion coefficients and strictly control the fiber core ellipticity in manufacture while making the overall mechanical performance of the optical fiber identical and with uniform and equal stresses. For example, material such as GeO_2-SiO_2 is used for the core while B_2O_3-SiO_2 is used for the cladding, their thermal expansion coefficients are equal. The relative refractive index difference is 0.34%. At the wavelength of 0.63 μm, the polarizing plane rotation is merely several degrees per meter while the beat length reaches above 100 m. This value is about three orders of magnitude lower than the birefringence of ordinary single mode optical fibers and is suitable for the polarization type fiber-optic sensor.

(2) The rotational optical fiber. This is a new-type optical fiber that still chooses fiber core and cladding materials of equal thermal expansion coefficients to control the base rod in fast rotation so as to make the intrinsic birefringence and stress birefringence of the optical fiber vary periodically along the length direction. In so doing, although the birefringence in each small segment is rather high, there is no birefringence for the entire piece of optical fiber on average, which is close to a piece of homogeneous optical fiber. For instance, a base rod with GeO_2-SiO_2 as the core, B_2O_3-SiO_2 as cladding, at a revolving speed of 1000 r/min during wire-drawing has yielded a rotational optical fiber with a birefringence that is almost zero.

2. High birefringence single mode optical fiber

Here, in sharp contrast to the low birefringence optical fiber, a birefringence as high as possible is needed. Like the low birefringence optical fiber, it's no easy matter to make the

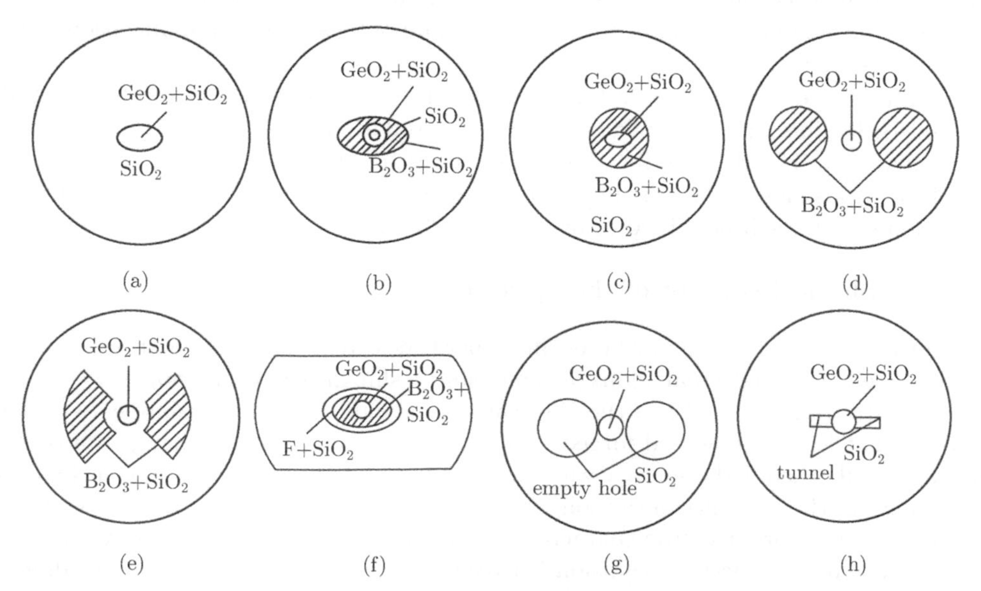

Fig. 8.4-4 Cross sections of several typical high birefringence optical fibers

optical fiber birefringence much higher than the birefringence of the ordinary optical fiber. Usually the B value is required to be in the range of $10^{-3}\sim10^{-4}$ orders of magnitude, which is equivalent to the optical fiber's beat length Λ in the mm order of magnitude. Since the optical fiber's rigidity is sufficient to stand the external interference of the mm order, the birefringence of the optical fiber is enhanced to make its beat length reduce to the mm order of magnitude. Then the coupling action generated by the external perturbation will be very small. Thus, over a rather long distance, the optical fiber will possess the characteristics of the linearly polarizing state transmission mode. As the linearly polarizing state transmission mode does not change in the main, the high birefringence is also known as the polarization maintaining optical fiber, or polarizing kept fiber for short.

Taken as a whole, the high birefringence optical fiber is subdivided into two classes, One class uses the non-axisymmetric structure to generate high birefringence, such as the elliptical core optical fiber, side pit, and side tunnel type optical fibers, as shown in Fig. 8.4-4(a), (c), (g), and (h), and the other uses the anisotropic distribution to generate stress birefringence, such as the elliptical cladding type, the panda type, the necktie type, and the flat cladding type optical fibers, as shown in Fig. 8.4-4(b), (d), (e), and (f).

8.5 The nonlinear effect in the optical fiber—the optical fiber soliton

We know from the nonlinear optical technology in Chapter 7 that, in essence, all media possess optical nonlinearity and the optical fiber is of course no exception. In particular, in the fiber-optic communication system, with the use of high strength lasers and ultralow loss single mode optical fibers, the influence of nonlinear optical effect in optical fibers is becoming increasingly serious. Since the field in the optical fiber is mainly constrained in the fiber core, and the core diameter of the optical fiber is very small, the field intensity in the fiber core is exceedingly high while the low loss of the optical fiber will lead to this kind of high optical intensity being maintained over a very long distance. The two aspects make the nonlinear optical effect in the optical fiber very obvious, further leading to the additional attenuation of the signals in the communication system. The crosstalk between adjacent channels and the physical damage affects the transmission performance of the system, restricting the transmitted optical power and the transmission distance. But this kind of effect can also be utilized to construct many useful signal transmitting and processing devices (e.g., the amplifiers and lasers) modulators and optic soliton communication, totally optical conversion, etc.

The nonlinear effect of the optical fiber includes the stimulated Raman scattering (SRS), stimulated Brillouin scattering (SBS), self-phase modulation (SPM) and the optic soliton, etc. We shall here mainly describe the basic concept of the latter two effects, especially optic soliton transmission, which, in the remote ultra-great capacity fiber-optic network transmission system, will be a kind of promising distortion-free transmission technology.

8.5.1 The physical concept of the optic soliton

The soliton was called the solitary wave in earlier times. In 1834, the British ship-building engineer Russell perceived a marvelous phenomenon: a ship drawn by two horses was making its way forward in a canal. When the ship came to a sudden halt, there appeared a large water peak that left the bow and forged ahead, its shape remaining unchanged during the advance. Russell called it the solitary wave and put forward the KdV equations in 1895 that could account for the existence of the solitary wave. In 1965, Bell Laboratories in the United States obtained high-order soliton characteristics by numerically solving the KdV equations. They also studied the process of collision between two solitary waves moving with different velocities and discovered that, except for the phase, the amplitude, shape, and movement characteristics of the two waves, all remained unchanged, manifested as the character of the particle. In 1976, Lamb proved in theory that self-induced transparency is a typical optic soliton phenomenon. People in 1980 proved experimentally that short pulses can also propagate in the optical fiber in the form of solitons, which they chose to call fiber-optic solitons.

Put more simply, the optic soliton is a particular form of propagation of energy or material, a kind of special electromagnetic wave capable of long–distance distortion-free transmission, each maintaining its independence after colliding with others.

The mechanism of the generation of the optic soliton in the optical fiber is a result of the constraint by two factors, the dispersion of the optical fiber and the optical fiber's self-phase modulation (SPM). The dispersion effect makes a pulse waveform scatter. The propagating velocities of the components at different frequencies of the waveform are different while the nonlinear effect of the optical fiber makes the front edge of the pulse become slow and the rear edge become quick. With the two constraining each other, it becomes possible for the pulse to propagate with its waveform remaining unchanged, forming the optic soliton.

Now consider the propagation of an optical pulse in an optical fiber, as shown in Fig. 8.5-1(a). Because of the nonlinear optical Kerr effect, the refractive index of the optical fiber can be written as

$$n = n_1 + n_2 I \tag{8.5-1}$$

where I is the light intensity, and n_2 is about $3.2 \times 10^{-16}\text{cm}^2/\text{W}$. Although the n_2 value is very small, as the propagation of the optical wave in the optical fiber is confined in the fiber core with an extremely small cross section, the optical field is considerably strong. Furthermore, as the optical fiber is very long, after propagating distance L, the nonlinear optical effect is generated, which is called the self-phase modulation of the optical wave in the optical fiber, the additional phase displacement being

$$\Delta\varphi = \frac{2\pi}{\lambda_0} n_2 I L \tag{8.5-2}$$

It can be seen that, for pulse components of different intensities, their phase velocities are different, and phase deviations are different. It's clear the red shift-frequency in the pulse leading edge part is reduced, but the blue shift-frequency in the pulse trailing edge part is increased, that is, the frequency domain of the pulse is broadened. Figure 8.5-1(b) shows the frequency deviation that has occurred.

Now let's consider the influence of dispersion. When the optical fiber possesses negative dispersion, $\partial v_g/\partial\lambda_0 < 0$ (i.e., the group velocity increases with the increase of frequency), the rear part of the pulse at a rather high frequency will be in the lead while the front of the pulse at a rather low frequency will lag behind, which will lead to the narrowing of the pulse,

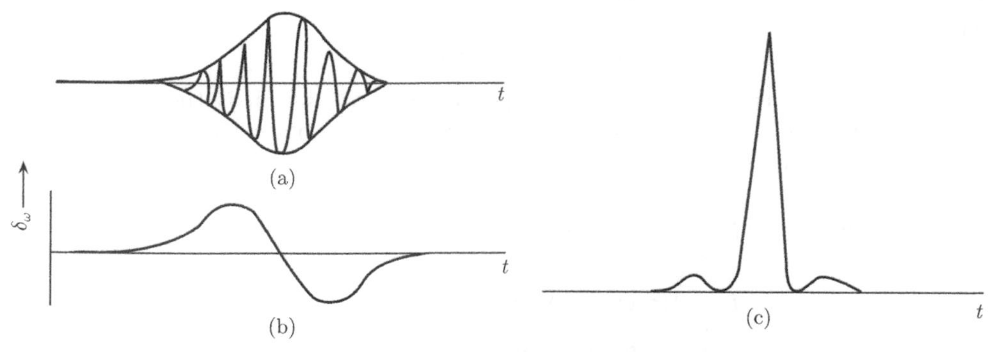

Fig. 8.5-1 The formation of optic soliton in the optical fiber

as shown in Fig. 8.5-1(c). Although the above-mentioned model is able to qualitatively explain the phenomenon of the narrowing of a pulse, it is insufficient to account for the mysterious process of pulse formation and the optic soliton. And it is necessary to write out and solve the nonlinear differential equation for describing the evolution of the pulse envelope shape.

8.5.2 The hyperbolic equation in a dispersion medium

First consider the propagation of the monochromatic plane wave in the single mode optical fiber. There exist the mutually orthogonal linearly polarized modes HE_{11}^{x} and HE_{11}^{y} in the single mode optical fiber, with the property of polarization of the actual transmission mode determined by the input excitation light. Suppose the mode of the input excitation light is polarized along the x direction, whose spatial relation can be written as

$$E_x = a(z)E(x,y) \tag{8.5-3}$$

where $E(x,y)$ is the power normalized eigenmode field distribution. $a(z)$ the phase amplitude coefficient, written as:

$$a(z) = A\mathrm{e}^{-\mathrm{j}\beta(\omega)z} \tag{8.5-4}$$

The propagation constant β is a function of the optical wave frequency ω. Thus, the differential equation for $a(z)$ can be written out:

$$\frac{\partial a}{\partial z} = -\mathrm{j}\beta a \tag{8.5-5}$$

Now consider the propagation of an optical pulse in a single mode optical fiber. Supposing the input excitation light possesses a narrow spectral width, with a central frequency ω_c. Then the transmission constant $\beta(\omega)$ can be expanded in the vicinity of $\omega = \omega_\mathrm{c}$ as

$$\beta(\omega) = \beta(\omega_\mathrm{c}) + \frac{\mathrm{d}\beta}{\mathrm{d}\omega}(\omega - \omega_\mathrm{c}) + \frac{1}{2}\frac{\mathrm{d}^2\beta}{\mathrm{d}\omega^2}(\omega - \omega_\mathrm{c})^2 \tag{8.5-6}$$

Thus Eq. (8.5-5) is written as

$$\frac{\partial a(z,\omega)}{\partial z} = -\mathrm{j}\left[\beta(\omega_\mathrm{c}) + \frac{\mathrm{d}\beta}{\mathrm{d}\omega}(\omega - \omega_\mathrm{c}) + \frac{1}{2}\frac{\mathrm{d}^2\beta}{\mathrm{d}\omega^2}(\omega - \omega_\mathrm{c})^2\right]a(z,\omega) \tag{8.5-7}$$

The space time relation of $a(z,\omega)$ is obtained from Fourier transform as

$$a(z,t) = \int_{-\infty}^{\infty} a(z,\omega)\mathrm{e}^{\mathrm{j}\omega t}\mathrm{d}\omega \tag{8.5-8}$$

If the frequency spectrum of $a(z,\omega)$ is very narrow, it can be conveniently expressed as the function of $(\omega - \omega_\mathrm{c})$ while introducing a quickly varying spatial phase factor $\mathrm{e}^{-\mathrm{j}\beta(\omega_\mathrm{c})z}$, written as

$$a(z,\omega) = A(z,\omega - \omega_\mathrm{c})\mathrm{e}^{-\mathrm{j}\beta(\omega_\mathrm{c})z} \tag{8.5-9}$$

in which $A(z,\omega - \omega_\mathrm{c})$ is the complex envelope of the wave, which is the slowly varying function of z. Taking Fourier transform of Eq. (8.5-9), we obtain the envelope function $A(z,t)$ of the space time relation, i.e.,

$$\begin{aligned} a(z,t) &= \int_{-\infty}^{\infty} a(z,\omega)\mathrm{e}^{\mathrm{j}\omega t}\mathrm{d}\omega \\ &= \int_{-\infty}^{\infty} A(z,\omega - \omega_\mathrm{c})\mathrm{e}^{-\mathrm{j}[\beta(\omega_\mathrm{c})z - \omega_\mathrm{c}t]}\mathrm{e}^{\mathrm{j}(\omega-\omega_\mathrm{c})t}\mathrm{d}(\omega - \omega_\mathrm{c}) \\ &\equiv A(z,t)\mathrm{e}^{-\mathrm{j}[\beta(\omega_\mathrm{c})z - \omega_\mathrm{c}t]} \end{aligned} \tag{8.5-10}$$

By use of the complex envelope $A(z, \omega-\omega_c)$ defined by Eq. (8.5-10), we obtain from Eq. (8.5-7)

$$\frac{\partial}{\partial z}A(z,\omega-\omega_c) = -\mathrm{j}\left[\frac{\mathrm{d}\beta}{\mathrm{d}\omega}(\omega-\omega_c)+\frac{1}{2}\frac{\mathrm{d}^2\beta}{\mathrm{d}\omega^2}(\omega-\omega_c)^2\right]A(z,\omega-\omega_c) \tag{8.5-11}$$

Then by using the Fourier transform relation equation

$$\mathrm{j}(\omega-\omega_c)^n A(z,\omega-\omega_c) = \mathrm{F}\cdot\mathrm{T}\left[\frac{\partial^2}{\partial t^2}A(z,t)\right] \tag{8.5-12}$$

the anti-Fourier transform of Eq. (8.5-11) is obtained:

$$\left(\frac{\partial}{\partial z}+\frac{1}{v_g}\frac{\partial}{\partial t}\right)A(z,t) = \frac{1}{2}\frac{\mathrm{d}^2\beta}{\mathrm{d}\omega^2}\frac{\partial^2 A(z,t)}{\partial t^2} \tag{8.5-13}$$

where $v_g^{-1} = \mathrm{d}\beta/\mathrm{d}\omega$ is the reciprocal of the group velocity. If the term to the right of Eq. (8.5-13) is zero, the envelope $A(z,t)$ will propagate with group velocity v_g without occurrence of distortion. But if this term is not zero (i.e., $\mathrm{d}^2\beta/\mathrm{d}\omega^2 \neq 0$), as it is an imaginary number, the propagation phase of the wave will be affected, which will lead to distortion of the pulse. Equation (8.6-13) is the famous hyperbolic equation, which is the basic equation for the study of the transmission of the optical pulse in a dispersion medium. Based on this equation, the problem of broadening or compression of an ultrashort pulse during propagation can be explained.

8.5.3 The nonlinear Schrödinger equation

Now suppose the dielectric constant of the single mode optical fiber is acted on by a certain perturbation so that the refractive index n_1 of the fiber core becomes $n = n_1 + \Delta n$. Thus the propagation constant of the conducting mode will correspondingly change $\Delta\beta$. Under the far from cut off condition, $\beta \approx k_0 n_1$, thus

$$\beta'(\omega) = \beta(\omega)+\Delta\beta(\omega) \approx k_0(n_1+\Delta n) = \beta(\omega)\left(1+\frac{\Delta n}{n_1}\right) \tag{8.5-14}$$

Then

$$\Delta\beta = \beta(\omega)\frac{\Delta n}{n_1} \tag{8.5-15}$$

Therefore, this term should be added to the expanded equation of the propagation constant. Thus we have

$$\beta = \beta(\omega_c)+(\omega-\omega_c)\frac{\mathrm{d}\beta}{\mathrm{d}\omega}+\frac{1}{2}(\omega-\omega_c)^2\frac{\mathrm{d}^2\beta}{\mathrm{d}\omega^2}+\beta(\omega_c)\frac{\Delta n}{n_1} \tag{8.5-16}$$

Here we replace ω with ω_c, and $\beta(\omega)$ with $\beta(\omega_c)$. Following the derivation of Eqs. (8.5-6)~(8.5-13), we introduce the envelope function $A(z,\omega-\omega_c)$ and use Eq. (8.5-12) for Fourier transform to obtain the nonlinear equation

$$\mathrm{j}\left(\frac{\partial}{\partial z}+\frac{1}{v_g}\frac{\partial}{\partial t}\right)A(z,t) = -\frac{1}{2}\frac{\mathrm{d}^2\beta}{\mathrm{d}\omega^2}\frac{\partial^2 A(z,t)}{\partial t^2}+\beta(\omega_c)\frac{\Delta n}{n_1}A(z,t) \tag{8.5-17}$$

In the above-mentioned process of derivation, it has not been indicated how the perturbation of the refractive index is generated. Now suppose Δn is induced by the optical pulse via the Kerr effect, i.e.,

$$\Delta n = n_2|A(z,t)|^2 \tag{8.5-18}$$

Thus Eq. (8.5-17) becomes

$$\mathrm{j}\left(\frac{\partial}{\partial z}+\frac{1}{v_g}\frac{\partial}{\partial t}\right)A(z,t)=-\frac{1}{2}\frac{\mathrm{d}^2\beta}{\mathrm{d}\omega^2}\frac{\partial^2 A(z,t)}{\partial t^2}+\beta(\omega_c)\frac{n_2}{n_1}\left|A(z,t)\right|^2 A(z,t) \tag{8.5-19}$$

Then, by means of variable replacement

$$\tau=t-\frac{z}{v_g},\ \xi=z,\ K=\beta(\omega_c)\frac{n_2}{n_1} \tag{8.5-20}$$

Thus Eq. (8.5-19) becomes the standard nonlinear Schrödinger equation:

$$\mathrm{j}\frac{\partial}{\partial \xi}-\frac{1}{2}\frac{\mathrm{d}^2\beta}{\mathrm{d}\omega^2}\frac{\partial^2 A}{\partial \tau^2}+K\left|A\right|^2 A=0 \tag{8.5-21}$$

where the second term is the dispersion term and the third is the nonlinear term.

Under the negative dispersion, or $(\mathrm{d}^2\omega/\mathrm{d}\omega^2) < 0$ condition, the allowable solution for Eq. (8.5-21) can be a stable hyperbolic secant type pulse. Since the pulse broadening due to optical fiber dispersion is offset by the variation of the medium's nonlinear refractive index, this pulse will not change its shape in the process of propagation, that is, it has formed an optic soliton, as shown in Fig. 8.5-2(a). Figure 8.5-2(b) and (c) are several higher-order solutions of Eq. (8.5-32) found with a computer. In order to obtain the first high-order soliton, the input pulse amplitude must be increased to twice the corresponding input amplitude of the fundamental soliton (or the power increase is fourfold); the optical pulse shape varies periodically in the course of propagation, with the pulse width compressed to minimum at the half-period and its original shape restored after a full period. For the next higher-order soliton, it is required that the input pulse have its amplitude increased threefold (or the optical power increased ninefold). The optimum pulse narrowing by compression

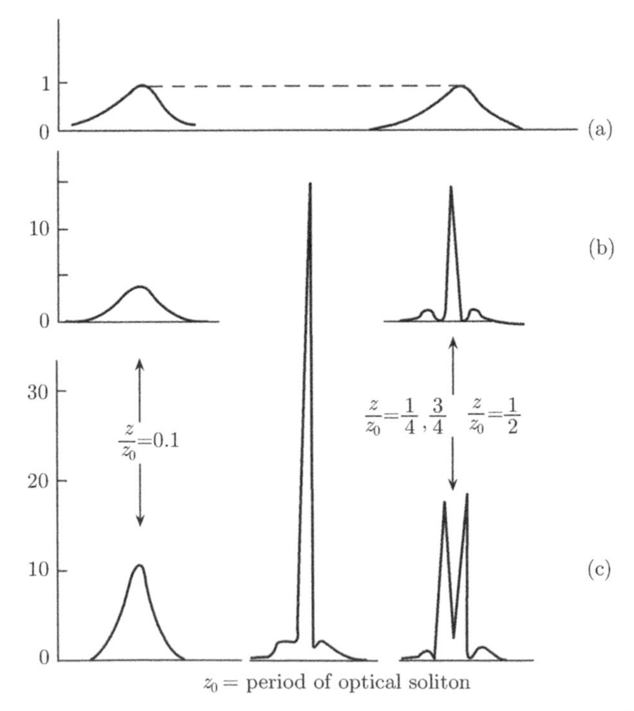

Fig. 8.5-2 Optical soliton in an optical fiber (computer solved)

occurs probably at 1/4 period and splits into two equal pulses. The case of still higher-order solitons would be even more complicated.

Under the positive dispersion or $(\mathrm{d}^2\beta/\mathrm{d}\omega^2) > 0$ condition, the allowable solution for Eq. (8.2-51) can be a stable hyperbolic tangent type pulse, called the dark soliton, "dark" referring to the background being bright, there being a "dark" pulse that propagates in a specified form. By contrast, the optical soliton under the negative dispersion condition is called the "bright soliton", or the background is dark, with a bright optical pulse that propagates in a specified shape.

8.6 The optical fiber joining and coupling technology

In the following we shall describe the technology of joining optical fibers with others and coupling the optical source with optical fibers.

8.6.1 The processing and joining of optical fibers

1. The fiber cutting method

When an optical fiber is being coupled with an optical source or a detector, to improve the coupling efficiency, the optical fiber end face should be polished as a mirror surface and be perpendicular to the fiber core's axial line. A simple method for performing end face slicing of an optical fiber is using the optical fiber cutting tool, shown in Fig. 8.6-1. Place a piece of naked optical fiber on a rigid body with radius R (in general several centimeters), the diamond knife pressing a slicing mark in the optical fiber in the direction perpendicular to the direction of the optical fiber. Then exert a tensile force on the optical fiber (to tighten the fiber with the "scar of wound"). Under the action of the bending stress and tensile force, the crack generated by the slicing mark will gradually expand, so that the optical fiber will be capable of being cut apart flat and smooth like a mirror's surface. Make sure that ordinary combination pliers is not used to cut optical fibers apart because, in so doing, owing to the brittleness of quartz, the fiber will break off with a rugged cross section that just cannot be used.

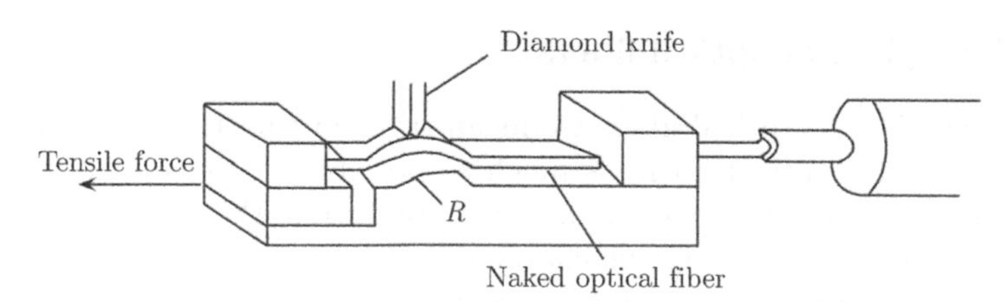

Fig. 8.6-1 A method for cutting naked optical fiber apart

2. The joining of optical fiber with optical fiber

The joining between optical fibers is divided into two kinds, the permanent joining and the flexible joining, the former in general subdivided into two modes, joining with adhesive and thermal melting joining. Whatever the mode, it is necessary to use the V-type groove or the precision sleeve tube, with the axial centers collimated. Put in the adhesive and make it solidified or, by the use of thermal sources such as a CO_2 laser or an electric arc discharge, melt the fiber counter-joining portions. When cooled, the optical fibers will be able to be joined, as shown in Fig. 8.6-2. The loss with this kind of end-to-end joining has already reached the 0.1-dB level.

It is pointed out in passing that, in optical fiber joining, should any one of the joining deviations shown in Fig. 8.6-3 appear, joining loss would ensue. The insertion losses introduced in (a), (c), (f), and (g) of the figure are maximum. The relevant calculation formulae or experimental data can be found in books or material on this subject.

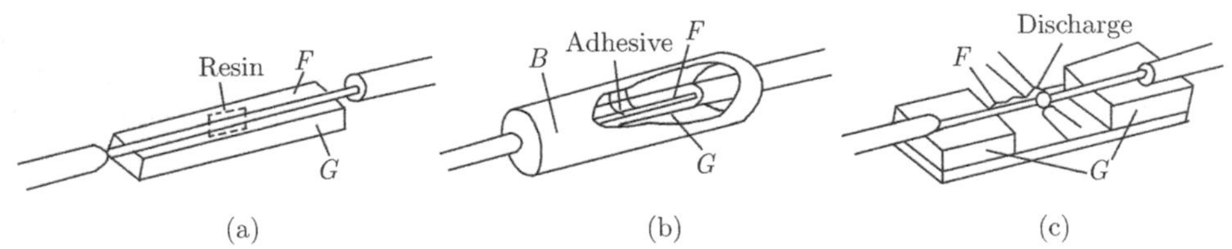

Fig. 8.6-2 Permanent joining of optical fibers:
(a) V-type groove method; (b) Sleeve method; (c) Thermal melting method (F–a naked of optical fiber, G–V-type groove, B–sealed sleeve (glass, ceramics, etc.)

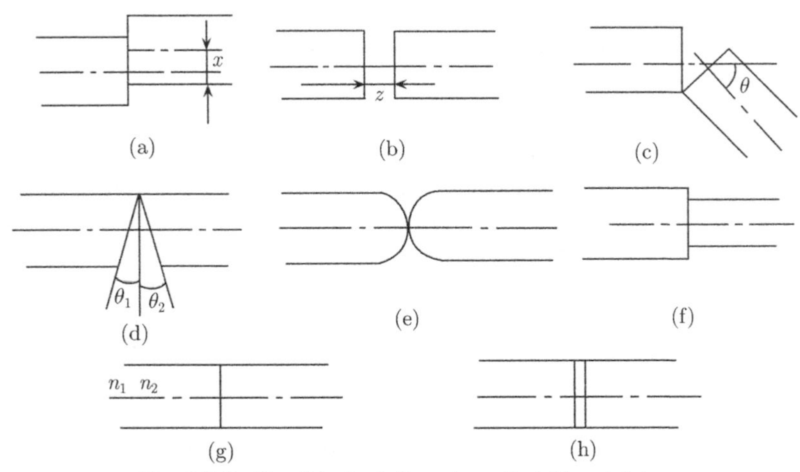

Fig. 8.6-3 Possible deviations in optical fiber joining:
(a) Axial deviation; (b) End face gap; (c) Axial included angle; (d) End face inclined; (e) End bent; (f) Deviation of core diameter; (g) Refractive index unmatched; (h) End face coarse

8.6.2 Optical coupling of optical fibers

By the optical coupling of optical fibers we mean transmitting the optical power emitted by the optical source into optical fibers as much as possible. This is a rather complicated issue involving the spatial distribution of light source radiation, the light emitting area of the power source, as well as the light-receiving characteristics and transmission characteristics of the optical fiber, and so on and so forth. Here we shall only describe some coupling methods and their practical techniques.

1. Direct coupling

By direct coupling we mean having a piece of optical fiber with an end face that is a plane directly placed close to the light-emitting face of the light source, as shown in Fig. 8.6-4. In the case of the optical fiber being definite, the coupling effect is closely related to the kind of the light source. If the light source is a semiconductor laser, as its luminous area is smaller than the area of the optical fiber's end face, so long as the light source is sufficiently close to the optical fiber's surface, the light emitted by the laser can all shed on to the end face of the optical fiber. Taking into consideration the degree of mismatch between the angle of divergence of the light source light beam and the optical fiber's receiving angle, the coupling efficiency is in general 20%, with 80% wasted. If the light source is a luminescent diode, then it will be even more serious since the angle of divergence of the luminescent diode is even larger, its

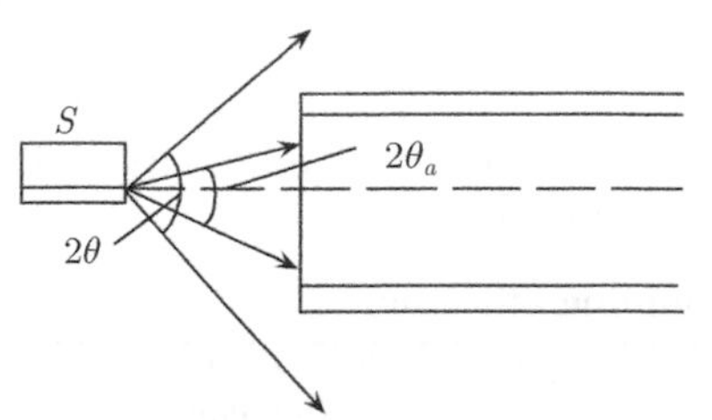

Fig. 8.6-4 Optical fiber directly coupled with light source
(S–light source; θ–light source light emitting field angle; θ_a–optical fiber receiving angle)

coupling efficiency being basically determined by the optical fiber's receiving angle, i.e.,

$$\eta = \frac{P}{P_0} \approx \mathrm{NA}^2 \tag{8.6-1}$$

For instance, if NA = 0.14, then $\eta = 2\%$.

In order to enhance the coupling efficiency, one method is inserting a lens between the light source and the optical fiber's end face, called lens coupling.

2. The lens coupling

Can the lens coupling method enhance the coupling efficiency or not? The answer is both in the affirmative and negative. There is here a problem of concept of the coupling efficiency. For the Lambert type light source (e.g., the luminescent diode), no matter what optical system is incorporated in it, its coupling efficiency can never exceed a maximum value:

$$\eta_{\mathrm{max}} = a\frac{S_{\mathrm{f}}}{S_{\mathrm{e}}}(NA)^2 \tag{8.6-2}$$

where a is a coefficient. The above equation shows that, when the light-emitting area S_{e} is larger than the optical fiber's receiving area S_{f}, incorporation of no optical system will be of any use. The maximum coupling efficiency can be obtained using the direct coupling method. When the light-emitting area S_{e} is smaller than the optical fiber's receiving area S_{f}, incorporation of an optical system will be of use and can increase the coupling efficiency. Moreover, the smaller the light-emitting area S_{e} is, the more will the coupling efficiency be increased. Under this criterion, there are the following lens coupling modes.

(1) The optical fiber end face spherical lens coupling

The simplest method of incorporating a lens is make the optical fiber end face semi-sphere-shaped to make it perform the function of a short focal distance lens, as shown in Fig. 8.6-5. It can be seen from the figure that the role of the end face spherical lens is to enhance the equivalent receiving angle of the optical fiber so that the coupling efficiency can be increased. A number of experimental results are as listed in Tab. 8.6-1, from which it can be seen that this kind of coupling method is fairly efficient for the step type optical fiber while not very good for the refractive index gradient type optical fiber.

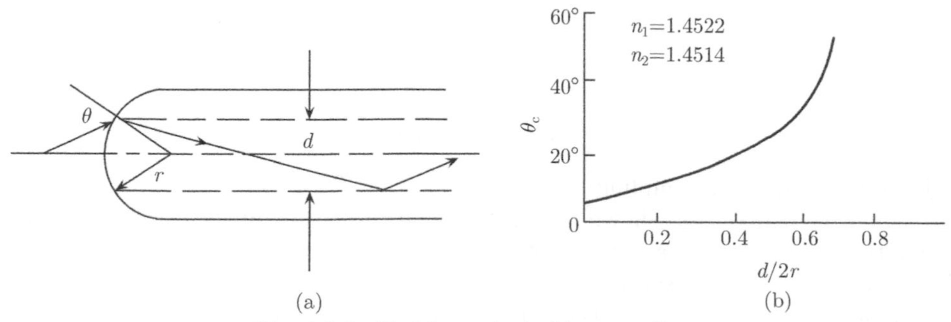

Fig. 8.6-5 End face spherical lens coupling:
(a) Schematic of end face spherical lens; (b) Relation of equivalent receiving angle θ_{c} with $d/2r$

(2) Cylindrical lens coupling

The light emitted by a semiconductor laser is asymmetric in space. The light beam is rather concentrated in the direction parallel to the pn junction ($2\theta_{//}$ is 5°~6°). In the direction perpendicular to the pn junction, divergence is rather great ($2\theta_{\perp}$ is 40°∼60°). So, the efficiency of direct coupling is not high. If effort can be made to have the light beam in the direction perpendicular to the pn junction compressed, the entire light spot will change from the slender ellipse to a shape close to a circle. However, if coupled with an optical fiber with a circular cross section, the coupling efficiency will be greatly increased.

Tab. 8.6-1 End face spherical lens coupling efficiency

Serial no. of optical fibers	type	NA	Core diameter /μm	Plane end coupling efficiency/%	Spherical lens end face coupling efficiency/%
1 B; SiO_2	Step	0.1707	55	24	63
2 B_2O_3; SiO_2	Part grad. type	0.148	50	13	19
3 GeO_2; SiO_2	Gradient type	0.1707	46	17	26
B_2O_3; SiO_2	Gradient type	0.162	42	10	15

The above-mentioned goal can be attained by means of the cylindrical lens, whose device is as shown in Fig. 8.6-6. A detailed investigation shows that, when the radius of the cylindrical lens is identical with that R of the optical fiber, the laser is located on the optical axis, and the lens surface is at $z = 0.3R$, the maximum coupling efficiency can be obtained, which is about 80%. If the location of the laser has deviated axially, then the coupling will be markedly reduced, that is to say, such a mode of coupling makes a very high requirement on the accuracy of the relative locations of the laser, cylindrical lens, and the optical fiber.

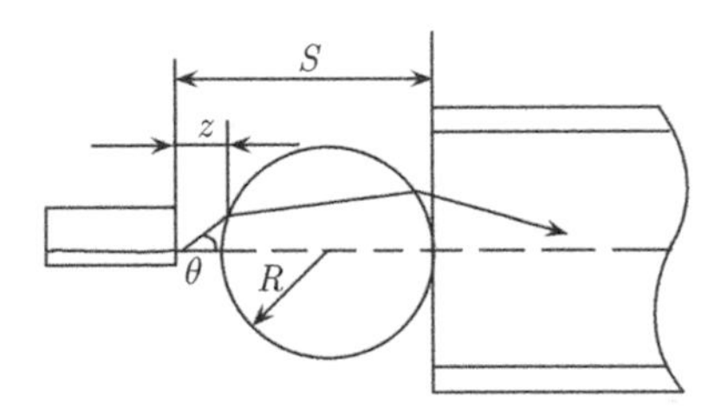

Fig. 8.6-6 The cylindrical lens coupling

(3) The convex lens coupling

First place the light source on the focus of the convex lens to make the light become parallel light. Then focus the parallel light onto the end face of the optical fiber with another lens, as shown in Fig. 8.6-7. This kind of coupler is composed of two parts, each containing a convex lens. Since the light is parallel, requirement on the accuracy of the joined parts is not high and adjustment and assembly are fairly easy, while the coupling efficiency can be as high as 80% and above.

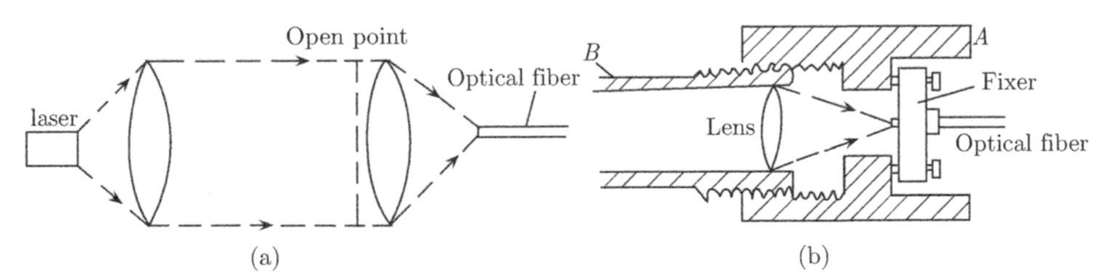

Fig. 8.6-7 Lens coupling

(4) The self-focusing lens coupling

By replacing the convex lens shown in Fig. 8.6-7 with a section of the self-focusing optical fiber of length $\Lambda/4$ (Λ is the beat length), we can also construct a coupler. It is customary to glue the optical fiber and the self-focusing optical fiber lens together. After getting into the self-focusing lens, all the parallel light enters the optical fiber upon being focused, as shown in Fig. 8.6-8. With its compact structure, stability, and reliability, this form of coupling is a fairly good form of coupling.

(5) The conical type lens coupling

Make the front end of an optical fiber into a gradually diminishing conical shape by corrosion as shown in Fig. 8.6-9(a), or have it scarified and drawn so as to make it thin and become a cone-shape as shown in Fig. 8.6-9(b). The radius of the front end is a_1, while that of the optical fiber itself is a_n. Light is injected into the optical fiber from the front end at an angle θ' and is emitted toward point A on the interface after being refracted at an angle r_1 as shown in Fig. 8.6-9(c). Since the interface is an inclined surface, $r_2 < r_1$. If the slope of the cone face is not abrupt, that is, the cone's length $l \gg (a_n - a_1)$, there is approximately

$$\frac{\sin r_{n-1}}{\sin r_n} = \frac{a_n}{a_{n-1}} \tag{8.6-3}$$

It can be proved that there is the following relation between the optical fiber's receiving angle θ_c' in the presence of a cone and the receiving angle θ_c when the end of the optical fiber is flat:

$$\frac{\sin \theta_c'}{\sin \theta_c} = \frac{a_n}{a_1} \tag{8.6-4}$$

The above equation shows that the numerical aperture of the optical fiber with a conical lens is a_n/a_1 times that of a flat end optical fiber.

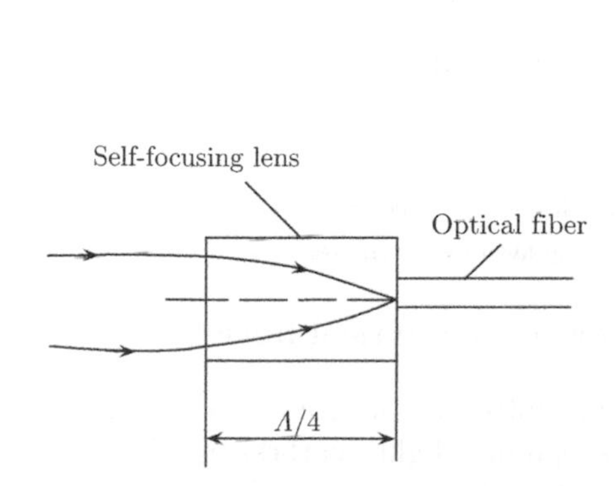

Fig. 8.6-8 Self-focusing lens coupling

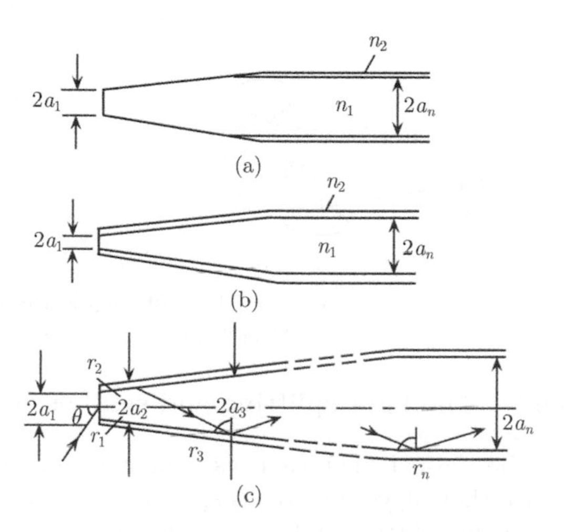

Fig. 8.6-9 Conical lens coupling

If only the diameter $2a_1$ of the front end face is greater than the area of the light source, the coupling of this kind of coupling mode can reach as high as 90% and above.

3. Fiber-optic holographic coupling

As the optical holographic film can convert one wavefront of light to another, it can be used as a fiber-optic coupler; the method of fabricating holographic coupler is as shown in Fig. 8.6-10(a). After passing through the optical fiber, laser becomes diffused light. As an object light I_F, the straight emitted light reflected by mirror M is used as the reference light. Heavy chromic acid gelatin or halosilver photographic emulsion film is used as the holographic recording medium. This holographic film is a fiber-optic coupler, as is shown in Fig. 8.6-10 (b). When in use, it is required that the laser beam that is in conjugation with the reference light when recording the hologram be used for illumination so as to gather the light beam for reproduction and coupling into the optical fiber. In principle, the coupling efficiency of this kind of coupling is very high. However, owing to the influence of the diffractive effect of the holographic film and the attenuation loss, the actual coupling efficiency is by no means superior to that of lens coupling. Nevertheless, its greatest advantage is that it can be applied as a multifunction optical component. Shown in Fig. 8.6-11 is a fiber-optic sensor system using holographic coupling devices. In Fig. 8.6-11(a), H_1 plays the role of two lenses and one beam splitter while H_2 two lenses and a light synthesizer. In Fig. 8.6-11(b), H has the role of two lenses, a spectroscope, and a light synthesizer, to play. Clearly, the holographic coupler can greatly simplify the conventional fiber-optic sensor system.

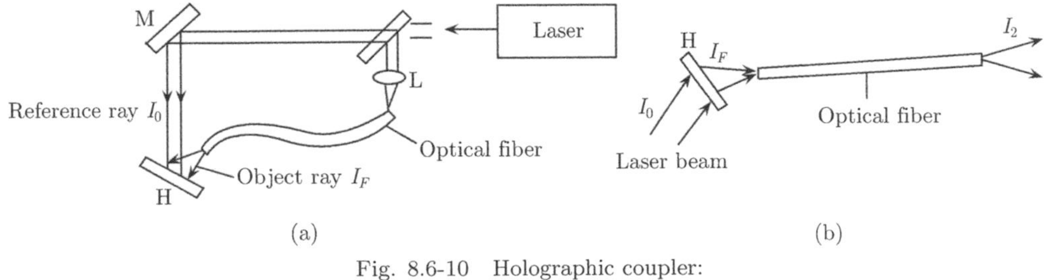

Fig. 8.6-10 Holographic coupler:
(a) Process of fabrication; (b) Role of coupling

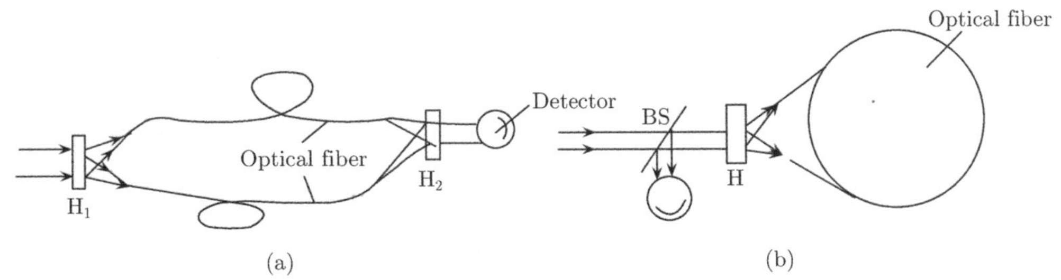

Fig. 8.6-11 A fiber-optic sensor using holographic coupler:
(a) Mach-Zehnder interferometer; (b) Sagnac interferometer

8.6.3 The light splitting and synthesizing devices for optical fibers

The optical fiber changes a light beam from one into two or from two into one. A device that plays this role is spoken of as a spectroscope or a light synthesizer. The optical fiber light splitting and synthesizing devices are helpful to the preliminary integration of the optical system. Shown in Fig. 8.6-12 are some recently developed optical fiber light splitting and synthesizing devices.

Shown in Fig. 8.6-12(a) and (b) is the light beam focusing type, the light synthesizing loss being 3 dB or so; shown in (c) and (d) is the half-transparent half-reflective type, the loss being 3.7 dB or so; shown in (e) and (f) is the waveguide coupling type, the loss at 5 dB or so; shown in (g) is the distributed coupling type, the loss at 3.7 dB or so, and shown in (h) and (i) is the partly reflective type, the loss at 4.7 dB or so.

8.7 Laser atmospheric and underwater transmission

The application of such technologies as atmospheric and underwater laser communication and detection usually takes the atmosphere and the underwater regions as the channel. Compared with optical fibers, the transmission characteristics of these channels are even more complicated and unstable. Compared with the microwave waveband, these problems appear to be more outstanding, restricting the many advantages in laser application from being brought into play and made use of. For this reason, the study of the transmission characteristics of laser in the atmosphere and underwater has already become a specialized research field. This chapter will present a brief description of a number of the basic concepts and problems.

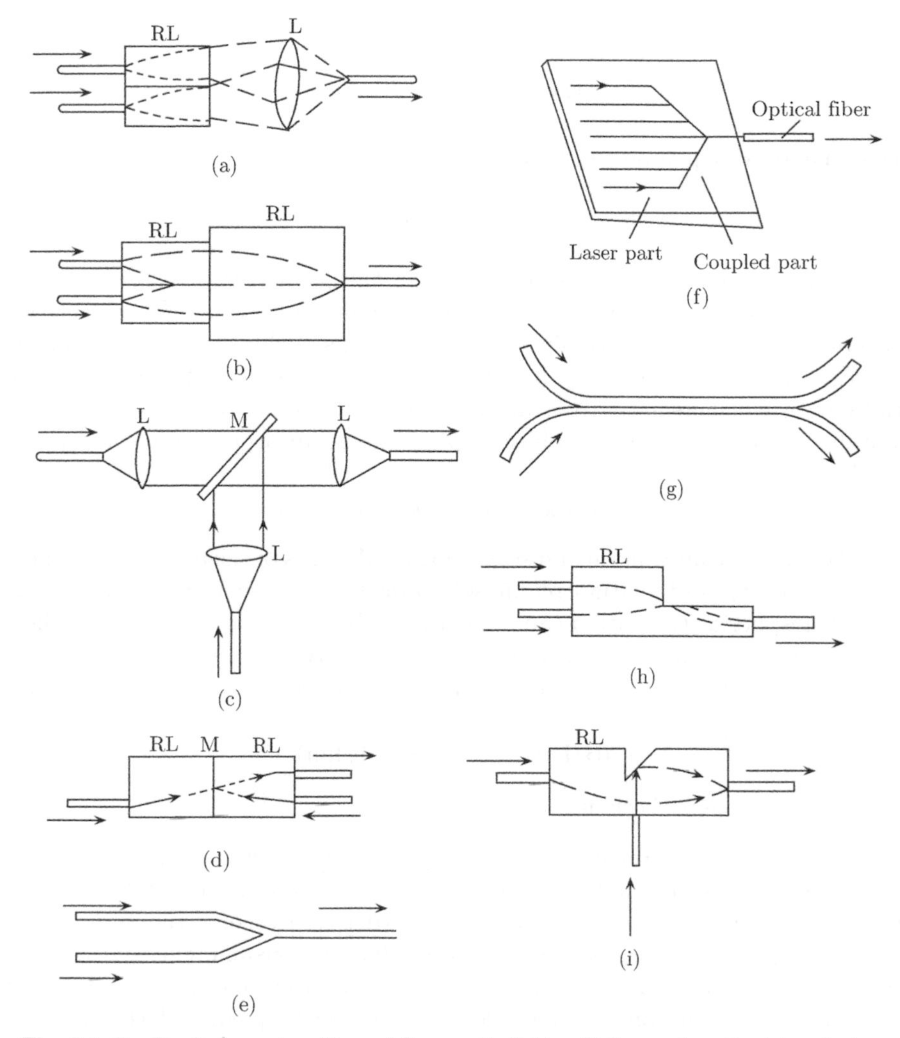

Fig. 8.6-12 Typical constructions of fiber-optic light splitting and synthesizing devices

8.7.1 Atmospheric attenuation

When laser radiation passes through the atmosphere, owing to the various gases and particles existing therein, such as dust, smoke, and fog, plus such weather changes as the blowing of a wind and raining, part of the optical radiation energy will be absorbed and turned into energy in other forms (such as thermal energy) and part is scattered and deviates from the original direction of propagation (i.e., the direction of radiation for redistribution). The total effect of absorption and scattering causes the intensity of the propagated optical radiation to be attenuated, which is what is spoken of as atmospheric attenuation.

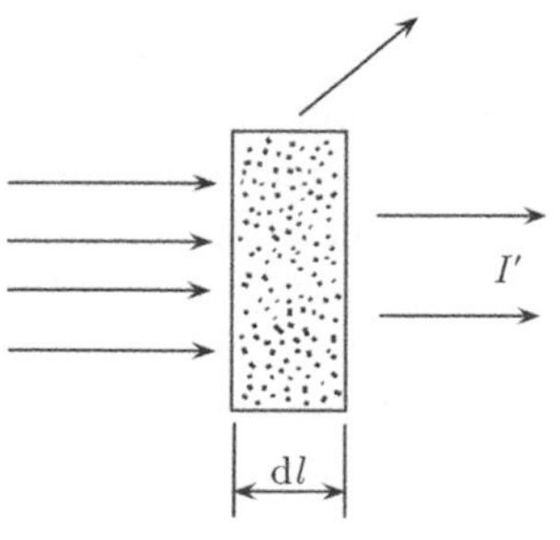

Fig. 8.7-1 Attenuation of optical radiation

Suppose a monochromatic optical radiation of intensity I passes through a thin layer of thickness $\mathrm{d}l$, as shown in Fig. 8.7-1. Prior to the emergence of the nonlinear effect, its amount of attenuation $\mathrm{d}I$ is proportional to I and $\mathrm{d}l$, that is, $\mathrm{d}I/I = (I' - I)/I = -\beta \mathrm{d}l$.

After integration, we have

$$T = I/I_0 = \exp\left(-\int_0^L \beta \mathrm{d}l\right) \tag{8.7-1}$$

Assume that the above equation can be reduced to

$$T = \exp(-\beta l) \tag{8.7-2}$$

where T is the atmospheric transmissivity (%) over the propagation distance L; I_0 and I are the light intensities before and after passing through the distance L, respectively; and β is the atmospheric attenuation coefficient (1/km). This is the Lambert law that depicts the atmospheric attenuation, showing that the light intensity appears to regularly attenuate exponentially with the increase in the transmission optical path. As the attenuation coefficient β contains the two independent physical processes of absorption and scattering, β can be expressed as

$$\beta = k_m + \sigma_m + k_a + \sigma_a \tag{8.7-3}$$

where k_m and σ_m are the absorption and scattering coefficients of the atmospheric molecules, respectively; k_a and σ_a are the atmospheric sol absorption and scattering coefficients. So a study of the atmospheric attenuation can obviously be boiled down to a study of the above-mentioned four basic attenuation parameters. In engineering application, the attenuation coefficient often uses (1/km) or (dB/km) as its unit; the conversion relation between the two is

$$\beta(\mathrm{dB/km}) = 4.343 \times \beta(1/\mathrm{km}) \tag{8.7-4}$$

1. The absorption by atmospheric molecules

When a light wave passes through the atmosphere, under the action of the light wave electric field, the atmospheric molecules generate polarization and cause the frequency of the incident light to make forced vibration. So, in order to overcome the internal damping force of the atmospheric molecules, the light wave has to consume energy, part of which will be transformed into other forms of energy (such as thermal energy), manifested as the absorption by the atmospheric molecules. When the frequency of the incident light is equal to the natural frequency of atmospheric molecules, resonant absorption will occur and there emerges the maximum value of atmospheric molecular absorption. The intrinsic absorption frequency of molecules is determined by the internal moving form of molecules. The internal movement of polarized molecules is in general made up of the electronic motion in molecules, molecular vibration, and the rotation of molecules round their mass center, the molecular resonant absorption frequency that results corresponding to the optical wave's UV and visible light region, near IR and intermediate IR, and far IR regions. Therefore, the absorption characteristics of molecules heavily rely on the frequency of the optical wave.

Owing to the difference in structure from molecule to molecule, totally different spectral absorption characteristics have been manifested. Most of the multiple atomic molecules that make up the atmosphere, with the exception of N_2, are polarized molecules, of which the O_2 molecules possess permanent magnetic dipole moment, while the remaining molecules all possess permanent electric dipole moment. Despite their greatest content (about 99%) in the atmosphere, the N_2 and O_2 molecules exhibit almost no absorption in the visible light and near IR regions. N_2 is a polarization-free molecule without a revolving structure and it is simply because of the action of the atmospheric pressure that it exhibits weak pressure induced absorption for 4.0 μm. Though with magnetic dipole moment, the O_2 molecule exhibits only very weak absorption of UV light (in the vicinity of 0.2 μm and 0.7 μm) and it shows very great absorption only for the far IR and microwave waveband. Therefore, in the visible light and near IR region, no consideration is given to its role of absorption. For molecules other than those of H_2O and CO_2, despite their spectacular absorption spectral line in the visible light and near IR region, since their content in the atmosphere is very small, their absorption role is in general not considered, either. However, high up in the sky, it is because all the remaining attenuation factors are already very weak that the absorption role of O_2 there is considered. The H_2O and CO_2 molecules, especially the former, have a broad vibration-rotation and pure rotation structure. Hence, they are the most important absorption molecules in the visible light and near IR region. This is the major factor for the attenuation of laser on a fine day. The central wavelengths of their main absorption spectral lines are listed in Tab. 8.7-1.

Tab. 8.7-1 Main absorption spectral lines of visible light and near IR region

Absorption molecules	Main absorption spectral line's central wavelengths/μm
H_2O	0.72 0.82 0.93 0.94 1.13 1.38 1.46 1.87 2.6 3.15 6.26 11.7 12.6 13.5 14.3
CO_2	1.4 1.6 2.05 4.3 6.2 9.4 10.4
O_2	4.7 9.6

It's not hard to see from Tab. 8.7-1 that, for certain particular wavelengths, the atmosphere exhibits extremely strong absorption, so much so that the optical wave just cannot pass through. In view of the selective absorption characteristic of the atmosphere, the near IR region is divided into eight sections while the wavebands of fairly high transmissivity are called the "atmospheric windows" (Fig. 8.7-2). In these windows, the atmospheric molecules appear to be weakly absorptive. The currently frequently used laser wavelengths are all located in these windows.

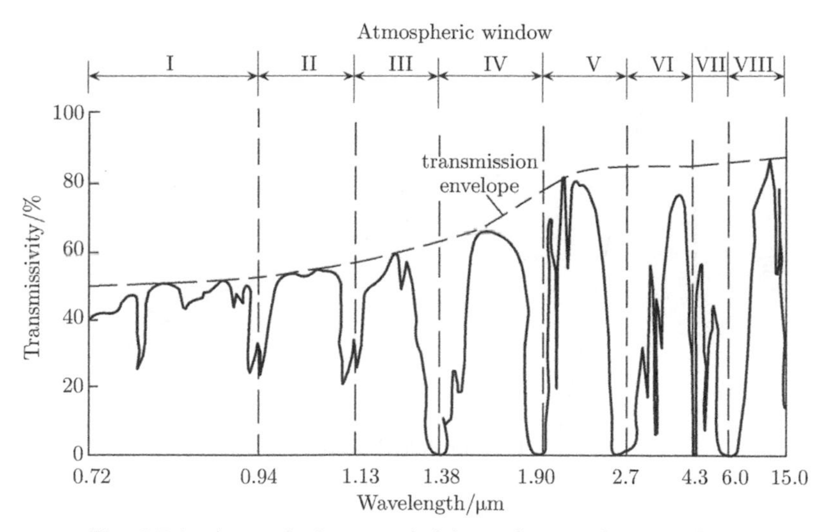

Fig. 8.7-2 Atmospheric transmissivity and atmospheric windows

2. The scattering of atmospheric molecules

When light passes through the atmosphere, the absorption and scattering by atmospheric molecules make the intensity of the transmitting light weaker while the electromagnetic field of the light wave makes the atmospheric molecules generate polarization to form vibrating dipoles, thereby emitting subwaves. If the atmosphere is optically homogeneous, the superposition of these subwaves will result in the light continuing to propagate only in the refractive direction while in the other directions, because of the interference by the subwaves, the light and subwave will offset each other, so no light appears. However, there always exists in the atmosphere statistical deviation of the local density and the average density—the density fluctuation destroys the homogeneity of the atmosphere, causing the coherence of the subwave to be destroyed. In addition, owing to the presence of various kinds of particles in the atmosphere, part of the radiated light will propagate in other directions, leading to the scattering of light in all directions.

As the dimension of the atmospheric molecule is very small (10^{-8} cm), at the visible light and near IR waveband, the radiated wavelength is always far greater than the dimension of the molecule. The scattering under such a condition is usually called Raleigh scattering. It is pointed out by the law of Raleigh scattering that the intensity of the scattered light is inversely proportional to the fourth power of the wavelength. Hence the molecular scattering coefficient is also inversely proportional to the fourth power of the wavelength.

The empirical formula of the Raleigh scattering coefficient is

$$\sigma_m = 0.827 \times N \times A^3/\lambda^4 \tag{8.7-5}$$

where σ_m is the Raleigh scattering coefficient (cm^{-1}); N is the molecules number (cm^{-1}) in a unit volume; A is the scattering section (cm^2); and λ is the optical wavelength (μm). It is known from this equation that the molecular scattering coefficient is proportional to the molecular density while inversely proportional to the fourth power of the wavelength. The greater the wavelength, the weaker the scattering; the shorter the wavelength, the more intensified the scattering. That's why the visible light scattering is more intensified than the IR scattering while the blue light scattering is more intensified than the red light scattering. So bright skies look azure. The relationship between the molecular scattering coefficient and the optical wavelength is as shown in Fig. 8.7-3.

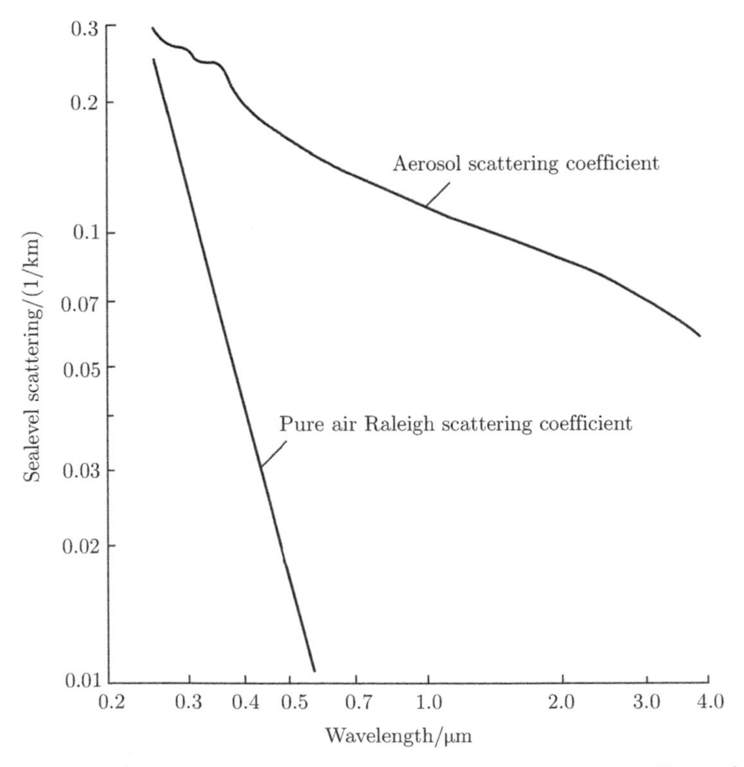

Fig. 8.7-3 Relationship of horizontal transmission of atmospheric scattering coefficient (calculated value) for standard bright atmosphere sea level with the optical wavelength

3. The attenuation of atmospheric aerosol

In reality, apart from atmospheric molecules in the atmosphere, there are large quantities of solid-state and liquid-state particles with a granularity of 0.03~2000 μm, the greater part of which are dirt, flue particles, micro-droplets, organic microorganisms, etc. Most solid-state particles not only directly make the atmosphere turbid (called haze), but also it is the center for the condensation of vapor that plays a very big role in forming clouds, fog, rain, and snow. Since the suspension of these particles in the atmosphere is in the sol state, it is usually called atmospheric aerosol.

The attenuation of the optical wave by aerosol includes the scattering and absorption of aerosol. It is pointed out by the theorem of light scattering that when the optical wavelength is far greater than the size of the scattering particles, Raleigh scattering will be generated while if the optical wavelength is equal to or smaller than the scattering particle size, Mie scattering will be generated. Raleigh scattering is strongly related to the optical wavelength while Mie scattering relies on the aerosol particle size and the distribution of density as well as such characteristics of aerosol as its refractive index. Its relation with the optical wavelength is far weaker than that of Raleigh scattering (Fig. 8.7-3).

The dimension distribution of aerosol is extremely complicated. In general there exist both the action of Raleigh scattering and that of Mie scattering at the same time. But it is shown after measurement that the dimension of aerosol is far greater than that of the atmospheric molecule. For instance, on a fine day with particularly good visibility and low humidity, the radius of the greater part of aerosol particles is between 0.1~1 μm while that of only a small part is between 1~10 μm. If it is cloudy, aerosol (mainly solid-state particles) is composed of particles with a radius of 0.03~0.2 μm. The radius of fog particles of various concentrations is between 3~ 60 μm, while that of the cloud particles is between 2~30 μm. The radius of rain particles is between 60~2000 μm. Therefore, for the aerosol, it's generally unnecessary to think about Raleigh scattering while mainly the role of Mie scattering should be considered.

(1) The attenuation in fine, cloudy, and foggy atmosphere

As far as engineering application is concerned, estimating atmospheric aerosol attenuation value using atmospheric visibility is more convenient and practical. This is because the main cause of visibility restriction is not the atmospheric molecule but atmospheric aerosol. So, it's quite natural to associate visibility with aerosol, especially for fine, cloudy, and foggy weather when this method is even more effective.

In meteorology, visibility is defined as: against the background of daytime horizontal sky for light of 0.55 μm (the wavelength for which the human eye is most sensitive), the farthest visual range within which a sufficiently big black body (target object) can be discerned. The determination of the visual range depends on the normalized comparison of the target object with the background and the visual perception threshold of the human eye. During an observation at distance R, if the target apparent brightness is $N_{\mathrm{T}R}$ and the background apparent brightness is $N_{\mathrm{b}R}$, then the contrast C_R of the target with the background is

$$C_R = (N_{\mathrm{T}R} - N_{\mathrm{b}R})/N_{\mathrm{b}R} \tag{8.7-6}$$

When $R = 0$ and $C_0 = (N_{\mathrm{T}R} - N_{\mathrm{b}0})/N_{\mathrm{b}0}$, it is called the inherent contrast of the target with the background while the contrast between C_R and C_0 ($\varepsilon = C_R/C_0$) is called the normalized contrast of the target with the background, that is, the observer's visual perception degree. A normal human eye's visual perception threshold $\varepsilon = 0.02$, meaning the visual perception degree when viewing a sufficiently big target in the farthest discernible distance with the normal human eye. Evidently, when $\varepsilon > 0.02$, it is possible to distinguish the target from the background whereas if $\varepsilon < 0.02$, it will be impossible to do so. Therefore, the visual range $R = V$ that satisfies the condition $C_{R=V}/C_0 = 0.02$ is called the visibility. According to the magnitude of visibility, the atmospheric condition can roughly be divided into ten levels (Tab. 8.7-2). As such a division is too sketchy, in actual engineering application the V value is directly used to represent visibility. For example, by the so-called standard fine day we mean the atmospheric state in which $V = 23.5$ km and the relative humidity is lower than 20%.

Tab. 8.7-2 Levels of visibility

Visibility level	Daytime visible distance/m	Daytime invisible distance/m	Remarks
0	< 50	$\geqslant 50$	Dense fog
1	$\geqslant 50$	$\geqslant 200$	Thick fog
	< 200		
2	$\geqslant 200$	$\geqslant 500$	Medium fog
	< 500		
3	$\geqslant 500$	$\geqslant'$ 1000	Light fog
	< 1000		
4	$\geqslant 1000$	$\geqslant 2000$	Thin fog
	< 2000		
5	$\geqslant 2000$	$\geqslant 4000$	Haze fog
	< 4000		
6	$\geqslant 4000$	$\geqslant 10000$	Light haze
	< 10000		
7	$\geqslant 10000$	$\geqslant 20000$	Fine
	< 20000		
8	$\geqslant 20000$	$\geqslant 50000$	Very fine
	< 50000		
9	$\geqslant 50000$		Particularly fine

According to the Lambert law of monochromatic radiation attenuation, under the atmospheric horizontal homogeneity condition, only the aerosol attenuation will be considered. Equation (8.7-2) can be rewritten as

$$T_\lambda = \exp(-\beta_{a\lambda} L) \tag{8.7-7}$$

where λ is the laser wavelength and L is the horizontal transmission distance. As $\beta_{a\lambda}$ is a function of the reciprocal of the wavelength, it can be written in the following form:

$$\beta_{a\lambda} = A\lambda^{-q} \tag{8.7-8}$$

By taking logarithm on both sides, we obtain $\ln\beta_{a\lambda} = \ln A - q\ln\lambda$. It can be seen that $(-q)$ is the slope of $\ln\beta_{a\lambda}$ with respect to the straight line $\ln\lambda$. The q value can be determined by experiment.

Additionally, it is known from $C_{R=V}/C_0 = 0.02$ that, if the apparent brightness of the target (radiation source) is far greater than that of the background at any observation distance, i.e., $N_{TR} \gg N_{bR}$ and the background apparent brightness does not vary with the distance of observation, or $N_{b0} = N_{bR}$, then the physical meaning represented by $C_{R=V}/C_0 = 0.02$ is the atmospheric transmissivity $T_{0.55} = 0.02$. Therefore,

$$T_{0.55} = \exp[-A(0.55)^{-q}V] = 0.02$$

where 0.55 is concerned with the 0.55 μm wavelength. After calculation, we can find $A = (3.91/V) \times (0.55)^q$ which, substituted into Eq. (8.7-8), gives

$$\beta_{a\lambda} = (3.91/V) \times (\lambda/0.55)^{-q} \tag{8.7-9}$$

where V is the visibility (km); λ is the laser wavelength (μm); and q is a constant closely related to the wavelength and visibility.

For visible light, $\lambda/0.551$, hence $\beta_{a\lambda} = 3.91/\mathrm{V}(\mathrm{km}^{-1})$.

For near IR light,

$$q = \begin{cases} 1.6 & \text{(when } V \text{ is very great)} \\ 1.3 & \text{(medium visibility)} \\ 0.585V^{1/3} & \text{(when } V \leqslant 6\text{km)} \end{cases}$$

For the sake of image intuition, Figs. 8.7-4 and 8.7-5 have been plotted based on Eq. (8.7-9) and $\beta_{a\lambda} = 3.91/V$, with a brief description of the weather given in the figures.

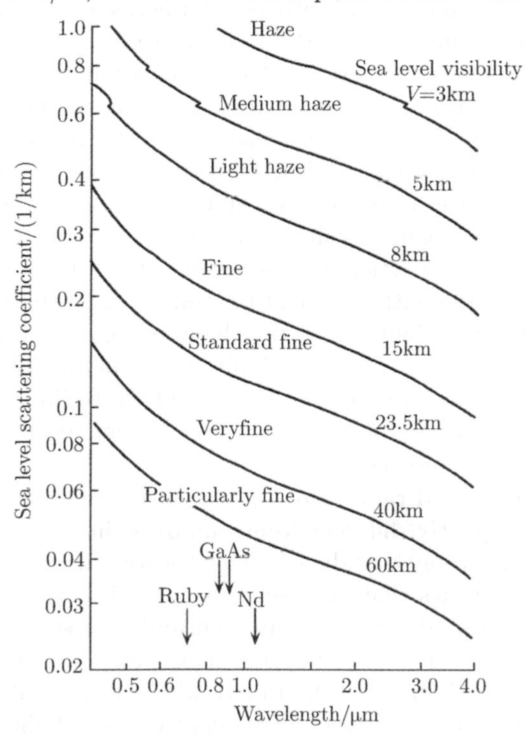

Fig. 8.7-4 Relation between aerosol attenuation coefficient and wavelength under different sea level horizontal visibilities

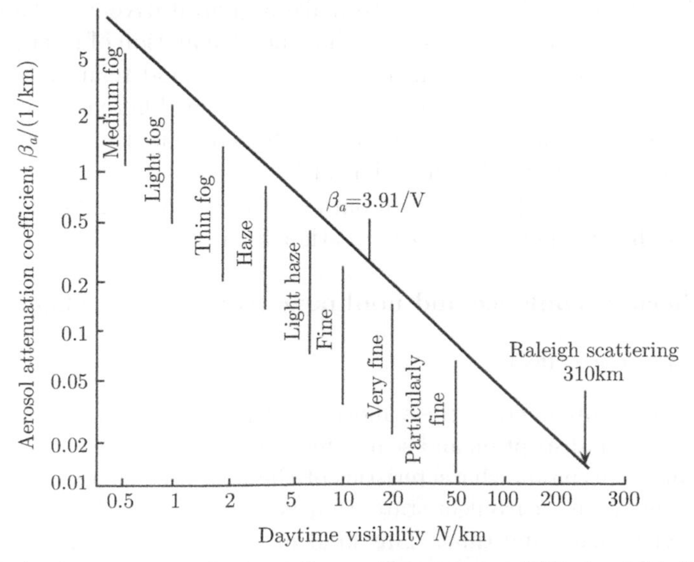

Fig. 8.7-5 Relation between aerosol attenuation coefficient and visibility in visible light waveband

As 1.06 μm and 10.6 μm laser is currently fairly widely used, we give a brief account for the applicability of Eq. (8.7-9) to the two wavelengths. For the 1.06-μm laser wavelength, there almost does not exist molecular absorption ($< 10^{-6}$ km^{-1}). The molecular scattering, too, is 1~2 orders of magnitude smaller than aerosol attenuation, the main attenuation being in its scattering and absorption. Many actual measurement results have proved that it's very applicable estimating the attenuation of 1.06-μm wavelength using Eq. (8.7-9). For the 10.6-μm laser wavelength, usually the molecular scattering can be neglected. The main attenuation is that of molecular absorption and aerosol attenuation. On a fine day, molecular absorption is dominant. With the worsening of weather, the action of aerosol attenuation will become increasingly great. Therefore, for hazy and foggy weather, Eq. (8.7-9) is also often used to estimate its attenuation value. It has been pointed out by both actual measurement and theoretical analysis that, in the foggy and hazy aerosol system, the attenuation of the 10.6-μm wavelength light is one order of magnitude smaller than the visible light and laser in the near IR waveband. The longer the radiation wavelength is, the smaller the aerosol attenuation will be. With the gradual increase in the granular size of aerosol particles (eg the particle radius greater than 10 μm), the evident relation between attenuation and wavelength will tend to vanish. So, the 10.6-μm laser possesses good fog- and haze-penetrating ability. This waveband is a considerably ideal transmission waveband. However, in a thick fog, eg when visibility is 200 m and 50 m, the 10.6-μm laser's attenuation coefficient can be as great as 20 dB/km and 50 dB/km, respectively. It's clear that the contribution of fog to the attenuation of laser is very serious.

(2) The attenuation by rain and snow

The difference between fog and rain is not simply in the difference in precipitation, but mainly the fog particle is greatly different from raindrops in size. It has been shown by research that, although on a rainy day the content of water (if it is 1.0g/m^3) in the atmosphere is in general over 10 times that of a dense fog (if it is 0.1g/m^3), the radius of a fog drop (in mm order of magnitude) is merely one-thousandth or so that of the raindrop, the raindrop gap being much greater. So the visibility is higher than in fog through which the optical wave can easily pass. Moreover, the raindrop's forward scattering effect is strong, which will markedly reduce the attenuation of straight emitted light beam resulting in the attenuation coefficient of rain being over two orders of magnitude smaller than that of fog.

As it is very hard to make a physical description of snow, what with the lack of data on the refractive index of snow, it's very hard to make a quantitative calculation for the time being. Some experimental investigations show that the attenuation of laser in snow is similar to that in the rain; the attenuation coefficient has a fairly good relation of correspondence with the intensity of snowfall. The attenuations of laser of different wavelengths in snow are not much different, but in terms of identical water content, the attenuation of snow is greater than that of rain but smaller than that of fog.

Tab. 8.7-3 lists the estimated attenuation data on three wavelengths of 0.5 μm, 1.06 μm, and 10.6 μm in fine, hazy, foggy, and snowy weather for reference.

8.7.2 Atmospheric turbulence and nonlinear transmission effect

1. Atmospheric turbulence effect

The problem of the components of the atmosphere that make optical wave energy attenuate through the action of absorption and scattering is discussed above. In our discussion, no mention is made of the dynamic characteristics of the atmospheric components. In reality, the atmosphere is always in a turbulent state; that is, the refractive index of the atmosphere varies irregularly with space and time. This kind of turbulent state will make laser radiation randomly change its optical wave parameters in the process of propagation, seriously

Tab. 8.7-3 Attenuation of light of three different wavelengths under different meteorological conditions

Meteorological condition	Wavelength/μm	Attenuation coefficient/(dB/km)
Fine weather	0.5 and 1.06	≈ 0.06
(humidity 20%)	10.6	0.54
Haze	0.5 and 1.06	1.4
0.5~10 μ heavy and small, 0.5 mg/m^3	10.6	0.66
Thin fog	0.5 and 1.06	9
0.5~10 μm heavy and light, 0.5 mg/m^3	10.6	0.9
Visibility ≈ 2 km		
Fog	0.5 and 1.06	18
0.5~10 μm heavy and light, 0.5 mg/m^3	10.6	1.9
Visibility ≈ 0.5 km		
Rain: 5 mm/h	0.5 and 1.06	0.6
25 mm/h	0.5 and 1.06	4.2
75 mm/h	0.5 and 1.06	7
Rain (1000 mm, 50 mg/m^3)	10.6	1.2
Snow: light snow	0.5 and 1.06	1.9
Heavy snow	0.5 and 1.06	6.9

affecting the quality of the light beam and leading to the emergence of such phenomena as what we call a light beam's intensity scintillation, bending and drift (also known as direction jitter), light beam diffusion and distortion, as well as the degeneration of spatial coherence or, taken as a whole, the effect of atmospheric turbulence. For instance, light beam scintillation (also known as atmospheric scintillation) will make laser signals undergo stochastic parasitic modulation to lead to the extra atmospheric turbulent flow noise causing the receiving signal-to-noise ratio to be reduced. This will lower the lidar's detection rate while increasing the undetected error rate and augmenting the simulation modulated atmospheric laser communication noise and increasing the bit error rate in digital laser communication. On the other hand, light beam direction jitter will make laser deviate from the receiving aperture, reducing the strength of the signal while light beam spatial coherence degeneration will lower the efficiency of laser heterodyne detection, etc. Therefore, investigations on atmospheric turbulence are receiving increasingly great recognition.

When studying the forms of movement of the atmosphere, it is usually considered that the atmosphere is a uniformly mixed single gaseous-state fluid, whose form of movement is divided into the laminar and turbulent movement. The former consists in the fluid particles doing regular and stable flow, with both the flow rate and direction in a thin layer being definite values and with no mixing occurring between layers in the course of movement while the latter is a kind of movement of an irregular vortex. The moving tracks of fluid particles are very complicated, there being both transverse motion and longitudinal motion, the moving velocity of every point in space randomly fluctuating around a certain average value.

It is known from experimental investigation that, in a certain volume of a gas or liquid when the ratio of the inertial force to the viscous force borne by the boundary of the volume exceeds a certain critical value, the regular laminar motion of the liquid or gas will lose its stability and shift to irregular turbulent motion. The value of this ratio is just the Reynold's number that represents the features of the moving state of a fluid.

$$Re = \rho \Delta v_l l / \eta \tag{8.7-10}$$

where ρ is the density (kg/m^3) of the fluid; l is a certain characteristic dimension (m); Δv_l is the amount of variation of the moving velocity over a distance of the l order of magnitude; and η is the viscosity coefficient (kg/m · s) of the fluid. The Reynold's number Re is a dimensionless number.

When Re is smaller than the critical value Re_{cr} (this value to be measured by experiment), the fluid is in stable laminar motion while when it is greater than Re_{cr}, it is in turbulent motion. As the viscosity coefficient of the gas is rather small, the motion of a gas is mostly a turbulent motion.

It can be verified by numerous examples that there does exist turbulent motion in the atmosphere. For example, when observing an object through air heated by the sunshine at the earth's surface, we can see the phenomenon of the "chatter of an object". The process of its formation and development is approximately as follows: in the moving process of the atmospheric laminar flow, stimulated by the illumination of the sun, the earth's surface radiation, and the action of friction between airflow and ground (also called the lower pad face), within a certain local range, an abrupt change in the airflow velocity takes place. The laminar flow loses its stability and generates maximum vortices (called level 1 turbulence). But their Reynold's number Re is rather large (both Δv_l and l are rather large), the $Re > Re_{cr}$ condition is very easy to satisfy, and the motion is extremely unstable. So the large vortices are decomposed into smaller ones (called level 2 turbulence). Now the variations of their corresponding dimension l and velocity Δv_l are both smaller than those of the preceding level, but the Reynold's number is still rather large and can still satisfy the condition $Re > Re_{cr}$. The motion is still unstable, so the smaller vortices continue to decompose into even smaller ones again and again until they become minimum vortices (or minimum turbulent air masses) that no longer satisfy the $Re > Re_{cr}$ condition. This process is in effect a kind of energy transfer. With the gradual diminution of the vortices, the influence (i.e., the thermal energy loss) of the gas viscosity coefficient becomes greater and greater. When the energy obtained by the small vortices is all supplied to the thermal energy loss produced due to the presence of the viscosity, the vortices will no longer decompose but become the tiniest vortices in the turbulent motion.

It's clear that the dimension l of the atmospheric turbulent air mass has an upper bound L_0 and a lower bound l_0, or $L_0 > l > l_0$; L_0 and l_0 are called the outer dimension and inner dimension of the turbulent air mass, respectively (Fig. 8.7-6). In the vicinity of the ground, l_0 is usually of the millimeter order of magnitude while L_0 is the altitude of the point of observation from the ground surface.

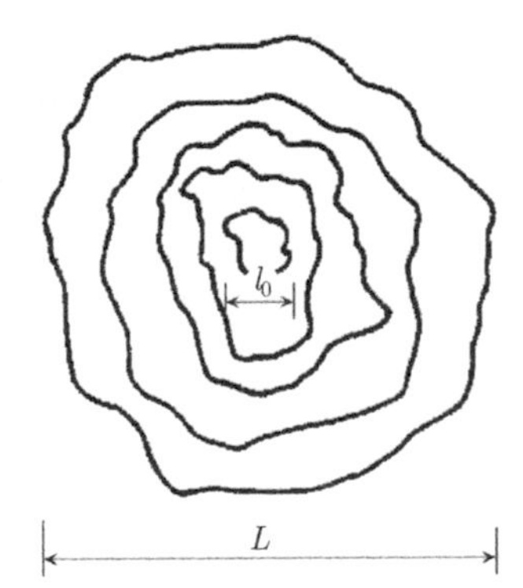

Fig. 8.7-6 The microstructure of atmospheric turbulent flow

As a result of the atmospheric turbulent flow movement, the atmospheric moving speed, temperature, and refractive index randomly rise and fall temporally and spatially, with the fluctuating field of the refractive index directly affecting the transmission characteristics of laser. By what we call the atmospheric turbulent flow effect of laser, we mean the effect of laser radiation during transmission in the fluctuating field of the refractive index.

The theory of atmospheric turbulent flow shows that the atmospheric velocity, temperature, and refractive index statistical characteristics obey the "2/3rds power law"; that is,

$$D_i(r) = \overline{(i_1 - i_2)^2} = C_i^2 r^{2/3} \quad (l_0 < r < L_0) \tag{8.7-11}$$

where i represents the velocity (v), temperature (T), and refractive index (n), respectively; D_i is the structure function of the corresponding field; r is the distance between points of examination; C_i is the structure constant of the corresponding field; and the unit is $\mathrm{m}^{-1/3}$.

The statistical characteristic of the refractive index of atmospheric turbulent flow directly influences the transmission characteristics of a laser beam. Usually the numerical value of the refractive index structure constant C_n is used to characterize the strength of a turbulent flow, i.e.,

Weak turbulent flow $C_n = 8 \times 10^{-9}\ (\mathrm{m}^{-1/3})$

Medium turbulent flow $C_n = 4 \times 10^{-8}\ (\mathrm{m}^{-1/3})$

Strong turbulent flow $C_n = 5 \times 10^{-7}\ (\mathrm{m}^{-1/3})$

The laser beam is a kind of finitely expanding light beam. The influence of atmospheric turbulent flow on light beam propagation is closely related to the ratio of the light beam diameter d_B to the dimension l of the turbulent flow. When $d_B/l \ll 1$, that is, the light beam diameter is much smaller than the turbulence dimension, the chief role of the turbulent flow is to make the light beam integrally randomly deflect. On a distant receiving plane, the projecting point of the center of light beam (or the position of the light spot), with a certain statistically average position as the center, there will occur rapid random jittering (whose frequency can be from several Hz to tens of Hz). This phenomenon is called light beam drift, whose numerical value can be denoted by the amount of drift or drift angle (the ratio of drift amount to the transmission distance). In addition, if a light beam is regarded as a whole, we shall discover in minutes that its average direction has evidently changed. This kind of slow drift is also spoken of as light beam bending. When $d_B/l \approx 1$, or the light beam diameter is equivalent to the dimension of the turbulent flow, the turbulent flow makes the wavefront of the light beam deflect randomly to form the fluctuation of the arriving angle on the receiving plane (the arriving angle is defined as the included angle between the wave normal and the normal of the optical axis receiving plane) so that image point chattering is generated on the focal plane of the receiving lens. When $d_B/l \gg 1$, or the diameter of the light beam is much greater than the dimension of the turbulent flow, the light beam cross section contains multiple turbulent flow vortices, each independently scattering and diffracting that part of the light beam illuminating it, bringing about random fluctuation of the light beam intensity spatially and temporally, the light intensity now great now small, hence the so-called light beam intensity scintillation. Meanwhile, light beam expansion and splitting is also generated. Even when the turbulent flow is very weak and the atmosphere is very stable, it is still possible to observe such variations as the distortion and twist of the light spot, its internal pattern structure, etc.

As the turbulent flow dimension l is continuously distributed between l_0 and L_0, and the diameter of the light beam steadily varies in the course of propagation, the above-mentioned turbulent flow effects always occur at the same time, the total effect being to make the light beam's spatial and temporal coherence markedly degenerate. Some principal research results are as follows:

(1) Atmospheric scintillation

The amplitude characteristic of atmospheric scintillation is characterized by the logarithmic intensity variance σ_I^2 of intensity I at a certain point on the receiving plane. By considering that $I = A^2$, there is

$$\sigma_I^2 = \overline{\left(\ln \frac{I}{I_0}\right)^2} = 4\overline{\left(\ln \frac{A}{A_0}\right)^2} = \overline{4\chi^2} \tag{8.7-12}$$

where $\overline{\chi}^2$ can be found through theoretical calculation while σ_I^2 can be obtained by actual measurement. Under the weak turbulent flow and uniform turbulent flow strength condition,

$$\sigma_I^2 = \overline{4\chi^2} = \begin{cases} 1.23C_n^2(2\pi/\lambda)^{7/6}L^{11/6} & (l \ll \sqrt{\lambda L} \ll L_0) \\ 12.8C_n^2(2\pi/\lambda)^{7/6}L^{11/6} & (l \gg \sqrt{\lambda L} \gg L_0) \end{cases} \text{For plane wave} \tag{8.7-13}$$

$$\begin{cases} 0.496C_n^2(2\pi/\lambda)^{7/6}L^{11/6} & (l \ll \sqrt{\lambda L} \ll L_0) \\ 1.28C_n^2(2\pi/\lambda)^{7/6}L^{11/6} & (l \gg \sqrt{\lambda L} \gg L_0) \end{cases} \text{For spherical wave} \tag{8.7-14}$$

where λ is the laser wavelength, L is the horizontal transmission distance, l_9 and L_0 are the turbulent flow's inner and outer dimension, $l_0^2/\lambda \gg L$ represents receiving in the near field region, and $l_0^2/\lambda \ll L$ represents receiving in the far field region. It is known from the above two equations that, with respect to the scintillation characteristic of the plane wave and spherical wave, except for their different coefficient values, they are governed by identical laws, the different coefficient values showing that the scintillation of the plane wave is greater than that of the spherical wave. But as a whole, the strength of scintillation is proportional to the strength of the turbulent flow and inversely proportional to the 7/6th power of the wavelength. When the wavelength is short, scintillation is strong whereas when it's long, scintillation will be weak. Meanwhile the magnitude of scintillation is proportional to the 11/6th power of the transmission distance.

However, it is shown by both theory and experiment that, when the intensity of the turbulent flow is increased to a definite extent or the transmission is increased to a definite limit, the scintillation variance will no longer continue to increase in accordance with the above-mentioned law. Rather, it is slightly decreased and appears to be saturated. So it's called the saturation effect of scintillation.

In addition, similar to radio signals, because of the random fluctuation of the refractive index, atmospheric scintillation also possesses frequency characteristics, its peak value frequency f_0 depending on the vertical component $v_\perp$ of the average wind velocity along the direction of the light beam propagation; that is,

$$f_0 = v_\perp / \sqrt{2\pi\lambda L} \tag{8.7-15}$$

(2) Bending and drift of a light beam

The phenomenon of bending and drift of a light beam is also called the phenomenon of astronomical refraction, which is mainly restricted by the fluctuation of the atmospheric refractive index. Bending is manifested as the slow variation of the statistical position of the light beam while drift is the rapid jump of the light beam around its average position. If the influence of humidity is neglected, the atmospheric refractive index n in the optical frequency band can be expressed by

$$n - 1 = 79.10^{-6} P/T \cdots \quad (\text{or } N = (n-1) \times 10^6 = 79P/T) \tag{8.7-16}$$

where P is the atmospheric pressure intensity, and T the atmospheric temperature (K). According to the law of refraction, in the case of horizontal transmission, it's not hard to prove the light beam curvature

$$c = \frac{1}{R_0} = -\frac{\mathrm{d}N}{\mathrm{d}h} = -\frac{79}{T}\frac{\mathrm{d}P}{\mathrm{d}h} + \frac{79P}{T^2}\frac{\mathrm{d}T}{\mathrm{d}h} \tag{8.7-17}$$

where R_0 is the curvature radius of light beam bending and it is specified when the light beam bends downward, the curve c should be positive; $\mathrm{d}N/\mathrm{d}h$ is the vertical gradient of atmosphere refraction. Under the sea level condition, $P = 101325$ Pa, $\mathrm{d}P/\mathrm{d}h = -12100$ Pa/km, and $T = 20°$C which, substituted into Eq. (8.7-17), yields $c = 32.2 + 0.93\,\mathrm{d}T/\mathrm{d}h$ (μrad/km). It is thus known that, when the temperature vertical gradient $\mathrm{d}T/\mathrm{d}h = 35°$C/km, $c = 0$, the light beam dose not bend; when $|\mathrm{d}T/\mathrm{d}h| < 35°$C/km, c is positive, and the light beam bends downwards; when the temperature vertical gradient $\mathrm{d}T/\mathrm{d}h > 35°$C/km, c is negative, and the light beam bends upward. It has been found by experiment that, in general, the light spot rises high in the daytime, and the light beam bends upward while at night, the light spot falls and the light beam bends downward. It's not hard to prove that, at the horizontal distance L, the distance the light spot deviates from its original position:

$$l = 0.5cL^2 \qquad (L \gg l) \tag{8.7-18}$$

With respect to the drift of a light beam, theoretical analysis shows that its angle of drift is closely related to the beam width W_0 of the light beam at the outlet of the emitting telescope; the mean square value of the angle of drift $\sigma_a^2 = 1.75C_n^2 L W_0^{-1/3}$ from which it can be seen that the thinner the light beam is, the greater will the drift be. Adoption of a wide light beam can reduce light beam drift. It has been found by experiment that, like scintillation, light beam drift exhibits obvious diurnal variation. When $C_n > 6.5 \times 10^{-7}$ m$^{-1/3}$/h, the c value is about 40 μrad and will no longer vary according to the expression $\sigma_a^2 = 1.75C_n^2 L W_0^{-1/3}$, showing that drift also has the saturation effect. The frequency spectrum of drift generally does not exceed 20 Hz, its peak value is below 5 Hz. The statistical distribution of drift obeys normal distribution. For instance, it has been found by measurement that $\sigma_a = 40$ μrad (equivalent to 8″), whose physical meaning is when laser is shot horizontally, because of the influence of the atmospheric turbulent flow, the probability of an error smaller than 8″ in hitting a target is 68%, and the probability of an error smaller than 16″ is 95%.

The above discussion shows that light beam bending and drift should not be mixed up. Around noontime, the drift of a light beam is very abrupt, yet the average position of the light spot is relatively stable. Conversely, in front of and behind the turning point of the temperature gradient (i.e., the turning point at which the light beam changes its direction), the light spot average position varies very quickly, but now the light beam drift is very small.

(3) The turbulent flow effect related to spatial phase fluctuation

If the receiving plane is not a target plane, but the receiving is carried out through a lens and on its focal plane, then we shall find that there are image points that are jittering. This can be explained as, while the light beam is found to drift, the light beam's arriving angle on the plane, affected by the turbulent flow, is also randomly rising and falling. That is, for the portion of the wavefront equivalent to the receiving aperture, random fluctuation is generated relative to the tilting of the receiving plane.

2. The atmospheric nonlinear transmission effect

The atmospheric turbulent flow effect mentioned above is a kind of linear transmission effect of laser atmospheric transmission at the low power level with no consideration given to the reaction generated by the laser energy action to the state of the atmosphere; that is, the state of the atmosphere has not changed its original turbulent flow parameters simply because of the laser transmission. When the laser energy is sufficiently great, significant variation will take place; that is, evident changes in the atmosphere's natural turbulent flow state will take place. Since such changes are caused by high energy laser beams, in the process of transmission, the optical beam quality of high energy laser beam is not only influenced by the linear transmission effect encountered by the low energy laser beam but also by a series of nonlinear transmission effects induced by high energy laser beams at the same time.

The atmospheric nonlinear transmission effect mainly refers to the phenomena of thermal halo and atmospheric breakdown blocking of the high energy laser beam in the transmission in the atmosphere. As the intensity of laser has exerted a distinct impact on the state of the atmosphere, the basic equation of describing light propagation has become a nonlinear equation. So the effect is called the nonlinear propagation effect. As nonlinear problems are rather complicated, only some basic phenomena are treated here.

(1) Atmospheric thermal halo

In the case of linear propagation, the absorption and scattering of the atmosphere result in the attenuation of laser energy. Under the high energy laser condition, the atmosphere's absorption process is markedly strengthened so that the atmosphere is locally heated, forming the increment of the pressure intensity of gases. The original thermal equilibrium state is destroyed. In order to regain their thermal equilibrium, the gases expand at the sound speed, leading to a change in the density of gases and further to changes in local refractive indices and hence the propagation characteristics of the light beam. When such changes in the refractive index accumulate to a definite extent, the atmosphere will become a nonuniformly divergent lens. Distinct distortion of the contour of the propagating light beam can occur. If it is the case of a convergent light beam, then the size of the light spot on the focal plane is clearly augmented while the illuminance is evidently reduced. Such a phenomenon is called the phenomenon of atmospheric thermal halo. Obviously, the atmospheric thermal blooming effect has restricted the high energy laser beam from maintaining high quality propagation.

It should be pointed out that thermal halo is also a kind of turbulent flow, but one not formed by the temperature gradient of the atmosphere itself. Rather, it's a special turbulent flow formed by the absorption of the energy of light beam in its path of propagation by the medium-thermal halo. Hence there exists the thermal halo threshold value P_{th}.

It can be proved by referring to atmospheric thermodynamics that there occurs the threshold value power of thermal halo in a transversely flowing gas:

$$P_{\mathrm{th}} \approx \frac{\pi}{4} \cdot \frac{\gamma p_0}{\gamma - 1} \cdot \frac{1}{\alpha(n_0 - 1)} \cdot \frac{\lambda^2 v}{a} \tag{8.7-19}$$

where p_0 is the atmospheric pressure intensity; n_0 is the initial atmospheric refractive index; a is the atmospheric absorption coefficient; $\gamma = C_p/C_v$; C_p and C_v are the molar specific heat at constant pressure and constant volume of the atmosphere, respectively; v is the transverse wind velocity; and a is the transverse dimension of the light beam. The above equation shows that the thermal halo threshold power of the continuous laser radiation is proportional to the square of the atmospheric pressure intensity and wavelength while inversely proportional to the absorption coefficient and the light beam transverse characteristic length. Therefore, other things being equal, thermal halo is more apt to occur for laser radiation of a short wavelength and great linear absorption coefficient. If the light beam is a convergent one, then thermal halo always occurs first at the focal point.

For CO_2 laser of 10.6 μm, take $p_0 = 10^5$ N/m^2; $n_0 - 1 = 3\times10^{-4}$; $\gamma = 1.4$; $\alpha = 3\times10^{-9}$/m; $a = 10^{-2}$ m; and $v = 0.3$ m/s. Then $P_{\mathrm{th}} \approx 10^3$ W.

It has been proved by both theoretical analysis and experimental research that in the case of continuous wave steady-state thermal halo the light beam's profile becomes distorted, its shape appearing as a "crescent", as shown in Fig. 8.7-7, in which the circular dotted line represents the initial light spot free from the influence of thermal halo and the solid line graph represents the equal intensity contour line of the distorted light beam. It can be seen from the figure that, when the laser power exceeds the threshold value of thermal halo the

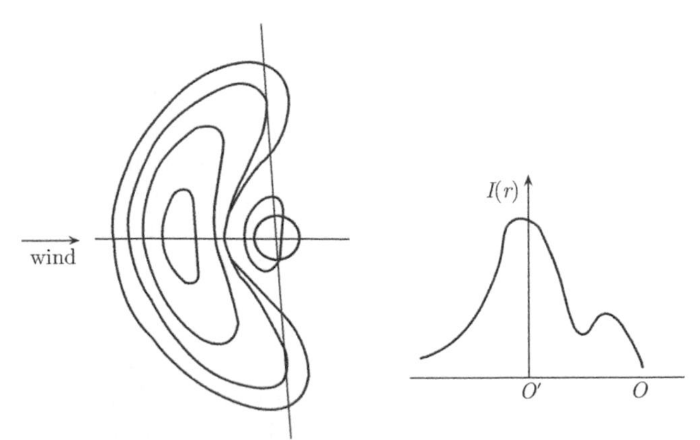

Fig. 8.7-7 Typical phenomena of thermal halo

shape of the cross section of the light beam markedly diffuses and becomes distorted. The center of gravity of the distorted light beam deviates from the initial point 0 to point 0′, that is, it inclines towards the direction of the transversely blowing wind. If such a light beam is focused, the light spot on the focal plane will become distorted likewise and the illuminance is markedly reduced also.

The problem of thermal halo in short pulse high energy laser propagation involves the transient process in which the thermal halo is set up. The case is even more complicated and there are some characteristics different from those of the continuous wave thermal halo.

For the single pulse thermal halo, the pulse being very short, the compression ratio very small, and the thermal halo threshold value power very high, the thermal halo effect is rather weak.

During multiple pulse propagation, under the condition of very high pulse repeatability and long pulse rows, a thermal halo similar to that of thc continuous wave will occur. This is because, although thc thermal halo of the pulse itself is rather weak, because of the effect of accumulation, however, the thermal halo of the subsequent pulses will become serious.

(2) Atmospheric breakdown

With laser of even higher power, when the laser beam propagates in the atmosphere, not only is there the problem of the generation of the thermal halo by heating of the atmosphere, but also, as the electric field in the light beam is extremely strong, there will be generated the phenomenon of light induced atmosphere ionization. This is called atmospheric breakdown. The atmosphere getting broken down itself is at the sacrifice of laser energy that is lost. In the passage of the already broken down atmosphere is formed the distribution of plasma. Further action between the laser beam and plasma will make the propagation of the plasma along the light beam rapidly expand radially and longitudinally. This will consume even more laser energy. When 90% of the laser energy is absorbed, scattered, and consumed by the plasma, the phenomenon is called passage blockage.

It is shown by research that atmospheric breakdown is mainly a result of the interaction between a laser beam and the atmospheric aerosol rather than one of the interaction with the molecules of such gases in the atmosphere as N_2, O_2, and CO_2. Atmospheric breakdown is first of all aerosol breakdown. Further action of laser with the aerosol plasma finally leads to clean atmospheric breakdown. The aerosol particles that play the principal role in atmospheric breakdown are droplets of water and solid-state particles. The action between a laser beam with ordinary energy and the aerosol particles is linear; that is, the absorption and scattering process of light does not change the state of the particles themselves; that is, under the condition of thermal halo, although changes have taken place in the distribution state of particles, the state of the particles themselves is still unchanged. It is only under the action of laser with a still higher power that the state of the aerosol particles will change (i.e., ionize), the process of forming ionization being roughly: irradiated by high energy laser, water drops are first heated, the heat energy absorbed makes the water drops vaporize, then pressure expansion, crushing, and diffusion are generated and accompanied by the blast wave and detonation wave. Vapor mixes with the surrounding air and continues to absorb energy, forming photo-plasma.

Research results show that the breakdown threshold value power of clean atmosphere (atmosphere not containing particles the size of 0.1 μm and above) is in agreement with the calculation values of the microwave breakdown theory, roughly 3×10^9 W/cm^2. The threshold value of the water drop is two to three orders of magnitude lower than that of the clean atmosphere, mainly depending on the dimension of water drop and the pulse width of laser.

8.7.3 Characteristics of laser underwater transmission

Of different kinds of waves propagating in water, the longitudinal wave (sound wave) has the least attenuation. So acoustic susceptance technology is widely adopted whereas the attenuation of the transverse wave (electromagnetic wave) is in general very serious so that it's almost impossible for the radio wave and microwave in wide application on land to be applied underwater. But the optical wave is an exception. In contrast to the microwave, its attenuation is rather small; in particular, the emergence of laser has made techniques such as ranging, collimation, illumination, photography, TV, etc. feasible within limited distances underwater. However, owing to the influence of underwater transmission characteristics, such applications are still greatly restricted and there are features different from those on land. This section will simply make a brief description of certain characteristics of underwater laser transmission.

1. The characteristic of attenuation

If the distance of transmission is rather short, like transmission in the atmosphere, the way attenuation of a beam of monochromatic parallel light goes on underwater also obeys Lambert's exponential law:

$$P = P_0 \exp(-\beta L) \tag{8.7-20}$$

where P_0 and P are the laser power (W) when the transmission distances are 0 and L, respectively; β is the attenuation coefficient (m^{-1}) that includes absorption and scattering.

Customarily, the attenuation length L_0 is used to represent the magnitude of attenuation underwater with the definition $L_0 = 1/\beta(\mathrm{m})$, the physical meaning of which is, over a distance of an attenuation length, the power of a laser beam will be reduced to $1/\beta$ that of the initial value. It's clear that the greater the attenuation, the shorter the attenuation length will be.

The attenuation coefficient β is closely related to the laser wavelength. Shown in Tab. 8.7-4 are the attenuation coefficient values over the laser wavelengths of running water measured in a water tank. The attenuation of the running water includes pure water (distilled water) absorption and particle scattering, while Fig. 8.7-8 shows the spectral absorption characteristics of distilled water. For different water qualities, the attenuation characteristics are greatly different. Figure 8.7-9 shows the variation of the attenuation length with the wavelength in different marine regions.

Tab. 8.7-4 Attenuation coefficients of running water

Attenuation coefficient / Wavelength/μm	Running water attenuation coefficient β/m^{-1}	Distilled water absorption coefficient/m^{-1}	Particle scattering coefficient/m^{-1}
0.4900	0.086	0.037	0.049
0.5200	0.099	0.041	0.058
0.5650	0.115	0.060	0.055
0.6000	0.243	0.197	0.046
0.6943	0.545	0.513	0.032

It is known from the above figures and tables that the optical attenuation in the infrared and ultraviolet wavebands is very great and just cannot be used underwater. In the entire visible light waveband, the attenuation of blue and green light is minimum. So this waveband is often spoken of as the "underwater window". It's not hard to find from the data listed in Tab. 8.7-4 that the attenuation lengths of light of wavelengths 0.4900 μm and 0.6943 μm are 11 m and 2 m, respectively. This shows that the transmission performance of blue light is much better than that of red light. Performing simple transform of Eq. (8.7-20), we can find the equation for the action distance of the optical pulse to be

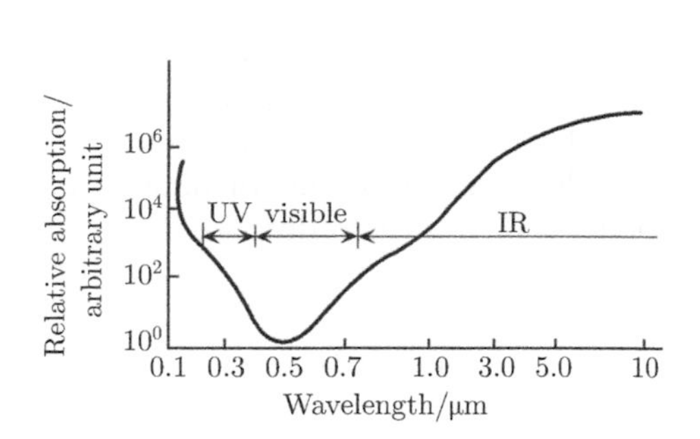

Fig. 8.7-8 Spectral absorption characteristics of distilled water

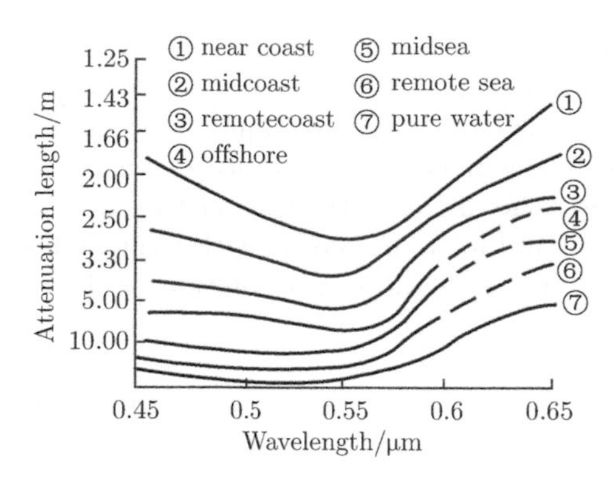

Fig. 8.7-9 Attenuation in different sea regions

$$L = -\frac{1}{\beta}\ln\left(\frac{P}{P_0}\right) = \frac{2.3}{\beta}\ln\left(\frac{P_0}{P}\right) \tag{8.7-21}$$

If P_0 and P in the equation are understood as the transmitting power and the detector's minimum detectable power, then L is the longest distance the optical pulse is capable of transmitting underwater. Take $P_0 = 10^6$ W, and $P = 10^{-14}$ W. For the light of wavelength 0.4900 μm, its action distance can reach as great as 500 m; for the light of wavelength 0.6943 μm, its action distance is merely 80 m. It can be seen that it's very hard to apply ruby underwater. In addition, for different water qualities, the attenuation characteristics are greatly different. It is known from Fig. 8.7-9 that the seawater in the remote marine region is clean and the attenuation length is rather great whereas the seawater in the offshore region is turbid and the attenuation length is greatly reduced.

2. Forward scattering

If the distance for measurement is increased, while suitably expanding the receiving area, then the measurement data will markedly deviate from the value calculated using Eq. (8.7-21) and the power received will be greater than the value estimated according to this equation. The reason lies in that the action of forward scattering has not been considered in Eq. (8.7-21). For convenience in discussion, illuminance is used to represent the magnitude of the optical radiation on the receiving plane. Illuminance is defined as the radiation power that falls on the unit receiving area or passes through the unit area (the unit is W/m^2).

The scattering of light in the direction of transmission is called forward scattering while that in the opposite direction is called backward scattering. Forward scattering involves a complicated process of scattering (Fig. 8.7-10). The radiation that scatters for the first time and deviates from the optical axis is scattered once again by another scatterer. For some cases this process is repeated many times. A considerable part of radiation in these cases enters the direction of the optical axis anew or slightly deviates from the optical axis to reach the receiving plane. This is called the multipath radiation while the radiation in the initial parallel optical beam that directly reaches the receiving plane is called the single path radiation. Thus, the total illuminance H_r on the receiving plane should be the sum of the single path illuminance H_r^0 and the multipath illuminance H_r^*; that is,

$$H_r = H_r^0 + H_r^*$$

where

$$H_\gamma^0 = (J/L^2)\exp(-\beta L)$$
$$H_\gamma^* = [(JK)/(4\pi L)]\exp(-KL)$$

where J is the radiation intensity (W/sr), K is the multipath attenuation coefficient (m^{-1}), and L is the distance of transmission (m). Some people have found $\beta = 0.6\ m^{-1}$ and $K = 0.187\ m^{-1}$ using 0.53 μm green light in lake water. It is known from this that forward scattering makes the distance of an optical beam transmission manifestly increase. The longer the distance, the greater the contribution by the forward scattering. This effect may be helpful for illumination with laser, but it's unfavorable to laser scanning or photography since it causes the scan resolution or the object background contrast to decrease.

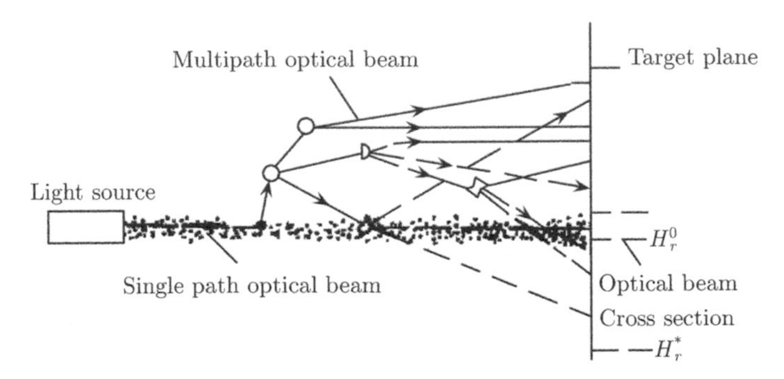

Fig. 8.7-10 Schematic of forward scattering

3. Backward scattering

Backward scattering is much stronger than forward scattering. This is a remarkable feature of underwater transmission. For instance, when driving in a heavy fog, an experienced driver always has the taillights on and the headlights off. With the taillights of the car before him, he can clearly see it. But if he drives with the headlights on, then the intensive backward scattering from the heavy fog will make him incapable of seeing anything. This is a convincing illustration of the influence of forward scattering. Underwater, the backward scattering is even more intensive. The greater the power, the greater the backward scattering. Powerful backward scattering often makes the receiver so saturated that it cannot receive any information. For this reason, for applications of underwater ranging, television, photography, etc., our chief concern is how to overcome the influence of backward scattering. Some measures are recommended below.

(1) Use filters and analyzers properly to distinguish between the irregularly polarized backward scattering and regularly polarized target reflection.

(2) Separate the transmitting light source from the receiver as much as possible.

(3) The most effective method is adopt the range gate technique (Fig. 8.7-11). When a laser pulse is propagating toward the target, the shutter of the receiver is closed. During this time, the continuous backward scattered light has no way to enter the receiver. However, when the optical pulse signal reflected by the target underwater returns to the receiver, the shutter of the receiver opens all of a sudden and records the target information. Thus the influence of the backward scattering underwater can be effectively overcome.

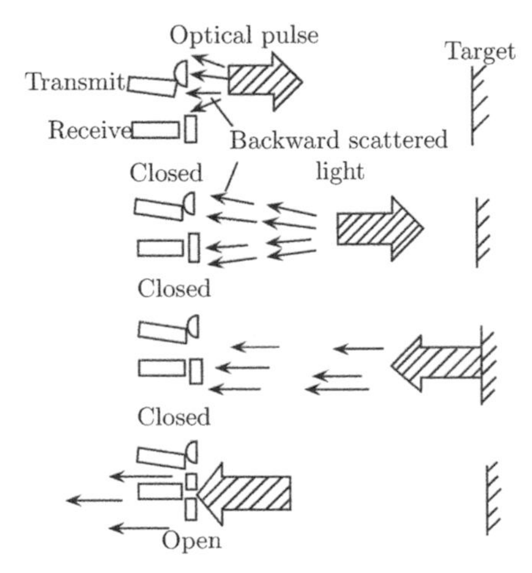

Fig. 8.7-11 Schematic of optical range gating

Exercises and problems for consideration

1. By referring to the theory of optical fibers' electromagnetic mode and the WKB method, sum up the significance of the fiber-optic weak conduction condition.

2. Proceeding from the ray Eq. (8.2-1), prove that the track of a ray in a homogeneous medium is a straight line and that the ray in a non-homogeneous medium is bound to incline toward where the refractive index is high.

3. Prove that the numerical aperture of the step optical fiber with a bending radius R is NA= $\left[n_1^2 - \left(1 + \frac{a}{R}\right)^2 n_2\right]^{1/2}$, and discuss the influence of R on NA; a in the equation is the radius of the fiber core.

4. Prove Eq. (8.2-60') using the method of transmission matrix.

5. There is now a self-focusing optical fiber with $L = \Lambda/4$. Try plotting the graphs of rays in the optical fiber and on the end face when a beam of parallel rays and convergent rays is injected into the end face and tell why.

6. For the use of wavelengths $\lambda_1 = 0.63$ μm, $\lambda_2 = 1.3$ μm, $\Delta = 0.02$, $n = 1.46$, find the condition for the single mode.

7. What is the relationship between fiber-optic dispersion, bandwidth, and pulse broadening? What influence will it exert on the fiber-optic transmission capacity?

8. Sum up the advantages of the 1.3-μm and 1.55-μm fiber-optic transmission systems.

9. What is the birefringence of single mode optical fibers? What influence does the phenomenon of fiber-optic birefringence exert on the fiber-optic transmission system?

10. What is polarization-maintaining optical fiber? How is its polarization-maintaining characteristic realized?

11. What does the optical soliton mean? What is the mechanism of its formation? what isits use?

12. Hold a discussion on what problems should be considered when an optical fiber is to be coupled with a detector. How is effective coupling realized?

13. What is the atmospheric window? Try to analyze the atmospheric attenuation factors of light radiation with the spectrum located in the atmospheric window.

14. What is the atmospheric turbulent flow effect? What does the 2/3 law mean? What influence does the atmospheric turbulent flow have on laser transmission?

15. How is the atmospheric thermal blooming effect generated? What influence does it exert on laser transmission?

16. What are the special problems of underwater laser transmission?

References

[1] Lan Xinju, Laser Technology, Wuhan, Huazhong University of Science and Technology Press, 1995.

[2] A. W. Snider, J. D. Lover, The Optical Waveguide Theory, translated by Zhou Youwei et al., Beijing, People's Posts and Telecommunications Press, 1991.

[3] Zhao Zisen, The Principles of Single Mode Fiber Communication System, Beijing, People's Posts and Telecommuinications Press, 1991.

[4] Peng Jiangde, Basics of Electronic Technology, Beijing, Tsinghua University Press, 1988.

[5] Yang Xianglin et al., The Fiber-optic Transmission System, Nanjing, Southeast University Press, 1991.

[6] Zeng Fuquan, Theory and Technology of Optical Fibers, Xi'an, Xi'an Jiaotong University Press, 1990.

[7] E. UDD, Fiber-optic Sensors, New York, John Wiley & Sons, Inc., 1991.

[8] Wu Jian, Wang Junbo, Hu Zhiping et al., High Energy Laser Massive Linear Propagation—Theory and Technology, Chengdu, College of Telecom Engineering Press, 1988.

[9] AD715270.

[10] AD671933.

Index

Printed and bound by CPI Group (UK) Ltd, Croydon, CR0 4YY

Printed and bound by CPI Group (UK) Ltd, Croydon, CR0 4YY
21/10/2024
01777100-0005

Optics, Lasers, and Photonics

Features Self-Contained, Independent Chapters for Flexible Use

As different laser technologies continue to make it possible to change laser parameters and improve beam quality and performance, a multidisciplinary theoretical knowledge and grasp of cutting-edge technological developments also become increasingly important. The revised and updated **Laser Technology, Second Edition** reviews the principles and basic physical laws of lasers needed to learn from past developments and solve the many technical problems arising in this challenging field.

The first edition of **Laser Technology** was classified by the Chinese National Education Committee as a "national-level key textbook." This edition presents the fundamentals of physical effects in technical devices and implementation methods to create a clear and systematic understanding of the physical processes of different laser technologies.

Logically presenting the various types of laser technology currently available, this updated second edition:

- Explores the transmission of information using optical waves with modulating technology
- Shows how beam energy or power can be greatly enhanced through Q switching, mode-locking, and amplification
- Explains how mode selection and frequency stabilizing technology make it possible to improve light beam directionality or monochromaticity
- Describes nonlinear optical technology that helps obtain new frequencies and light waves
- Covers transmission in the atmosphere and underwater

Technical improvements to enhance laser performance in different applications have given rise to new physical phenomena. These have resulted in a series of new laser branches and fields of applied technologies, such as laser physics, nonlinear optics, laser spectroscopy, laser medicine, and information optoelectronic technology. This book analyzes this growth, stressing basic principles but also including key technical methods and examples where needed to properly combine practical and theoretical coverage of this distinct area.

CRC Press
Taylor & Francis Group
an **informa** business
www.crcpress.com

ISBN 978-1-138-37276-4
9 781138 372764